BUILDING CONTRACT CLAIMS AND DISPUTES

2nd edition

DENNIS F. TURNER B.A. (Hons), F.R.I.C.S., M.C.I.O.B.
AND ALAN TURNER J.P., DIP. ARB., F.R.I.C.S., F.C.I.ARB.

Addison Wesley Longman Limited
Edinburgh Gate
Harlow
Essex CM20 2JE
England
and Associated Companies throughout the world

Visit Addison Wesley Longman on the world wide web at:
http://www.awl-he.com

First edition published 1989
Second impression 1990
This edition published 1999

ISBN 0 582 28511 9

British Library Cataloguing-in-Publication Data
A catalogue record for this book is available from the British Library

Typeset by 35 in 10/12pt Ehrhardt
Produced by Addison Wesley Longman Singapore (Pte) Ltd.,
Printed in Singapore

CONTENTS

LIST OF FIGURES

LIST OF TABLES

INTRODUCTION

In coming to a second edition of what was *Building Contract Disputes*, and renaming it *Building Contract Claims and Disputes*, we are aware of the change of emphasis which this implies. In fact, this change has come about, not by shifting ballast, but by a considerable expansion of payload in the treatment of claims, backed up by a substantial body of further legal cases. There has also been a change by taking account in detail of a distinctly wider range of contracts and approaches than before, as a perusal of the contents will reveal. This has been accompanied by a restructuring of the chapters dealing with delay and loss and expense to develop this and emphasise its thrust.

Although the previous edition deliberately stopped short of last-resort matters and sought to major on prior resolution by consent, the trends towards adjudication and other tribunals arising out of Latham and the Housing Grants, Construction and Regeneration Act 1996 has made it desirable to deal with resolution methods which have come more into prominence. Because these are seeking to promote by persuasion or stronger means the earlier and more gradual solution of problems and disputes, as the earlier edition of this book advocated, they are to be welcomed. Quite an amount of space has thus been given to the contractual provisions and the wider apparatus to which they relate. It is hoped that the resulting blend of good practice procedures, contractual provisions, legal cases and resolution processes together with explanation of good claims methods and the case studies will contribute to awareness of where the rights and wrongs lie, not only of entitlements and defences, but of balanced settlement.

The contents seek to meet the needs of various disciplines within construction, so that at most points the treatment is not usually addressed explicitly to those on one side only of disputed matters. The client, client's adviser, contractor and subcontractor alike should be able to read and interpret for his own purposes. The general level has therefore been set with two types of reader in mind: those who are seeking explanation of basic principles, and those who are looking for more extended treatment of issues with which they are already acquainted. In more precise terms, this dual approach is intended for those who are in the late stages of qualifying in their disciplines and also those concerned with continuing professional development – or simply pursued by the need to know more! It is hoped that a coherent approach has been achieved and that the types of subject-matter are reasonably distinguishable by the divisions within as well as between chapters.

Extended use has been made of case studies in the latter stages of the book to apply the more generalised discussion of earlier parts to concrete situations, and occasionally to supplement previous discussion. It may be surmised that behind some of the examples lie real life cases, from which extrapolations and adaptations have been made for present didactic purposes. Some of the prototypes might have difficulty in recognising themselves, which is probably just as well! The amount of detail has been varied both within and between the studies to avoid impracticable length and also to cover a representative selection of situations and resolutions. As a result, the case studies are not worked examples which may be followed slavishly (usually a dangerous policy, as actual conditions may vary so much), but illustrations of what may be done, combined with discussion of the limitations that often attach to specific solutions.

The aim in the case studies is to use an approach to financial settlements based upon the obvious intention of the contracts that settlement should proceed progressively, rather than occur entirely at the end in the form of a 'big bang'. As a result, there is not the setting up of a target claim, subsequently to be shredded by a series of missiles. In fact, generally the term 'claim' has largely been avoided, as again not representing the intention of the contracts – within a contract framework there are 'entitlements'. If this emphasis helps just a little to dispel the emotive side of many disputes, it can but have succeeded. In recognition of the need on occasions for an 'end of contract claim', some expanded guidelines on and an example of such a document are given, so that other chapters may be turned to and applied to it when necessary. To help with the problem of looking at disputes in something of a vacuum, as is inevitable in a book, the studies given are each linked closely to a site layout and programme to indicate more expressly how design delays and disturbances affect progress and performance. In one study, an analysis of the contract bills of quantities has been included to give 'scale' to what is happening.

No apology is made for considerable cross-referencing and a limited degree of repetition of key ideas between chapters. It is realised that no authors are sufficiently gripping in style or precise in structure or content as to compel every reader to read such subject-matter hypnotically from beginning to end: many will wish to dip. Given also that readers have differing needs, this is inevitable. But dipping in and out carries the risk of missing some essential point covered elsewhere. Repetition also allows the reader to start with either the more theoretical, earlier chapters or the more practical, later ones, and then turn to the other set. The order used in the book suggests what is thought to be best, but dictates of time often lead to some compromise, and so there is, we repeat, some repetition.

Although the strict principles have been kept in view, discussion has recognised the difficulties of staying exactly with theory all the time. It is hoped that the text as it has emerged will not cause too much offence to either purist or pragmatist. Reality is, after all, rather like that as well.

We have found this edition an interesting exercise in the logistics of working for most of the time at more than arm's length, but for much more. In case any readers should wonder, we do happen to be brothers, on which fraternal note we commit you all to the text as it emerges.

Thanks are due to Michael Battye, partner of Masons solicitors, for allowing access to their excellent reference library, to Roy Gordon, honorary secretary of the Arbrix Club Construction Group, for an extended loan of a set of *Building Law Reports*, to Nicholas Gould and Michael Cohen for bringing to our attention the research they had carried out into alternative dispute resolution, to Rena Monks for assistance in preparation of the law case table and case index, to Ken Bristow for an introduction to the aid of document scanning and to our families for the usual forbearance when such a minor task as a publisher's deadline replaces other activities.

Dennis F. Turner
Alan Turner *July 1998*

ACKNOWLEDGEMENTS

We are grateful to the following for permission to reproduce copyright material:

Sweet & Maxwell for Tables 27.1 and 27.2 from 'ADR: appropriate dispute resolution in the UK construction industry', *Civil Justice Quarterly*; and Macmillan Press Ltd for a figure (figure 5.3) from *Building Procurement* (2e) by Alan Turner (1997).

ABBREVIATIONS AND FURTHER READING

ABBREVIATIONS USED IN TEXT

The following are the main abbreviations used in the text. The full names of documents are not necessarily the precise titles, but are what might otherwise have been used in the text. In the case of the JCT documents, several revisions have been introduced. The latest considered are those of April 1998. Where the dating is significant, it is indicated in the text.

JCT contracts and related documentation

JCT 80 contract/form	JCT Standard Form of Building Contract, 1980 edition, including amendments 1 to 18. This is published in several variants. Unless qualified, a reference may be read as being to any variant.
JCT 'clause'	A clause of the foregoing contract.
IFC 84 contract/form	JCT Intermediate Form of Building Contract, 1984 edition including amendments 1 to 12.
IFC 'clause'	A clause of the foregoing contract.
NSC/W	JCT Employer/Nominated Subcontractor Warranty, 1991 edition.
NSC/C	JCT Nominated Subcontract, 1991 edition.
DOM/1	Domestic subcontract to JCT forms, 1998 revision.
Green Form	Nominated subcontract relating to the JCT 1963 contract.
Blue Form	Domestic subcontract relating to the JCT 1963 contract.

Government contract GC/Works/1

General Conditions of Contract GC/Works/1 for Building and Civil Engineering major works with quantities, as prepared for central government works, 1998 edition.

Institution of Civil Engineers (ICE) Contract

ICE Conditions of Contract, Sixth Edition, 1991, with Corrigenda (August 1993), Guidance Note (March 1995) and Amendments (Reference ICE/6th Edition/Tax/February 1998), with amendment ICE/6th Edition/HGCR/March 1998.

New Engineering Contract (NEC)

The New Engineering Contract, 1994 edition with addendum NEC/ECC/Y/2/April 1998 to take account of the HGCR Act.

Method of measurement SMM7

Standard Method of Measurement of Building Works, Seventh Edition.

Institutions and Associations

ACE	Association of Consulting Engineers
BEF	British Property Federation
CDM	Construction (Design and Management) Regulations 1994
CIA	Chartered Institute of Arbitrators
ICE	Institution of Civil Engineers
JCT	Joint Contracts Tribunal
RIBA	Royal Institute of British Architects
RICS	Royal Institution of Chartered Surveyors

Other abbreviations

ADR	Alternative dispute resolution
PFI	Private Finance Initiative
BOOT	Build, own, operate and transfer

FURTHER READING

These works are mentioned as affording both expansions of topics covered within the context of the present volume and extensions into surrounding areas. As such, they are referred to when they have an immediate relevance; they may also be consulted to obtain a more general view and a deeper treatment of their subject-matter at the points concerned or beyond.

Building Contracts: A Practical Guide
By Dennis F. Turner

This book gives an extended survey of and commentary on well over forty of the most widely used standard forms of contract in the United Kingdom construction industry. It provides a detailed examination of the private edition of JCT standard form with quantities and a comparative treatment of the other JCT main contract forms, including the intermediate and minor works forms, and those for design and build work, prime cost and management contracting. It also deals with the related subcontracts, tender forms and warranties, extending to those for domestic subcontracts beyond the strict ambit of the JCT. It also covers the GC/Works/1 contract and the ICE main form.

In its approach the book considers all aspects of the contracts etc in a rounded and detailed way that extends beyond the selective treatment of particular themes in the present volume. It places special emphasis on the explanation of those aspects which are of immediate relevance to those working with the documents, while backing this up with comment on the legal aspects of the provisions and reference to a body of decided cases. It runs to almost six hundred pages and is in its fifth edition.

Design and Build Contract Practice
By Dennis F. Turner

This book is a treatment of the principles of the design and build system of producing construction work and examines its procurement and operation in detail. It relates practical applications of design and financial methodology and contractual aspects throughout the successive stages of projects, from conception to final occupation and residual liabilities.

Throughout there is consideration of the terms of the JCT with contractor's design form of contract, especially those which are particular to design and build. Separate chapters extend this to cover similar and distinctive provisions of the JCT design portion supplement, performance-specified work, supplementary provisions related to the BPF system and nominated subcontracts. There are also chapters on the BPF system, the GC/Works/1 design and build contract and the ICE design and construct contract. These aspects are related to one another to afford a comprehensive survey of the field for clients, designers and other professionals in the construction industry and other interested persons. The book is in its second edition and is over three hundred pages long.

Building Procurement
By Alan Turner

This book explains the methods of building procurement commonly used in construction. It is widely accepted that choosing the right method of procurement is crucial to the success of any project. Seldom is there an automatic choice of how to obtain construction works, and this book begins with first principles: Why does a client want to build? What quality does he want? When is the project to be finished? And how much does the client want to spend? Only after these minor matters are settled will construction contracts be discussed and drawn up. If wrong decisions are taken at the early stages of a project, then very often disappointment and disputes will occur later on, as examined in the present volume.

The book examines reasons for building, relationships of parties, their contractors and designers, how to choose the right route in particular circumstances. It discusses PFI and BOOT procurement and the recommendations of Latham. Discussions of contracts for consultants and contractors is followed by fourteen case studies that examine how and why clients made their procurement choices and how successful they were. The book, now in its second edition at over two hundred pages in length, is aimed to appeal to both students and practitioners.

PART 1

PRECONTRACT ACTIVITIES

CHAPTER 1

BACKGROUND TO CONTRACT DISPUTES

- A complex process
- Clients and the industry
- Matters underlying disputes
- Outline practical approaches

This book is about building *contract* claims and disputes in particular, rather than about building disputes in general. This is a topic which has exercised the minds of many within and beyond the industry for years and it conjures up a number of discrete subtopics, numbers of which are explored in the detailed sections of this book. Because of the proverbial relation between wood and trees, it is useful to survey some principles which underlie much of the detail. To the initiated, they may savour of nothing new, but it is hoped that any reader will find something of value among them.

A COMPLEX PROCESS

It is perhaps surprising that more does not go wrong, even today, with the production of buildings when the intrinsically complicated and hazardous nature of the activity is reviewed. This is most obvious in terms of accidents and so forth during construction, while problems of securing quality and ensuring that designs do not fail come close behind. But the operational process is at least as subject to problems, even if they are not so apparent to the outsider. What happens in this process is the spawning-ground for the disputes that this book considers.

It has been said that factory production is a line of work going through people, whereas site production is a line of people going through work. This difference arises from the need to install the work on an individual and perhaps difficult site, with all the problems of coordination and the added ingredient of the weather. The work is usually also individual in content, even if it repeats the character of other buildings. This is not new, and down the centuries problems of time and money overrun have been found. Several of the old law cases testify to this, as also, for example, do the records of the building of London churches two hundred years ago. Although today's refined systems for controlling the process undoubtedly have improved so many aspects, they also show up more starkly things that do go wrong. This reflects the premium on the rapid completion of buildings which are often inherently more complex, and can be much larger than in the past. What can be done to improve affairs? This theme has occupied numerous top-level committees, as well as those in humbler positions in the industry. The findings of the Latham Report, *Constructing the Team*, found largely the same problems existing in the industry in 1994. In particular, Latham concluded that the number of disputes that arose in the industry was a major factor which perpetuated poor relationships and poor performance in the industry. It is not the claim of the present work to provide the answers, rather it seeks to make some sticking-plaster available for application when an injury occurs. Even so, its basic propositions may be oversimplified as follows:

(a) Avoid problems whenever it is possible to do so without excessive expenditure of resources.
(b) Accommodate problems in the most efficient way, when (a) is not possible, accepting some disturbance as inevitable.
(c) Resolve any disputes which arise out of the compromise of (b) as smoothly as possible in the circumstances resulting.

This 'council of imperfection' is suggested, not as something new, but as something consciously to adopt. It is an approximation to what usually happens, but perhaps by default rather than intent. It is not a call to be careless, but to weigh the options and select the optimum and not strive for the unattainable.

CLIENTS AND THE INDUSTRY

Construction clients all want the right building, at the right time and for the right price. They vary in how far they are equipped to achieve these objectives, even if the structure of the industry and prevailing conditions give them the chance. The reasons for building and the priorities and differing objectives that different types of clients and contractors may have is explored in *Building Procurement*. (Books referred to are listed at the front of this book.)

The seeds of possible future conflict are often sown in the early days of a project and it is essential for a client and his advisers to establish their objectives from the outset. If the appropriate procurement path is chosen it is more likely that future conflict can be managed within the risks that the procurement method will contain. No method is risk-free for client and contractor and some place most of the risks on only one of the parties. If this is appreciated from the outset it may make dispute and possible conflict more understandable, so that contingency in time and/or money has been allowed for by the parties.

The entirely 'lay' client, that is one unversed in the procurement process, faces a number of special obstacles. He may not know very clearly what it is that he wants to have produced. Even if he does, he may not know how to present a brief over design or programme. He may well be puzzled by the split between design and construction organisations, how they relate to him and to each other. He may be tempted to opt for the 'cheap' or the 'quick' way to meet his need, without seeing the pitfalls that there may be. In particular, he may be too optimistic about what can be done within the money and time available. When things are rolling forward, he may not realise the effects that his changes of mind will have, even those that embody flashes of near-genius.

All clients face numbers of internal problems. Their budgets are restricted by total limits, by when the money is available and by the return (in any way) that the project must achieve. Time may be pressing to meet markets, to provide social satisfaction or whatever is the target. Technological changes in requirements may affect the solutions that are embodied in a building. These effects may be in conflict, but they may not be foreseen until work is under way. They may be compounded by the activity of building being on already occupied sites or, worse, in occupied buildings. The individual nature of many buildings has already been touched upon.

The response of the industry to clients as a whole shows first of all in a structure that has evolved over the decades. In the order in which clients may become aware of their existence, there are design consultants, cost consultants, contractors, subcontractors and suppliers, leaving aside all the more peripheral personnel such as inspectors. It is a tenet of systems theory that organisations tend to become rigid according to the functions that they have been performing and not always to be responsive to change. In addition the structure may make the flow of activities and information more tortuous than a single operation might necessitate.

In the case of construction, the process of providing a building filters through this structure, which largely facilitates it. But because typically there are several organisations involved, the possibility of gaps in communication or coordination is present. Indeed it may be said that most disputes, which are not essentially technical or due to changes of mind, occur because of disjunction at interfaces. Some of these are due to misinformation etc and some to misunderstanding of what is required. Even when there is some form of integration of activities, as in design and build arrangements, the same gaps are possible, although liability for them is shifted.

In more traditional patterns, the major contractual arrangements are set up on the basis that the client and contractor hardly speak to each other, as the nexus is to be found in the consultants for the project. It may then be argued that the objectives of the latter are not entirely identical with those of the two parties to the building contract, that is the client and the contractor. This is not to suggest lack of professional concern, but simply that they are constrained by their own upbringing. Thus an architect may not be primarily concerned with information flow, as against aesthetics and function of the finished project, whereas a quantity surveyor may not see clearly the importance of cash flow to a contractor, as against ultimate settlement at an uncertain date. Some, but not all, of the newer patterns and contracts address the communications issue (as it may partially be termed) and this is brought out on occasions in this volume.

MATTERS UNDERLYING DISPUTES

At risk of some repetition of points just made, some of the more common, if not always more obvious, causes leading to disputes of the types considered most often in this volume may be set out. Many of them are avoidable at some cost, which may or may not be justified. Sometimes a matter may best be left as a risk, and not everything can be foreseen every time. The causes are present, whether they should be pre-empted is a matter of policy, but the decision should be made consciously when the cause and risk are known to exist.

Decision making and results

A number of early decision areas always exist. It is a fact of most adventures in life that the early decisions are usually the most critical, because they condition those that follow and because they are often irreversible without an unacceptable cost. Fundamental is the care taken with the initial concept of the scheme (at whatever quality level is sought) and the design brief into which it is transformed. Hard behind comes the time allowed for both design (with accompanying planning permission etc) and construction. Inseparable from quality and programme is the product of the two, namely cost. A balance will need to be struck between the competing trio of quality, programme and cost, and the constituents of each should be examined critically at the earliest stages of a project. An optimum will arise and it will seldom be possible to obtain a client's desired quality to the programme and cost that he wants. The client's critical requirements must be matched with his resource provisions.

There are also several basic conditions which may be unavoidable and so should be evaluated closely. The site and existing buildings may constrain what can be achieved, both as an end-product and also on the way in terms of phasing and so on. The nature of the scheme may bring in particularly complex constructional or organisational methods, because it is unusually integrated or innovative. External constraints may be present, or archaeological interest may be anticipated.

Implementation of the concept divides into two areas: decisions and achievement. Decisions are tied in with development of the brief, perhaps happening in part post-contractually, and also with setting up the contract. It lays itself open to the possibility of seeking to pinch pennies while introducing undue risk for the client, for instance by making inadequate provision for elements in the programme needing time or expenditure. An obvious area of doubt is in the legal provisions, where ambiguities or gaps may be left over points of procedure. But conversely, the attempt to be oversmart and cover every option may create fresh loopholes that never existed before. Post-contractually, this type of approach may turn into lack of tolerance that stirs up its own protective reaction. Special care is needed when overlapping contractual arrangements are used, individually near-impeccable, but together causing discrepancies.

Implementation, by activities to produce the work, follows on what has been sketched in the last paragraph. Inadequate consideration of the management system for the whole operation is a likely cause of weakness. Overcoming this does not necessarily mean introducing something complicated, but rather ensuring there is a logical, unambiguous allocation of responsibilities, with a clear decision path through the whole programme and watertight procedures for communicating decisions to those who need to know. It is often argued that people produce information to suit their own perceptions of what is needed, not to give what others need to use. Rationalisation of information to suit others might actually mean less unneeded work carried out by the producers for themselves! Common problems over information are that it is incomplete, unclear, error-ridden or just plain late in arriving. The rise of some forms of contracting over recent years may be as much due to the attempt to dispose of these failures as to secure any radical difference in the way in which work is organised on the site and paid for by the client.

Even with the appropriate systems, there are still weaknesses that may creep into their use. Changes of mind are not necessarily a sign of indecision: they may show a willingness to reflect and not be blinkered, but they can spread havoc if they are not controlled. They are another example of optimisation: at a certain level of cost, a good idea ceases to be worth implementing. Furthermore, although the end results may be similar, there are no proper excuses for putting off decisions until too late or for indulging in panic action when a difficult situation does come to light. These instances may look as though they are mainly to be laid at the doors of clients and consultants, but may be the responsibility of contractors also, who can fail to foresee problems or to organise their own activities. This may aggravate a loss situation, or even lead to one which cannot be charged to the client.

Team interactions

The correct system may have the wrong team members. Consultants, it has been suggested, may be bad organisers, even worse a minority are bad at designing. It is not unknown for them to be unreasonable to a greater or lesser degree. Contractors and subcontractors too may not be well organised, and they may be of the wrong size (either way) for the work they undertake. They may lack expertise in particular types of project – sometimes in any type, it appears! It may be their fault that they accept the work, but it may be the precedent fault of others that they were ever selected to tender. They may be commercially inept on particular contracts, something that it may be difficult to foresee will happen. Tenders may be too low for profitability, work may be performed uneconomically, or (with or without these features) firms may be 'claims conscious'. Sometimes they just do not know how to secure their normal entitlements in the final account.

This survey is cast in a deliberately gloomy vein, to make its point. But when all this is said, many of those operating in the construction field are doing their level best to perform the miraculous today and the impossible not long after. It is also pertinent that authors who live in ivory towers should not drop bricks. There are also a number of factors which arise from quarters other than clients and those with whom they contract. There are the actions or inactions of local authorities and others with powers of approval or who perform work under statutory powers, there are unions and there is that special British institution, the weather. Not all of these sources produce effects over which the client may be liable to meet the extra cost arising, although he may be faced with the resulting delay to his building without means of redress. Eventually clients pay for such hazards as an addition to tenders for jobs in general. It is a matter of policy which risks are borne by the single contract and which are spread.

Position of the architect or other design team leader

Whoever leads the design effort for a project carries a particular responsibility, not only for the design itself, but for numbers of the issues which have been outlined so far. This is true even if there is some other person who is acting as project manager and who coordinates the total execution.

As a minimum, the architect (as he will usually be called hereafter) has three powers:

(a) To inspect the works with a view to approving them and so perhaps to disapproving them.
(b) To issue instructions which change what the project costs both the client and the contractor.
(c) To issue certificates about the contractor's performance, expressing approval over work and other matters, and also to enable the contractor to be paid.

These duties he performs under his contract with the client, so that he is prima facie liable to the client for any breach of his contract. But he may also be liable to the contractor in tort for negligence in some circumstances. The courts are tending to enlarge the areas of tortious liability and it seems that architects are no more immune than other mortals.

OUTLINE PRACTICAL APPROACHES

It is desirable to wind up this chapter on a fairly positive note, as something of a corrective to the tone so far. The central message has been that it is realistic to seek an optimum level of certainty in decision making and implementation, rather than some perhaps elusive maximum, the cost of which in time or money may not be justified.

This target must be set against the background of what it is practicable to achieve, given the existing structure and working of the construction industry, and the difficulty of evading these constraints at the level of the single project. It is seldom possible to tie down a scheme to give absolute fixity before the contract is placed, so that no change can deflect it from its planned path of serenity and certainty. This means there must be suitable ways of building in flexibility over the physical works, the financial arrangements and the programme from concept to completion. There are the contractual mechanisms available to satisfy most needs, and these in particular are surveyed in the next chapter. The more common aims may be outlined here.

Design freezing or development

Primary will be the degree of fixity in the scope and design of the works. In an ideal world, design would be done once, in advance, and the works would be built as designed. The reasons why this does not happen need not be rehearsed further. As stated above, a balance will need to be struck between the often competing criteria of cost certainty and programme, which may require that tenders are sought when design is not complete. A client will then be committed to expenditure when cost certainty may be relatively small. Early assessment of the degree of cost predictability and control required and achievable by a client for the type of project concerned is unavoidable. This is explored in detail in *Building Procurement*.

The practicalities of life mean that ways of accommodating changes and of settling the unsettled are needed. Too often it is assumed that the simple availability of a clause about variations gives adequate licence for anything, whereas it is intended to cover only fairly limited changes, introduced into the programme of construction work in an orderly manner and with plenty of warning to avoid disturbance of production. If more drastic alteration or late establishing of intentions is expected, a move to a cost reimbursement contract becomes desirable at the very least. Even without drastic change of intention, such factors as delay in giving information or variations out of proper time need more than a variation clause to allow adequate valuation to be made within the contract.

Four basic state-of-design situations may be isolated in relation to the time of tendering, and so establishing the financial basis:

(a) *Design complete and frozen at tender*: This has the maximum advantage for price keenness and a straight run at construction, avoiding extra costs then. It also has the longest pre-tender time and, rigidly applied, excludes flexibility over design development.

(b) *Design complete but not frozen at tender*: This has the price advantage, but then loses this during construction, according to how disturbing variations are to progress, quite apart from their own direct costs. There may be both cost and time overruns.

(c) *Design incomplete at tender and declared to be so*: This is indicated by a basis of approximate quantities or prime cost. The pre-tender time is shorter, because less design is done and tendering documents are less precise. Pricing is higher, but an element of later uncertainty is discounted in this, although the contractor is not obliged to accept dilatoriness from the client's side without recompense.

(d) *Design incomplete at tender but not declared to be so*: This is concealed by a basis of firm quantities or (with more difficulty) by one of drawings and specification. The concealment occurs because drawings in any detail are not issued for quantity tendering, and the bills do not usually look any different when the uncertainty exists. This may secure the maximum price advantage, unless detected, and may be used to shorten the pre-tender time. It leads to problems later, usually intensified to the extent that they were unexpected by the contractor.

There are situations in which clients need to rush into construction on site as soon as possible, but the simple question of starting on site does not always mean an earlier completion. Even if it does (and clients may be very pressing on their consultants), the cost of achieving it may outweigh the advantage gained. A rudimentary approach to this problem is shown under 'Client's risk analysis' below.

Contract financial bases

Contract basis is developed in the next chapter and is fundamental to much of the rest of this book, but is set in context here.

A contract basis usually contains three strands relevant to this issue. One is certainty over the amount payable, so far as this is can be achieved, for reasons already indicated. A second is some sort of a formula upon which payment is based. This may range from 'this much for the whole works', through some form of measured quantities and unit prices to payment by costs incurred. In the case of both quantities and a cost basis, the formula awaits the insertion of detailed values into its spaces to give the final amount due. The third contract strand is leeway to negotiate over such elements of excess expenditure as are due to matters which can be attributed to default of the client or his advisers. This is largely the burden of this book. Somewhere between are the effects of uncertainties which are hardly anyone's fault in many cases, such as ground conditions, the condition of existing structures, market changes and technical development.

There are risks of a commercial nature in the market, mostly widely understood and accepted and others of a less predictable nature, which may or may not be insurable. It is usually the client's advisers who set the pattern of who carries each risk, and this should be properly assessed in any contract with peculiar features. Otherwise, tenders will be loaded to cover excessive risks or clients will face higher settlements to cover the risks which they have assumed. For most regular situations, the standard contracts provide a sharing out of risks which the industry has become accustomed to view as 'fair', but the position should be watched.

When all this has been done, there is a need for mutual tolerance between those operating a building contract to avoid the onset of constant niggling over small issues of departing from 'the book'. Given the will, much can be settled by judicious give and take, while still reaching a sensible conclusion. Attitudes of scoring points or splitting hairs over legalistic interpretation can harden into lack of will to settle at all. Even in the matters considered in this book, where some of the hairs must be split, there are often issues of no moment where common sense can steer affairs through to a safe haven. It is sometimes said that most disputes are settled at the right amount, even if for the wrong reasons. The thought is comforting, although it may be suspected of providing doubtful reassurance for those who do the settling!

Client's risk analysis

The risk of delaying design and the possibilities of serious revisions are not simple matters to assess against the hope of saving time by an early start on-site. The following model does not give a 'rule of thumb' for deciding, as there is not a standard disturbance situation. It indicates some major factors which often enter into the picture and which should be weighed in deciding whether a trade-off is likely to be beneficial. When claims situations develop, it is time to consider whether the risk paid off!

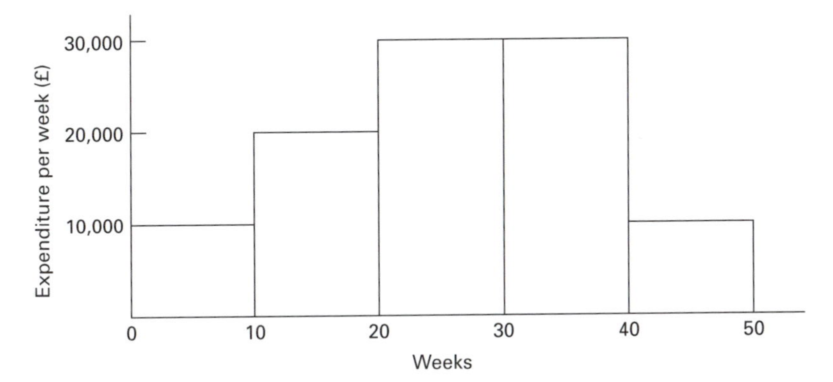

Figure 1.1 Anticipated expenditure pattern

A project is assumed with the characteristics below. The terms and ideas used in what follows are developed in later chapters, so the model here may be left until those chapters have been read, if the extra detail is needed first. Here is some data to begin with:

(a) Contract value £1,000,000.
(b) Contract period 50 weeks, whether this is compressed or not is irrelevant.
(c) Work starts 10 weeks before design is completed, the financial basis being unimportant within the broad-brush picture.
(d) Liquidated damages are £4,000 per week, and it is assumed that the benefit to the client of 10 weeks time-saving is therefore at least £40,000.

The rate of construction and pattern of expenditure anticipated are shown in Figure 1.1.

Present concern is limited to claims due to design not proceeding as intended when information is not supplied on time or is inaccurate and needing correction. Other matters leading to reimbursement are ignored, although what they are and when they occur within the total pattern will affect payment. Often they increase it, perhaps by compounding an effect. Occasionally, taken in combination, they produce less effect than they would have done separately.

It is assumed that a disturbance effect occurs only in *any one* of the five periods, so the effects tabulated are *alternatives*, not parts of a cumulative series. The effects, which are treated extensively throughout this book, are divided into two categories:

(a) *Prolongation*: Simple extension of the contract period of a nature qualifying for extra payment.
(b) *Disruption*: Occurrence of uneconomical or abortive work, whether or not the programme is extended.

It is also assumed that a disturbance in any one time period affects the next two (if present) in both these ways, although at a reduced rate, and that the results are then overcome by corrective action, paid for in the figures shown. According to the expenditure rate in a given period and those following, the effects may be rated descriptively as follows.

Period	Immediate effect	Knock-on effect
A	low	high
B	medium	high
C	high	high
D	high	medium
E	low	nil

These suggest that there is greater loss and expense during a period when work is proceeding more rapidly and when there is a greater quantity to follow, as might be expected. Subject to these considerations, there is also more risk of knock-on effects when the initial impact occurs early, although the scope for recovery of time is also greater, if the effect can be overcome. Much depends on how an effect can be contained. If a disturbance occurs in a clean-cut way, there is more likelihood of deferring work without much disruption, although prolongation will still take place.

Period	Expenditure (£)	Immediate effect (£)[a]	Knock-on effect (£)[b]	Prolongation (£)	Total (£)
A	100,000	5,000	(2%) 10,000	(3 wks) 15,000	30,000
B	200,000	10,000	(2½%) 15,000	(5 wks) 25,000	50,000
C	300,000	15,000	(3%) 12,000	(6 wks) 30,000	57,000
D	300,000	15,000	(3%) 3,000	(4 wks) 20,000	38,000
E	100,000	5,000	–	(2 wks) 10,000	15,000

[a] The figures in this column are 5% of the corresponding entries in the expenditure column.
[b] The percentages in parentheses are applied to the expenditure figures for the next two periods, e.g. the 2% knock-on effect for period A is applied to the expenditures for periods B and C to give 0.02 (£200,000 + £300,000) = £10,000.

The prolongation rate is given as including the loss of liquidated damages as well as reimbursement of the contractor's prolongation costs, so each of the totals has to be set against the £40,000 saving due to the initial overlap of design and construction time by ten weeks. The occurrence of two sets of disturbance, or a slow dribbling disturbance can have heavier effects, especially in prolongation. As the figures stand, the isolated disturbances in Periods B and C each cost more than the value of the project time they save the client.

CHAPTER 2

CONTRACT SOLUTIONS AND RISK DISTRIBUTION

- More traditional contract bases
- Assumptions underlying the more traditional bases
- Less traditional contract bases

Several responses are possible in the light of the problematic nature of organising building contracts. It is possible to decide to do better next time, by being on one's toes rather more often and treading on other people's less often. It is possible to ignore the situation and just press on. Neither of these takes a positive account of where the problems might lie. A more realistic response is to accept that the situation is indeterminate in some ways, but that it is possible to respond with such precautions as are available by way of appropriate contract arrangements – as well as doing a bit better! This chapter reviews the common alternatives for procurement methods and forms of contract that are available for a chosen procurement route in terms of their strengths and weaknesses for present purposes, so that some may be expanded in later chapters. Others are given to afford some basic comparisons.

A more comprehensive discussion of contracts and procurement methods is given, in *Building Contracts: A Practical Guide* and in *Building Procurement*. Before a client and his advisers consider the type of contract that they require, they should analyse carefully their building needs and agree the priorities for finished product, overall programme, and all-in price for everything they will need to spend to achieve their objectives. This is explained fully in *Building Procurement* and it should be understood that the contract solution will arise from the procurement method, not the other way around. The distribution of risks between contractors, consultants and client will be part of the criteria used to determine which procurement route is chosen. Figure 2.1 shows a broad assessment of the risks between the three types of procurement, using the assessment criteria listed in the Building Round Table report *Thinking about Building*.

The more traditional arrangements described in *Building Procurement* and below have evolved over a long period and have become those which clients, or their advisers more often, put forward to contractors as the contract basis. By contrast, the newer methods of design and build, construction management and management contracting mentioned have appeared as specific reactions, it must be suspected, principally from contractors against failure to operate the older methods properly. As often happens, out of such reaction have come methods valuable in themselves, so that clients are embracing and suggesting them quite often to contractors, sometimes despite their advisers in the first instance. Probably these roots give the

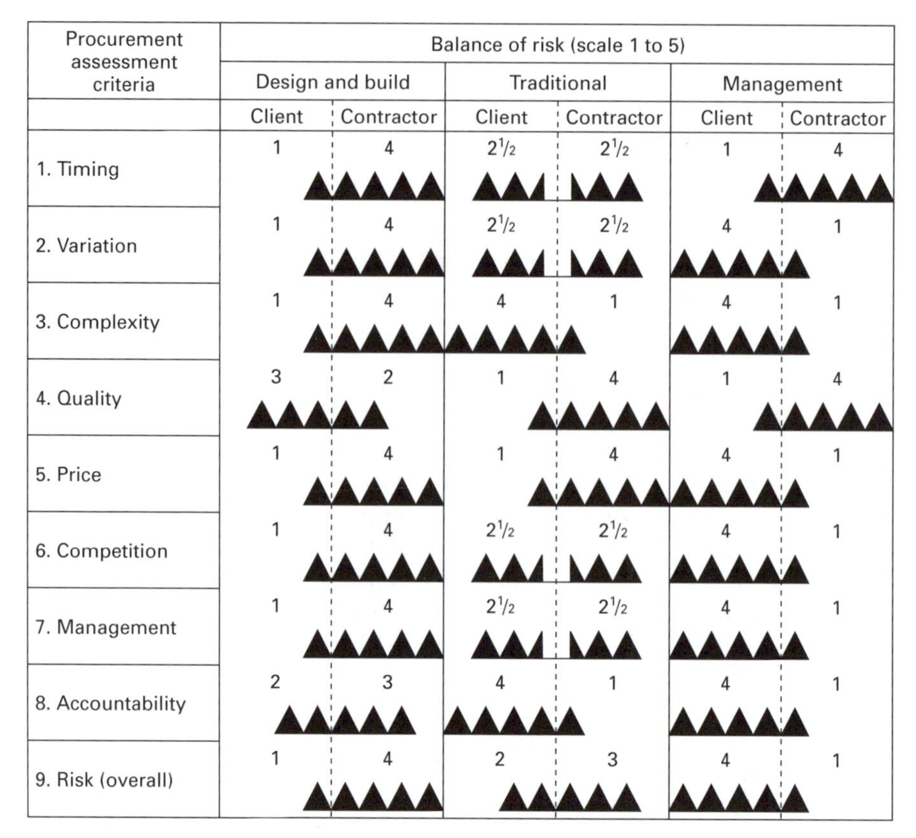

Procurement assessment criteria	Design and build: Client	Design and build: Contractor	Traditional: Client	Traditional: Contractor	Management: Client	Management: Contractor
	Balance of risk (scale 1 to 5)					
1. Timing	1	4	2½	2½	1	4
2. Variation	1	4	2½	2½	4	1
3. Complexity	1	4	4	1	4	1
4. Quality	3	2	1	4	1	4
5. Price	1	4	1	4	4	1
6. Competition	1	4	2½	2½	4	1
7. Management	1	4	2½	2½	4	1
8. Accountability	2	3	4	1	4	1
9. Risk (overall)	1	4	2	3	4	1

Figure 2.1 Assessement of risks between the procurement routes

fundamental reason why the newer approaches need less consideration in a book on claims and other disputes: they are adopted in part to avoid the contentious areas caused by failure to use the other methods properly.

Construction covers many technical types of work with great variation of scale and where parties seek different risk distributions. They are affected by many factors, including the desired speed, timing of design and information communication, and the extent and nature of subcontracting. There is room for all the methods of contracting in the complexity of today, provided that a suitable one is chosen in any instance and not abused.

MORE TRADITIONAL CONTRACT BASES

Here 'traditional' simply indicates that these bases (where design is separate from construction) have been in use in a substantial way throughout the twentieth century at least (how times change), so that the bulk of work has been performed under one or other variant of them, and still is, as opposed to under design and build or construction management/management contracting. Incidentally, JCT Practice Note 20 gives advice on considering the appropriate form of contract for building work from within the JCT range of forms.

Drawings and specification

Used to give a financial basis of a simple lump sum, drawings and specification are the prevailing method for a fairly small, uncomplicated project, such as a single house or a straightforward shed-like building. Legally it gives an entire contract, as described later in this chapter, in that the basic commitment is to pay the contract sum for the contract works, neither more nor less. It is intended for projects that have been fully designed before the contract is placed, because it contains minimal provisions for pricing variations. There may be no provisions, but desirably there should be an unquantified schedule of rates (described more fully in Chapter 6) to give details for incidental variations that occur. Such a schedule is usually derived from the contractor's build-up of his tender, that is his own 'builder's quantities' structured and priced however he may choose, rather than in relation to any standard method of measurement. Some editing of the wording of the schedule may be negotiated by the architect or the quantity surveyor before the tender is accepted, to make it more suitable for its purpose, but it is essentially proffered by the contractor. As such it is open to be construed against him on the contra proferentem principle discussed in several later chapters, as an independently prepared set of quantities is not (see next heading), in the event of disagreement over its meaning.

This type of contract works very well for those jobs for which it is intended, by cutting down on unwanted detail. It allows marginal variations to be priced from the schedule, but introduces risks over substantial variations. There are two reasons for this, the first being that the contractor carries the risk in his tender of errors in his quantities, because only the prices out of his build-up become part of the contract. Of course, this risk works either way, although major overestimation of quantity is likely to lose him the job in the first place. The client is also at risk, by the same token. If then a major variation occurs, say by way of omission of the whole of an item, it will be priced at the unit rate in the schedule, but measured at the actual quantity involved, rather than that on which the contract sum was based. It is presumably the case that elsewhere in the contract sum, the contractor will have amounts effectively compensating for such over- or underprovision, but he may find himself in difficulties if the quantity error is substantial.

The second risk arising out of such a schedule is that of deliberate manipulation of the schedule pricing. This is related to the former risk. Because the quantities do not form part of the contract, the contractor may be inclined to adjust his normal level of pricing for particular items, according to whether he expects an increase or decrease due to variations. This is a dubious practice and can easily rebound, if the direction of variations has been misjudged, as considered further in Chapter 3. It is also an option that may occur to whoever is 'editing' the schedule before acceptance, especially in the light of inside knowledge of pending variations. The contractor is quite entitled to resist editing that seeks to change prices which are genuine, even if high in the opinion of others.

The overarching risk with this form of contract is that it is used for projects which are too large and/or complex to be based on such limited apparatus to cover variations. These may be extensive in scope or large in value or, worse, both. Any weaknesses in the schedule content will be magnified on the larger contract. In a highly competitive environment, there are pressures to use this system far beyond normal limits. Much depends on the relative expertise of the parties as to who comes best out of the resulting encounter, but this is hardly a commendation of the practice leading up to it. There is evidence that the JCT minor works contract and the IFC contract (the latter discussed extensively hereafter), are both suffering from this abuse.

Should there be such occurrences as disturbance of progress leading to a claims situation, this type of contract can only be more vulnerable. These occurrences are less likely in smaller contracts, or less likely to be made an issue because they are intrinsically smaller, but the basic contract provisions are the same, while backed up by less financial detail in the schedule.

Firm bills of quantities

Firm bills of quantities remain the most common way of arranging the financial sides of the larger contracts, which was intended to mean over about £150,000 or rather more for simple works. Inflation has moved this figure on somewhat but it should be a sound basis for consideration. They are still lump sum or entire contracts. The quantities are part of the contract, but are present to explain what is in the contract sum and not to offer an opportunity to either party to select which parts of the works the contractor will perform.

The regular contents of bills of quantities may be divided into three types. There are preliminaries, so called because they come first in the documents, but more properly relating in large part to what are the overheads or what the economist would call the fixed costs of the project. The bulk of the content is in the actual quantities, which give the amount of work to be produced, and so are more the equivalent of variable costs, in that a fairly small alteration in quantity does not produce a corresponding alteration in the preliminaries. They are variable, but only in ways which the contract sets down. Then there are several types of item which may be grouped together for present purposes: prime cost sums, provisional sums and daywork. These all represent allowances for work still to be instructed when the contract is entered into, and so somewhat uncertain for the contractor. In fact, some of the allowances may never be needed at all and, except perhaps for daywork, the tenderers have to include the same sums in their tenders. All of these receive further consideration in various chapters.

The pricing of quantities lays itself open to similar practices as may occur with schedules of rates, in drawings and specification contracts and elsewhere. Where they receive distinct criticism (although it is really a question of degree) is in the facility they afford to have variations valued. This may seem a strange criticism, and put like this it is, for it is a great strength that the means of controlling changes in the sum payable are so readily at hand. But what those concerned with the design of buildings for the client can be lulled into believing is that variations can be introduced willy-nilly, with the contract formula for their valuation taking account of all the change in cost that is incurred by the contractor. But enough individually innocuous variation instructions can add up to a radical change in the works and their conditions of execution, so that provisions about loss and expense due to variations are also needed. These allow the excess cost to be reimbursed, but lead some designers at least to feel under attack when claims are presented. They may be right.

Even worse is the fact that 'firm' quantities can be used as a cloak for indecision. Designers can cherish an illusion that everything will come out right in the end. Quantities may be so presented as to look firm, but really be at best an intelligent guess at what the designers will eventually come up with. There is an old saying about a crystal ball. There are even more variations to contend with, while the contractor may have been misled into assumptions over precision about the work when pricing these treacherous wolves in sheep's quantities. The ground is prepared for uncertainty as to what was required and what has been priced, which can hardly help in settling what may be a difficult contract anyway.

The lesson here is that firm quantities should be used when conditions are firm: the client knowing what he wants, and the design team having designed it fully. Uncertain elements can be identified as provisional, but otherwise the contractor can trust what he gets. Human errors in quantities are always possible; these will be corrected. If the use of such quantities disciplines early firm design, and this is possible, there can only be gain. If this is not attainable, it is better to think about the next option.

Bills of approximate quantities

It must be admitted that simply labelling bills as approximate does not dispose of the problems – these are just admitted from the beginning. These bills are sometimes attacked because they lead to higher pricing by tenderers, but is this not reasonable? Anyone entering into a less certain commitment is entitled to expect more for doing so, and possibly will be less awkward about settlement as a result. It is true that approximate quantities are an invitation to defer decisions, but when decisions cannot be made they have their strengths. Equally though, the facility to measure what is actually done once when it happens, rather than to start with something and amend it later, does not dispose of the extra costs of disturbance or working out of sequence.

If bills are made approximate, only the measured sections should be affected. The preliminaries still need to be definite, otherwise many of the ground rules for the project will be uncertain. Nothing is impossible here, but it is the area where the greatest effort to be precise is needed.

In the main, discussion of quantities in this book is related to an assumption of firm quantities, as what is said applies often to all quantities. Where necessary the distinctions over approximate quantities are drawn. When the contract is in being, they become effectively a schedule of rates (see below) for final valuation, but one where the basis of pricing is acknowledged to be affected by the weighting and proportioning of the original quantities. It is difficult to assign any confidence level to these quantities, although tenderers must make some assumptions – and put in the higher prices just mentioned! Differences between the parties over the assumptions which should have been made must be expected and resolved against the built-in uncertainties.

This basis is still that of a lump sum, but with the sum actually calculated at the end of the period.

Schedules of rates

Schedules of rates are mentioned above, as an adjunct to drawings and specification contracts. The present reference is to contracts without any initial lump sum, but based upon a schedule of rates without quantities, so leading to complete measurement of work as executed. In this case the schedule is prepared by the client's quantity surveyor and is based upon defined rules of measurement, such as those in a standard method of measurement. Effectively, the schedule is used just like bills of approximate quantities once the contract is running.

The difference is that when tendering takes place, there are no quantities available, just the unquantified schedule. Tenderers therefore must decide as best they can, the disposition of work in terms of the proportions of work sections or trades. They may not even have a totally clear picture of the nature of the work or its scale, although in such cases some additional formula may be included to allow adjustment for at least the variation of scale. Usually they are pricing by taking a prepriced schedule and quoting percentages on or off sections within the schedule. This means that on top of uncertainty about the work, they are forced to accept given differentials between items within a section, which may actually cover quite a range of work. This prepriced approach is the most practical way of obtaining tenders that can be compared with reasonable ease in the absence of common quantities to give immediate comparability, the alternative being to compare every price in an initially unpriced schedule between tenders. It does not help tenderers though, unless they are used to the particular schedule and know its idiosyncrasies.

These schedules lead to higher prices and to more negotiation of additional prices during valuation of work under them, because of the uncertainty that surrounds their use. Usually they are restricted to cases in which the whole contract is beset with uncertainty, such as term contracts for a running programme of small works and maintenance. Here work crops up over a period of two or three years and is ordered and paid for as a series of single jobs. Commonly, there is a break clause in the basic contract allowing either party to get out at a few months' notice, and this not infrequently happens. This arrangement is thus convenient for what it covers, rather than particularly economical. Because of the small sections of work and the option to break off, it does not often lead to disputes of the types considered in this book, as the problems of operation which arise can usually be accommodated in the provisions for adjusting preliminaries and other allowances for overheads. The basis is not considered further within the scope of this book.

Prime cost

The prime cost arrangement is otherwise known as cost reimbursement. It is most frequently used when the nature of work is uncertain, as well as the extent, and when prior estimation of a lump sum price or of unit prices is therefore impracticable. It may also be used when speed precludes other approaches, or when the client has a special relationship with a given contractor. Its essence is that the contractor is paid for most of his direct expenditure on site at the level incurred, such as hours of labour and plant and quantities of materials. Site overhead costs, head office expenditure and profit are covered either by a lump sum or by a percentage on the directly reimbursed expenditure. The rates for labour and plant and the allowances to be made with materials for elements such as discounts, must be defined and dealt with precisely. Competition between tenderers is limited to these items and no overall tender amounts are obtained, so that often only one contractor is invited to quote at all. It is more relevant to secure a reliable firm which will work efficiently and economically, than apparently competitive rates from uncertain firms.

In operation, these contracts do not lead to disputes in the same way as contracts related to the other bases. This is because they give 'payment for what is spent' in the main, rather than 'payment for what is produced' as do the others, again in the main. If there are disputes over this part of the contractor's remuneration, they are usually over whether he has been wasteful in his use of resources, and whether the client should therefore meet the full amount.

There are usually provisions in the contract for the architect to decline payment for excessive levels, or even to order labour etc to be taken off site. What is excessive can be contentious, as it may well impinge on the contractor's organisation of the work and timely completion, which remain his responsibility. Such a disagreement could be about the substitution of plant for labour, for instance, and lead to division of responsibility over the level of total expenditure at the end of the work. Under other contract arrangements, the architect has even less power to intervene, a point taken up several times in later chapters.

Alternatively, the amount paid on the prime cost basis is regulated by some 'target' cost for the project, agreed in advance against an estimate based on shadow bills of quantities. Deviation from this target results in an adjustment of what the contractor actually receives. This may be set so that he shares the difference up or down with the client or, not infrequently, so that he concedes the whole difference if this is a saving. The estimate then becomes a maximum. The salient area of dispute here can be whether the works have been varied, while the estimate has not been adjusted sufficiently to take account of this. Although variation instructions are needed in such a contract, they may mistakenly be regarded as of less importance, as the contractor is being paid primarily for what he spends. Proper attention to both legs of the financial system is therefore crucial.

It is, however, the case that quite a substantial part of the price payable is derived from the addition made to the actual prime cost for overheads etc. When this is by way of a percentage, it clearly varies as a direct function of the other inputs. This is the accepted contract basis, so there can usually be no argument over its adequacy. When it occurs as a lump sum and there is a gross distortion of the prime cost from what was anticipated, there is room for readjustment even without any element of disturbance of work. It follows similar reasoning to that for the adjustment of preliminaries (Chapters 6 and 31).

ASSUMPTIONS UNDERLYING THE MORE TRADITIONAL BASES

Although they do not apply uniformly in each of the bases, several common assumptions apply in most instances. Even in the case of prime cost, most of them are relevant.

Entirety principle

The contract is held to be 'entire' – there is an obligation on each party to perform his part in return for that of the other, and each may expect the other to undertake the whole of his obligation. Whether or not a contract is an entire one is a matter of interpretation of the parties; it depends on what the parties agreed. 'Complete performance' of the contract by the contractor is a condition precedent to the liability of the employer to pay. Thus, the contractor may not break off the works part way through without some penalty – in a contract without any relieving provisions, he will forfeit the right to any payment for what he has done. Equally the client may not choose to have the works terminated early without facing an action for breach, again unless there are other provisions. In particular, he may not extract work from the contract and give it to another to perform, at least not while the remainder of the works is under way. To operate a contract under these precise assumptions, it is necessary to have complete definition of the works, in nature, scope and detail.

All the forms of contract referred to later in this book include clauses that modify the rigorous position. All have provisions for effecting variations to the works, that is changes in what is to be built. Some have provisions requiring the contractor to accept that defined sections of work will be executed by subcontractors, who are sometimes to be chosen by the architect. The variations arrangement expressly, and the subcontract arrangement implicitly, allow the possibility of introducing incremental changes which may affect considerably in aggregate what the contractor produces. In all cases it is intended that the changes shall be instructed in such a way as not to disturb regular progress. Otherwise, extra payment arises.

Design and construction

It follows from these points that design must be done entirely or substantially before the contract is awarded, both to allow proper tendering and for work to proceed in an organised and, hopefully, profitable manner. Even a prime cost contract constituted like the JCT prime cost contract, where there may be a lump sum but adjustable fee or a percentage fee, assumes a reasonable definition to make the fee a feasible option, even if prime cost is not on wider grounds. All the above arrangements are based upon design by an architect or other consultants acting with him and under his coordination. The contractor has no design responsibility expressly given to him, and it is a matter of doubt whether any can be passed to him by roundabout means under these contracts.

The construction methods needed to produce the designed works are, however, entirely the contractor's prerogative, unless any stipulations are made in the contract documents. The programme details are his concern also, so long as he meets the end date and any intermediate dates over phasing that may have been given.

The various provisions over these elements delineate the major responsibilities reasonably clearly. Other risks over injury and damage are also allocated, while there are provisions allowing or excluding the possibility of the contractor recouping expense due to market fluctuations in his costs. It is generally true that time is important in the contracts, so there are provisions over liquidated damages for delay in completing, but money appears to be more important. Thus, it is common to find that extra time can be granted, but earlier completion (even to make up lost time and with extra payment) cannot be insisted upon by the client.

There is some imbalance in the way the parties are cast in the contracts, due to the traditions of the industry. The client has very much a back seat, and even those matters which he may well initiate are communicated to the contractor by the architect. He is silent unless affairs reach the possibility of a determination (Chapter 18). By contrast, the contractor is in the front line and highly active, as he must be. The balance is held by the architect, with the quantity surveyor in a mopping-up role. Although both are engaged by, or even on the staff of the client, they are required by the building contract to act between the parties and hold the scales over differences between them. The architect even has to hold the scales when he is being weighed in the balance – not the easiest piece of acrobatics! These issues affect a number of points discussed hereafter.

LESS TRADITIONAL CONTRACT BASES

Although they need less consideration later, the position of the newer bases may be set out here. In many respects, they do not diverge from the others, which itself is worth stressing. If they are less prone to disputes, it is because of the way in which the parties approach them, rather than anything constitutional about the contractual patterns.

Design and build

Design and build, otherwise known as package contracting or the all-in service, may be seen as a variant of the drawings and specification contract already described. Its major feature is the transfer of design responsibility from the client's side to the contractor himself. It is common for the client to provide quite a large amount of preliminary design work as part of his brief, perhaps up to sketch design stage. The contractor takes over and develops as much or as little as he gets and is responsible for the whole, subject to the possibility that he may not have been able to check some of the assumptions on which he was instructed to proceed.

Current practice is mainly for a client, or his designer, to produce a scope design and then require a contractor to 'develop and construct' from that scope. Very little pure design and build is currently carried out where the contractors compete on submitting different designs in answer to the client's requirements. This is because it is expensive to have, say, four contractors each design different solutions and there are then difficulties for the client in comparing tenders based on different designs. In 'develop and construct' a reasonable amount of construction details, including foundations, structural solutions and engineering services may be left to a contractor to finalise, but generally he will have to seek approval for his 'development'. It is therefore common for a contractor to present a design which is not, indeed cannot be, finalised in all of its details at tender stage, although even before the contract is made, the design should be complete enough for the client to see what he is going to get. The client may still require changes to the design after the tender has been accepted, although how they are implemented is primarily the contractor's concern. Because of requirements for a contractor to obtain approval to the way in which he has developed his design, there may be plenty of opportunities for 'claims' to arise over responsibilities for delay and/or extra costs arising to a contract, if the employer's approval process does not meet a contractor's assumptions made at tender stage. Several high-profile projects have recently experienced protracted approval times for design development, and these have led to claims.

The contract sum is again a lump sum, as perforce it must be when what lies behind it in design comes about in the manner indicated. The contractor gives a contract price which is fixed even though the design is incomplete, although the price will be amended to accommodate the client's later changes. To aid in these calculations, there is usually a contract sum analysis, corresponding to the schedule of rates under the drawings and specification pattern. Its form is variable, but for various reasons (including the potential development of design) it best takes the form of a number of lump sums, breaking down the main sum for the whole contract.

Many of the provisions of such a contract over progress, disturbance and payment are very similar in structure, and sometimes detail, to those in other contracts, needing comparatively little independent consideration here. (Chapter 24 covers some significant points; more detailed consideration may also be found in *Design and Build Contract Practice* and *Building Procurement*.)

Multiple contracts

The term 'multiple contracts' is used to cover the pattern in which there are two or more direct contractors to the client working on a site in operational equality, in that no one of them is the 'main' contractor with possession and control of the site throughout his contract period. This is distinct from what is intended to be the relatively incidental presence of other 'persons' who come on site and perform work with the contractor's agreement and in circumstances to which he is also in agreement, but not under contract with him. This is what the JCT contracts usually provide and the contractor still has an otherwise full and unimpeded possession of the site in the regular sense. A multiple contract position may be viewed as equivalent to that of subcontractors when there is a main contractor, but with the client taking the place of the main contractor so far as control of site operations goes.

An arrangement like this is used by clients who are having major works performed in which there may be several radically different types of construction or installation, when there is no clear candidate for general contractor either in terms of expertise or bulk of work within the project. It is judged more advantageous for the client or his consultants to provide the overall coordination. It may also suit cases in which no contractor is on site throughout the whole construction programme, but some of the several contractors overlap in execution while others perform their work more or less end to end, and in which the development of the project design, or even of its concept, is spread over a lengthy period, so that compression is desired. This approach is suited to projects like refinery, power station and other complex engineering constructions, where there is a client of sufficient experience and sophistication to take a very active part in proceedings on a day-to-day basis. It may be that considerable portions of the design are performed by the contractors who carry out the fabrication and installation. Any suggestion that the arrangement may be used on a regular building project of normal size, as a way of cutting out a layer of overheads by removing a general contractor's supervision should be resisted by the client's advisers; it is usually false economy.

An approach of this nature usually arises out of interaction between scale, specialisation, design, time and uncertainty; it may, and usually does, lead to all manner of procedural and organisational complications, some of which are outlined in Chapter 13. It clearly needs contract terms that are in many respects quite distinct from those generally used in this book, terms whereby the client picks up many of the unresolved or contingent costs and other responsibilities.

The pattern of management contracting to be considered in the next section represents something of a halfway house, in that there is effectively one direct contractual link for the client, although there is a rather hybrid system of works contractors, in the JCT form at least. A related but full multiple-contracts version for building work comes about with construction management (also noted below). With any multiple-contractor approach, a highly developed form of project management (also noted below) becomes a necessity, as may be gathered from the points made in Chapter 13. All of these options represent something of a return to the trades contracts system in use in England during the nineteenth century and more recently in Scotland.

Management contracting and construction management

The essence of a management contract, or the now more popular system of construction management, is that design is still generally performed by the client's consultants, but that an additional operator, the management contractor or construction manager, is introduced into the basic structure. The details of this arrangement are open to numerous variations, but generally a contractor or manager performs no work on site other than to provide a team to manage the construction that happens there. There will also be the advantage that the contractor or manager will be able to arrange and manage preconstruction activities and to preorder materials and contracts. All work is carried out by a series of separate contractors in a direct relationship with the contractor or in direct relationship with the client, as under the construction management system, but coordinated by the management contractor or construction manager. He also coordinates the interface between these numerous contractors and the design team, who manage their own internal affairs as usual, subject to pressures put upon them by the management contractor or construction manager to deliver information.

Construction management is closely related to management contracting in that there is a series of separate contractors performing the physical works (in this case including the provision of temporary facilities), but the various contractors are in direct contract with the client, who also provides overall coordination. It may be regarded as the full form of the multiple-contract approach outlined above and considered for particular reasons in Chapter 13.

How each of the site contractors is paid depends on the details of the contract with the client, although these will have common elements. The quantity surveyor generally remains responsible for settlement overall, but there is also a responsibility on the management contractor to ensure that the budget which he agrees with the quantity surveyor is not exceeded without authority. The management contractor usually cooperates with the design team to ensure that the design is developed with a view to ease of physical construction, but he does not assume any design responsibility, nor can he dictate on matters of design to those with responsibility.

The JCT management contract in fact consists of a group of contracts and other agreements and has had comparatively little exposure in practice or in the courts. Following its issue, the popularity of the management contracting system waned, to be replaced in part by construction management. It uses the concept of works contractors to deal with those persons in a contractual relationship with the employer and the management contractor, but controlled by the latter. At present no standard form of contract exists for construction management.

The emphasis in management contracting and construction management is upon professional management offered by the manager, who is acting for the client alone. It is for this reason that it is better for the contractor/manager not to perform any parts of the work. If he did, he might be faced with a conflict of interests between the overall management and cost and what suits his own share of the work. It is argued that this contract approach gives economic, buildable design, with shorter design and construction programme times, both on site and from initial concept.

If any dispute arises between the client and his management contractor or construction manager, it can be seen that it will differ from those with a contractor in the traditional role because of the management contractor's responsibility for coordination, and total progress and budget: all partly shared with others. Because of some of these difficulties, construction management has now almost replaced management contracting in popularity. Difficult areas of dispute in these rather specialised contracts are beyond the scope of this book but general principles will often be applicable. In the case of the other contractors, who are basically subcontractors, such as the JCT works contractors, disputes are more likely to fall into the traditional pattern: they are probably paid directly by the client, but are operating rather like subcontractors in other arrangements. They may therefore cause delay or disturbance to others, or suffer similar problems themselves, so that claims etc of the types discussed in Parts 3, 4 and 5 of this book are possible and covered in the contracts to allow for settlement and also any contra-charges which may arise.

Arguably, management contracting and construction management require a sophisticated client and project management capability acting for him in order to be most effective. This procurement system is discussed in more detail in *Building Procurement*.

Project management

Project management may be distinguished in two forms. In one form it involves the traditional pattern of responsibilities and contracting between client and contractor, but with the introduction of a project manager within the client's team. The project manager will often be responsible for management of the contractor and the consultants. He may have a background as a contractor, but may well have come from one of the design professions. As in management contracting, his role is to coordinate the whole matter of design and construction to secure the best integration, but not to organise the interaction of the various persons who may be performing work on-site. This, and responsibility for their payment, remains with the contractor as usual.

This pattern does not change any contractual relations, other than to give an extra one – the project manager's own! The project manager may be regarded as an extension of the client: his own man up front, not needing to be mentioned in the traditional forms of contract. Alternatively, he may be seen as crystallising what the architect has traditionally done in coordinating the functions of the rest of the consultants, while also being the primary dealer with the contractor. For present purposes, project management does not therefore introduce any new dimensions. Nor does it dispose of any old ones, although the aim is to avoid some of the interface problems that lead to disputes.

A project manager is given named status in GC/Works/1 Contract (1998) and the New Engineering Contract (NEC). In both forms the function of design is deliberately separated from the design function.

In the second form, the project manager performs similar functions, but in relation to a construction management arrangement or some other variant on the multiple-contract pattern. Here he also acts to coordinate the various contractors, so stepping into the organisational role of the management contractor. There are numerous ways of setting up this approach to allow for the variety of situations it is intended to meet.

CHAPTER 3

TENDERING: CONTRACTOR'S POLICY, EXAMINATION AND CONTRACT

- Contractor's policy
- Examination and acceptance of the tender
- Offers, statements and representations
- Resulting contract (or not)

The first two chapters have reviewed the circumstances of disputes and the types of contract arrangements, more commonly emanating from the client's area, which have evolved to give workable arrangements. This chapter looks at the way in which tenders and their assessment can affect subsequent actions over disputes. This is very much the stage at which the potential contractor is making the running, but when the client's advisers can also sow the seeds for future trouble, or sift them out. Whenever 'contractor' is mentioned, 'subcontractor' may be substituted, in relation to either the client or contractor as appropriate.

For simplicity, bills of quantities (usually firm) are assumed, and competitive tendering is considered rather than negotiation. As these conditions give the most complicated information by way of detailed pricing, they also provide most opportunities for the interplay of different tactics in structuring the pricing and in disentangling any snags before settlement is in prospect. Other patterns are not devoid of these matters, and the corresponding possibilities may be inferred.

CONTRACTOR'S POLICY

Tendering

Tendering is broader than pricing and covers the higher-level decisions about a particular contract. These include preliminary decisions about whether to tender at all.

For all contractors, these are conditioned by the nature of the project. This means attention to the type of work, its scale and location, and such matters as the time given for its construction. It is also important to weigh up the client himself, whether he is stable, known for changing his mind or for being a bad payer, frequently late. Some of these things lend themselves to counteraction by way of extra reimbursement: late payments do not usually do so (Chapter 8) but have determination as a last resort, whereas insolvency can be countered hardly at all. Unless he makes allowance for those matters that do not completely put him off tendering, the contractor can expect a hard time in attempting to make up losses that he did not provide for and that are without remedy.

With these issues cleared, the contractor still has to evaluate the risk elements of the contract itself, such as the market conditions nationally or locally when set against the fluctuations provisions, and whether any undue risks of a contractual nature are to be assumed or may be negotiated. In all these lie his profit margin – or the lack of it! This chapter is not a treatise on tendering as such, but simply illustrates areas where future disputes may germinate.

Normal pricing

Unless there is an extreme deviation, the effects of the decisions so far mentioned are difficult to detect, whether made well or poorly. They are blanketed by the overall tender and do not show in the detailed pricing. Discussion of the latter now is related to net pricing policy, that is exclusive of the risk and profit levels set as a board decision. If ascertainment of loss and expense in particular is related to some analysis of the contract bills (by no means always the case; see Chapters 11 and 33), then the structure of pricing can influence the result, or at least how easy it is to arrive at it.

The more straightforward matters are those aspects of measured items which usually are not required to be stated in some special way under the rules of a standard method of measurement, but which affect the price level, sometimes substantially. These include the quantity of an item or group of items of similar work, as scale especially affects the use that may be obtained from plant. The JCT provisions specifically refer to this aspect as one applying to ordinary variation valuation (Chapter 6). Allied to this and not expressly mentioned in this way, but implied in the term 'similar character', are the questions of repetition and intricacy, the presence or absence of either of which can affect the cost of 'identical' quantities of work. This is especially true of alteration work, or other 'piecemeal' situations. Phasing or the timing of work within the overall contract period can affect what work will cost. Sometimes these considerations will influence whether to sublet the whole or even part of the work given together in the bills.

This is a difficult area in general for pricing, and standard methods are wary about such detail. This is understandable, as projects vary so much in their detailed character. It is also the case that most contractors would readily admit that they cannot obtain site costing information in sufficient detail to make it sensible to try to price all of these factors, even if they were made known, and if estimators then had the time and the aptitude to take them into account. One major national contractor is reputed to maintain that it is practicable to isolate seven cost centres only, without achieving any subsidiary detail.

These considerations mean that the approach when tendering in most cases is to assess the general incidence of these unspecified factors and make some allowance in prices on a fairly broad basis. As a result, disputes over the effect of disturbance and so forth start from a basis of uncertainty over what conditions the tender really anticipated. Furthermore, such matters as loss and expense are not to be argued from the tender basis, but on the actual deviation from what should have happened. They must therefore be argued from an average set of conditions, related to items actually more or less costly in the first place and then affected by events during progress! This theme recurs later (Chapter 11).

Abnormal pricing

Deliberate attempts to distort the inevitable mechanics of pricing can also be made. A common area of difference of opinion is how to deal with preliminaries. The extreme split is between those who price them separately and those who spread their amounts as a percentage or other allowance over the prices for measured items. If preliminaries represent costs which are incurred in a way not related directly to the remaining costs of the project, the former method has the advantage by simple logic. Such costs are examples of the economist's so-called fixed costs, which may remain fixed for some variation in other costs (the variable ones) within a band, and then change outside the band in either direction at a rate or in discrete jumps not bearing a direct relationship to the variable costs.

It may be suspected that those who do not segregate these costs when tendering often do not know with any precision how they are incurred, and so take the easy approach. This may work reasonably well with work of closely similar character, but may lead to disaster with unusual work having an unusual concentration of site overheads etc. Whether separate pricing occurs or not, the prices given in the contract bills are not to be used in the ascertainment of loss and expense, whereas those for preliminaries in particular are to be adjusted in relation to variations on quite distinct principles from those used for measured items. However, realistic values can offer useful guidance in some circumstances of investigation (Chapters 11 and 33). It is dubious to suggest that spreading the value of preliminaries over the measured amounts helps the contractor to keep his options open when negotiating extra amounts. Such practices tend to rebound.

More common manipulation occurs over pricing measured work, and two aims may be distinguished. One is to enhance cash flow by front loading the pricing. This means that those sections of work performed earlier, such as excavations, are priced higher than true cost, whereas others like finishings are priced lower. The result is that the job provides extra working capital early on. This may be fine, so long as the fact is not forgotten by the contractor part way through and too optimistic a view of profitability is not taken. It is even less happy for the client if insolvency of the contractor comes, and there has been an overpayment in the earlier stages. This point recurs hereafter: the significance here is that it may affect not only cash flow, but also the pricing of variations and settlement in some circumstances.

The other aim of manipulation is to try to anticipate the way in which the project is going to shift in terms of variations. Correct spotting of items due to increase in quantity and higher pricing than usual in consequence, will bear obvious fruit. This is also true of spotting decreases and pricing accordingly. It is correspondingly untrue when the wrong forecast is made and related action taken. Such provisions as those in the JCT forms link variation valuation directly to the prices in the contract bills, so that the consequences follow. As already hinted, the ascertainment of loss and expense does not follow directly, making attempts at disturbance spotting an unproductive activity. But again, price distortion for any reason can have unexpected effects if it becomes necessary to seek collateral information from the pricing.

EXAMINATION AND ACCEPTANCE OF THE TENDER

Pricing

Yet another avenue of distortion may come about not by the contractor's planning at all. This is alteration of pricing during examination of the tender. This may be urged by the quantity surveyor to secure consistency between items or to correct imbalances, real or supposed, of the types already discussed. It is sometimes proposed as a way of taking up errors in the arithmetic in the bills, as an alternative to making a percentage adjustment to the total on the summary. This may be convenient for minor amounts, especially if balancing items can be found. But it can sow the seeds of future problems if used for major amounts, and especially if the preliminaries are altered – always a temptation because of the lump sum pricing. It is always the contractor's right to insist that his pricing remains as he has originally given it, even in respect of arithmetical errors on the face of the bills. This may not be wise for practical reasons, or for diplomatic reasons, but it is valuable if a point of significance is at stake. Once an alteration has been incorporated, it is as binding as any other contractual figure.

There are more general issues over financial detail, usually resting with the quantity surveyor to clear, as he has structured the tendering document. Initially, he may well look at how the tender compares with his own cost plan or other estimate – and then at early retirement! Inside the bills as tendered, he will be looking at the elements mentioned above as affecting the contractor's policy, especially those which suggest some manipulation of pricing, because they have a mirror effect for the client.

The contractor should remember that the quantity surveyor, who is to become the impartial valuer between the parties once the contract is in being, is acting for the client during assessment of tenders to ensure a commercial deal for him. This is quite proper at this stage, it being assumed that the contractor will look after his own interests adequately. He must therefore be prepared to do so, and not assume that the quantity surveyor will take his part, although still pointing out any clear errors. It remains that any errors of pricing or arithmetic which go through into the contract bills cannot be corrected in favour of either party, but are held to have been accepted.

Several particular situations may arise at assessment. There can be effects of the confidence level in the design as developed, both in the documents in general available to tenderers and in the bills in particular. This may show in amendment bills produced for pricing after tenders have been received, or even in alternatives requested in the original tendering. From the contractor's side, there may be qualifications which he puts forward to the tendering basis in the invitation, over such questions as fluctuations, the contract period or phasing. These are not contentious in themselves, but need care in ensuring that the secondary document is not priced in such a way that, if embodied in the contract, it will produce anomalies in the price structure or, worse, lead to the wrong total tender on the amended basis.

This danger is always possible when such elements as plant and temporary facilities are to be used differently in the amended scheme and are not identified separately for pricing purposes. Even more, when bills are approximate there is a tendency to treat them as schedules, as though the very fact of being able to select items and then measure the final quantity imparts some elasticity into what the unit prices then cover. The essential point here is that when such a revision has been agreed and incorporated into the contract, the opportunity for either party to obtain any difference in what is payable has passed. This may then prejudice the course of later negotiations over important changes in the balance of quantities, when prices for them are based upon the earlier structure.

Programme

Alongside the pricing, the programme is another aspect of the tender which may need attention. Standard contracts are somewhat ambivalent over the provision and status of contractors' programmes. The JCT position is to allow for an optional master programme post-contractually and for it then to be kept up to date, but without any significance being assigned to it (Chapter 4). Sometimes tendering documents ask for a programme to be supplied with the tender, perhaps again not giving a specific reason for this, or perhaps disclaiming any responsibility from the client's side for it when examined.

Such a programme may be useful in assessing the realism of the contractor's approach, especially in an awkward project, and in checking that he has taken account of any intermediate dates due to phasing etc. It also warns the architect about when he is likely to have to produce details and schedules for the contractor (*Merton* v. *Leach* in Chapter 15 illustrates aspects of this). It remains necessary for the architect to hold to the position of non-approval during any discussions over any such programme, otherwise he may find he has assumed some stance which is beyond his recognised position, so committing his client in the event of a dispute related to the data embodied in the programme.

Whatever may be left unsaid and by implication as not being the case, the provision of a programme (even when not requested) is some prima facie evidence of the contractor's intentions. It may be questioned later as to its realism or seriousness, but it can be useful to one party or the other in a dispute. Therefore, it is wise for the architect, or a contractor dealing with a subcontractor, to look at it carefully and to ask, without commitment, for any clarification during tender assessment, when there should be no special interpretation in prospect, although this cannot be automatically assumed. In the case of *Piggott Foundations Limited* v. *Shepherd Construction Limited* (1994) it was established (before an Official Referee and the contract was on DOM/1) that provided the subcontract works do not unreasonably interfere with other works being carried out and provided they are completed within the subcontract programme, there is no obligation on a subcontractor to plan the sequence and programme of its works to coincide with the programme of a main contractor's works. The post-contract aspects of programme are discussed further in Chapter 4.

OFFERS, STATEMENTS AND REPRESENTATIONS

A number of issues hover around the relatively firm area of documentation, which constitutes the essence of any contract. Some of these issues arise at the time of tendering and in subsequent discussion.

First there is the question of the offer itself. It is established law that there need to be both offer and acceptance to give a valid contract. Usually, the contractor's offer is based directly and simply upon the documents issued to tenderers, which themselves constitute an offer to treat. Occasionally, he will put forward an offer which qualifies the documentation issued and this too will be embodied in a further document, probably a letter. Whether the contractor qualifies his tender or not, if the client accepts it as put forward, there is a binding contract. On the other hand, if it is not taken into the written contract, it may well be of no effect. Thus, in *Davis Contractors Ltd* v. *Fareham Urban District Council* (1956) a letter of the contractor's stating that his offer was subject to adequate labour and material being available was not incorporated. The contractor was unsuccessful in his action over the effects of prolonged delay due to shortages – the employer suffered the delay and the contractor suffered the expense. This must be distinguished from a representation made by one party which induces the other to contract, as discussed below.

If, in either of these instances, the client purports to accept the offer as made, but subject to certain amendments, no contract immediately results. This is true even if the client says in relation to the contractor's qualified offer, 'No, I do not want that, but I accept your tender on the basis of the original terms issued.' The contractor has never made such an offer, so it does not exist to be accepted. All the client can do is to put forward a counter-offer, which is then open for the contractor to accept.

Much vaguer situations can come into existence through discussions arising out of queries during tendering, or during more formal pre-tender meetings, perhaps before tenders are actually invited. Some of them directly or indirectly produce effects considered in Chapter 4. It is quite possible for either party to enter into the contract on the understanding that matters not delineated in the documents are nevertheless after a certain fashion. This is especially the case over arrangements about the site or over its character, about the programme and the provision of facilities. The architect may, for instance, state that an owner of adjoining land has agreed to make an area available to the contractor as a stockyard for the duration of the contract, and that the client is foregoing the rent. As a result, the contractor's tender is lower than it would otherwise have been. Alternatively, the contractor may state that he will complete some section of the works ahead of the rest in the absence of any phasing provisions. This may even be the factor that secures him the contract, against the competition of others. In neither case is anything in writing – what is the contractual position?

In the first place, a building contract need not be in writing at all to be valid, although the question of proving its terms becomes daunting if it is of even average complexity. What is true of a contract as a whole, is true of a part of it.

Secondly, there is the distinction between a 'condition' and a 'warranty'. For this purpose at law, a 'condition' is not the equivalent of a clause in a contract form itself, as in common practical usage. It is some representation which induces a party to contract, and which is fundamental to the contract. By distinction, a warranty is a statement that does not go to the root of the contract, although the effects of it not turning out to be true may still be serious. Whereas a breach of condition can lead to damages or to repudiation or rescission of the contract, a breach of warranty leads only to damages, but not rescission, assuming in each case that money will give adequate recompense. In the examples given, it is unlikely that their breach would be serious enough for them to be regarded as conditions. Even if they were, but were not discovered before work started up (quite likely in these examples), the aggrieved party would possibly settle for damages, rather than have to break off the work.

When the breach is of a condition and so fundamental in its effect, it may entitle the aggrieved party to treat the contract during or before progress as repudiated by the other, so as to bring it to an end, or to treat the precise and apparently binding terms of the contract as overruled. The courts are careful to restrict recognition of fundamental breach. Thus, in *Hong Kong Fir Shipping Co. Ltd* v. *Kawasaki Kisen Kaisha Ltd* (1962), a case in which such breach was alleged over the non-availability of a hired ship due to the defendant's fault for twenty weeks out of a total hire period of twenty-four months, it was held that there was not fundamental breach, as the plaintiff could still obtain a large part of the hire benefit. (This case introduced the concept of 'intermediate' or 'inominate' terms, namely that 'terms' in a contract had to be read in the context of that contract and of the nature of an event giving rise to an alleged breach of contract in a case. In order to determine whether or not a term or condition is a 'fundamental term' or a 'core condition' in a contract it is not sufficient only to look at what such a 'term' or 'condition' may be called in that contract. Generally, a contract is only repudiated when a contract-breaker has renounced his obligations under a contract or rendered them impossible of performance.)

Underlying this discussion is the question of misrepresentation – making an untrue representation that induces the other party to enter into the contract, but which is not then embodied in the contract. Misrepresentation is of fact alone, and not in giving an opinion or statement of the law (the position may be different over giving professional advice). This is a complex area of law, with several divisions of the topic. In many cases today, the defendant in an action is placed in the position of having to demonstrate that he made a statement with 'reasonable ground to believe' it true and that he in fact did so believe, this being the requirement of the Misrepresentation Act 1967. A misrepresentation gives the representee the right to treat the contract as voidable, that is an option to declare it void or to affirm it. It follows that a misrepresentation may become a matter with the standing of a condition.

In the case of *Howard Marine and Dredging Co. Ltd* v. *A. Ogden & Sons (Excavations) Ltd* (1977), the defendants had based their tender for work upon statements by the plaintiffs about the capacity of barges which they were to hire. These turned out to be excessive, although based upon usually reliable data. It was held by a majority that the plaintiffs should have gone back to the basic ship data, so it was deemed they had not been sufficiently careful. Judgment was given for the defendants, who had already repudiated the contract and counterclaimed for damages.

But it is not only the parties to the contract who may be liable to legal action over misrepresentation. When a duty of care in giving advice exists, a person who gives advice without due care may be sued in tort for failure to exercise reasonable care. The classic statement of this position arose in *Hedley Byrne & Co. Ltd* v. *Heller & Partners Ltd* (1963). Here a bank gave a credit reference about a client, which turned out to be negligent when the client went into liquidation soon after. The bank avoided liability only because of a rider that the reference was given 'without responsibility'. The House of Lords was clear that liability would otherwise have lain. There are implications here for consultants who proffer advice, perhaps gratuitously, during precontract discussions. Usually, they are acting as agents for their client, but the possibility of other relationships needs to be watched. Even as agents, they may need to look over their shoulders at their clients, who may seek recompense from them after action from the contractor, claiming that he has been misled.

It is also possible for misrepresentation to occur by silence, and this is perhaps an area of greater concern to consultants. It is not that silence constitutes misrepresentation in isolation, but that it may constitute misrepresentation if it occurs in such a way as to distort a positive representation, or when an earlier statement is found to be untrue. There is always the feeling that it is up to the contractor to make up his own mind, and this is true. But he cannot decide what he does not know about.

These are areas on which contract forms are silent, because they are concerned with the operation of the contract as agreed, and not with what led up to it. They are therefore usually areas where the parties must come to some agreement over differences themselves. The architect and the quantity surveyor, if named in the particular contract, can do nothing on their own initiative, because they have no powers given to them, even if they effectively caused the problem in the first place. In the event of failure to agree, the parties have to consider legal action or arbitration.

A thorny situation would occur if a subcontract tenderer put in a tender which was far too low, the contractor embodied it in his main tender, the main tender was accepted and then the subcontractor withdrew before he was accepted, as such persons may sometimes do. There would be no breach, as there would be no subcontract, but the contractor would have relied on the subtender and maybe could not obtain another anywhere near so low. Much would turn on how aware the contractor should have been of the inaccuracy (a question perhaps of how specialised was the work), but it has been suggested that an action for tort might lie, in view of the subcontractor's possible negligence. This has yet to be tested.

RESULTING CONTRACT (OR NOT)

Completeness

Several comments on the status of elements in a building contract flow from the immediately preceding discussion. Although there is always the question of evidence (perhaps needing both judge and jury, if queried), the recording of an agreement can take several forms:

(a) Entirely oral, that is spoken or parol.
(b) Entirely written, which includes any form of document, such as a drawing.
(c) A mixture of (a) and (b).

For a building contract of any substance, there is usually a sizeable body of documents. In any contract so evidenced, there is then usually a presumption that the whole intention of the parties is contained and expressed in what has been recorded. If a contract is a speciality, it becomes particularly important to ensure that what is evidenced by this formal method is complete. Even in the absence of this, the presumption may be critical in later discussion. It will stand against an assertion by one party that he did not intend what he has signed, whether he contends for oversight in reading it or just that he made some other form of mistake in approaching the agreement. Mutual mistake is another question, and normally the parties will seek to rectify this anyway.

The general rule is that a document is both 'exclusive and conclusive' over its own terms, as has been supported in a long series of cases, particularly with reference to leases. These include *Angell* v. *Duke* (1875) where the provision of additional, unmentioned furniture was held to be excluded, and *Henderson* v. *Arthur* (1907) where variant terms about payment were excluded. However, the courts may find that a contract does not express the whole intention of the parties, even though it may appear to do so. Thus in another lease case, *Walker Property Investments (Brighton) Ltd* v. *Walker* (1947), the absence of mention that the tenant might use certain extra storage did not debar inclusion of this accommodation in the scope. Without access to the exact documents, it is rather unclear how the distinctions were defined.

The status of statements in such cases is usually assessed evidentially by when they were made during the course of negotiations; broadly speaking, the later the statements, the more cogent they will appear, because affairs will have become clearer overall. It will also be taken into account whether any part of their immediate subject-matter is in writing and sheds any light on what is not, and whether they were based upon specialised information available to one party only, so that the other had to rely upon it. These statements then have to be weighed against whether the written contract is complete and precise in itself, without ambiguity, obvious error or misunderstanding.

In the absence of clear cases on building contracts here, it is probably safe to say that the primary presumption is likely to hold in the face of silence. If a situation then occurs like one of the examples already given about extra stockyard space or phasing of work, the answer is to be explicit. It is possible that the production of correspondence or minutes not incorporated into the contract would constitute acceptable evidence, but this sort of document is usually vague on the details unless edited and incorporated. In any case, unless the aggrieved party is prepared for proceedings, he cannot find out in the face of opposition whether this is so.

If a contract is patently incomplete, say by the omission of a completion date, the courts will imply a reasonable provision if the parties cannot agree. Similarly, they will act to give coherence to a contract containing discordant terms, as may arise when a standard printed document is incorporated alongside other elements specially composed for the individual contract. Here the normal rule is for the special document to overrule the standard, as this presumably is what the parties intended.

When considering the completeness of a contract, the courts will generally be very reluctant to interpose matters that the parties have failed to include. Courts start from the position that parties to a contract have made their bargain, however unsatisfactory that bargain may appear to one or both of the parties when a dispute arises, or however unsatisfactory it may appear to an outsider coming fresh to their bargain. Implied terms are one of the rare areas for possible court interpretation. Contracts rely on their express terms, and in the case of construction contracts they will invariably be written and involve use of a standard form of contract. Where there is no adequate express term, it may be that a court *may* decide, solely in order to make a contract work, to *imply a term* into a contract.

In *The London Borough of Merton* v. *Stanley Hugh Leach* (1985) it was implied that (i) Merton would not hinder or prevent Leach from carrying out its obligations under the contract, (ii) there was an implied undertaking that Merton would ensure that its architect would carry out those things that it must do to enable Leach to carry out the construction works and (iii) in particular, that the architect needed to provide the correct information to enable Leach to build to and that there was no contractual duty imposed on Leach to check the accuracy of information for divergencies and discrepancies.

In *Davy Offshore* v. *Emerald Field Contracting* (1991) the judge restated five basic points required for implication of a term into a contract, as (i) it must be reasonable and equitable to imply a term, (ii) it must be necessary to imply in order to give business efficacy to a contract (so no term will be implied if a contract is effective without the implied term), (iii) it must be so obvious that 'it goes without saying' that the contract needs the implied term and that was what the parties would have done had they realised its 'absence', (iv) it must be capable of clear expression and (v) it must not contradict any express term in the contract.

In *J. and J. Fee* v. *The Express Lift Co. Limited* (1993) it was held, under DOM/2 conditions of subcontract, that the following terms would be implied: (i) the contractor would not hinder or prevent the subcontractor from carrying its contract obligations, (ii) take all steps reasonably necessary to enable the subcontractor to carry out its obligations and (iii) because there was nothing expressly in DOM/2 to require the contractor to supply information to a subcontractor a term should be implied for this which required the contractor to provide correct information at such times as was reasonably necessary for the subcontractor to fulfil its contractual obligations.

Consistency

Some standard forms of building contract contain provisions which reverse this last rule of interpretation, in particular the JCT group, so guarding against mutilation of their own close-knit terms by ill-considered additions, but possibly also leading to effects not intended by at least one party when special terms are modified by the standard. Numbers of actions have been brought on these matters, with a variety of detail. *English Industrial Estates Corporation* v. *George Wimpey & Co. Ltd* (1973) and *M. J. Gleeson (Contractors) Ltd* v. *Hillingdon London Borough Council* (1970) may be consulted as examples. The proposed introduction of extra terms of a contractual nature by either party therefore needs careful examination, and possibly legal advice. They tend to appear in the preliminaries of the bills, perhaps as statements of intention to change the standard contract wording, perhaps as seemingly innocuous clauses in their own right. Their effect is the same, however they appear or are expressed, unless the standard wording is directly amended in the actual contract form. It is possible to delete just the standard clause that leads to this situation, and some authorities would recommend this. If so, redoubled care is needed, as then the specially written will prevail in all cases.

It is also a principle of interpretation that in the event of a dispute, a document in a contract will be construed against the party who put it forward, even though the other may have expressed no opposition to it at the time. This contra proferentem principle applies to a purpose-written document or to a ready-printed one, if it is proffered by one party. Here 'proffered' means that the one party has produced a document for use which has originated with him or with his side of the client/contractor syndrome.

The series of contract forms produced by the Joint Contracts Tribunal (JCT) do not fall into this group, because they have been negotiated by representatives of both clients and contractors, among others. They thus do not count as having been put up unilaterally, even though the individual client and contractor are very unlikely to have been on the JCT, and even though one of them will have initiated the proposal to use the form for the immediate contract. In contrast, forms such as those produced by the Association of Consultant Architects and by central government are unilateral documents, so they are subject to the contra proferentem rule; the same goes for the New Engineering Contract.

Care

Attention to detail when entering into the contract will be well repaid if there is a difference later over content or interpretation. Both parties should check their records of telephone conversations, meetings and correspondence to ensure that all relevant items have been drawn into the final agreement. Thereafter it is usually too late: documents so embracing as those for a building contract do not admit of being readily circumvented by 'Well, I meant . . . and you know it.' This may mean amending some of the documents as used for tendering. If so, amendments should be initialled by the parties themselves to avoid doubt or ineffectiveness. Other points may warrant gathering up into a consolidating document, which is then referred into the main list of documents.

Another place for care is whether a contract has come into being at all. This is usually quite obvious: even if the formal contract has not been drawn up and signed, there will usually be a letter of acceptance. This will suffice if it refers to all the documents already mentioned, although it is to be regretted that greater promptitude in executing a contract is not always observed. But it may be quite a distinct matter if the contractor receives simply a letter stating that the client intends to enter into a contract and so would he, the contractor, please proceed with certain work or ordering of materials to save time. A letter of intent is often meant to give comfort to both parties but it can be the source of later problems if not executed correctly. A letter of intent is really an intimation of a separate contract and should include an undertaking by or on behalf of the client to pay separately for what is being ordered if the main contract does not go ahead.

On occasions, large organisations enter into construction activity perhaps without using the necessary care, in some cases without formalising contracts. Speed of operation is sometimes a driving force and it is argued that 'the formalities can catch up later'. In *A. Monk Building and Civil Engineering Limited* v. *Norwich Union Life Insurance Society* (1991) there was perhaps an extension of the approach taken in British Steel Corporation (see below), but it was not successful. A letter of intent was issued by Norwich Union's construction managers which authorised Monk to proceed up to a maximum expenditure of £100,000. Work continued under 'cover' of the letter of intent on the basis that a contract would be awarded. Subsequently work of over £11 million was carried out and over £7 million was paid for, but the balance was disputed. Norwich contended that the letter was an 'if' contract, whereby if no contract was concluded then Monk's costs would be limited to 'proven costs'. It was held that as the work was carried out at the request of Norwich Union then the basis of remuneration should be a quantum meruit, as it was held there was no contract with terms concluded. Indeed, one of the reasons was that the construction manager had no authority to contract on behalf of Norwich Union. This case emphasises that it is imperative to examine the original authority (in this case an agent's letter) which is said to give rise to a contractual liability. It follows that any person dealing with an agent should ensure that the agent has the requisite authority to commit its apparent client. This is particularly applicable today when roles played within construction teams are complex, not always made clear to all parties and sometimes change from project to project from those thought to be 'traditional', whatever that term used to mean.

A previous case was *British Steel Corporation* v. *Cleveland Bridge & Engineering Co. Ltd* (1981), where it was held that a letter of intent from the defendants to enter into a subcontract with the plaintiffs asking them to go ahead with work, followed by the plaintiffs performing work, did not constitute a subcontract with terms. They were entitled to be paid on a quantum meruit. By contrast, in *Trollope & Colls Ltd and Holland, Hannen & Cubitt Ltd* v. *Atomic Power Constructions Ltd* (1962) a contract concluded some months after work had started was held to have retrospective effect, covering inter alia its payment terms.

As always, these cases confirm that circumstances determine a finding in a particular case. If a contractor can successfully argue that although he has carried out the work there was no contract in existence, he may be able to establish that he had no obligation to complete the works on time, or even at all, and that he is entitled to be paid a quantum meruit. Although there is little detailed authority to show how a quantum meruit will be assessed, it is perceived in the majority of cases that recovery on that basis will exceed the amount which a contractor would recover under the alleged contract. In *Mitsui Babcock Energy Ltd* v. *John Brown Engineering Ltd* (1996) John Brown proceeded on a letter of intent. There were disputes over tolerances of work and Babcock Energy argued that there was no concluded contract. The Official Referee concluded that there was a contract. The essential point was that if the parties intended to enter into a contract but later disagreed over some of the terms, that did not prevent the contract being binding unless, without the missing terms (those over which they disagreed), the contract could not be enforced. In this case payment on a quantum meruit was not substituted for the payment provisions of the contract. In *VHE Construction PLC* v. *Alfred McAlpine Construction Ltd* (1997) the Official Referee observed:

> It is remarkable that this question (was there a contract between the parties, and, if so, what were its terms) is probably the most frequent issue raised in the construction industry. On projects involving thousands and sometimes millions of pounds, when a dispute arises over payment, the first issue very often is to decide whether there was a contract and, if so, what were the terms of the contract, if any.

The parties conducted themselves as if a contract was in fact in place (often conduct can be relevant to whether a contract has been formed but this is not always so) and the Official Referee concluded that 'a contract was completed by a telephone call between representatives of the parties . . . and was further agreed to by the parties by their conduct.' The content of oral evidence in vital telephone calls was hotly disputed at trial and underlines the obvious that crucial matters on contracts are surely so much better committed to writing.

PART 2

CONTRACT ADMINISTRATION: MAINLY RELATED TO JCT CONTRACTS

CHAPTER 4

PHYSICAL AND TIME PROBLEMS

- Programme and method statements
- Site possession, access, ground and other conditions

The chapters in Part 2 deal with the principles of matters which may become the subjects of disputes during and following the progress of a building contract. Some of their causes may be seen to have their roots in elements set out in Part 1, whereas others arise directly from events or non-events during progress. Numbers of these provisions are illustrated in the case studies in Part 4.

The significance of the provisions of individual contract forms becomes important in these discussions. The emphasis is placed upon JCT 80 in its 'with quantities' edition and upon IFC 84, although other editions of JCT 80 can usually be substituted without any change of meaning, other than the obvious change from the contract bills to the contract specification. Much of what is stated applies to other forms of contract, but should be read against the specific provisions of a particular form. For instance, the GC/Works/1 (1998) contract (Chapter 20) gives the programme a more prominent place, whereas the New Engineering Contract (Chapter 22) deals quite distinctly with whole tracts of procedures on the need for the contractor to issue revised programmes to demonstrate to the employer that the completion date is still attainable. Beyond these lie other, largely non-standard, contracts for such situations as construction management and multiple direct contracts for separate elements of one project, as outlined in Chapter 13. Only occasional reference to any of these is made in this chapter.

The client, as he has been termed throughout Part 1, is given his title of 'the employer' from now on, following the terminology of the JCT and most standard forms.

With entry into the contract, both parties pass from the area of negotiation, with the possibility of withdrawal at a cost to their separate selves mainly of goodwill. Instead they are bound to actions of various sorts, constrained within the contract timescale. These actions start with the activity of the contractor to get work under way, even if off-site, of the employer to make the site available and of the architect on his behalf to provide initial information. The importance of prompt action all through cannot be overemphasised, with so many causes of dispute flowing from its breach.

Many of the more 'routine' problems encountered during contract progress fall into the group looked at in this chapter. They happen with depressing frequency across the range of building work, and they affect the smooth flow of work which the contractor is entitled to expect from the employer under many contract arrangements. Often, but not always, they are symptomatic of insufficient planning before construction begins or of undue changes of mind by the employer's team during construction. *Building Procurement* discusses all the methods of procurement used in the construction industry and the risks that each one contains for client and contractor. Sometimes the problems represent and arise from a deliberately assumed risk of partial planning to allow an early start to be made, followed by some unevenness in progress, for which the employer knows he will then have to pay. If programme speed and an early start, with presumably an earlier completion, are wanted then objectives of cost control and price certainty will need to become subservient to programme. These risks may be contrasted with the less frequent risks and hazards set out in Chapter 18 and the more specialised difficulties discussed in Parts 3, 4 and 5.

Because many of these problems stem from the employer or his agents, standard contracts deal with them in such ways as to relieve the contractor of the more obvious ill-effects. However, there are others which remain his responsibility almost entirely, so far as cost and time go. These include the wider state of the economy and the effects of government actions, whenever fluctuations provisions and closely circumscribed extension of time clauses do not help. He is cushioned against the weather only to a limited extent, and against such matters as traffic conditions not at all. Although he may gain extra time to offset strike losses, he receives no further payment and nothing else to compensate for unexpected union actions. The effects of variations in the regional workload of the industry are entirely his, for better or for worse. Sometimes these items may overlap in effect with those over which he is afforded some protection (Chapter 15).

PROGRAMME AND METHOD STATEMENTS

Contract stipulations

It is a matter of precontract activity to examine and evaluate any statements put forward by the contractor in support of his tender, but they are considered here in view of their later effects. These statements may have been specifically required of the contractor, perhaps even being an elaboration of outline requirements given by the employer in the enquiry documents. They should have been checked so far as practicable for reasonableness, and the contractor should have been questioned about how far he really expected to stand by what they contained. It is important that no one on the employer's behalf should have actually accepted them as binding upon the employer, or have given any explicit approval of them. At most, there should be an expression that they have been 'noted' for information only.

If it is necessary to include some requirements over the contract programme in the tender information, this should be limited to the essentials, so that the contractor has the maximum scope to organise his operations himself for greatest efficiency. Usually only the start and finish dates will be important to the employer, and even these may be expressed as so long after the date of acceptance of the tender. Provided acceptance is not delayed unduly, a little flexibility here is not objectionable, and the contractor can protect himself against too much by stipulating how long his tender remains open for acceptance when he submits it. As a group, standard contracts more often than not ignore the possibility of a staggered commencement of the works, but usually have some provision (perhaps in a supplement) for staggered completion. Any such options should be clearly incorporated into the tender, paying special regard to the possible pitfalls over liquidated damages for delayed completion (Chapter 9).

Inside the end dates for the contract, there may be dates that are critical for the employer's wider operations. These intermediate dates may be fixed dates by which certain work is to be completed. Although they may not lead to the handing over of the work concerned to the employer, they may be required to allow direct contractors to enter upon site work (Chapter 13). Other dates may not be absolute stipulations in themselves, but may act as constraints within which alone the contractor may perform defined parts of the works. As a general rule, the more such dates are given, the more causes of difficulty may arise, either from failure to give all relevant information about what are essentially complex matters or from affairs miscarrying in practice. There is also a peculiar problem in entering enforceable contract provisions against the contractor here, when the standard provisions relate only to final overrun of the programme and special provisions can easily conflict (see *English Estates* v. *Wimpey* in Chapter 3 and the discussion under 'Payment' below).

Working methods are even more difficult, or even dangerous, to delineate. JCT forms refer to several particular obligations and restrictions which the architect may require to be varied, but give no guidance over how to introduce them initially (see next section). Some of these are more in the category of programming points than working methods proper, although this is an area where some overlap exists. It is best not to require the contractor to perform work by particular methods, but simply to restrict positively those which he may employ when, for instance, noise or pollution is unacceptable. A requirement that he must carry out demolition or alteration work in defined ways technically is a likely recipe for shared or transferred responsibility when something goes wrong (see *Bowmer & Kirkland* v. *Wilson Bowden Properties* in Chapter 5).

Even more, the drafting into contracts of terms that require the contractor to carry out the works in a workmanlike manner enters in broad spirit into how he works, while leaving surprise that to proceed in such a manner was not already an implied contract term.

Status of documents post-contractually

When the employer has put forward specific requirements for tendering and these are incorporated into the contract, the position is clear. There is also no doubt if the contractor has made any stipulations over such matters as the contract period, and these in turn are incorporated.

The least clear position under contracts in general exists over the contractor's own master programme setting out how he proposes to perform the works. If this document is not proffered before the contract is awarded, it remains the contractor's private document for his own use, with no contractual status. JCT clause 5.3.1.2 is specific and requires the contractor to 'provide the Architect (unless he shall have been previously so provided) with 2 copies of his master programme' and also to provide amendments to take account of extensions of time granted by the architect. This clause is singularly unhelpful by virtue of what it leaves unsaid.

In the first place, the level of detail in the programme is not given: a master programme certainly indicates no more than major dates, which will give the architect just a broad forecast of the contractor's progress. Secondly, the 'unless' provision means that no contractual significance can be assigned to a document that may not have been produced (in any sense of the word) before entering into the contract. Furthermore, the document has no definite post-contractual function, and this is presently under review. There are several principal options and gaps over a programme under the JCT contract; and the IFC contract is similar in effect, but without mentioning a programme at all:

(a) *As information for the architect's use when supplying information in time for the contractor's work to proceed*: Clauses 25 and 26 (which see) refer to the contractor requesting information closer to the event in most cases than 'so soon as possible after the execution of this Contract'. Furthermore, the programme is hardly detailed enough for guidance in this way.
(b) *As information for the architect's use when deciding upon claims and extensions of time*: This is not said anywhere, and only an assumed connection can be deduced over extensions of time from the reference to incorporating the effect of extensions when granted.
(c) *As a binding initial statement of what the contractor is going to do*: This is nowhere stated and the contractor has no greater (or lesser) obligation than to 'carry out and complete the Works' in accordance with the contract under clause 2.1.
(d) *As a regularly revised statement of what the contractor intends to do*: Except for the reference to revisions due to extensions of time, there is no requirement for the contractor to amend the programme at all, quite apart from working to it. It may therefore be amended for extensions and at the same time be quite out of step with reality and misleading.

The GC/Works/1 contract takes a much firmer line over these issues so far as the extent and review of information goes, while still relying on the usual methods of getting back on programme or dealing with the consequences of being behind. The NEC contract is strongly built around the programme being a significant part of the management of the construction works. It requires the contractor to notify the project manager as soon as he considers any matter will affect the programme – the 'early warning' procedure. Failure to do this may result in the contractor not being given a 'compensation event' which will entitle him to payment for the effect of any instructions that have caused a change to the last issued programme. The NEC requires the contractor to submit a revised programme each time he considers the completion date will not be met. The revised programme will need to show how he proposes to maintain the completion date last agreed and what measures and cost will be involved in keeping to that programme. The NEC approach arises from the philosophy that the contractor knows most and is best at controlling his construction programme. A cost is likely to be paid for this if the changes that require reprogramming have resulted from instructions of the project manager.

Use of documents post-contractually

The only indisputable purpose of method statements and programmes in the absence of special provisions is to help the contractor in performing his own contractual obligation, noted under (c) above. This is to proceed to completion by or ahead of the contract date, and in all other respects as contract. How the contractor does this is entirely his own concern, so long as he is not bound by any subsidiary stipulations of the types already discussed. It follows, therefore, that he may vary the detail given in his programme in any way that does not prevent him achieving the objective. A possible point of doubt is over his further obligation that he shall 'regularly and diligently proceed' under JCT clause 23.1 or IFC clause 2.1. This appears to require him to maintain regular progress even when he has time to take a break and still meet the completion date, although *GLC* v. *Cleveland Bridge* (see under 'Payment' in this chapter) throws a query over this when there is an excessively long contract period. The case is in the context of fluctuations, though it does appear to relate. Subsequent cases below appear to confirm this.

It is nevertheless the case that the contractor's programme and other data prepared for his domestic use may provide prima facie and perhaps even sole information from which to assess what should have happened when some form of disturbance occurs and leads to a dispute. For this reason, there is something to be said for the architect or another adviser seeing what the contractor has available precontractually, even if not formally requested and however rudimentary it may be. In fact, the less polished it is, the more may it be taken as an unbiased statement of intent. The longer it has been around, the more it may have been 'adjusted' by a devious contractor!

If the contractor has been working to a programme fairly closely during the earlier stages of the contract, there is again a strong presumption that it represents a fair projection of what was intended for the rest of the time. In all cases, as close an examination as possible should be made to establish whether the programme should be used as part of the basis for a settlement, or whether a fresh construction is effectively needed for this purpose.

Such an assessment should take account of several basic factors:

(a) The logic of the programme in respect of sequence and interrelation of sections of work.
(b) The durations assigned to each section in relation to the overall programme time.

(c) The reasonableness in amount and character of the resources allocated to each section, and of those resources allocated to be available for wider purposes throughout the project.
(d) Whether factors other than those adduced as causing the disturbance under discussion have affected the programme, adversely or otherwise.
(e) To what extent the contractor has chosen to revise the programme, and whether this has led to greater efficiency.

The last point returns to the contractor's right to amend any programme at his sole discretion, within any constraints placed upon him, such as when there is closer control or at least monitoring of the programme, for instance as is provided in the GC/Works/1 (1998) contract and the importance placed on the programme in the NEC contract (Chapters 20 and 22). It may arise out of consideration by the contractor of the preceding points in part, or simply represent wider second thoughts.

The architect or the quantity surveyor and now sometimes a programme consultant, as most likely to be involved in discussions, should therefore regard a programme as useful, but not infallible. He is entitled to disregard it where it is faulty for his purposes, and substitute some other basis for his calculations. This becomes a delicate matter when the contractor indicates that his programme is faulty, but that he had plans to revise it in its later stages to compensate for this. He then further indicates that these plans have been frustrated by the causes of disturbance that he now alleges. It is no part of a concession of extra time or money to make good the contractor's deficiencies for him. On the other hand, there is no presumption that any contractor is always going to execute his work with perfection if left to himself. The starting-point for adjustment should therefore be a reasonable, if occasionally fallible, plan of action by the contractor, but not one which is heavily flawed or deliberately distorted.

When the contract has been awarded, there is clearly no power to require the contractor to produce any programme or method statement beyond the contract requirements. Similarly, what can be required cannot be asked for during general progress in any special form (such as a network), unless the contract documents are explicit. If a dispute occurs, the contractor is obliged to give necessary information to those assessing the extent of compensation. Although they cannot insist on any particular presentation, it will obviously be in the contractor's interests to do what he can to ease their understanding of his case, and so comply with reasonable requests.

A number of cases relate to the use of contractor's programmes for post-contract disputes. In *Glenlion Construction Ltd* v. *The Guinness Trust* (1987) the contractor was required to provide a programme and this showed that he planned to finish early, which the JCT contract allowed. He was delayed by lack of information from the architect. It was held that he had a right to finish early, whether or not he had produced a programme that showed this. There was no corresponding duty on the employer to provide information that enabled him to finish early and so the architect was not obliged to provide information in time to permit an early finish. The employer must not prevent the contractor from finishing early but he is not bound to take positive measures to implement the contractor's privilege or liberty.

There was no automatic link between delay to the programme for an early finish, an extension of time and loss and/or expense.

In *Piggott Foundations Limited* v. *Shepherd Construction Limited* (1994) it was established that provided the subcontract works do not unreasonably interfere with other work being carried out and that the subcontract works are completed within the subcontract programme, there is no obligation on the subcontractor to plan the sequence and programme of its works to coincide with the programme of the main contractor's works. The subcontract was on DOM/1 and the case was decided before an Official Referee. Once a construction period becomes delayed and/or disrupted, it is important for each party to keep records of the events and actions that it took. Without knowing the circumstances of a particular project, it is not clear what use or relevance programmes and method statements may have, but it is advisable for contractors and consultants to retain any master programme with each update, the level of resources that were expected to be used and which resources support each programme. Then a comparison of that programme with as-built records and the resources that were used to support the as-built programme may be crucial to negotiating extensions of time and/or loss and expense entitlements. In *Aries Powerplant Ltd* v. *ECE Systems Ltd* (1995) a preliminary issue was to decide when Aries was to complete its work and whether or not time was of the essence in the contract. The documentation was often confusing and the judge held that the contract should be construed on the basis that where later provisions were not consistent with earlier ones then the parties must have intended the later ones to prevail (take heed, inconsistent document producers). It was stated that, in construction contracts, the normal rule is to say time is not of the essence. Furthermore, it was held that time could only be made of the essence provided the work was the same as had been tendered for and that final drawings and specifications had been produced at tender stage. It is an implied term that if delivery was not required by any particular agreed or estimated date then completion would be within a reasonable time upon reasonable notice being given. This suggests that unless an employer/main contractor is able to do everything required to allow work to be done by a subcontractor (or contractor) then it is most unlikely that time will be of the essence, and any attempt to make it so will fail and be to the disadvantage of the employer, the one trying to enforce the essence.

Special programme features

Control

Several elements can affect the operation of the programme. One is the incidence of nominated or named subcontractors under the appropriate contract form. Here it may be necessary for both the architect and the contractor to be involved to different degrees in agreeing the subcontract programme with the subcontractor. Special considerations may obtrude if a particular subcontractor is cheaper on price but cannot meet the programme which the contractor requires. Even more special considerations crop up if renomination becomes necessary. These aspects are looked at in Chapter 7. Here it is simply noted that, in the employer's interests, a 'trade-off' calculation may be relevant; the trade-off is between what is arranged and what becomes payable to the contractor or to the subcontractor, according to the solution.

It is also necessary for a contractor and his subcontractor to take care over the terms on which they subcontract. In *Martin Grant & Co. Ltd* v. *Sir Lindsay Parkinson & Co. Ltd* (1984), a subcontractor had carried out work under a non-standard domestic subcontract to a JCT form which contained a provision 'at such time or times and in such manner as the Contractor shall direct . . . the Sub-Contractor shall proceed with the said works expeditiously and punctually to the requirements of the Contractor so as not to hinder hamper or delay the work'. It was held that the contractor was not obliged, as the subcontractor put it, to 'make sufficient work available to [the subcontractors] to enable them to maintain reasonable progress and to execute their work in an efficient and economic manner'. The terms were sufficiently clear and express to preclude an implied term, such as the subcontractor sought. In *Piggott Foundations* v. *Shepherd Construction Ltd*, it was held that the subcontractor was not required to comply with the main contractor's programme, on interpretation of DOM/1 subcontract. The words 'the progress of the works' in the opinion of the judge meant that the subcontractor had to carry out the works in a manner as would not unreasonably interfere with the actual carrying out of any other works which could conveniently be carried out at the same time. There was no contractual obligation upon the subcontractor to carry out his work in any particular order or at any specified rate of progress. This appears to mean that a contractor may have some difficulty in maintaining control over his programme and the subcontractor.

Another pressing situation is where a contract leads to considerable factory production of specially designed units. Here a disturbance of the programme can be more drastic, as well as more difficult to evaluate because of its off-site effects. It may cause dislocation to factory production for more than the immediate contract, because of the difficulties of absorbing a build-up of units not yet wanted on-site. The alternatives become whether to deliver items to the site and use up what may be very tight storage, whether to store at works and use up the normal buffer space quite rapidly, or whether to seek storage elsewhere off-site. These are quite distinct from the problems caused by disturbance of the programme entirely on-site. They may be easier or harder to evaluate, according to whether or not they are self-contained effects (see case studies in Chapters 30 and 32).

More subtly, the effect of a disturbance will be changed according to the extent to which the contractor sublets work and varies his flexibility. What is serious for him, with a whole programme to arrange, may have little effect on a subcontractor who just comes and goes. Equally, the reverse may be the case, because it is the complete commitment of the subcontractor for that project.

Payment

Given a generous contract period and no intermediate constraints, the contractor may choose to perform a heavy weighting of his work early or late in the period, rather than spread it evenly. If he performs early, the employer pays early but receives no benefit until practical completion. If he performs late, the employer pays late but probably pays more if the contract is on a fluctuating price recovery basis. It was held in *Greater London Council* v. *Cleveland Bridge & Engineering Co. Ltd* that the choice remained the contractor's. Here the contractor could have manufactured gates for the Thames Barrier some three years before installation and have saved increased costs for the employer, but then have borne financing costs until he was paid on installation. He was within his rights to postpone manufacture, despite statements of intent about manufacture in his programme. It was stated that, as a general rule, a contractor was not obliged to operate the programme to his maximum detriment to obtain the maximum benefit for an employer. Here specifically, this was despite a provision that the contractor was 'to execute the works with due diligence and expedition', the equivalent of 'regularly and diligently proceed with [the Works]' contained in the JCT and IFC contracts.

Those latter provisions continue with 'shall complete the same on or before' the completion date, so that the contractor appears to have the right to finish early and be paid accordingly, perhaps by proceeding very diligently. It is difficult to modify contracts to avoid these effects, as most stipulations can be circumvented by stratagems of timing work. Only numerous intermediate dates may be successful and this policy has been criticised earlier on other grounds. The best answer is a realistic contract period.

SITE POSSESSION, ACCESS, GROUND AND OTHER CONDITIONS

Contract provisions and their limitations

JCT contracts are no exception to the general range of building contracts in the ways in which they treat matters relating to the site of the works. In fact, they say very little about the site itself, by implication 'the site of the Works'. They usually refer to 'the Works' as the locus where it all happens, once the contractor has been given possession of the site. This blurred distinction is commonly of no consequence, but may be unfortunate in some circumstances, as the site is conceivably more extensive than the strict works.

References to the site are made specifically in respect of the employer giving possession to the contractor and of any obligations and restrictions (discussed hereunder) which the employer may impose over use. These are given respectively in JCT clauses 23.1 and 13.1.2 and in IFC clauses 2.1 and 3.6.2, and their intentions are clear. It would appear that a single, undivided site is envisaged by the contracts, although there is no difficulty in having a site in parts physically separated, provided this is adequately described in the preliminaries of the contract bills or specification.

There is no provision for the contractor to be given possession by instalments (as occurs in the case study of Chapter 32), and this would need amendment of the relevant contract in both clause and appendix and inclusion of details in the preliminaries. JCT 80 and IFC 84 have provision in clauses 23.1.2 and 2.2 respectively for the employer to defer giving possession for up to six weeks beyond the contract date for possession, provided an insertion in the appendix gives a specific period. Delay beyond any defined period under those contracts, or any delay at all under other contracts, constitutes breach of contract. This is a post-contract deferment, but reasonably the contractor might claim for loss and expense due to inadequate notice. This allows the possibility of action for damages and also, in a serious case, of the contractor treating the contract as repudiated (Chapter 9). The contractor must use his discretion here, as he needs to consider goodwill and, even when matters are seriously disturbed, whether he wishes to proceed with the contract. This of course he is still entitled to do, and negotiation of a sum (the need for which rises with inflation) is likely to sour continuing relationships less than an initial resort to legal action.

Beyond these points, the contracts refer to the works for all major purposes. These include the distinction between materials 'delivered to or adjacent to the Works' and materials 'before delivery' to these places, affecting issues of payment and responsibility treated elsewhere (Chapter 8).

The term 'adjacent to the Works' is usually interpreted as something like 'not too far away', rather than actually contiguous, which is not at all helpful. It appears to be the intention that 'Works' here is equivalent to 'site'. If so, anywhere adjacent is not intended to be the site. Suppose the employer makes an area belonging to him and detached from the site available for storage, workshops etc, then it becomes necessary to define it either as part of the site or as not. It is not strictly the site of the works, but this distinction is only an inference. The site is best regarded as what the employer makes available for the contractor to perform the works. If the contractor himself obtains the use of an adjacent area of land, this should not be designated as part of the site, since questions like ownership of materials will not be determined by location on this area.

The JCT contracts are also silent (as the GC/Works/1 and ICE forms are not) about conditions around the site, such as private roads, accesses through areas belonging to the employer and not part of the site, and temporary easements and risks of trespass to land or air space (but see questions of ingress and egress in later chapters affecting extension of time and loss and expense). These variable matters need careful consideration and legal drafting in many cases.

Normally the site will be obviously larger than the works in extent, or coextensive with it, but there is a particular need for definition of working areas in the case of alteration work contained entirely within an existing building or (potential confusion) spreading without possession across an existing functional site of the employer. Here the question of phased possession rears itself, perhaps in a multitude of small instances, and is best dealt with by provisions about 'by mutual arrangement between the employer and the contractor, from time to time.' Each of these goes beyond the plain contemplation of the contract, but needs care nonetheless. The routine pattern of sole possession by the contractor will not apply, and this should be spelt out. The problems faced by a subcontractor under provisions with some similarity, but more compulsion about phasing are seen in

Martin Grant v. *Lindsay Parkinson* and *Piggott Foundations* v. *Shepherd Construction*, as discussed above and in Chapter 7.

Site conditions: normally encountered problems

It is relevant to emphasise that the contractor faces a variety of difficulties to be overcome on many sites, even without the employer imposing any special ones. Many disturbance situations involve the compounding of these problems or the subversion of well-laid plans.

Apart from the disposition of the workforce to secure best results, there are considerations of site layout which can affect the costs of storing and handling materials, the grouping of temporary accommodation and the effectiveness of plant. These can mean problems in securing efficiency in combining the use of these facilities, or selecting the optimum sizes of items. Disturbance of the 'best fit' cases can therefore involve extra costs permeating many operations. What effect there is depends very much upon the site configuration and the restrictions which are introduced by the employer, quite apart from the intrinsic nature of what is being constructed.

Again, these elements are intensified when the works are fragmented by consisting of alterations, or are located in several places within a large occupied site. It may be very difficult to obtain high utilisation of facilities, even without disturbance of what has been planned. What is the best sequence for, say, plant may be at variance with the dictates of site offices or supervision.

These points are obvious enough to any contractor, but are mentioned briefly as what may escape the notice of those usually concerned more with the finished product than with the means of producing it. Sometimes their incidence should deter those responsible from introducing even quite desirable variations, in view of the resultant costs linked to the stage at which they are introduced.

A highly specific aspect of the site characteristics is in the ground, which deserves a section all to itself.

Ground conditions

What is contained within the substrata of the site is of peculiar interest in many cases, because the contractor not only works within its confines, but does such things as digging out, supporting, filling back and taking away materials and disposing of water! This interest is intensified by the uncertainty as to the precise nature of the substrata until it is actually exposed. Some contract clauses are more reserved than others about ground conditions, the JCT contracts are silent (even the contract for design and build work, where more than excavating is at stake; see Chapter 24), whereas the GC/Works/1 and ICE contracts use as their standard the judgment of an experienced contractor (Chapters 20 and 21). The NEC contract also uses the 'experienced contractor' standard (Chapter 22). The Standard Method of Measurement of Building Works, Seventh Edition (SMM7) is by no means so reserved, a pattern to be found in other methods that may be regarded as broadly similar in this present discussion, which does not seek to interpret them in detail.

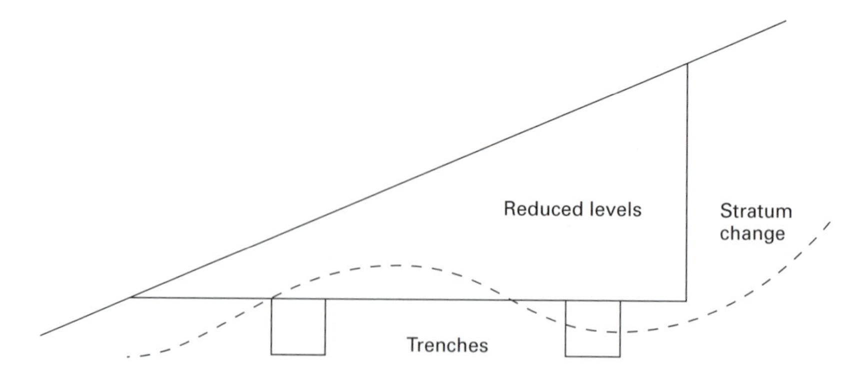

Figure 4.1 Excavation conditions and measurement

The traditional practice when tendering has been to place on the contractor as much risk as is reasonable over the costs of excavation and associated works. This has sometimes meant all risks, with transgressions of reasonableness not unknown. Because conditions can vary quite radically within short distances horizontally or vertically on some sites, rules have emerged within methods of measurement to provide as much data as is available and to cope with the diversity.

A key principle is to provide borehole or trial-pit information about ground conditions to tenderers, perhaps with pits left open during tendering. As an extension of this, a full report encompassing close technical analysis of the soil and of water conditions is regarded as essential for any major project, and no contractor should sensibly proceed without one. Even with this provided, there remains the question of its contractual status. If it is erroneous at the points and times at which it was obtained, then the contractor is entitled to any redress for being misled. If it is unrepresentative of the conditions throughout the site, the position is fairly complex. Under SMM7, with some divergence from the sometime position that the contractor is responsible for making up his mind upon the data given, or in any other way, his pricing is then binding upon him in all these respects.

SMM7 leaves it to the contractor to assess, from the detail given, the practical nature of what he will excavate and subsequently handle, so he cannot assert that a properly described gravel stratum is more difficult to deal with than he had anticipated. It is well known that the rules of measurement ignore the position in plan of excavations on the site, hence their relation to individual boreholes or other similar information. They are preoccupied with depths below starting levels, which are to be existing ground levels and which are nearly always required to be stated. As a result, there may not be a precise correlation between the quantities and the ground survey, so this remains a 'contractor's risk' area.

Figure 4.1 throws up several possibilities, assuming that the stratum change shown is to a more difficult material, but not one which has to be given separately within the rules. The reduced levels of excavation shown in Figure 4.1 is to be classified by its maximum depth, whereas part of it which is not near maximum depth (and might have been in a shallower depth classification) is in the more difficult material. Furthermore, the trench nearer ground level is entirely in the more difficult material, and the trench deeper down is only partly in the more difficult material (although again the proportion will not be apparent). These obscure questions are usually of interest (if that is the word) to quantity surveyors only, as also may be what is intended in SMM7 by the term 'existing ground level'. The main point is that the required depth classifications do not lead to the descriptions attached to the quantities giving sufficiently precise information every time.

SMM7 requires 'the pre-contract [ground] water level' to be stated in the bills and re-established during progress, so there has to be a remeasurement of the quantity of excavation below the level. The pricing of the one measured 'extra over' water item itself must take account of the effect of the water on the material being excavated, and on its handling characteristics thereafter, unless it falls into one of the categories of 'soft' materials mentioned below as to be kept separate in any case.

If information about subsoil conditions is not available, assumptions about ground and water should be made in the contract bills, upon which pricing and quantities are then held to be based. These assumptions will then be corrected by remeasurement, so as to reflect any revised balance of soil conditions. It is also implicit in this arrangement that the prices may be amended for different constituents, as they are not when actual data has been provided and the contractor makes his own assessment.

It is long-standing policy that particularly hard and soft ground conditions merit separation in the contract bills. The contractor won his contention that rock should have been measured (and so priced) separately in *C. Bryant & Son Ltd* v. *Birmingham Hospital Saturday Fund* (1938), as the then current SMM required. The existence of trial holes showing rock, but not made plainly available to tenderers, was not sufficient for him to assess his obligations properly. Thus rock is given separately for excavation but not support, usually as a provisional item subject to remeasurement. The main problem is one of definition, and SMM7 still holds to a long-standing definition which, by its use of 'can be removed only by', is alone in making the measurement criterion the method of execution and not the finished product (a hole in the ground). Even then, there is some doubt in the expression used: 'can only' is not the same in effect as a term with 'practicably' included – will the positing of enough plastic spoons to wear away the stratum count to avoid paying for rock? Soft excavation means unstable ground, such as running silt and running sand. Here the position is reversed and the excavation is not distinguished, but the support is to be. No definition is given, so that one should be supplied.

When SMM7 does not apply, these rules may be changed by the applicable method and the effect should be noted closely. When the contract is based upon a specification, a schedule of work or other unquantified document, some corresponding statement about ground conditions may be needed when the project involves extensive groundworks or the conditions are expected to be difficult. This should take account of any ground exploration, or make assumptions, so that tenders are not unnecessarily inflated to cover major uncertainties. Any schedule of rates provided by the contractor will need similar qualification in complex cases.

It will be clear that anyone not familiar with the concepts involved under this heading, should seek advice before drafting any provisions on the topic. Although some risks for the contractor are reduced by the expedients introduced, quite extensive areas of residual risk may be discerned and fresh ones may be created. In themselves, the rules relate under all contracts and according to the method of measurement in use (if any) to the actual process of excavation and the activities of earthwork support and disposal of water, spoil etc. When read in conjunction with provisions such as those in the GC/Works/1 and NEC forms about unforeseeable ground conditions, they stretch to relate to other work within or otherwise affected by those conditions, as well as such wider contingencies as the programme (Chapters 20 and 22). Contractors who act on incorrect subsoil survey information and suffer loss may be able to demonstrate that the survey information had become a warranty or condition in the contract. This was so in *Bacal Construction* v. *Northampton Development Corporation* (1975), where the contractor had designed substructure works together with bills of quantities for ground conditions. During the work, the soil was found to be very different from the survey information supplied and the contractor was compensated with damages.

Some sites are in areas where the discovery of antiquities during excavation work is highly likely. It is not practicable to add anything in the contract bills or specification to ease the way through. The contractor is entitled to extra time and payment in severe enough cases, and in advance the employer can only take account of his own risk and then hold his breath. This type of situation is illustrated in Chapter 30.

Site conditions: imposed obligations and restrictions

It is quite common for the employer to impose particular requirements on the contractor, some positive and some negative. These may well overlap with points discussed under the preceding heading and with the wider contract programme discussed before that. It is therefore important to ensure that any such provisions are not at odds with contract clauses or with other standard clauses in the preliminaries. If they do conflict with any contract clauses, the contract clauses will prevail by reason of items such as JCT 80 clause 2.2.1 and IFC clause 1.3. If they conflict with other documents, there is no certain way of establishing priority and the documents will be read as a whole to obtain the presumed meaning. If other ready-printed documents are included in the bundle, then the specially prepared clauses are to be given precedence over them.

There is no restriction on what the employer may introduce, should he have special circumstances on or around the site. He may be concerned about noise, dirt, smell, traffic and a host of other possibilities, but must be reasonably specific. It is not enough to say that the contractor must observe any regulations which may be imposed upon him during progress. He is entitled to tender on the basis of definite conditions, either stated or gathered from an investigation of the site while he is tendering. This does not necessarily mean an absolutely clean run when within an occupied and operating area, but it should be consistent with the right of possession that he has in the absence of information to the contrary.

Standard contracts are by definition not suitable vehicles for material about the introduction or bounds of such tailor-made stipulations. JCT clause 13.1.2 and IFC clause 3.6.2 both allow the introduction and variation of obligations and restrictions in four categories:

(a) Access to the site or use of any specific parts of the site.
(b) Limitations of working space.
(c) Limitations of working hours.
(d) Execution or completion of the work in any specific order.

The express inclusion of these permitted categories may be seen effectively as restricting the matters on which the architect may instruct variations, beyond physical changes in the works, while not restricting the sorts of matters over which obligations and restrictions may be introduced in the initial contract. This is because the architect's power to issue instructions at all is limited to those matters expressly given in the conditions, variations as defined being one such matter.

This suggests that particular care is needed in selecting and describing any obligations and restrictions in the first place, as any included in the contract bills but not falling within this list are strictly not eligible for variation, whereas others not covered by the contract bills cannot be introduced post-contractually. This is hardly a practical or necessarily fair position in many circumstances, but appears to be legally correct. Some leeway may be gained by including matters which are as close a fit as possible to what is anticipated and then varying them, which may be done if and only if the initial description has been worded so as to fall at least partly within one or more of the four categories listed above.

Direct contractors and others

Standard contracts are usually linked to the concept of the contractor being given full and uninterrupted possession of the site for the duration of the contract works. But they also introduce the possibility of the employer engaging others to perform work outside the contract, but going on concurrently with it on the site, or even of the employer himself performing such work. Such cases occur in JCT clause 29 and IFC clause 3.11, although curiously the former omits any reference to work being carried out actually on site.

The intention of the two JCT contracts is that any other work shall be relatively incidental in scale as against the works within the building contract; this is the basis of the following discussion. In passing, however, there is nothing contractual which expressly requires this to be so, or even for the other work itself not to be building work. Other standard contracts, such as the GC/Works/1 form treat this matter and tend to imply (but no more) the possibility of more major work being performed. All of these contracts look to the building contractor as the dominant presence, often providing facilities of various sorts to the others present. In the context of what Chapter 13 terms multiple contracts, the various contractors present may in several cases be almost equal in size within the project. Here the considerations advanced under the present heading may have some relevance, but the wider issues outlined in Chapter 13 will predominate and hopefully be dealt with in the contract conditions for the project. If there is a need for a large number of contractors to be employed directly by a client, it may be that construction management is the more appropriate procurement method (see *Building Procurement* and Chapter 2 for explanation and discussion of construction management).

The two JCT clauses both take account of two situations in which such direct contractors, independent of the building contractor, may intrude into what is otherwise his undoubted right of sole possession of the site during his contract period. One situation is where the contractor has notice when tendering that such contractors will be present; the other situation is where he is advised during the contract period.

In the former case, it is required by JCT clause 29.1 that the contractor is to be given in the contract bills or specification 'such information as is necessary to enable the Contractor to carry out and complete the Works in accordance with the Conditions'. IFC clause 3.11 refers more briefly to 'Where the Contract Documents provide for work not forming part of this Contract', thus allowing a more simple identification of the work within the context of a more simple contract.

The JCT requirement has two sides. There must be enough information given in the documents to allow the contractor to perform his own work in accordance with his own contract. This presumably is intended to mean enough information for the contractor when tendering to price the effects of any direct contractor being present and working, rather than for him actually to perform his own work without any hindrance at all. There must be enough information in due course for the contractor to proceed with site work.

Although neither category of information need be entirely contained within the other, it is likely that the information required for site work will be the more detailed, by covering such features as precise setting-out data. This the contractor does not need when tendering, and its absence could hardly be held to invalidate the provisions of the clause. (He will expect, at least with quantities contracts, to receive much positional and similar information during progress in respect of his own work, as the contracts permit.) To avoid contention and doubt, it is useful to agree during tender examination or negotiation that enough information has been made available for pricing within the contract documents, and that further information is properly recorded for issue later as part of the supplementary detail to be given to the contractor when the contract is formalised.

The other side, inferred in the IFC clause and made express in JCT clause 29.3, is that the contractor has no responsibility towards these persons as subcontractors. This affects the extent of his liability over indemnities and insurances and removes from him any responsibility for their work as performed.

In the present context, significantly, it means that he is not responsible for organising their operations; this is the concern of the employer or the architect. The contractor should look to the architect for technical and programme information, while acting to coordinate his precise activities by day-to-day consultation with direct contractors to the extent set out in the contract bills etc. This will probably be given in a manner broadly similar to that applying over subcontractors, but the contractor is not responsible by any implication for more than is expressly laid upon him.

In particular, the contractor has no responsibility to perform attendance or afford temporary facilities unless these have been itemised adequately for pricing. He should only deal with these elements under a variation instruction, and should pay especial attention to any transfer of responsibilities which may be bound up in an apparently simple mandating of extra work.

The second way in which direct contractors may be introduced, by post-contractual action of the employer, is covered in JCT clause 29.2 and, again, IFC clause 3.11. Here the contractor has a right to withhold his consent on reasonable grounds, which might be valid if the proposed extra work would interfere drastically with the works as such. It is to be hoped that any direct work proposed during progress would in fact be relatively incidental or free-standing, and not extend to something of as great a consequence as the structural frame, which should have been foreseen! If something quite critical did arise, without which the finished works would be severely diminished in value to the employer, it is likely that it would be held to be unreasonable for the contractor to withhold consent, even though his own work would be greatly affected. Beyond this question of consent, the clauses say nothing, and the extra work falls into the same category as that under the previous clauses in all other respects.

Several factors come into play to safeguard the contractor against the effects of extra direct work introduced during progress, or indeed against any unexpected effects of any work already described in the contract. There will be variation instructions to cover any attendance, coordination, changed finished work and the like. Then the provisions for extension of time and loss and expense are available when appropriate. There is even the big stick of determination over delay by such contractors. Each of these aspects is discussed in later chapters.

No aspect of these provisions gives the employer the right, through the architect or otherwise, to take work out of the contract and give it concurrently to direct contractors. If it is desired to remove work from the contractor, it must be omitted by variation instruction, with the routine effects on valuation, and be left out until after the contractor has achieved practical completion. Only then is the employer free to give it to another without being in breach of contract.

CHAPTER 5

INFORMATION, INSTRUCTIONS AND APPROVALS

- Functions of the employer
- Authority of the architect
- Issue of instructions by the architect
- Mistakes in contract and other documents
- Approval of the works

This chapter surveys the matters contained in its title, working from the basis of the JCT and IFC contracts for the most part. Before this, it needs also to set out the positions and operations of those originating such matters: the employer and the architect.

FUNCTIONS OF THE EMPLOYER

Although the functions of the contractor are in general too obvious to need separate mention by way of introduction, those of the employer are not so obvious. This is because he actually has very few functions in contractual terms, as the architect acts in relation to the contractor for most purposes, although the alternative positions created under the various contracts discussed in this book should be examined, especially those compared in Part 6, where the client may have distinctly greater powers to act directly. In fact, the employer mainly pays when advised to do so and takes certain actions in emergencies. His other activity takes place behind the scenes in conveying his wishes about the building works to the architect.

This division is not unimportant, as is seen from the functions allocated to the architect and introduced in this chapter. It courts confusion for the employer to take an active role when the architect acts in a manner like that of a separate project manager and carries the primary design responsibility as well. The employer should therefore avoid giving direct instructions to the contractor about any matter and always see that they are issued by the architect. This helps to avoid discrepancies and conflicts, and the possibility of gaps in communication. The contractor for his part is advised to decline politely to receive such instructions, as is his contractual right, by referring the employer back to the architect as the proper channel. Faced with a persistent employer, he should discuss the position with the architect, who will not wish to see such strains occurring, especially if they affect design matters.

Table 5.1 Major obligations and rights of employer

	JCT	IFC
Obligations		
CDM Regulations	1.5	1.12
	6A	5.7
Agreeing supplementary domestic subcontractors	19.3	–
Insuring works etc as contract option	22B, 22C	6.3B
		6.3C
Meeting amounts interim and final certificates	30.1.1.1	4.2
	30.8	
Paying nominated subcontractors direct when contractor defaults	35.13.5	–
Rights		
Receipt of completed works from contractor	17	2.9
In entirety	2.1	1.1
To specified standard	2.1, 14.1	1.1, 1.2
On time	23.1	2.1
Defects and defaults remedied		
By contractor	17.2, 17.3	2.10
By others, contractor charged	4.1.2	3.5.1
Indemnities over patent rights	9.1	–
Indemnities over injury	20	6.1
Liquidated damages for late completion	24.2	7.1–7.3
Determination of contractor's employment	27.1, 27.2	7.1–7.3
	28A.1	7.8
Completing works after last and recovery of costs	27.4, 28A.3	7.4, 7.9
Collateral agreements with nominated subcontractors over design, delay, loss and expense, determination	NSC/W	

Note: The list of clauses covers major elements only, and should not be read as exhaustive.

The employer himself may wonder about his power to query what he receives from the contractor in terms of quality, and what he pays for it, when his hands are so tied. His direct contractual avenue is to resort to arbitration or court action against the contractor himself. His other avenue is to proceed against the architect over matters in which he considers that the architect has failed in what he has allowed the contractor to produce physically and in time. Here the employer is acting under his own contract with the architect, rather than through the building contract, just as he would over a design fault or other lapse of the architect that does not overlap with the contractor's responsibilities. He may even choose to act against both at once on matters arising out of the contractor's activity, so that he may obtain satisfaction from them both in some proportion or other.

A list of the major obligations and rights of the employer under or in relation to JCT 80 and IFC 84 is given in Table 5.1.

AUTHORITY OF THE ARCHITECT

Procuring the works

Procuring the works is the ultimate responsibility of the client or employer. This has been explored in detail in *Building Procurement*, where the options for 'design separate from construction' are compared with the options for 'design combined with construction'. Under the latter arrangement, contractors will take on design responsibility and will be responsible for control of information to suit their programme. This is discussed in Chapter 24 and in more detail in *Design and Build Contract Practice*, where the implications of contractors carrying out design are considered.

A key feature of the standard forms taken in this volume is the naming of the architect or an equivalent person to exercise special functions within the building contract, in addition to the design functions which he performs under his agreement with the client. These special functions exist partly to allow him to carry his design concept through to realisation, when built by another with whom he has no contract. They also allow him to stand between the parties and decide various matters, as is convenient when he is already in that mediating position. There is no intrinsic connection between all of these functions. In all, they do give him considerable power and responsibility, but they are strictly defined and so are limited. He has been described in these respects as 'the creature of the contract'.

Directly related to his design function, the architect (as he may be termed alone for simplicity) has powers to ensure that his design is what is built. These powers cover inspection to see that the building is correct in layout and disposition overall, having the right quality and sufficient detail. Subservient to these is his power to have tests performed on materials and work. But in the way that building design and tendering proceeds, it is usually necessary for supplementary information about the design to be fed to the contractor during progress to 'explain' the requirements. This involves working to a time schedule and also implies that no change will be introduced by what is provided as explanation. However, since changes are almost inevitable, the architect may also introduce them into his design in a regulated way, hence changing the amount payable to the contractor. In each of these areas, there are corresponding provisions to protect the contractor from laxity or abuse of power on the part of the architect.

Taking decisions

The further role of the architect under JCT contracts is very much that of keeping the parties in as harmonious a relation as possible. He is the focus for communication in both directions, and the contract machinery provides for him to issue and receive most notices and to be the one who issues instructions and certificates on a range of matters. All these items are listed in Table 5.2.

Among them are a number of issues over which the architect is to act as decider in matters of potential dispute. Some are day-to-day elements such as whether work

Table 5.2 Architect's instructions and certificates

	JCT	IFC
Instructions		
Discrepancies or divergences	2.3	1.4
Discrepancies or divergences, performance-specified work	2.4	–
Statutory obligations	6.1.3	5.4.3
Setting out works	7	3.9
Opening up, testing, removal, excluding persons	8.3–8.6	3.12–3.14
Variations and provisional sum expenditure	13.2, 13.3	3.6, 3.8
Contractor's variation quotation	13A	–
Schedule of defects	17.2	–
Defects during liability period	17.3	2.10
Insurance of other property	21.2.1	–
Insurance of liquidated damages	22D.1	–
Postponement of work	23.2	3.15
Account documents to quantity surveyor	30.6.1.1	4.5
Antiquities	34.2	–
Nominated or named subcontractors	35.5.2, 35.6, 35.8 35.9.2, 35.18.1.1 35.24.6, 35.24.6.3 35.24.7.3, 35.27.7.4 35.24.8.1, 35.24.8.2	3.3.1–3.3.3
Nominated suppliers	36.2	
Integration of performance-specified work	42.14	–
Certificates		
Practical completion of works	17.1	2.9
Completion of making good defects thereafter	17.4	2.10
Completion of making good defects after partial possession	18.1.2	–
Payment of insurance monies after damage	22A.4.4	6.3A.4.4
Failure to complete works	24.1	2.6
Expenses etc after determination by employer (optional)	27.6.4.2	–
Interim certificates generally	30.1.1.1	4.2
Interim certificate to pay off nominated subcontractors	30.7, 35.17	–
Final certificate	30.8	4.6
Failure to prove payment of nominated subcontractors	35.13.5.1	–
Failure of nominated subcontractors to complete	35.15.1	–
Practical completion of nominated subcontract works	35.16	–
Interim certificates (three) following nominated subcontractor determination	35.26	–

is up to standard, but others are questions of whether the contractor is entitled to an extension of time or to extra payment for disturbance of progress. This leaves room for some tension, as the architect may be acting as judge and defendant in the same matter, to the extent that what is under consideration may be the result of his own dilatoriness or error. This point recurs in latter chapters explicitly, and underlies some of the discussion here. It may be suggested too that the architect is in an awkward position, because he is ultimately responsible for the cost of the project to the employer. So too is the quantity surveyor, in that he has prepared the cost control information, but he does not then make decisions that add to the cost in the way the architect does. At worst, he is to be blamed for a bad forecast, whereas the architect has actually incurred the expenditure for the employer. The contractor may be forgiven for the occasional feeling of unease when contentious matters arise.

Limits to powers

There are, though, distinct limits to the architect's powers under the contract, because he has only those powers that the contract gives him. For instance, he can issue instructions only on those matters stated, and the contractor can challenge anything apparently out of line (see 'Issuing instructions'). More widely, the contractor has the right to seek adjudication, arbitration or possibly litigation over the effect of the architect's decisions. The employer too is protected by this appeal to adjudication etc, and both parties have the possibility of court action. The costs, procedures and scope differ between the channels, so which one constitutes the more desirable channel depends upon the individual situation. Too extensive to be considered in this chapter, these ideas are discussed in Chapter 27. As most disputes come down to money, the powers and duties of quantity surveyors should be understood in contracts (accepting that in ICE and NEC contracts a quantity surveyor's functions are not separately recognised by a named role in the contract). However, in building contracts most of the arrangements for agreeing final accounts rest in the hands of the quantity surveyor, and there is doubt as to whether the architect can interfere with them, provided the quantity surveyor has proper contractual authority for what he includes. Also, the architect is not able to intervene in the contractor's construction organisation to direct how he shall perform the works, in such terms as how he disposes his equipment about the site, the sequence of a contractor's working (as opposed to sequence of the works), how much labour and plant he uses, and so on. The contractor's overriding obligation is to 'carry out and complete the Works' (so both JCT 80 and IFC 84), that is to produce the finished article within the time given. For another to interfere with this activity would be to undermine the basic contractual responsibility. This is a clear area where the contractor may protect himself by asking the architect to substantiate his own authority under the contract. This immunity of the contractor from meddling, as it might become, extends to the detailed means of producing an individual piece of work within the whole. Contrast this with the philosophy adopted in the NEC contract, where the project manager (PM) is able to instruct the contractor to require a change to be implemented in the way that the project manager wants. It is suggested that the PM will not lightly become involved in construction sequencing and methods of construction but the NEC contract provides for that to happen if circumstances require it. Compensation for such an entry into the domain of the contractor is provided for by means of a 'compensation event', if time cost or quality has been affected beyond that previously contracted for by the contractor.

In each case the implied restriction on the architect is subject to any specific provision to the contrary in the contract documents. Thus, there may be requirements over phasing or restrictions of when the contractor can obtain access to sections of the site. In detail, there may be precise specification clauses about how he is to mix concrete, the weather conditions in which he may lay it or how he is to protect it, giving a more restricted situation in some respects than under a purely design-mix specification.

There is a distinction among these examples. There is no authority for the architect in JCT contracts in particular to instruct over the level of productive resources, such as labour, that the contractor may deploy, including whether the hours per week should be increased (or for that matter decreased). This is irrespective of whether extra payment is held out to the contractor for such changes. Only in prime cost contracts do such powers intrude to some extent.

By contrast, the architect has some authority to instruct over amending phasing, access and cognate matters, at least when the contract contains some initial statements about them. He also has authority to instruct postponement (see the next section and also Chapter 15). He may clearly amend the specification, even to the extent of changing the basic philosophy of how work and materials are covered to produce roughly the same finished product. But exercise of these powers carries with it a balancing right for the contractor to seek an upward adjustment of the contract sum, or concomitantly for a downward adjustment to be obtained from him in appropriate cases.

ISSUE OF INSTRUCTIONS BY THE ARCHITECT

There are three aspects to an architect (broadly, this will apply also to an engineer or project manager) issuing instructions.

Construction contracts are predicated on the theory that tender information, which becomes contract information, contains by far the majority of the information that a contractor needs to construct the project (provided the contract is one for 'firm' quantities or not otherwise a contract indicating that it is on provisional information to a large degree). Otherwise, a contractor has not been provided with the information on which to give a 'lump sum tender'.

When asked what information, drawings, specification etc they need to prepare a firm bill of quantities to obtain a firm tender, quantity surveyors have been known to reply, 'Just the same information that the contractor will need to use for construction.' Therefore, the information that an architect may need to issue as 'instructions' during the contract period can be categorised in three ways:

(1) Information that is not, for whatever reason, complete and given in the tender but is inherently known to be required by the scope and nature of the job as set out in the tender documents.
(2) Odds and ends of information that almost inevitably occur during the course of any contract – little points to do with finishing, colours etc.
(3) Variations, which by their nature are changes to the contract.

Category (1) is discussed in relation to possible claims if not handled adroitly, particularly in relation to work under JCT contracts. Categories (2) and (3) are not of concern in this instance. Category (2) should be minor matters, by definition, and category (3) is dealt with in Chapter 6.

Situations often arise during progress on-site when the architect must instruct over what has to be done, subject to the limitation of his powers discussed immediately above (see 'Limits to powers'). Under the standard forms, the scope of his instructions is restricted to areas specifically authorised, as listed in Table 5.2, so the contractor has a right to challenge whether an instruction comes within this list under JCT clause 4.2 or IFC clause 3.5.2. If it does not, he need not comply. He should follow the procedure carefully and be clear on his position first. Equally, the contractor need not comply at once with an instruction on any subject which the architect 'purports to issue . . . other than in writing', as JCT clause 4.3.2 has it.

That clause allows, but does not require, the contractor to confirm oral instructions back to the architect within seven days of issue and then to act on them 'within 7 days from receipt' by the architect, unless the architect has dissented. This is a slightly precarious arrangement in a closely timed instance, as the contractor does not know exactly when the architect receives the confirmation. If the contractor places an urgent order quickly, its cancellation due to the architect's slow reaction may lead to an extra payment. Strictly, the architect cannot dissent outside the limit. If there is any doubt, he should therefore issue a further instruction cancelling the contractor's written confirmation of an oral instruction. This gives the contractor authority for payment.

Most contractors operate the 'confirm back' system in the interests of progress. It also allows them to give their version of a dubious instruction, particularly over some matter other than a variation. It is, however, strictly in order for the contractor to ignore an oral instruction, even if its later proper issue causes extra cost. Subject to questions of goodwill, this may even be worth considering with an architect who is persistently remiss. The burden of proof is his and, in the nature of the case, he is hardly likely to have good records.

Alternatively, the contractor may choose to press on with unconfirmed instructions and rely on the provision for the architect to confirm at any time before the issue of the final certificate. This can be risky when faced by an architect with a bad memory or bad records, especially if the final account is high. This risk runs either way, according to who can make out the stronger or more unscrupulous case. The provision is always useful to deal with the items which are overlooked by everyone until, for instance, the quantity surveyor digs them out during his tedious final working through.

Information release schedules

In relation to instructions that fall under category (1), as discussed at the beginning of the previous section, important changes occurred in JCT contracts in 1998, brought about under amendment 18 to JCT 80 and amendment 12 to IFC 84, issued in April 1998.

These make contract provisions for the architect to state what information (i.e. information that has not been provided in tender/contract documents but which is inherently known to be necessary when the job starts on site, discussed under category (1) in the previous section of this chapter). The architect knows what work he has not drawn or specified and that it will be required for construction. As explained above, in theory the contract expects that all the information provided for tender will be all the information that the contractor will need for construction, but in reality the industry knows this to be incorrect. (Sadly, on too many projects the information provided at tender stage is not only incorrect and incomplete but is sometimes untrue; this can lead to circumstances where claims easily occur.) By the provision of new recitals in 1998, and an amended clause on instructions in the JCT contracts, for instance in JCT 80 in the Sixth Recital and clause 5.4, the architect is now committed to supply information referred to in the information release schedule at the time stated in the schedule (subject to any amendment agreed between the contractor and the employer, a proviso that allows the parties to vary such time by a further agreement, and if there is disagreement between the parties on reaching such agreement then the matter can be referred to an adjudicator). Although there is still a requirement for a contractor to notify the architect when he foresees that information will be required, there is an absolute requirement for the architect to state, in the information release schedule, when he will provide outstanding information, a bone of contention at the very heart of many construction disputes.

This is a significant change from previous JCT provisions and carries with it sanctions of extensions of time and loss and expense if the late issue of information, beyond that indicated in the information release schedule, causes delay or disturbance. Flow of information is an area that has always provided fertile grounds for disputes. Now the JCT amendment puts good practice into the JCT contracts as a provision. It remains to be seen if this lessens the likelihood for disputes and claims or gives more specific opportunity for them when, as before, information is provided late and incomplete. The nature of clients, designers and contractors needs to change to meet the standards required by these contract requirements.

Instructing subcontractors

It is axiomatic to the operation of a normal building contract, as with other areas of commercial life, that the contractor's subcontractors are his responsibility. This statement must be qualified, because standard contracts provide for various types of subcontractor:

(a) *Domestic subcontractor*: The freedom of the contractor under common law to choose a subcontractor to perform some part of the works prima facie to be performed by him directly is constrained by a requirement for him to obtain the architect's consent to any subletting. Conversely, the contract documents may require him to sublet a specified part of the works to one person out of a number listed by name in the documents. Under either arrangement, the contractor is dealt with and paid as though he performs the work concerned, and in turn deals with the subcontractor entirely as though no limitation had been placed upon him during selection.

(b) *Nominated subcontractor*: The contract documents may reserve some part of the works to be performed by a subcontractor whom the architect nominates to the contractor after entry into the contract. Subject to safeguards, the contractor must otherwise accept the nominee as his own subcontractor. At the tender stage, the contractor will have included a prime cost sum given to him to cover the amount of the subcontractor's account, rather than an amount affecting his own competitiveness. Thereafter, the subcontractor is paid through the contractor an amount related to his own terms and set against the original sum. Although he is a subcontractor, he receives special treatment over a number of matters relating to his programme and other matters defined in the contract.

(c) *Named subcontractor*: The contract documents may state that a named person is to perform some part of the works as priced by the contractor. At the tender stage, the contractor bases his tender for the part concerned upon the price of the subcontractor supplied to him as part of the tender documentation. Thereafter, the contractor is dealt with and paid for the work as though he himself performs it. The subcontractor receives some special treatment over given matters, but these are less extensive than when there is nomination.

The three types of subcontractor are taken in detail in Chapter 7 and other specialised aspects in other chapters. The important facts for present purposes are that they are subcontractors and that this status is modified only to the extent provided in the contract concerned. The architect must therefore conduct his dealings post-contractually with each class of subcontractor entirely through the contractor over all matters affecting the execution of the works, including such aspects as quality, programme and variation instructions.

This philosophy maintains the position of the contractor as controller of the construction itself and avoids the classic confusion situation arising when lines of communication are incomplete or crossed. There is no single method of creating havoc quite as effectively as passing information or instructions randomly among the participants in the building process, except perhaps by deliberately ignoring the proper channels! The contractor is perfectly entitled to ignore any communications directed at his subcontractors of all varieties, but not sent through him. He should ensure that a proper pattern is established from the first day and, so far as possible, that it is not broken. The standard contract provisions, and those of subcontracts, make no allowance for direct communication between architect and subcontractor.

Subcontractors themselves should ensure that their own positions are clear and protected, by receiving instructions through, and only through the contractor. Otherwise they risk performing work that is not coordinated with the rest, so alterations are needed, for which they will not be paid. Even worse, they may not be paid for the work itself as 'instructed'. Whether the contractor receives proper instructions and confirmation is for him to sort out; subcontractors are entitled to be paid by the contractor for what he requires them to do.

The standard subcontract forms oblige the contractor to pass any relevant instructions of the architect to subcontractors. They distinguish 'directions' of the contractor, which are requirements he originates. When they relate to general site operations, such as where to position temporary works, the contractor's directions do not concern the architect. Alternatively, they may be supplementary aspects of what the architect has instructed, relating to precisely how subcontractors carry out their activities. Subcontractors are advised to confirm back to the contractor immediately whether, in their view, any of these matters carry financial implications over and above what is evident from any architect's instruction leading to them.

These extra costs may trace back to the terms of the subcontract or to delay in passing on the architect's instruction, so that work has progressed meanwhile to a stage beyond where it was when the architect acted. There is no slack time allowed by the contracts for the contractor to act, and he must ensure that subcontractors are notified immediately of any urgent matter. This is an area in which three-way disputes can easily arise, unless very careful routing and timing of requirements is observed and corresponding confirmations are issued and records kept. The tendency is for subcontractors, especially if they are smaller organisations operating across several contracts, to be the ones less well equipped and so unprotected at the time of settling accounts. As they are on the end of the line for payment, the lesson should be obvious.

If there are elements of work which subcontractors have undertaken to perform, they should watch that the contractor does not perform them instead and then render a contra-account. This can be particularly contentious if there is work which the architect instructs to be removed (perhaps under JCT clause 8.4) and so, by implication, replaced. If the work is integral with work performed by others, subcontractors have little option about the principle of removal, whereas they should expect to deal with it physically themselves where practicable to reduce the effect on their remaining work. When there is work by several persons in question, the simple instruction to remove does not settle whether any one person's work is what is actually defective. Action and records at the time are needed so that reimbursement for the reinstatement can be sought where appropriate.

The distinctions between the several types of subcontractor do not affect the procedures and underlying positions outlined above. Even in the case of nominated subcontractors, just the same channelling of information remains correct, despite the modified position over some programme and other matters. However, there is one area of difference which must be considered under the next heading for the various subcontract types.

Design by subcontractors

Two situations may be distinguished over the initiation and transmission of 'design', as distinct from responsibility for its adequacy (a subject discussed in

Design and Build Contract Practice and *Building Procurement*). One occurs under the main contract umbrella and the other beyond it. The former situation is where a domestic subcontractor is required to develop some details of design, and so may the contractor. The size of screws for a fixing is strictly an aspect of 'design', but here are some more reasonable examples:

(a) Layouts of final subcircuits for an electrical installation.
(b) Choice of materials or components to fulfil a performance specification.
(c) Design of connections for structural steelwork.

Of these, (a) usually happens on site as work progresses and (b) may be little different. On the other hand, (b) and (c) may be subject to approval of the proposals by the architect or another consultant. Although it is generally held that no design liability can be placed on the contractor under a JCT or IFC arrangement, the design must still pass through the contractor if it originates with a subcontractor, and approval must return the same way. This allows the contractor to make any comment which affects his responsibilities. It is submitted that approval in this case transfers responsibility for the design adequacy to the architect. The contractor should be satisfied with nothing less under these contracts.

The last situation is where the architect requires a subcontractor to perform design by parcelling out to him work which either he, as architect, or some other consultant might be expected to do. This is permissible under the forms concerned only when the subcontractor is nominated or named. Although the subcontractor is paid for the design as part of his subcontract price, he performs the design quite separately under the terms of the appropriate employer/subcontractor agreement. This means there is no reference to the design work in either main contract or subcontract, and no liability attaches to the contractor for the design or any consequence of its lateness or error.

In this situation the contractor is not responsible for the transmission of information between subcontractor and architect, and it should pass directly between them. Indeed, it may pass before the contractor or a nominated subcontractor is appointed. In the case of a named subcontractor, it always does so. If the contractor becomes involved in passing information, he acquires an added responsibility over the timescale, and may acquire one by implication over some aspect of the design. For the purposes of the contract, such information becomes part of the design detail which, when it is settled, the contractor is entitled to receive as part of the architect's design, even when he then passes it to the subcontractor who originated it, and whether or not he himself needs to utilise any of it anywhere else in the works. As much as any other information, it ranks within the provisions about errors in information, extension of time, loss and expense, and so forth. Any redress which the employer may seek for errors or the consequences of late supply of information from the subcontractor, which holds up any part of the works, will lie against the subcontractor under any employer/subcontractor agreement.

MISTAKES IN CONTRACT AND OTHER DOCUMENTS

With the general positions of the various persons delineated, the categories of mistakes that crop up in the documents for a building contract may be considered. It is primarily the contract documents that are treated here, but others issued during progress and related to the contract set are included, particularly instructions, as these often modify the contents of the contract documents.

Chapter 3 concluded by showing how the contract is expressed in the documents alone, unless peculiar circumstances exist that require examination of any representations. There now arises the question of misstatement within the documents. This is distinct from misunderstanding of what they actually say, where the result of misapprehension will lie with the party under it, unless the question of ambiguity leads to invocation of the contra proferentem rule (Chapter 3). In the case of misstatement, this same rule comes into play, in that the party composing a document is prima facie responsible for its contents. There are three statements about mistake that may be distinguished within contracts based upon JCT 80 or IFC 84:

(a) The contract conditions will prevail over all other documents, by virtue of JCT clause 2.2.1 or IFC clause 1.3. A plea of accidental clash will not usually stand to reverse this, although a case like *English Industrial Estates* v. *George Wimpey* (Chapter 3) was otherwise decided on special facts and should be noted. As any clash is likely to be verbal, rather than arithmetical, it will be corrected as it emanates from the employer's side. If necessary, adjustment in the final account will follow.
(b) Mistakes in or between the text or graphics of documents are to be corrected and financial adjustment will follow on the same reasoning. The situations leading to this position are considered below.
(c) Mistakes in pricing or in arithmetic on the face of the documents are not to be corrected, so they are borne by the contractor in either direction. This follows explicitly from JCT clause 14.2 and implicitly from the agreement to pay the contract sum in IFC article 2.

The leading clauses relating to (b) are JCT clause 2.3 and IFC clause 1.4, which are broadly similar.

Two editions of JCT 80 need considering with IFC 84 here: that 'with quantities' and that 'without quantities', the latter being with the specification as one of the contract documents. These give the following documents to go with the conditions and the drawings, which always form part of the contract. In the case of IFC 84, the options are mutually exclusive.

(a) The contract bills as mandatory for JCT 80 with quantities, and optional for IFC 84.
(b) The schedules of work as optional for IFC 84.
(c) The specification as mandatory for JCT 80 without quantities, and optional for IFC 84.
(d) 'Numbered documents' as optional for JCT 80 in any version, to allow for nominated subcontract documents as required.

For work definition, each case therefore has the contract drawings and one other document, or several parts of the same document, and perhaps the numbered documents. There may be what JCT 80 terms 'discrepancy in or divergence between' documents, or 'inconsistency in or between' them, which is the umbrella equivalent term in IFC 84 (see next section). All of these amount to mistake, either within a document or in terms of what has been given in two or more documents, perhaps by two or more persons, such as the drafting consultants. There may also be a mistake in the sense of a single error, say of a figure, which is not at odds with anything else in the documents, and which may therefore be more difficult to discover.

These various locations of mistakes have several effects, notably at the tender stage and during progress, that is pre- and post-contract. They may lead to adjustments of the contract sum, and perhaps also to rectification of information or even of work performed.

Mistakes affecting the tender and the contract sum

Although aspects of tendering have been outlined in Chapter 3, the effects of mistakes in the documents supplied to the contractor may not appear until the contract is under way. As the contractor's tender sum, with any adjustment during examination, becomes the contract sum, any undiscovered mistakes of this type will be carried forward. Contract drawings are not mentioned separately here, but they are related to the documents which are mentioned, by containing the other element of any divergence or by containing a discrepancy that will have been interpreted in some way in the documents mentioned. The numbered documents for nominated subcontractors should not have any effect, as prime cost sums cover their value at this stage to the extent that there is a directly financial effect.

Contract bills

When there are contract bills, JCT clause 14.1 and IFC clause 1.2, in its third paragraph, both state that 'the quality and quantity of the work included in the Contract Sum' are those in the contract bills. This means that the bills prevail for financial purposes under the contract, as well as for giving the specification. JCT clause 2.2.2.1 and IFC clause 1.5 give the Standard Method of Measurement of Building Works (currently SMM7) as applying to these bills, unless something contrary is stated in 'any specified item or items'. There must therefore be an explicit and particular indication of any change to the method of measurement underlying the bill items. This method requires various matters of description to be covered, as well as essentially computational points.

The means of indicating a change are left open. Here are some of the possible ways:

(a) Special explanatory endorsement of an item or group.
(b) Clear description of an item or group which shows without emphasis that it includes or excludes some liability which is normally treated in the opposite way.
(c) Definitions in the preambles governing the group.
(d) Measurement in a unit different from that prescribed and in some way highlighted.

These are listed in what is usually seen as descending order of clarity to the estimator, whatever may be said about his responsibility to read every word, no matter where and how it may be presented to him. Departures from the norm like these should be made extra plain.

There are in principle two ways in which the contract bills may be in error: by not agreeing with the contract drawings over quantity or description, and by having some mistake originating inside themselves. Non-compliance with the standard method, without saying so, would come into this latter category. The two forms of error may effectively have the same result in many cases: an incorrect amount is included in the contract sum or the contractor is misled about what to produce, or even both.

Each contract provides that any of these errors is to be dealt with by correction, as though a variation instruction has been issued (see later in this chapter).

Schedules of work

If there are schedules of work under IFC 84, the same principles apply, except there is no recognised method of measurement to support any quantities given. For this reason, the provision of quantities is of dubious value. There need not be any quantities, a schedule can simply be a specification arranged in such a way as to allow it to be priced paragraph by paragraph. This contrasts with the contractor taking off and pricing quantities which relate more loosely to a traditional specification, so that he may draw upon several specification items for the content of a single measured item. With schedules of work, the contractor may need to take off some form of quantities to price each item, but his effort is contained within the item in each case.

If supplied with quantities in such schedules for tendering and ultimately for contract purposes, the contractor therefore has to read what he is given with particular care, in case anything unusual has been done. To the extent that definition is uncertain, he would be well advised to endorse his tender as being based upon the understanding that the usual standard method of measurement has been followed so far as relevant and unless some other principle has been made explicit for any item of work. Clause 1.2 (third paragraph) gives these schedules the same status as contract bills, and clause 1.4 allows them to be corrected in the same way, except for silence over anything to do with the method of measurement.

Specification

The use of a specification without quantities is the last option, applying under JCT 80 without quantities and under IFC 84. There is no standard method of specification writing giving rules to be followed and deemed to underlie an actual specification, although there are published documents which are widely used, such as the National Building Specification, and the Common Arrangement system has given standardisation a further impetus. Both writer and estimator are therefore on their own to exercise reasonable judgment to a large extent. JCT clause 14.1 states that what is in the contract sum is what is 'shown upon the Contract Drawings or described in the Specification', without establishing either as prevailing. Presumably work may appear in both places or only one. IFC clause 1.2, in its first paragraph, gives the content of the contract sum as being what is in the contract documents 'taken together', but with the contract drawings prevailing if there is inconsistency. This is more helpful than silence.

When there is silence, the contractor is left with doubt. If something is in one document only, he must include for it, unless there is some strong reason against it being required. Thus he should include for an unspecified cupboard shown upon the drawings, subject to querying what its construction is to be, if he lacks information on the drawings. Equally, he should include for two roof ventilators if they are clearly specified, even though their exact positions are unsure. But if he finds a specification for 'the basement tanking' when there is no basement, he may reasonably not include the tanking in his tender. If a cupboard is not shown and presumably cannot be inferred, again the contractor cannot be expected to include it in his tender. If he finds that he has some complete inconsistency, such as differing numbers of floors on plans and sections, he can do nothing without having the position cleared up by the architect. The possible illustrations are unlimited.

This is a difficult area, because there are some fairly old legal cases which support the view that the contractor must allow when tendering for whatever is reasonably necessary to complete the works in an entire contract. This is what a building contract usually is and in particular is in the forms being considered, with some modifications not affecting the point at issue. For instance, in *Williams* v. *Fitzmaurice* (1858), it was held that the contractor had to supply without further charge floorboarding which had not been given in the specification and, it appears, not shown on the drawings for a house. Apparently, the view was taken that the floorboarding was a necessary part of the house.

The doubt that lies over these decisions for present-day purposes is that the technology of construction today is more complex and varied, so the contractor will not necessarily know as standard practice what has been omitted and needs to be supplied, while contract documents have also developed to show more detail as regular practice. Industry-wide standard documents, such as methods of measurement, drawings codes and specifications, using any common arrangement system, set standards of communication which the courts are increasingly likely to accept as normative. Even so, at the level of smallest detail, the contractor must ensure that he covers what reasonably is left to be inferred.

If, however, matters do not go so far as legal proceedings (and they are not very likely to warrant it for the smaller contract), it is likely that an inconsistency will be resolved in favour of the contractor who did not proffer the faulty documents. The contract provision under JCT clause 2.3 and IFC clause 1.4 is again for the architect to issue an instruction resulting in a variation. In general, it is necessary for a specification-based contract to be tied up more thoroughly than a quantities-based contract, because it cannot contain major uncertainty, as the other may do, explicitly or concealed. Only by the use of a provisional sum can undecided work be covered, as provisional quantities are not available. This removes competitiveness from this portion and so should be restricted to work of minor value.

Although it is not strictly within the area of error, the principle of the architect being able to issue such supplementary information under the JCT without quantities form is highly suspect. The same doubt occurs when the IFC form is used without quantities. This is because the contractor, when tendering in the absence of quantities which are to form part of the contract, is preparing his own quantities without redress if they are incorrect due to his own error. He therefore needs virtually all the construction information as tendering information, to eliminate uncertainty over what is required, which may lead him to make injudicious guesses during the hurry of tendering. He is unable to tell which rooms are to have painted walls and which wallpapered without a schedule or annotated drawings. He may, on the other hand, be indifferent when tendering to the precise position of a door in a given wall, so long as he knows all of its characteristics.

There should therefore be far less need for the contractor to receive information when drawings and specification are in use. If this happens frequently, there are probably variations or errors underlying the situation.

Mistakes affecting performance of work

Many of the mistakes considered above may also affect work on-site, because they lead to erroneous understanding of what is required. What is said is therefore applicable here as well.

Relationship of documents

The documents forming part of the contract with the several sets of conditions have been set out, but are repeated in less detail with the other post-contract documents, which are mentioned in JCT clause 2.3 and IFC clause 1.4 as possibly not agreeing:

(a) Contract drawings.
(b) Contract bills, specification or schedules of work (but not a contract sum analysis or schedule of rates, however, which does not or should not convey information to the contractor).
(c) 'Numbered documents', such as nominated subcontract documents.
(d) Instructions issued during progress, other than those introducing variations (the latter are bound to be 'out of step' with the contract documents, because they change something).
(e) Descriptive schedules or other similar documents 'necessary for use'.
(f) Further drawings or details 'to explain or amplify the Contract Drawings or to enable the Contractor to carry out and complete the Works'.
(g) Levels and setting-out information.

All of these are intended to fit together and not change what is to be produced, or any obligations and restrictions governing activity. Bearing in mind that the contractor has tendered and obtained the contract on a defined and adequate basis (it is to be assumed) without (e), (f) and (g) in particular (which are not given as introducing variations), their inclusion as contractually recognised may occasionally bring something of a chill to the spine. This is especially so when the contract is based on a specification which, with the drawings, purports to tell the contractor everything in the absence of quantities.

When there are contract bills, the reasonable intention of these extra items is to give the contractor information of a positional nature or detailing small parts of the works. Most, but not all, of this information was needed by the quantity surveyor for the preparation of what have now become the contract bills. That he does not always receive it or that it is not always subsequently based on what he included in his bills, is a well-known cause of many of the discrepancies and divergences being discussed. It is information which the contractor did not need when tendering, because the quantities contained the relevant quintessence of it.

Whether the basis is contract bills or specification, it is important that any supplementary information is just that: supplementary. The danger stage comes when it introduces changes which should be covered by variation instructions. It is often tempting to combine small variations into supplementary details, but this must be strenuously avoided by the architect or resisted by the contractor to escape confusion. When uncertainty over what is in the contract and what is variation is combined with what is mistake, the potential for chaos assumes unfortunately common proportions. There is no answer, other than a pedantic adherence to a rigid procedure.

Relationship of mistakes

Even assuming a straight run through the issues just raised, there may be errors in the documents emerging. Two categories of mistake are mentioned in the JCT clause:

(a) Discrepancies within individual documents.
(b) Divergences between two or more documents.

The IFC clause rolls these together as 'inconsistency in or between' documents. The effect is the same, and much of the earlier discussion also relates to these points. A further area of potential divergence exists over statutory requirements, as JCT clause 6.1.2 and IFC clause 5.2 recognise. This area lies outside the sphere of purely interdocumental differences, but extends to include possible clashes between variation instructions and these requirements. In all cases, the contractor is required to give notice to the architect, so that the problem may be resolved.

There is no automatic 'this document prevails' option, as mistakes do not come like that when considering what is actually required to be performed. This must be distinguished from the position of the conditions prevailing in contractual interpretation, and the position of the contract bills (if any) prevailing over what has been included financially within the contract. These two principles can reasonably establish certainty in their own areas, so that adjustment can be made from a declared base.

The architect is to give instructions to resolve the tension. These are not defined as variation instructions, nor as instructions that are deemed to constitute variations, except over statutory matters. This is because they may not introduce a variation but, by the choice of the one alternative, may simply confirm what is already included in the contract sum, be it in bills or otherwise. If, therefore, a variation is being introduced, it must be covered by an instruction in suitable form. If the effect is to show an error in the contract bills, or if it can be construed that way, it is possible to rely on JCT clause 14.2 or IFC clause 1.4 for the adjustment to be treated as a variation, deemed in the former case and actual in the latter.

In the case of statutory requirements, the conflict is between some section of the contract documents or a variation and an outside set of stipulations. If the requirements prevail, in the sense of a waiver not being available, a variation or further variation is almost inevitable. The instruction issued by the architect in response to the contractor's notice is therefore to be treated as though it were a variation instruction.

Responsibility to discover mistakes

Both JCT clauses use the term 'If the Contractor shall find', whereas the IFC clause on statutory matters uses the term 'If the Contractor finds', as an accommodation to movement in the language. The IFC clause on general inconsistencies has no such expression, but simply starts where the others continue, by saying 'the architect shall issue instructions in regard to' the various matters listed. It is suggested that the silence is as helpful as the statements. It also leaves affairs open to cover the situation in which the architect finds inconsistencies unaided.

The problem with the three clauses, and it is there by implication in the fourth, is in the 'If'. This is sometimes held to mean 'If there are any mistakes' or, more reasonably, 'If the contractor finds them'. A contractor has a clearly established duty to warn of any errors of design etc which he notices, and this would extend to any clashes or deficiencies in the documentation. In undertaking construction work, he holds himself out to be an experienced and competent builder, as distinct from an expert in matters of design which lie within the special sphere of an architect or engineer. The courts are tending to take stricter views of the liabilities of skilled persons in general, and designers and constructors are no exception. But this does not mean that constructors are to be as expert as the person whose work they receive as constituting what they are to construct. It is one thing for the contractor to notice that reinforcement has been omitted from a concrete beam, it is another for him to realise that the detailed reinforcement does not allow an adequate factor of safety by its quantity or positioning (see the cases at the end of this section).

This is an example of a mistake upon the face of a document apparent to the specially knowledgeable, but not to all. More difficult is a divergence between two documents, even if it does not require special expertise to discern it. If a nominated subcontractor indicates in sufficient time a need for holes for pipes through a concrete upper floor, this may mean rearrangement of reinforcement designed by the structural consultant, who should be advised and coordinated by the architect. If the point is missed by the architect, it may then be that the contractor also does not notice when setting about construction and working from the structural drawings. In principle, it is fine to say to him, 'You should have combed every drawing, as you knew that such holes were required in places,' but it does not always come about that way in the sequencing of information supply to him. Another example might be in coordinating specialist window details with the structural frame. If the same person, say the architect, produces the divergent information, it is often all the more difficult to suggest that the contractor is at fault. The inclusion of the numbered documents category in JCT 80 heightens this problem.

A clash with statutory requirements is a rather distinct issue. These divide most commonly between matters of planning control and matters of building control. Matters of planning control should be cleared before work proceeds and are generally at too high a level of concept for the contractor to be likely to fall foul of them. On the other hand, building regulations are his own backyard. They pose a distinct issue, because they stand outside the project as separately constituted, but bearing on it. Usually, therefore, a clash with them should be more apparent to the contractor than other discrepancies which could either be missed or which might be mistaken for special requirements of the designer. With this said, the clauses in both contracts do not place an explicitly greater responsibility upon the contractor to find and report.

In the end, the contractor is responsible for sorting out what he is building, but the original responsibility for getting it right lies with the designers. There is a whole area of uncertainty lurking in the shadows. Given the 'If' of the clauses and the realities of operations on and about sites, there is room for a deal of tolerance in treating these inconsistencies. The designers cannot pass on to constructors the responsibility for discovering the results of skimped work, whereas constructors cannot avoid all responsibility and they must perform routine and sensible checking. When they do find something, they must report it. If they do not, and this can be established, they are liable for the consequences of passing it by.

Two differing cases illustrate these principles and the way in which intricate technical questions have a bearing, while varying from case to case. In *Equitable Debenture Assets Corporation Ltd* v. *William Moss Group Ltd and Others* (1984) the issue was some leaking curtain walling designed by a nominated subcontractor. In addition to the subcontractor's liability, it was held that the architect was negligent over acceptance of the subcontractor's design and in other ways. The contractor should not have been expected to check the design in advance, when the architect had spent some time on this, but he should have reported the faults in the curtain walling that became apparent during erection, so he was liable in this respect. A term was implied in the contract by the court to this effect, as none was express in the JCT 1963 form. Design apart, the contractor was also liable for the subcontractor's bad workmanship on-site and for breach of the building regulations.

In *Victoria University of Manchester* v. *Hugh Wilson & Lewis Womersley (a firm) and Pochin (Contractors) Ltd* (1984), the problem was with external ceramic wall tiling applied to irregular concrete surfaces, a method untried by several of those involved. The architect had failed to warn the employer of the risks and the design was defective. The subcontractor had failed to fix the tiles adequately. Again, there was an implied term that the contractor would warn of any design defects believed to exist (something less than a complete duty to search for and find defects), but he was not liable for breach of this duty. This appeared to be based largely on his lack of previous experience with such tiling and therefore practical inability to warn. He was, however, liable for failure to supervise his subcontractor adequately, despite the contributory negligence of architect and clerk of works. Several other distinctions of liability were drawn, making the case more widely interesting reading.

The 'duty to warn' question appears to be heading towards another instance of what is reasonable, as the courts develop a more structured attitude to a problem which is growing with the intricacy of buildings. In *Bowmer & Kirkland* v. *Wilson Bowden Properties* (1996) it was held that it should be within the knowledge of a skilled person (which includes builders and contractors as stated below), when asked to deal with a specific problem, that problem may cause other problems and therefore the skilled person should check for such other problems or at least draw the other person's attention to the need for such a check. This will readily apply to demolition, central heating, compliance with CDM Regulations etc. It was held that

> when a skilled person is asked to deal with a specific problem, it is not enough for him to deal solely with that problem and walk away. It should be within his knowledge that the existence of that problem may have caused other problems, he should either check for the existence of other problems or, at least, tell his employer or client that a check ought to be made. The principle must apply to all skilled people, whether they be surgeons, lawyers, mechanics, builders, electricians or plumbers.

Then once work is constructed incorrectly, because the contractor did not notice, there is the direct cost of its correction and, associated with it, the contingent matters of extension of time and loss and expense which the contract provisions allow as appropriate. Even if the matter is raised early enough for redesign to prevent abortive construction, it may not be early enough to avoid delay and extra expense. The question then becomes not only should the contractor have noticed, but how soon should he have noticed? See *Bowmer & Kirkland* above.

There is no clear provision for the contractor to be paid for excess supervisory costs of sorting out the mistakes and shortcomings of others, and often it is difficult to segregate this from routine checking and cooperation. In a major instance it could well fall to be reimbursed as part of the managerial expenses incurred, following the reasoning in *Tate & Lyle* v. *Greater London Council* set out in Chapter 12.

APPROVAL OF THE WORKS

The overall scheme of approval

The standard contracts make limited reference to the question of approval of the works by the architect, and none to approval by the employer. The main scheme calls for little comment here and may be summarised to highlight responsibilities and show where problems might occur:

(a) The contractor undertakes to 'carry out and complete the Works in accordance with the Contract Documents', as the IFC version more succinctly puts it. The primary responsibility is upon him to perform, rather than upon the architect to ensure that he does and has.
(b) During progress, the architect has a right of access to the site and other places where work is being performed. The purpose of this is not stated, but may be inferred as to allow inspection and checking to take place.
(c) Also during progress, the employer may appoint a clerk of works as inspector under the directions of the architect. The clerk of works also has a right of access, but no power to issue instructions, which thus precludes him from instructing the removal of defective work (see below). Anything which he 'directs' must be confirmed by the architect.
(d) The architect is not obliged by the contract provisions to give approval to any work during progress. This is particularly true over quality, but would be difficult to sustain over grossly incorrect setting out of work, so that the works were wrong in position or scale.
(e) The architect may instruct the removal of defective work or materials during progress (see below).
(f) When the architect forms the opinion that practical completion has occurred, he is to certify this, and the employer may take possession of the works. There are subsidiary provisions over phased completion in various ways, but these may be ignored here. The architect is not required to have checked the detailed quality or completeness of the works at this stage.
(g) During the defects liability period following practical completion and usually of six months duration, defects may be found by the employer or the architect, it may be assumed, and notified to the contractor within fourteen days of that period ending. The contractor is obliged to make good any such defects at his own expense.
(h) Following making good of defects and up to the issue of the final certificate, the contractor is still liable for the cost of making good defects, but is not obliged to put them right himself. It is good law that he should be given the option of so doing, before others are engaged to do it, as applies during progress and the defects period.
(i) During the whole period before the issue of the final certificate, no certificate is conclusive over compliance of work with the contract requirements. This includes certificates of payment which include the work in question.
(j) The final certificate ends the operation of the contract, leaving the parties with their residual rights and liabilities during the limitation period which applies, and subject to the Latent Defects Act 1986.
(k) The final certificate provides that it is conclusive over those elements of the works which were expressed in the contract documents as to be 'to the reasonable satisfaction of the Architect', so that there is no further redress against the contractor, except for fraudulent concealment. It otherwise does not express that the architect is satisfied that the contractor has complied with the contract, although the architect does not issue the certificate until he is happy that the works are not showing any patent defect at the time of issue.

The effect of these several steps is that the architect has kept the works under observation during progress and for some time after, so avoiding the complication of having to take action against the contractor over every individual defect as it is found. Also the contractor remains liable for the works in respect of latent defects during the usual statutory period applying to contracts.

Defects during progress

The contracts are more explicit over defects during progress than over any other aspects listed above. Both the JCT and IFC forms contain the same basic provisions, but the IFC form goes further with extra provisions.

Common provisions

The basic, common provisions are in JCT clause 8 and IFC clauses 3.12 and 3.14. Their scheme is for the architect to be able to instruct the opening up of any work or testing of any materials, whether or not there are clauses in the contract documents mentioning specific tests etc. There may be such clauses in respect of such matters as regular testing of concrete cubes or runs of pipework. It is then usually possible for the cost of tests to be included in the contract sum, as they form a predictable proportion of total costs. Whatever may have been said or not said, the contractor is obliged as usual to comply with instructions given. If the results of tests show that the items were not in accordance with the contract quality, he must bear the cost of tests (if not in the contract sum) and replacement. If the items are in order, the contractor is reimbursed his costs. In a major case, he may find that he also has a right to extension of time or loss and expense payment, because of the disturbance factor.

It would be possible for the architect to run riot in using his powers, but it is not likely that he will, in view of the penalties incurred if he is wrong. There is no time limit after work, if so it be, is performed within which he must act. This makes sense, as defects may be suspected at any time after work is carried out. As the architect is under no liability to clear inspections as work proceeds, this aspect needs reasonable application to avoid upheavals long after work is installed and perhaps encased by other substantial and expensive work. Many critical features of work are covered progressively by the distinct inspection of the local authority, whose powers to condemn are separate from those of the architect.

The contractor is advised in any case of difference which cannot be resolved to take sufficient evidence for future use, as he is obliged to comply with the architect's instruction. Under the JCT contracts there is the possibility of immediate adjudication or arbitration, although some areas of dispute may not be referred to arbitration until after final completion. The evidence may be samples, photographs, independent reports or whatever is suitable. If the contractor refuses to remove anything after further notice, the employer has powers under JCT clause 4.1.2 or IFC clause 3.5.1 to engage others to do the work of removal and counter-charge the contractor. It is reasonable in such an impasse to allow the contractor time to take his evidence first.

In an extreme case of dispute, if the works are 'materially affected' by a refusal to remove, the employer has a right of determination against the contractor (Chapter 16). This should be reserved for such situations as major foundation problems which, if left, would prejudice the whole structure and cause extensive costs and lost building use. Even damages might not adequately cover this in terms of the employer's disturbance aspects, especially during occupation.

Special IFC provisions

Clause 3.13 of the form contains additional provisions to deal with the situation in which some work of a category is found to be defective and there is the question of whether other work in that category is defective. It is of course possible under the clauses already discussed to have successive sections of work opened up and tested, with the attendant risks over costs. The present clause aims to allow an exploratory procedure, which may stop short of destructive testing, but it also places the cost with the contractor in the first instance. The steps in the procedure are as follows:

(a) Work or material is discovered to be defective, presumably under the basic procedure in the preceding clause. This is also put right under that clause.
(b) The contractor is to inform the architect of what action he 'will immediately take at no cost to the Employer' to check that no further similar defects exist.
(c) The architect may instruct the contractor to open up work or test materials at no cost to establish that there are no similar defects, if one of three contingencies exists:
    (i) The contractor provides no statement within seven days.
    (ii) The architect is not satisfied with the proposals.
    (iii) Safety or statutory factors give great urgency to matters.
(d) The contractor is to comply with the instruction 'forthwith'.
(e) Without prejudice to compliance, the contractor may object to compliance and, if the architect does not withdraw his instruction, arbitration automatically ensues over the 'nature and extent' of what is being required.
(f) If and so far as the arbitrator decides in favour of the contractor, the contractor is to be paid his compliance costs.

The procedure therefore puts the contractor under an obligation automatically to act under (b) every time there is a discovery under (a), in case the architect acts under (c). It then allows the architect the option over whether to accept what, if anything, is proposed but, either way, the contractor has to act without charge, unless he can persuade the architect to change his mind completely. There are the additional hazards of the contractor not acting quickly enough (and the matter may require prolonged investigation) or of urgency existing. His only relief over the time pressure lies in putting forward a provisional proposal for action, to be followed by another when he knows more about the problem. Beyond this he must hope for the prospect of a favourable arbitration, if all else fails. The use of the procedure lapses at practical completion, whereas it could be useful during the diagnosis of defects thereafter.

The contractor is therefore being placed at risk, when one defect is discovered, of having to pay for what could become an endless series of investigations, with only adjudication and maybe arbitration to help him, and this coming in willy-nilly if he but complains and the architect remains silent. It is also possible for uncertainty to arise as to which clause, basic or continuation, is being invoked, as matters are not always as clearly segregated in time and place as discussion on paper may suggest. This leads to doubt over who is to act in what way, and over who is paying for the actions taken. There appears to be no power for the architect to agree to 'split the difference' with the contractor or otherwise to apportion responsibility. Only the arbitrator may review this, but often it will be after completion of the works and he will have to decide on evidence, rather than 'split the difference'.

It is to be hoped that, in practice, the architect and contractor will usually discuss procedures when a first defect has been found, before either of them rushes into a process which could be far more arduous than expected. The clause is a useful attempt at pointing a way through a common maze, but it does create its own dead ends.

Final certificate

Each standard form of contract has provisions for issue of a final certificate. This is the last chance that an employer has to be sure that the works are to the specification required. The conclusiveness and therefore importance of the final certificate concerning defects was brought out in *John Mowlem and Co. Limited* v. *Eagle Star Insurance Company Limited and Others* (1995). JCT contracts were amended in 1995 to recognise the 'conclusiveness' that was all-embracing before, namely that the architect was signifying by his final certificate that 'where the quality of materials or the standard of workmanship are to be to the reasonable satisfaction of the architect the same are to such satisfaction'. The provision, introduced by amendment 15, has the effect that materials and workmanship expressly described in drawings/specifications or schedules of work as having to be to the satisfaction of the architect are so to be and are, by the issue of the final certificate, certified as having been to the architect's satisfaction but that the certificate is not conclusive evidence that such other materials or workmanship also comply with any other requirement or term of the contract.

CHAPTER 6

VARIATIONS, PROVISIONAL SUMS ETC: MAINLY IN JCT 80 AND IFC 84 CONTRACTS

- Authority and instructions
- Definitions
- Procedures for valuation
- Principles of valuation
- Rules for valuation
- Special provisions for valuation
- Valuation provisions under various JCT contracts

This subject is extremely important for present purposes, because building works of even moderate scale are habitually beset by continual instructions resulting from changes of mind by the employer or his consultants, quite apart from those forced upon the parties by outside agencies, ground conditions and a selection of other hazards to progress. The result is often quite drastic amendment of what was to have been produced under the original contract documents, with equally drastic amendment of the contract sum, due to the cost of the actual variations and of their consequences in terms of disturbance and delay.

This chapter first examines the contract provisions of the chosen contracts over variations and then considers how they are to be implemented financially in their own right. This leads on to their relationship with the provisions over loss and expense for disturbance of regular progress.

Essentially similar provisions are contained in the JCT 80 and IFC 84 contracts to deal with variations, those in the IFC 84 contract being rather more simply expressed. Reference is made initially to JCT clause 13 in the 'with quantities' edition. Separate discussion is given of the differences arising when the other editions of JCT 80 are used, while discussion on IFC 84 is given where most suitable. The other contracts considered in this book have their provisions for variations discussed in the respective chapters in Part 6, where comparison with the present provisions may be made.

It may be noted, however, that the JCT and IFC contracts now have a greater measure of affinity with the GC/Works/1 contract in particular (Chapter 20), but with significant distinctions, over the treatment of variations and any related loss and expense (or expense only under that form). That contract requires that valuations of variations are to be made either by recourse to the prices in the bills of quantities or similar, or by a quotation provided by the contractor at the project manager's request and agreed by the quantity surveyor. The valuations are always to include for any expense which there may be due to disturbance, with no provision elsewhere in the contract for calculating such expense in a more overall way to allow for disturbance which may be spread more widely across the works than the scope of a single variation or even grouped variations. The questions raised are mentioned in Chapter 20.

The present contracts provide three options over variations, all of which are considered below. One is the traditional method of valuation of a variation after its instruction, based upon the prices in the contract bills or similar document, with loss and expense dealt with completely separately, as discussed in detail throughout much of this book. The other two are recent and both involve the provision of a quotation by the contractor for a variation. One of them occurs after the issue of any instruction, when the contractor may choose to put forward a quotation before the traditional method of valuation is considered; this method is used only if the quotation cannot be agreed. The other is for a quotation to be required from the contractor before a variation is instructed, allowing the possibility of considering its financial implications before proceeding. In each case of a quotation (but not otherwise), the amounts are to take account of loss and expense. All these factors are discussed hereafter.

A comparison of the several sets of provisions should be made carefully whenever changing use to a less familiar contract out of the range. A further difference is that the GC/Works/1 provisions do not spell out so many small 'ifs and buts'. They may well be all the better to deal with for this.

AUTHORITY AND INSTRUCTIONS

Variations

It is necessary under JCT clause 13.2 and the first paragraph of IFC clause 3.6 to give the architect power to instruct variations, so that the entire contract principle (Chapter 2) may be modified. The power is not given to the employer, to any consultant other than the architect, to the quantity surveyor or to the clerk of works.

It is possible, but not often desirable, for the architect to delegate some part of his powers to another here, as indeed it is necessary for him to designate which members of his own staff have any power to instruct variations or communicate with the contractor in other ways. Usually the effect of other consultants acting other than through the architect is confusion, whereas the effect of the employer acting direct may be anarchy. The quantity surveyor cannot act or even accept any variation not properly instructed by the architect, but must seek confirmation, even when it leads to an omission.

There is some case for the clerk of works dealing with small items, such as minor details or the positioning of subsidiary parts of services, subject to confirmation by the architect. This, among other things, is what JCT clause 12 regulates quite closely; in fact, its time limit of '2 working days' for the architect to confirm will often be impracticable. There remains the danger that small variations will have large consequences in the wider works, so that rapid checks are needed.

The contractor is within his rights not to comply with any so-called instruction issued by anyone but the architect, subject to the points just made. Instructions are discussed in Chapter 5, and the comments about the contractor confirming oral 'instructions' back to the architect before complying are particularly relevant here.

In addition to the contract provisions for confirming instructions in general, there is also JCT clause 13.2 allowing the architect to sanction any variation in particular which the contractor may introduce 'otherwise than pursuant to an instruction' proper, what may be termed 'deemed variations'. No procedure or timing is attached to this, so the contractor cannot rely on any actions of his own to make the option binding. It is useful to cover such cases as requests of the employer with which the contractor has complied, or even variations which the contractor has introduced himself. It is entirely at the architect's discretion whether to follow it, so the contractor cannot expect to abuse the provision.

In the nature of variations, instructions in writing include graphic data, which should be accompanied by some form of signed authority. Even a sketch or amendment endorsed on the one copy of a drawing being used on the site is acceptable as an instruction, provided it is authenticated by a signature. The main disadvantage here is that no one else may have a precise copy.

Although all of the foregoing applies under the JCT form, IFC clause 3.5 does not legislate for any system of confirming oral instructions or those issued by an unauthorised person, except by having the sanctioning provision mentioned above. Under the IFC form, the contractor should therefore when appropriate use a system which does not 'confirm back', but which requires the architect to put his purported instruction in writing, say by returning the contractor's form signed. Otherwise the contractor is best advised not to conform.

Strictly only adjustments of firm quantity items constitute variations, but usually the term and its derivatives are applied throughout this chapter to provisional sums and approximate quantities (mentioned under the next two headings); this point is repeated hereunder for certainty.

Provisional sums

Both contract forms require the architect to issue instructions about 'the expenditure of provisional sums'. (The JCT form includes a reference to sums in subcontracts, which need no special discussion here.) Such sums are commonly used to cover work which cannot be described and given in measured or other items, such as in a specification. When SMM7 is used, as it is with quantities under either contract form, it expressly permits provisional sums to be included in the tender and contract sum, dividing them into sums for defined work and for undefined work under general rules 10.3 and 10.5 respectively, according to the amount of information which is or is not given in accordance with those rules. In all other cases the power to include sums depends on custom, as does an understanding of what they are.

Following on its definitions of the two types of provisional sums, SMM7 adds further points in general rules 10.4 and 10.6, which have an effect beyond the immediate area of the bills of quantities in which they appear. These points are that in the case of defined sums, the contractor 'will be deemed to have made due allowance in programming, planning and pricing Preliminaries', while in the case of undefined sums he 'will be deemed not to have made any allowance'. The force of these further rules is firstly to affect the way in which provisional sums are expended, as discussed under 'Subsidiary provisions over valuation' later in this chapter Secondly, they also intrude into areas of contract provision that are outside the scope of a method of measurement, by speaking of programming and planning, which cannot be read as solely affecting pricing matters (see 'Programme effects' under 'Compliance with architect's instructions' in Chapter 15).

When once provisional sums exist, the architect does not have a discretion over whether to deal with them, as with variations: they must become the subject of instructions, even if only to say 'leave them out'. The contractor in turn should apply for instructions within the time band laid down in the clauses about extension of time and loss and expense, provided the sums are described clearly enough for him to know when they are needed. If they are described in a misleading way, this very fact may lead to a sustainable claim, if the contractor has reasonably made wrong assumptions about how the work portrayed affects other work.

Subject to these comments, there is nothing special to remark about provisional sums. They are effectively a premeditated variation waiting to be activated and valued like any other.

In the case of *St Modwen Developments* v. *Bowmer & Kirkland Ltd.* (1996) costs associated with the expenditure of provisional sums were held to include site preliminaries costs, supervision at site and attendance costs at site. (See 'Valuation of provisional sum expenditure' for more details on this case.)

Approximate quantities

These may be included in bills of otherwise firm quantities to cover work for which design is incomplete or the quantity is uncertain (say for groundworks in variable conditions), but where the general nature of the construction is sufficiently determined to allow indicative quantities with descriptions to be included and priced as part of the contract. These are omitted when the work is performed and the work is remeasured and priced, taking account of any differences along the lines of what happens with complete remeasurement under an approximate quantities contract (see later in relation to the JCT approximate quantities version). They may be seen as falling between firm quantities and provisional sums, by giving price detail as with the former and a provisional contract allowance as with the latter. They cause no special problems in direct valuation and are not generally distinguished in discussion hereafter. If they are sufficiently inaccurate, they may lead to extension of time or loss and expense as related adjustments.

DEFINITIONS

Provisional sums have been defined above, so it is only variations that need be dealt with here. For present purposes, the term 'variations' is used to embrace also the expenditure of provisional sums and the finalisation of work covered by approximate quantities.

Variations and variation of contract

The JCT and IFC contracts contain the same definitions of variations. These fall into two parts: physical variations and variations to site conditions. Both parts are subject to the proviso that 'no variation . . . shall vitiate this Contract' (JCT wording). This venerable expression is stating that variations may be indulged in without upsetting the integrity of the contract.

Beyond some point, variations shade into 'variation of the contract', that is the substitution of something other than the works originally in the contemplation of the parties. Where the boundary is crossed is a matter of fact, depending upon the individual situation in the contract. A drastic 'all-in-one' change, such as the inclusion of a fourth block of flats in a contract for three blocks, or turning a contract for a factory into one for sheltered housing would certainly constitute 'variation' in the grand sense. A change of this magnitude and character is something which the contractor is entitled to decline, because the subject-matter of the contract has been changed. A mass of smaller 'variations' could produce such a situation as an end result, but it is far less likely that the contractor could object in time, as the process would be incremental. Some major reassessment of the basis of payment might be due in retrospect, but this lies beyond the present scope. All of this points with much more likelihood to physical variations taken next, than to the other sort taken hereafter.

Variations of physical work

Physical variations are given in JCT clause 13.1.1 and IFC clause 3.6.1. They cover variation to the 'design, quality or quantity' of what is in the contract, and these terms are amplified by way of illustration, but not limitation, in respect of amounts or standards of work, including the removal of what is already there (other than for defects). These categories cover all the usual changes that are so familiar, without need to split hairs over the elements. Importantly, no limit is placed on when the architect may issue such instructions, so the possibilities of disturbed sequence etc are present in embryo.

The distinction between the valuation of variations and the ascertainment of loss and expense is important with both types of variation. It is usually easier to maintain with physical variations, although care is needed to follow the principles set out in this chapter and in those dealing with loss and expense.

Variations of obligations and restrictions

Variations affecting site conditions are dealt with differently, in JCT clause 13.1.2 and IFC clause 3.6.2, as the lists given are not illustrative, but specific and limited.

An example occurs in the case study in Chapter 32. There are four categories, giving the precise wording:

(a) Access to the site or use of any specific parts of the site.
(b) Limitations of working space.
(c) Limitations of working hours.
(d) Execution or completion of the work in any specific order.

There is therefore no authority for the architect to vary any 'obligations and restrictions', as these elements are termed, other than what is given here. He may instruct 'the imposition . . . addition, alteration or omission' of any of these only, and so cannot amend other matters of the contractor's working methods, such as health and safety. This does not prevent the inclusion of other obligations and restrictions in the contract initially, only their post-contract variation. SMM7 clause A35, for instance, gives the present list in substance, and also several other categories of less radical obligations and restrictions.

This type of variation to the contractor's general working arrangements on the site is especially likely either to overlap with matters leading to loss and expense payments, or to cause such a payment because the very variation throws progress out of normal. The initial remarks made under the heading 'Principles of valuation' about choice of reimbursement methods become critical in such a situation.

PROCEDURES FOR VALUATION

Who values and when

Both JCT and IFC clauses allocate the underlying duty of determining valuation in accordance with the rules to the quantity surveyor, who at this point is acting independently of the architect for authority. The recently introduced procedures discussed below allow for the contractor in defined situations putting forward his own valuations, but for the quantity surveyor still to have responsibility for checking and agreeing them or otherwise. There is vague provision for him to consult with the architect over issues about time and disturbance (see below) which may be appended to the contractor's figures. There is therefore a distinction from loss and expense proper, where the quantity surveyor can act only if the architect requires him to do so.

The traditional course of events has been for the quantity surveyor to value variations progressively as work proceeds or when drawing up the final account, and this is still possible under both clauses in the absence of any timetable for the quantity surveyor's work. This is a weakness of the system, as neither party knows where he is and procrastination of settlement is encouraged on both sides. On the other hand, it is not always easy to settle variations at the time, as they may be interrelated with other uncertain work or be affected by questions of sequence verging towards disturbance of progress.

Both of the clauses have recently introduced similar provisions over advance valuation. The mainstream clause 13 now permits the contractor in any instance of an already instructed variation to put forward a 'price statement' for its valuation, whereas the following clause 13A allows the employer the option of seeking a quotation from the contractor for a variation which is still only contemplated. These procedures are outlined below. It is now possible to decide that the desirability of a variation for the employer is conditional upon what the price will be (the 'value' question outlined below) and so seek to agree it in advance. But the contractor may not refuse a variation just because a price suggested by either process is not high enough. On the other hand, it does not force him to agree a price with the employer in advance, as there can be no binding 'agreement to agree' in law. If there is no prior agreement, with or without the attempt being made, the variation can still be instructed (if this has not already been done) and the contractor must then proceed with it and be paid on the basis of the quantity surveyor's valuation. This is always subject to adjudication, if the contractor and the quantity surveyor cannot agree on its amount, as are most matters.

As with many dealings, it is usually better to come to an agreement when once the hand has been put to that particular plough, otherwise relations may be soured. The alternative is not to embark on that approach at all. It is fair to suggest that major variations of obligations and restrictions are those most likely to create a need for prior agreement, at least over a basis of valuation. They are also the one area of instructions to which the contractor is allowed to lodge 'a reasonable objection' under JCT clause 4.1.1 or IFC clause 3.5.1. Although such an objection should not be based directly upon lack of price agreement, but upon subversion of the contractor's programme or method of working, prior agreement of an acceptable price or basis is likely to remove the possibility of an objection arising at all.

Three routes to valuing a variation

The variations provisions of the two contracts allow for three alternative ways to valuation, with a number of subsidiary points within each. They are best outlined without reference to every clause by number or to every small detail. The main flow of the clauses is illustrated in Figure 6.1, and key points are explained in the following list; the paragraph lettering corresponds to the letters in Figure 6.1.

(a) The architect may instruct under clause 13.2 over variations (and this takes in 'deemed' variations under various clauses) and must instruct over the expenditure of provisional sums and work covered by approximate quantities.
(b) The contractor may object to accepting such instructions when he has reasonable grounds under clause 4.1.1 over obligations and restrictions.
(c) The contractor by implication should not introduce changes of his own, including acting on the direct requirements of the employer, but the architect may sanction these as deemed variations under (a).
(d) The contractor may put forward a price statement under clause 13.4.1.2 in response to a variation instruction under clause 13.2 and based upon the valuation rules of clause 13.5 discussed hereafter. This may have requirements attached relating to the completion time and to disturbance. The quantity

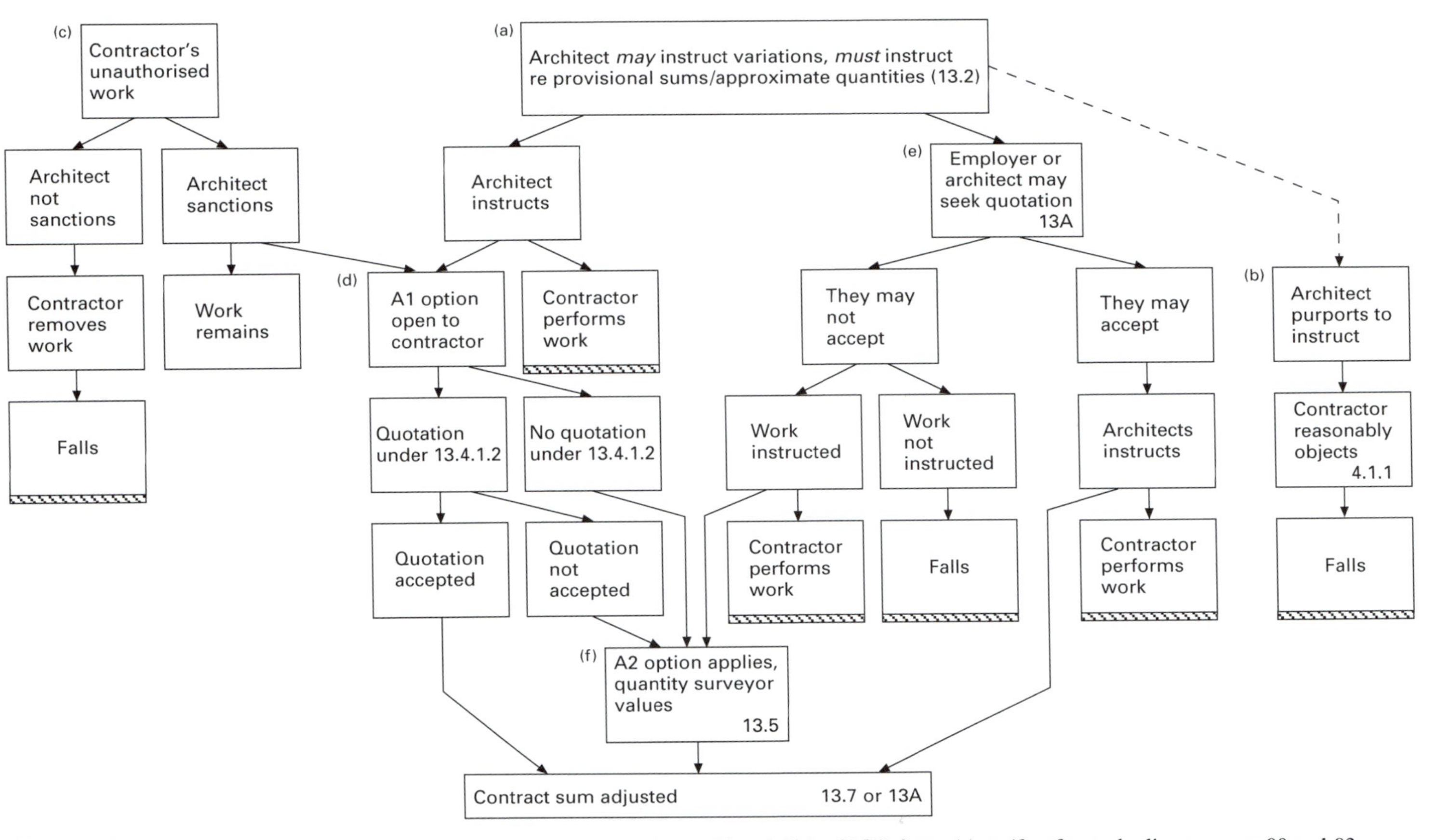

Figure 6.1 Variation instructions and valuation procedures under clauses 13 and 13A of JCT form: (a) to (f) refer to the list on pages 90 and 92

surveyor (liaising with the architect) may accept all or any parts of this submission and otherwise (f) applies. The contractor must comply with the instruction in any case.

(e) As an alternative to (d), the contractor may be asked to put forward a quotation under clause 13A, before an instruction is issued. This is to be related 'where relevant' to the contract prices, to allow for any preliminaries adjustment, to include for the cost of its own preparation (reimbursable whether it is accepted or not, provided it is reasonably prepared) and to have any time and disturbance requirements attached. Here the employer (no doubt advised) is the person to accept or reject the quotation. The alternatives following are (i) for the variation to be instructed upon acceptance of the quotation, (ii) for there to be a variation instruction and the work valued under (f), or (iii) for the work not to be carried out.

(f) As the fall-back from the operation of (d) or (e), a variation is to be valued in due course under the rules of clause 13.5. These are the traditional rules for valuation discussed below, while (d) and (e) are the newcomers within the contracts.

The use of the words 'may' and 'must' and other terms giving various degrees of discretion in the foregoing list should be noted closely, as governing the availability or closing of options. There are various quite tight time constraints on the use of (d) and (e) (periods of twenty-one days are mentioned in places), as these are intended to afford the opportunity to decide ahead of execution what the cost implications are. Even though an instruction under (d) has already been instructed, there remains the possibility of modifying or even withdrawing it.

The 'uncommitted' arrangement under (e) appears to be best suited to the possible introduction of a very large variation, particularly one including work of a different nature from that in the existing works, so that different pricing may apply and the other considerations given are also relevant (e.g. preliminaries adjustment). The mention of nominated subcontractors is significant. Something of the character of the infusion plant introduced into the project considered in Chapter 30 might have the type of work (particularly services) and sufficient programme disturbance to qualify, although it is on the large side for pricing in detail in a short time. The solution there was to arrive at a suitable pricing basis and measure as work proceeded.

Throughout the major part of the rest of this chapter, the traditional methods of valuation are addressed directly. However, the systems used of bill and *pro rata* prices, daywork etc are relevant to the two newer approaches set out in the clauses, and these may now be outlined in principle before going on to valuation detail.

Contractor's priced statement

JCT clause 13.4.1 and its IFC cousin each provide two routes towards valuing the results of instructions, either of which may be used in a particular case. The instructions concerned are those over formal variations, deemed variations arising out of other actions or inactions of the architect, the expenditure of provisional sums and the final definition of work covered originally by approximate quantities. All of these effectively change or clarify what the contractor has to produce.

The route given as of first instance is that subheaded 'Alternative A: Contractor's Price Statement'. This relates to an instruction already firmly given and with which the contractor must comply, be that necessarily soon or distant and with prejudice to whether agreement is actually achieved under alternative A. The alternative provides that the contractor *may*, but need not, put forward his 'Price Statement' for the variation within twenty-one days of the instruction and he 'may also separately attach' two non-price requirements, giving therefore up to three elements when required:

(a) The price for the work, based on the valuation rules provisions of clause 13. These are as discussed under the next main heading.
(b) Any amount in lieu of loss and expense not included elsewhere. This is not therefore necessarily calculated or ascertained as set out in Chapters 11 and 12, although the amount is described as 'in lieu of any ascertainment', suggesting that it should be reasonably comparable as a minimum.
(c) Any adjustment to the time for completion not included elsewhere and conceivably to make it earlier or later. This presumably is to be arrived at in broadly the same way as an extension under the rules looked at in Chapter 16.

There is a set of procedures for the quantity surveyor in checking the price statement, including taking 'consultation with the Architect'; this most obviously relates to the question of time adjustment, which the architect handles in all other cases. It is also relevant to the loss and expense equivalent, as the architect deals with at least the stage of agreeing in principle about strict loss and expense before (usually) passing it over to the quantity surveyor. The procedures then allow for the quantity surveyor accepting the price statement in whole or part or not at all, with follow-on procedures to work through to a possible amendment and acceptance. They also allow for him notifying the contractor (after consultation with the architect) as to whether his requirements over any loss and expense equivalent or time adjustment have been accepted. In the event that any of three elements is not accepted and bearing in mind that the variation instruction is already binding on the parties, that element falls to be determined under the otherwise existing provisions of the contract.

In the case of the price itself, there follows alternative B which provides that the valuation shall be made by the quantity surveyor following the traditional procedures in accordance with the rules in following parts of clause 13.

Contractor's quotation

Clause 13A and its IFC cousin provide for the possibility of the employer testing the water in a significant instance by obtaining a quotation from the contractor before a variation is instructed. The method set up follows that for the contractor's priced statement where suitable and it relies on several procedural elements:

(a) The contractor being prepared to receive an instruction to prepare such a quotation.
(b) Provision of sufficient information in time for the contractor to prepare the quotation and put it to the quantity surveyor within a period of twenty-one days, inclusive of any part to be performed by nominated subcontractors.
(c) Acceptance by the employer through the architect within a further seven days from receipt, and by implication checking, by the quantity surveyor.
(d) If the employer does not accept within the time, the architect is to instruct either that the variation is to be carried out and valued in accordance with the contract valuation provisions or that the variation is not to be perfomed at all.

The quotation is to consist of several elements:

(a) The same three as given above for a priced statement: the price itself, any amount in lieu of loss and expense and any adjustment to the time for completion. The price is to be supported 'by reference, where relevant' to the contract prices in the bills, which offers rather more leeway than the traditional rules, as well as implying that the work may well be of a distinct character.
(b) The cost of preparation of the quotation, which is to be reimbursed to the contractor within fair and reasonable limits even if the quotation is not accepted.
(c) If required, 'indicative information' on any additional resources required.
(d) Similar information if required on the method of carrying out the variation.

These several elements give a measure of fairness as between the parties, although after the procedure is exhausted the employer may still have his work performed on the contract price basis. It may moderate undue use of the process by the provision that the cost of the quotation is reimbursed. Earlier comments under 'Who values and when' are particularly apposite here.

PRINCIPLES OF VALUATION

Scope of valuation

Before dealing with the more detailed aspects, the distinction between these clauses and those covering loss and expense may be considered. The present clauses authorise the adjustment of the contract sum to cover the immediate valuation of variations themselves. This immediate valuation allows for differences in the quantity of finished physical work, and also for any special feature in the conditions under which the parcel of work is executed. These include such elements as performance out of normal sequence, but still in the context of regular progress, a term discussed in Chapter 10. But they do not include any consequential effects in the wider programme, which may mean that unvaried work is subject to disturbance of its regular progress, the criterion for meeting loss and expense.

If this distinction were simply one of using different clauses as the authorisation for adjusting the contract sum by the same mechanisms, little would be at stake.

The significance is that loss and expense is to be dealt with on a basis similar to that for common law damages, so the contractor's actual loss is ascertained and he is put back into the position in which he would have been had the disturbance not occurred. There are therefore elements of cost for which the rules are closely circumscribed (Chapter 12), while profit on the extra work caused is not allowable, although it may be that loss of profit on other work outside the contract does fall to be reimbursed. There are also strict rules about giving notices over disturbance occurring. The valuation of variations on the other hand is to be made on a basis similar to that used for the items in the contract sum, inclusive of the same elements and calculated at a similar level by using the *pro rata* or 'fair valuation' approaches described in this section.

Despite the distinctions of definition, it is not always absolutely clear in practice into which category some items of reimbursement fall. There is some room for manoeuvre between the parties in settlement, affecting how much may be payable. This arises because the basis is actual loss in the one case, and in the other a level related to that of the original contract, fully inclusive of head office overheads and profit (Chapter 12). How desirable it may be to use the contract level will depend upon how fiercely competitive it was and upon which party is viewing the desirability. Sometimes settlement based upon the contract level can be achieved more rapidly and therefore helps cash flow. It also then has the virtue of representing a 'bird in the hand', whereas a loss and expense claim can be subject to more dispute, owing both to its constituent calculations and to its often more emotive origin. If the quantity surveyor is dealing with variations and the architect with loss and expense, care is needed to avoid duplication.

The most recent amendments to the forms appear to have been introduced to bring these two routes closer together when the parties choose to allow this, by providing the alternatives described below.

Basis of valuation

The term 'valuation' is not altogether happy in the context of variations, and the less explicit term 'pricing' would have some advantages. What is being calculated is not necessarily the value of a variation to either party. The employer may value it in economic terms far higher than suggested by the additional or reduced price payable. Presumably he would not want it at all if he did not perceive, or think that he perceived, some balance of value in his favour. It is a fact of life, though, that he often does not know the cost effect and hence the economic effect of his decisions, or his architect's, until too late. Sometimes he never knows, among the mass or lack of detail which he receives.

From the other angle, it must be wondered whether the contractor sees a variation as a thing of value. He often expresses it rather differently! Maybe its main 'value' is as a lever for demonstrating that chaos has arrived, so he now deserves extra financial consideration. But equally, he does not receive his 'cost' in most instances. What he is paid is some construct based largely upon the prices in the contract bills, or whatever document underlies his contract sum, and this may differ noticeably from his costs as they arise. This is one of the points about tendering policy and pricing described in Chapter 3.

The rules given in the standard clauses leave little room as has been noted for any basis of calculation other than by relation to the contract bills etc for what are usually the bulk of variations. In the case of omissions, little else would be suitable anyway. As noted above, the results are inclusive of overheads and profit, automatically in most instances.

IFC clause 3.7 is less detailed than its JCT counterpart, but will usually lead to a similar result. Its lack of some details in fact allows it to cover in one contract the options which JCT clause 13 gives in its separate editions, but this then requires the other contract documents to be more explicit over rules. Failure here may lead to doubt and dispute at critical junctures, perhaps leading to documents being construed contra proferentem. This will commonly mean against the employer. Discussion of the rules is related mainly to the JCT with quantities pattern, with differences covered later.

RULES FOR VALUATION

Valuation of additions by measurement

This heading is broad enough to cover variations, provisional sum expenditure and approximate quantities. Only the first is referred to for brevity.

In the various contract rules, the JCT clause alone uses the expression 'valued by measurement', which is hardly precise, as measurement does not in itself value. A similar process may be read into IFC clause 3.7.3 (discussed hereunder), when quantities form part of the contract with clause 1.5 also applying.

JCT clause 13.5.1 gives three rules, which are to be applied in turn to a piece of work in a variation to see whether they fit, until the suitable one is reached:

(a) Work may be of 'similar character', may be 'executed under similar conditions' and may not 'significantly change the quantity' in the contract bills, presumably of the work to which it is similar. If all three criteria are met, the contract rates are to 'determine the Valuation'.
(b) Work may be of 'similar character', but the conditions or quantity, or both may not accord with (a). Here 'a fair allowance' is to be made for differences.
(c) Work may not be of 'similar character', which by definition changes the other two criteria. Then there is to be valuation 'at fair rates and prices'.

Several comments are needed on the three criteria themselves:

(a) *Similar character* is sometimes taken to mean 'identical in every respect', such as material, application, size and any other feature which SMM7 requires to be stated in an item description. If this is accepted, then under the first rule 'determine' really means 'use the bill rates exactly as they are', so that no pro rata pricing arises. It may be seen not to arise under the second rule either on this argument, and it clearly has no place in the third. It is more in accord with normal usage for 'similar' to mean 'like', while not excluding the identical (although pedantically perhaps it does). On this basis a reasonable scope for the term is 'work of the same material, perhaps differing in size and with some difference in application'. This will be recognised by the initiated as allowing pro rata pricing, as well as direct bill prices.

(b) *Similar conditions* may be taken to mean broadly those aspects of a piece of work which SMM7 does not expressly require to be stated. Thus, when deciding similarity or dissimilarity, location on a ceiling rather than a wall often falls to be considered under (a), whereas location on a first-floor ceiling rather than a tenth-floor ceiling may fall here. Again pro rata pricing can deal with the results. It is best to adopt a 'give and take' approach to much work in these respects, and to reserve adjustment to large cases, as almost any piece of work can be shown to be different from another in some minor way.

(c) *Significant change [in] . . . quantity* may occur quite independently of change in conditions, although one may shade into the other. Its more obvious effects are matters like different gang sizes, plant types or sizes and scale or phasing of delivery of materials. These effects are susceptible of pro rata pricing again, with some more give and take mixed in. (But see also under 'Changed conditions for unvaried work' hereafter.)

In practice the precise distinctions between these criteria and the rules in which they are embodied are not very important. Although they may be viewed as classic examples of quantity surveyors' desire to dance on the point of a needle, they do have the virtue of covering most options by a sort of blanket bombardment of overlapping shellfire.

When work goes outside the first criterion, the basis of valuation is to be 'fair'. This is to be interpreted as fair to both parties, so that the contractor receives a level of payment which covers the labour etc which he reasonably uses or should use in the operations concerned. Very often the work will be common enough in character, but just has not been included in the contract. The prices are then generally known and accepted within the industry within a fairly narrow band as reflecting inputs for work performed under 'average' conditions etc, and can be adjusted for any actual differences of conditions and quantity.

This approach means that the contractor does not receive his whole cost when he has been inefficient, but also that he may receive distinctly more if he has been particularly efficient. If the approach is used quite impartially, it should not even be necessary for the quantity surveyor to enquire what the level of efficiency has been. (It is also related more dubiously to abnormal pricing; see Chapter 3.) The ideal becomes more difficult to achieve when the conditions become far from normal, so that routine data are not available. Then some review of actual costs, modified by considerations of conceivable efficiency, may be needed.

A departure should also be considered if the contractor chooses to use some technique or piece of plant that is innovatory and the use of which is not reflected in standard data. This is a difficult area, as the contractor may contend that he, and not the employer, stands to gain from his own initiative. The test may be, What would he have done if pricing similar work in competition? And the answer may be, He would have priced somewhere between the old and the new if he thought his level of competitiveness would carry it. If so, a 'fair price' for a variation may be pitched at a similar level.

Added to these direct inputs must be overheads and profit. Both should be at the level included in the contract sum. There is no argument of substance for adjusting the profit level in particular to something higher or lower, because current tenders in general are coming in at a different level of competitiveness, or even that one from the same contractor has done this. Fair prices should be fair in relation to the contract bargain as struck, so keeping in step with pro rata pricing.

Another argument, which may be regarded as a special case of profit margin, but bigger because it edges into negative profit margin, concerns the use of erroneous prices in pricing variations, either directly or as the basis of pro rata prices. It is not always easy to say with confidence that a price is in error in either direction, but even when it is, the price should still be used. The clause being considered states plainly that 'the rates and prices for the work so set out' (that is in the contract bills) 'shall determine the Valuation' of 'additional or substituted work'. In fact, the present clause is the only one in the contract which specifically requires the use of the contract prices at all, so their primary purpose must be seen to be as a schedule of rates for valuing variations. The provisions about interim valuations do not require the use of the contract bills, except indirectly in that the calculation of formula fluctuations so requires.

It is sometimes argued that a contractor is not obliged to perform more work at an erroneous rate than the quantity which he priced in his tender. Put thus crudely, it ignores the direction of the error or what should happen if the quantity goes down. This is a four-edged sword which the employer might want to wield in some circumstances. It is also affected by the possibility of 'variation spotting' (Chapter 3) and of a gamble not succeeding. In the case of *Dudley Corporation* v. *Parsons & Morrin Ltd* (1959) the contractor had priced rock excavation at some 10% of an economic rate, the bill item being provisional. The actual quantity was about three times the provisional. The Court of Appeal held that the rate must stand, after considering the various proportions of rock and other excavation. It was said 'one must assume that [the contractor] chose to take the risk of greatly under-pricing an item which might not arise.'

Valuation of provisional sum expenditure

The rules for valuing additions variations will apply in most cases to work arising under this head, as most of it is likely to be measured and priced or else be daywork. Problems can occur over evaluating such things as 'obligations and restrictions' and attendances when their impact has been inadequately forecast in the bill provisions. The question of defined and undefined provisional sums under SMM7 (discussed in this chapter) may be important here, as the requirements must be adhered to carefully.

An example of valuation of work that was related to provisional sums occurred in *St Modwen Developments* v. *Bowmer & Kirkland Ltd* (1996), where it was decided by an arbitrator and upheld by an Official Referee on appeal that the contractor should be paid for preliminaries, supervision and attendance relating to provisional sums executed by domestic subcontractors. The bills of quantities in the contract documents included provisional sums for this work. It was found that attendant items, not valued by the employer in the final account, were due because the method of dealing with them in the bills of quantities departed from the Standard Method of Measurement by placing what, in effect, were nominated subcontracts without using the procedures contained in JCT 80 for nomination. The case illustrates that departures from standard documentation should be done with great care and caution and may well be taken against the party arguing for it (the contra proferentem rule).

Valuation of additions by daywork

This heading again relies upon the peculiar use of 'valuation' within the contract, this time in JCT clause 13.5.4. What is really being done here is to use the contractor's cost as the basis of payment, in circumstances where measurement or subsequent pricing is impracticable, and often where relatively expensive working occurs. The rules given require the incorporation into the contract of a schedule of percentage rates related to one or more standard definitions of daywork. Daywork is chargeable at current market rates, that is adjusted for fluctuations in market costs, with the addition of the percentage appropriate to the category of cost.

To operate this method of valuation, little more is needed beyond a record of the inputs incurred by the contractor, to which the costs and percentages may be applied. The contract requires the contractor to deliver 'vouchers . . . for verification' within the week following work being performed, and it is mainly this matter of vouchers which needs comment in the present context.

The stipulated procedure assumes that the architect himself can deal adequately with what may be a flow of sheets. Alternatively, it allows them to be routed to the architect's 'authorised representative', usually meaning the clerk of works as the person on the spot. This is the norm, as only someone present can properly keep a close eye on events. There is no statement that the appropriate signature authorises that valuation shall be by daywork or that the rates upon the sheet are correct, it merely verifies the time and other quantities given. Daywork is used only when 'work . . . cannot properly be valued by measurement' and the decision over this rests with the quantity surveyor. Even before the work is performed, the architect should not authorise that it is to be performed on daywork, and it must be questioned whether he can, even if he tries. It would be possible for the employer to do so under the arrangement already discussed, but an agreement by the architect would appear to be invalidated by the rules here as a whole.

It is a mixed blessing, but really desirable to have a valid agreement in advance by the quantity surveyor for daywork to be allowed for defined work. It means that uncertainty is removed and should ensure that records will be kept. On the other hand, it removes the incentive to perform work as economically as possible, in case daywork is not finally allowed.

It is common practice for daywork records of work to be kept 'for record purposes' when prior authorisation has not been given, and this can be useful so long as it does proliferate and get out of hand. Sometimes the practice is followed of signing all daywork sheets as 'for record purposes only', even when daywork has been agreed in principle. This does no harm, or good, because the sheets are always just records.

Once the sheets are signed, the quantity surveyor has two options only over hours etc stated: to accept them for use in daywork payment or to decline them and proceed with measured valuation instead. He cannot decide to reduce amounts because he thinks they are too high. With this said, he may still find that the evaluation of work performed in unusual circumstances (even of loss and expense) is helped by light shed from such records, although the dividing line between the two approaches is thin.

A different situation may arise when daywork overlaps with measured work. The two may be very difficult to segregate in the complexities of site working, but some split must be made, rather than having the whole area of work dealt with by daywork.

Valuation of omissions

It will be obvious that when work is omitted from the contract, the means for its evaluation lie to hand in bill items etc. Nothing more need be added over this direct evaluation prescribed in JCT clause 13.5.2, although the comments below on JCT clause 13.5.5 may be noted regarding remaining work.

Subsidiary provisions over valuation

JCT clause 13.5.3 gives three stipulations, two of which are innocent enough. These two are for measurement of variations to follow the same principles as in the contract bills (anything else is a recipe for disorganisation at the least), and for any percentage or lump sum adjustments of the contract sum to be reflected in valuation. Lump sum adjustments are usually the result of action during examination of the tender.

The third relates to the possibility of adjusting the amounts of items in the preliminaries of the contract bills. This does not occur as a direct percentage of the additions and omissions, but on assessment of what effect variations have on the sorts of items given in the preliminaries. The principles of adjustment are considered in Chapter 15. For fairly marginal amounts of variations, very often little or no adjustment will occur. Expenditure of provisional sums may not lead to any adjustment, as their amount should have been taken into account in assessing the contract preliminaries. If the amount expended is significantly different, some adjustment may lie. The exception is for sums for 'undefined work' under SMM7, mentioned earlier in this chapter, for which the amounts are *not* to be taken into account in assessing the contract preliminaries, so that an adjustment at the final account stage is more likely. Because of this category of adjustment, it is important that values are included in the contract bills for appropriate items, rather than just a total for the whole bill of preliminaries or, worse, that the value is spread over the measured rates at large.

The clause wording is that 'allowance, where appropriate, shall be made', rather than that 'the rates and prices . . . shall determine the Valuation', as for measured valuation. This can have two significances. It may be recognising that preliminaries are priced almost invariably as lump sums, and so these sums will be adjusted by some 'allowance'. Alternatively, it may be permitting the evaluation of the true extent of adjustment, even when the full pricing has not been shown in the contract. On balance, it appears reasonable to accept both of these possibilities together, rather than as alternatives. Acceptance of the former only would limit or remove adjustment in the case of partial or no separation of pricing. If an adjustment of some calculated value of preliminaries is then made, it needs to be remembered that some level of adjustment is automatically being made as part of the net effect of measured variations. This question is illustrated in Chapter 12.

SPECIAL PROVISIONS FOR VALUATION

There are several provisions, hardly precise enough to be rules, set out in the latter part of JCT clause 13. These both extend and limit the main rules in special circumstances when straightforward valuation of the variations etc is not adequate to deal with all the financial aspects.

Changed conditions for unvaried work

Sometimes the exact localised effects of a variation can be dealt with by measurement, for instance, without covering all the consequences. Other work may not have been subject to the instruction, but is nevertheless affected in its cost of execution to some significant degree. JCT clause 13.5.5 allows a revaluation of such work 'as if it had been the subject of an instruction . . . requiring a variation', with the rules as already explained covering revaluation. Use of the clause is subject to the qualification that the variation 'substantially changes the conditions' for the unvaried work.

An example of this class of effect would be the omission of work requiring the same scaffolding or other facility as other unvaried work, although whether it would justify a financial reappraisal would depend on how the scaffolding had been priced in the contract bills. If priced in the several items, varied and not, it would need an adjustment to reinstate the part cost of scaffolding otherwise being omitted. The addition of work sharing a scaffold would not create this problem, as it would be priced without scaffolding, unless it were an item already in the contract bills, when adjustment might be needed to avoid an overinclusion for the scaffolding. It might, however, lead to less continuity in performing the work which originally had the scaffold to itself, so increasing its cost.

This example is one of many which are really matters of access or other facility in performing work. It could be matched by others in which the dominant feature was sequencing, repetition or continuity, either in programme or physical disposition of work. Even without these aspects, there are ways in which a shift in cost can be caused by an addition or omission of the whole of one type of work at one end of a programme or the other, while leaving the rest of the programme intact. Rather similarly, a part of some work originally spread evenly at the various levels in a building can become spread unevenly high or low due to additions or omissions, so changing the amount of hoisting costs as averaged in the original pricing.

Two comments are called for here. Firstly, clause 13.5.5 refers specifically to 'conditions', which expression in JCT clause 13.5.1 is differentiated from 'character' and 'quantity'. As the work in question is itself unvaried, no question of character arises. But addition to or omission from work already in the contract bills may change substantially the overall quantity. Although that clause authorises pricing of the varied work (provided it is an addition) to cover this effect, it does not allow repricing of unvaried work to this end. The present clause does not do so either, so that strictly there is a gap in the provisions, as noted earlier. One solution which may be tried is to price the whole effect in the additional work, if it is additional, which is clumsy in arithmetical detail. The other is to interpret 'conditions' as equivalent to 'quantity', which comes close to some of the examples in the preceding paragraphs, and so reprice all the work much more rationally. Neither solution is really supported by the clauses, but something has to be done, and the second is suggested as filling the gap in contractual logic.

Secondly, like the rest of JCT clause 13, this clause is to be used to price work executed under changed conditions that can be foreseen before the work is performed, so the programme is adjusted to allow for as economical working as practicable. No question of disturbance arises here, as discussed below.

Fair valuation of variations

Although many variations can be dealt with by measured valuation or daywork, supplemented by some ingenuity, others resist these means completely. In these cases an undefined concept of 'fair valuation' is introduced by JCT clause 13.5.6. This is obviously intended to be some calculation not primarily dependent on measured quantities of the traditional variety, but assessed at a level that is reasonably consistent with the level of the contract sum, in terms of allowance for overheads and profit.

Two divisions are given: where the 'Valuation does not relate to the execution of . . . or the omission of work', and where some part of a valuation otherwise performed in accordance with the main rules cannot be dealt with in that way. The former division is particularly applicable to the valuation of variations in obligations and restrictions, where some amounts arise from broad analysis of a range of prices or from considerations needing lump sum adjustments. It is also related to adjustment of preliminaries; in fact, the two classes overlap very often, and some of their parts may overlap with the second division.

This second division may relate to cases in which variations change elements of cost like a central concrete batching plant or a tower crane, without changing drastically what is happening to each unit of work at the point of depositing or fixing. To arrive at revised measured prices for such work, it is often necessary in any case to calculate the cost of the central element and then spread it across the unit prices. It makes sense not to perform the spreading operation, which at the best may have to be delayed until the total quantities are available. This provision also covers any case of special supervision or staff costs, due to intensive working, reordering or other repeated work etc.

Again, this is an area which tends to lie close to disturbance of progress, so an overlap must be avoided.

Distinction from loss and expense

The variation provisions are not to take account of loss and expense, something made explicit by JCT clause 13.5.6. It is suggested that the dividing line between the two camps is not always clear in practice, although the theoretical distinction is there.

The primary difference is that loss and expense under JCT clause 26 flows from disturbance of regular progress of the works. Although variations in themselves may lead to work being rescheduled into patterns and sequences which were not envisaged in the original programme (be it that used for the tender or drawn up for construction), they may still be instructed in such a rhythm as to allow the rescheduling to be done well ahead of work on- or off-site, so that maximum productivity is achieved. This may of course mean dearer, or cheaper working, than would have occurred – which is the point of having clause 13 at all!

The difference financially has been touched on earlier in this chapter and is explored more fully in Chapter 14. Where there is room for genuine doubt over which theoretical category fits, the practical decision may be tempered by the likely financial outcome and the ease of achieving it.

Timetable for valuing variations etc

No timetable is given for agreement of variations, nor is there any explicit statement that they will be agreed in quantity or price. This is implicit in JCT clause 13.6, which simply allows the contractor to be present to take notes when the quantity surveyor is measuring. It also shows in the scheme for preparing and presenting the final account, followed as it is by the issue of the final certificate and the option of arbitration.

In the interests of progress, the quantity surveyor and the contractor may well arrange some form of progressive agreement of measurements and prices, as well as of the less difficult daywork sheets. Even when they do reach some extent of agreement, it is as well for it to be understood that, on prices in particular, there must be room for review of agreements if unexpected factors emerge. There may have been the agreement of prices related to one scale of working, when further variations undermine the basis used. Alternatively, it may turn out that a minor price agreed without too much consideration of its level suddenly becomes the fulcrum on which major negotiations hinge.

It is a well-established principle of law that prices used for interim payments on account do not bind the parties in final settlement, as to amount or even as to whether an item should be included at all (Chapter 8). These facts should reassure those agreeing interim valuations that they have a right to review, while not encouraging them to be rash in what they include. Even when variations are not accurately agreed, they should be included in payments on account when performed, as JCT clause 13.7 requires, as well as JCT clause 30 dealing with preparation of the payments. This is important for the contractor's cash flow, although his redress if not satisfied is rather limited (Chapter 8).

VALUATION PROVISIONS UNDER VARIOUS JCT CONTRACTS

Summary of valuation provisions under contracts other than JCT 80

In view of the length of JCT clause 13 and the foregoing comment upon it, the more important features are listed in outline, to help comparison of the regular valuation provisions, leaving aside the question of prices and quotations given in advance:

(a) Where possible, measured valuation is to be used for additions, with prices derived from those in the contract bills, either directly or by the pro rata method, to take account of differences in the varied work.
(b) If these prices are not suitable, other prices of a similar level of competitiveness are to be applied to the measured quantities.
(c) If measurement and pricing are not suitable at all, daywork is to be used for additional work.
(d) Measured valuation is to be used directly from the contract bills for all omissions.
(e) Elements of original pricing, such as preliminaries and percentage adjustments, are to be taken into account where suitable.
(f) The effect of variations on conditions in which unvaried work is performed is to be allowed.
(g) Special fair valuation may be used for elements of cost not susceptible to any of the foregoing methods, in whole or in part.
(h) Loss and expense due to disturbance of regular progress is not to be allowed as part of the valuation of variations.

None of the complications or inadequacies already described are mentioned above, and 'variations' should be read as including 'provisional sum expenditure'.

Provisions under IFC 84

Although this contract is identical in its effects to JCT 80, there are a few points to note. For ease of reference, here are the parts of the two clauses dealing with mainline valuation. Most of them indicate straightforward rearrangement of order, needing no further mention.

JCT 80	*IFC 84*
13.4.1	3.7.1
13.5.1.1	3.7.3
13.5.1.2	3.7.3
13.5.1.3	3.7.4
13.5.2	3.7.2
13.5.3	3.7.6
13.5.4	3.7.5
13.5.5	3.7.8
13.5.7	3.7.4
13.5 (proviso)	3.7.7
–	3.7.9

The priced document

IFC clause 3.7.3 refers simply to 'the priced document' and to 'the relevant values therein', rather than to 'the Contract Bills' and to 'the rates and prices . . . so set out', as does JCT clause 13.5.1 in its several parts. The nature of the priced document depends on the financial basis of the contract, indicated in the IFC recitals, to which IFC clause 3.7.1 refers back. The recitals give alternatives to choose from:

(a) A fully priced specification, schedules of works or contract bills. Any one of these implies a complete breakdown of the contract sum.
(b) A specification supported by a contract sum analysis or a schedule of rates. Either of these implies a limited amount of data from which to price variations etc.

Of these, contract bills and a schedule of rates both give measured items for reaching a quantified valuation. Contract bills are discussed in the main section on valuation in this chapter; the schedule of rates is discussed in a later section on JCT 80 without quantities. The schedule of rates must contain its own rules for measurement, as the usual method of measurement cannot be assumed. IFC clause 3.7.3 should be sufficiently detailed over pricing aspects, even though more compressed.

A priced specification and priced schedules of work are by no means clearly different, as both envisage a division of the works into a series of items each permitting self-contained pricing. For a specification to be laid out in any way that helps valuation of variations, or for that matter the original estimating on the basis indicated, it must be broken into items related to parcels of work, rather than to trade divisions which may spread all over the project. The only reasonable distinction is that schedules of work can be used alongside a trade-based specification to give a rearrangement of material for pricing purposes without great detail in the works.

When this has been said, the distinction between schedules of work and a contract sum analysis is also not absolute, except that the employer supplies the schedules of work and the contractor supplies the contract sum analysis. It may be inferred that the schedules are intended to be divided in some detail, whereas the analysis may not be. The term 'contract sum analysis' has come into regular currency through the JCT with contractor's design form, where it can mean many things (see *Building Contracts: A Practical Guide* and *Design and Build Contract Practice*). It is always desirable that the architect or the quantity surveyor should provide the main divisions of the analysis and a clear guide to the level of detail for the contractor before he tenders. Otherwise, less than enough detail is more likely than the opposite, as well as the possibility of unsuitable divisions. There will then be a need for recasting and editing, as can happen with a schedule of rates (see later).

Which of these options to use is a matter to be decided in the light of the individual project. The intention is that IFC 84 is used for 'works of simple content', as is endorsed on its cover, but this is not entirely the way it is working out in practice. Furthermore, even simple works can be subjected to undue torture by way of variations, so their likely extent should also influence the decision. It makes no sense to use none of them, that is just to have a specification and no pricing, indeed the contract then becomes unworkable by uncertainty and possibly an analysis would have to be imported later. It may be even more damaging to try to use more than one of them, as all manner of clashes and doubts over precedence can arise. One is needed, but only one.

The possibility of the contract sum analysis or the schedule of rates not providing all the data needed for valuation of variations is recognised in IFC clause 3.7.9, which allows a fair valuation in those circumstances, as becomes inevitable. This option may be needed in the case of omissions as well as additions, as in other contracts. It will then be reasonable to look into detail in the contractor's estimating papers, although these do not form part of the contract and are not binding. Subject to this and to any errors or special adjustments made within it during tendering, the detail may be useful in conjunction with other information and judgment. Subcontract NSC/C allows the revaluation of subcontract work affected in this way by main contract variations.

Subsidiary provisions

Nothing is said in this contract about the treatment when valuing variations of lump sum or percentage adjustments given in the priced document. This should nevertheless follow the principles of the JCT clause, which is spelling out normal practice. The smaller the contract sum, the less likely is it that any lump sum will be taken into account.

Although the provisions for daywork terms are given in IFC clause 3.7.5, there is no procedure for verifying records or forwarding them to the architect, within a given time or at all. This should be dealt with in the specification, the contract bills or whichever document applies.

Provisions under JCT 80 without quantities

Under this contract, the only possibility offered as standard for valuation prior to the 1987 revisions is a schedule of rates. That revision also introduced schedules of work as in IFC 84 just discussed, and the comments may be read to apply here. The original JCT clause 5.3.1.3 requires the schedule of rates to be provided by the contractor post-contractually, if it has not been provided already. The advantage always lies with it being provided before the contract is entered into, thus saving an area of possible disagreement later. It would be possible, say, for the quantity surveyor to prepare the schedule unpriced, so avoiding any problems of wording and layout. It is usually considered sufficient, with the small projects envisaged for this form of contract, to allow the contractor to use his own work, subject to any editing of the wording.

This is so, because the routine procedure is for the contractor to take the build-up which he prepares for his tender and use it as the basis of the schedule of rates. In conjunction with the quantity surveyor, if any, he will delete all the quantities and extensions, so that just item descriptions and unit prices remain. With any editing for presentation purposes, the result is the contract schedule, with the advantage that it is clearly related to the price level of the tender, and hence the contract sum. Not only are the quantities and extensions removed, but they are denied any significance over accuracy, so the contract is based upon a lump sum without quantities, but with a schedule for variation purposes only (which is strictly how contract bills are used also). This position must be adhered to rigidly during all post-contract work.

With the schedule so provided, the present version of JCT clause 13 runs extremely close to its 'with quantities' cousin. Rather like the one in the IFC form, it ignores lump sum and percentage adjustments, but the same comments apply. It does not give the provision about fair valuation if the schedule does not cover the variation, because it is more detailed in other respects. If the schedule has been prepared as outlined above, at least omissions should be fully covered.

The main problem areas with this contract are in deciding how to apply the clauses about errors, or rather to decide when errors may exist, so affecting the starting-point for adjustment of variations. Errors in quantity included in the contract sum lie with the contractor, as given above. JCT clause 14.1 states that both 'quality and quantity of the work included in the Contract Sum shall be deemed to be that which is shown upon the Contract Drawings or described in the Specification'. The use of 'deemed' shows that the contractor's own quantities are not relevant. The 'shown . . . or described' provision means that work need be shown only in one place or the other for it to count as in the contract sum, so that the contractor has read the documents together.

JCT clause 2.2.2 clouds this position somewhat by dealing with 'error in description or quantity or omission of items' in the contract drawings 'and/or' the specification. It requires any of these aspects to be corrected as though variation has been instructed. Thus, whereas JCT clause 14.1 allows an item to occur in either document, the present clause seems to be considering the possibility that omission from one only will constitute an error. It is suggested that the correct interpretation must be that clause 14.1 overrules this clause, which then refers to errors in description in one document alone which cannot be seen to be in error when the other is read. Omissions must be seen in the same light: there must be an item which should by normal practice be in one document only and which cannot be seen to be missing, or the item must be missing from both, where it would normally appear. Error in quantity is less easy to reconcile: it can best be interpreted as a drawing error such as the wrong height of a wall, or a specification error such as giving the wrong number of brackets for a shelf. The discussion of discrepancies and divergences in this type of contract in Chapter 5 is also relevant here.

The drawings and the specification are to be read together, without either taking precedence. JCT clause 2.3 requires the contractor to refer any discrepancy or divergence to the architect for resolution. Whether or not a variation instruction is needed, hence any difference in payment, will depend upon how the architect resolves matters. This same clause places 'further drawings or details . . . to explain or amplify' within its orbit, which is reassuring, as the contractor should have received all of them in order to place the tender! This provision applies of course to removal of doubt over what has to be done, as well as over what is to be paid, a matter discussed for the 'with quantities' form in Chapter 5.

JCT clause 14.3 affirms that any quantities supplied to the contractor 'at any time . . . shall not form part of this Contract', which can only reasonably be interpreted as supplied beforehand and presumably for tendering. Any practice of supplying such quantities cannot be too strongly condemned: if no one is prepared to stand by them, they should not be used.

Provisions under JCT 80 with approximate quantities

Several features in this form build up the background picture of the one distinctive feature here, the uncertainty about the original quantities and the necessity for complete remeasurement and valuation for the final account. The uncertainty under any contract based on this approach occurs at three levels:

(a) The extent to which the design is incomplete, so that the scope and detailed nature of the scheme is in doubt. Only rarely will the design be complete, with time alone forcing the use of approximate quantities.
(b) How accurately the quantities represent the scheme as designed, be the design advanced or otherwise.
(c) The uncertainty of the contractor's pricing, in view of his uncertainty over what it is for which he is tendering.

These levels have a cumulative effect. The first affects the contractor's view of his preliminaries and of how far to reflect such features as repetition and buildability in his unit prices, as well as determining the best that the quantity surveyor might achieve by way of accuracy. The second is the result of the first and also depends on how long a time is available to prepare the quantities. One feature of approximate quantities is that they can be of any degree of approximation, allowing very rapid preparation when the need arises, but less coincidence with reality as a result. The third level is that not only are the quantities uncertain, but the prices attached to them are based on an uncertain assessment.

Approximate quantities are very useful to the employer in some circumstances, but they do throw more responsibility than usual on the contractor. Some of this he will tend to absorb by including a greater margin in his pricing for the greater risk he takes, but the rest is what he may expect to come to him in settlement, not just in correct, final quantities, but in review of the pricing. The less accurate the original basis, the more sweeping this review may need to be. The employer should be prepared by his advisers for the financial result, even if he is less than aware of how it will be incurred.

Against this backcloth, the JCT form aims to achieve middle ground. This is indicated by the note at the head of the articles of agreement, but not forming part of them or of the rest of the contract, that the form is intended for works which have been 'substantially designed but not completely detailed'. The note also mentions the approximation and the need for remeasurement. In principle, there is nothing within the contract which specifically limits its use to so high a level of prior design, while it is fairly difficult to produce a form which facilitates a far less well-developed scheme. The major problem is dealing with price adjustment, which is more a question of how the bills of quantities are composed, than of what the contract says. It can only allow leeway, as this one does.

Within the articles, the reference is made to 'Drawings and Bills of Approximate Quantities showing and describing, and intended to set out a reasonably accurate forecast of the quantity of, the work to be done'. It is a clear, grammatical deduction from the order of this rather compact form of words, that both sets of documents are to show and describe the work and both are intended to forecast the quantity of work. Whether this is precisely what the drafters intended may be wondered, as it is at odds with normal practice.

The functions of drawings and quantities are somewhat distinct in that usually the former 'shows' the major dispositions of work, whereas the latter 'describes' the detail, if not its positioning. In this sense, they may be seen as performing the two functions jointly but essentially apportioned between them, and the wording given may be accepted as tolerably correct. The inadequate alternative would be to deny the drawings any part in what follows and to allocate all the functions to the bills, which do not properly 'show' work. Such a construction would also run across what is the obvious interpretation of the rest of the wording in this paragraph of the articles.

The concept of apportioned functions recombining to give the total picture cannot be applied to the one function of 'setting out' the quantities. It is accepted practice, when bills form part of the contract (as they do here), that they are to be taken as alone embodying the quantities without need for recourse to the drawings. The exception lies with such elements as scaffolding and other temporary items, which are either not measured at all or are given in more rudimentary quantitative form than is otherwise the practice. These are likely to vary from one contractor to the next, according to detailed site method, and so are commonly assessed by inspection of the configurations shown on the drawings. The resulting sums must then be allocated against the quantities, including preliminaries items. For the quantities as 'set out' expressly, the present wording is unfortunate if the two types of document diverge considerably in the picture of scale which they paint.

JCT clause 13.1 states the usual basis that the contract bills give both the quality and quantity of the work in the contract, which should be held to override the uncertain wording just discussed. It must be held then, in general, that only the bills can supply quantities, whereas the drawings continue to indicate character, complexity and other features of the work as usual. It can be seen, however, that the possibility of dispute is latent here, if the drawings and bills diverge widely, particularly over pricing plant and similar items.

JCT clause 14 deals with 'Measurement and valuation of work, including variations and provisional sums'. The mention of variations might be unexpected in an all-remeasurement contract, if it did not cover variations of obligations and restrictions, as distinct from physical, designed work. Even so, it is intended to cover both varieties of variation. The contract has three sets of information which need mention here:

(a) *Contract drawings*: These are recorded in the articles and may not be complete, but are to be assumed to be correct so far as they go. They are not automatically laid aside when the contract is formalised, so it is necessary to vary them if they are altered in some way. In a contract in which they are simply rough drawings, they should be omitted by a single variation instruction, stating that all working drawings will be issued thereafter under separate instructions. If they really are capable of development from where they are, it may be possible to issue instructions varying them. Often the reality will lie between these options. Usually, it is safest to assume the former situation and to omit all of the original drawings at the beginning, even if a few are reinstated later.

(b) *Drawings and details*: These are to explain and amplify under clause 5.4. They have a definite place in a firm quantities contract, where only a few drawings become contract drawings out of those which were available for preparation of the contract bills. In such a contract, they are issued without an instruction, because they vary nothing. In the present contract, their purpose is the same, but the execution may be faulty in that what is in the contract for them to explain and amplify is none too clear, as between drawings and bills. If they are issued without an instruction, the quantity surveyor has no authority to use them, if the contract drawings which they purport to amplify are removed. Even with those drawings still current, the supplementary drawings are confusing in status, so the facility to issue them as such is best not used. A variation is always to be preferred.

(c) *Variation instructions*: These include drawings, under clause 14. If the advice under (a) and (b) is followed, they become the embodiment of everything to which the contractor works, so avoiding doubt.

As far as valuation goes, the provisions follow the other JCT variants closely, allowing for the absence of omission measurement. In place of the reference to a significant change in quantity leading to a change in rate, there is 'Where the quantity of the work was not reasonably accurately forecast'. It is to be expected that this provision will have greater importance in this form of contract. It really covers at least three situations: where an individual item or set of items is affected, where the whole content of the project varies, and where a swing in the distribution of trades affects their balance and interaction in the progress of the whole.

As ever, there is a need to keep a sense of proportion over what constitutes a serious departure from a reasonably accurate forecast. For example, what is serious for the subcontractor when the contractor is looking at a sublet trade, may look quite different when looking at the job as a whole. The contractor may then be conceding a change in prices in the subcontract, while it is not made in the main contract but absorbed by the wider compensating adjustments. On the other hand, an adjustment of preliminaries is more likely to occur in a remeasurement contract. There is no special mention of change in quantity in clause 26 here as leading to loss and expense, as the effect of the variations clause is to take up the various differences, leaving disturbance of programme as a distinct issue.

CHAPTER 7

SUBCONTRACTORS: APPOINTMENT AND CONTROL

- Basic distinctions in subcontracting
- Domestic subcontracts
- Nominated subcontracts
- Named subcontracts

This chapter deals with the contract control of the various types of subcontractors within a JCT context and as listed immediately below, especially obtaining them initially and replacing them in special circumstances later. As between contracts, the pattern varies quite extensively, especially over nomination and renomination, the JCT arrangements being in most respects the most detailed (some might say fussy) and tending to give more responsibility and potential liability to the employer. Other aspects of subcontracts are given in Chapter 5 and in chapters hereafter, with the same aspects of main contract work. Discussion of multi-way claims in Chapter 13 is relevant to many aspects of subcontract operations.

BASIC DISTINCTIONS IN SUBCONTRACTING

Standard arrangements

There are four readily distinguishable treatments of subcontracting allocated between the JCT and IFC contracts, which have some parallels under other contracts, but with quite extensive differences in application which should be checked in each case. The JCT and IFC options cover most principles and are given here in some detail to indicate the major issues:

(a) *Domestic subcontracts at the contractor's own initiative*: These are cases in which he chooses entirely for his own reasons to pass out some portion of the works to another. This is a process also termed subletting within the contracts, while within the industry the results are also known as private subcontracts. This treatment occurs in both contracts.
(b) *Domestic subcontracts into which the contractor is required to enter*: The permitted choice of subcontractors is given to him when tendering. The alternative terms again apply. This treatment occurs only in the JCT contract.
(c) *Nominated subcontracts into which the contractor is required to enter*: A sum of money is stated in each case for inclusion in his tender and the actual choice of subcontractor is notified to him by the architect, usually after the contract is in existence. This treatment also occurs only in the JCT contract.

Table 7.1 Comparison of main contract clauses dealing with nomination and naming of subcontractors[a]

	JCT	IFC
Local authority etc not subcontractor	6.3	3.3.8
Definition	35.1	3.3.2
Contractor's tender	35.2	3.3.2
Procedure introduction	35.3	–
Documents relating to subcontract	35.4	–
Contractor's objection to subcontractor	35.5	3.3.2
Architect's instruction	35.6	–
Contractor's obligation on receipt	35.7	–
Non-compliance of subcontractor, notice	35.8	–
Architect's duty on receipt	35.9	3.3.1
Number not used	35.10	–
Number not used	35.11	–
Number not used	35.12	–
Interim payment of subcontractor	35.13.1	–
Direct payment of subcontractor	35.13.3	–
Prenomination payment of subcontractor	35.13.6	–
Extension of subcontract period	35.14	–
Failure to complete subcontract works	35.15	–
Practical completion of subcontract works	35.16	–
Early final payment of subcontractor and defects	35.17–35.19	–
Position of employer and contractor	35.20–35.21	3.3.7
Restrictions in contracts	35.22	–
Number not used	35.23	–
Circumstances where renomination/naming	35.24	3.3.3–3.3.6
Determination of subcontract/employment	35.25–35.26	3.3.3

[a] The clauses are listed by primary reference to the JCT nomination subject-matter, so the IFC references give only partial correspondence, in view of the differences between the two subcontracting methods.

(d) *Named subcontracts into which the contractor is required to enter*: A single subcontractor is named in each case to whom he is to go for a subtender on which to base his own tender. This treatment occurs only in the IFC contract, although there is a minor reference which has similarities in JCT clause 35.1.

A comparison of the main contract provisions over nominated and named subcontractors is given in Table 7.1.

Non-standard arrangements

There are two salient ways of departing from standard practice. One is to set up a subcontract which is not one of the four standard arrangements listed above: perhaps by creating a hybrid, perhaps by producing a new strain. This is not very likely, but may occur on the initiative of the employer or his advisers. The subcontractor should look carefully for the underlying motive, as it usually increases his responsibilities, probably over design and other liabilities towards the employer by way of some collateral warranty. The arrangement may well be faulty legally and also introduce procedural flaws.

The second departure is to use an amended or non-standard subcontract for a standard arrangement. This can be done properly in any JCT situation only for a domestic subcontractor, as the other forms are obligatory under the related main contracts. Even small amendments are likely to need corresponding main contract amendments. The contractor may well put forward his own domestic subcontract conditions, which again the subcontractor should check closely before acceptance. Most of the chapter headings in this book suggest areas worth looking at, particularly those of programme, valuation and payment (such as 'pay when paid' and set-off). The individual subcontractor often cannot secure standard conditions for an individual subcontract, and has to rely more widely on the broader pressures from his trade association or similar body. He has to weigh the terms offered against the need to obtain the work within his own bargaining strength, but should never sign without understanding. Strictly, nothing can be done from the employer's side and usually nothing will be in practice.

Both these departures are undesirable. They may be seen as mutually variants one of the other and, to some extent, the comments made apply to both. As they are so variable, they are not considered elsewhere, but their significance should be assessed when they are encountered. One case indicates the degree of care which the subcontractor should employ in reading the terms offered.

In *Martin Grant & Co. Ltd* v. *Sir Lindsay Parkinson & Co. Ltd* (1984) the subcontractor had carried out work under a non-standard domestic subcontract to a JCT form which contained a provision 'at such time or times and in such manner as the Contractor shall direct . . . the Sub-Contractor shall proceed with the said works expeditiously and punctually to the requirements of the Contractor so as not to hinder hamper or delay the work'. It was held that the contractor was *not* obliged, as the subcontractor put it, to 'make sufficient work available to the [subcontractors] to enable them to maintain reasonable progress and to execute their work in an efficient and economic manner'. The terms were sufficiently clear and express to preclude an implied term, such as the subcontractor sought. When in doubt, a subcontractor should seek advice before commitment.

Control of subcontracting

Within the broad legal position affecting subcontracts, a contractor may subcontract or sublet any part of his work, unless he is restricted by the terms of his own contract. This latter is what the JCT and IFC contracts do (and others considered later in this book), to varying extents according to the type of subcontract involved. Here are some of the major reasons for controlling aspects of subletting:

(a) *Quality of what is produced*: Although the specification may be identical, different firms may perform differently against it, but still within its limits. Even if they fall short, indifferent performers are not always easy to pin down by reference to the written word. If possible, it is better to eliminate them in advance.
(b) *Programme compliance*: This is another aspect of performance, usually more sensitive in the way in which it affects performance of elements not within the subcontract, but offering somewhat more scope about what may be tied down in writing.
(c) *Design*: With specialised work, it is frequently the subcontractor who designs it, so the design cannot be performed on a consultancy basis for execution by another.
(d) *Price*: This may be the other side of all three coins so far listed, so that some form of direct involvement from the employer's side becomes desirable.

None of these considerations, or even all of them in combination, make contractual control indispensable, but it is often held to help. They may also be the seed-bed for later disputes, which is the reason for stating them here.

Although the methods of subcontracting vary to quite an extent in their detail, they all result in *subcontracts* as such. The subcontract documents accompanying each are more similar than distinctions at main contract level might suggest.

DOMESTIC SUBCONTRACTS

The root principle of any domestic subcontract is that the contractor selects and appoints the subcontractor to perform work for which he, the contractor, has tendered as part of the main contract, so that several key factors emerge:

(a) *Detailed terms of the subcontract*: These are entirely between the parties, unless the main contract imposes any obligation on the contractor to include any particular terms. This is not done in the contracts here considered, except by stipulations that fluctuations be in keeping with those in the main contract and that there are provisions about property in materials passing from subcontractor to employer when the contractor is paid. There are standard forms available to produce a harmonious result over all issues, but their use is optional.
(b) *Method of tendering*: This is left to the contractor, as is the financial basis (subject to the fluctuations point) and the nature of all contract documents, including drawings and specifications.
(c) *Subcontract price*: This is between the parties alone. The price payable by the employer to the contractor for the work concerned is not affected by any difference between the two prices, in either direction. Settlement of the final prices payable is similarly separate.
(d) *Control of the subcontractor over performance*: Matters of time, site organisation and quality rest with the contractor, who is responsible to the employer for any fault under the main contract.

(e) *Design responsibility*: Unless any responsibility is placed on the contractor (as there is not under the JCT and IFC contracts), none can be placed on the subcontractor via the subcontract. By the time that the subcontractor is appointed by the contractor, it is too late for the employer to arrange anything by another route, as may occur with the other subcontract arrangements.
(f) *Mutual commercial and other responsibilities and liabilities*: Over matters such as injury and damage, these responsibilities rest between the parties. Responsibility for some insurances are, however, governed by the main contract provisions.
(g) *Stipulations about appointing another subcontractor*: None are given if the first is lost for any reason. The contractor must settle the consequences himself and either complete outstanding work himself or obtain consent to another subletting.

These features flow from the general principle that privity of contract is solely between the parties to a contract, even when some terms (as here) are dictated by another. They are equally applicable when the contractor has decided to sublet or when the employer has required it. In most respects, they also apply under the other forms of subcontracting taken below.

Contractor's initiative

When the contractor wishes to sublet, he is required by JCT clause 19.2 or IFC clause 3.2 to obtain the architect's consent, which is not to be unreasonably withheld. Reasonable grounds would be that the proposed subcontractor regularly performs inadequately over quality, but also over programme if he is to perform a critical section of work. Immediate signs of serious instability would also qualify, as would a strong reason for the employer to insist that he had chosen the particular contractor for special skill in the type of work concerned, so debarring any subletting. The giving of consent in no way relieves the contractor of his responsibilities over the work concerned, nor does it prevent him taking it back on himself.

It is implicit in this arrangement that the architect would have to demonstrate his reasonableness to the contractor, although if the architect did not demonstrate anything, the position would be the same. Ultimately, 'unreasonably' in the clauses has reference to what the courts or an arbitrator would hold. In the event of deadlock, recourse to one of these tribunals would be necessary. Reference to arbitration, probably now after adjucation, during progress is possible on any issue under the IFC form, but only on a few issues under the JCT form, of which the present is not one. Court action is not so constrained, but is not notorious for its speed, so a quicker solution is unlikely.

If then there is disagreement, but the contractor does not wish to or cannot push his rights at once, he must accept the architect's decision under protest and seek another subcontractor. This may run him into delay finding a replacement or extra expense because of a higher price. These points should be recorded to the architect, so that suitable redress here and on any other consequential issues may be sought later.

Even if the architect gives consent to a particular subletting, he may do so too late to meet the contractor's programme. Here the provisions over extension of time and loss and expense do not apply, as they relate only to delay by the architect in providing 'instructions, drawings, details or levels' (Chapter 16). Early action is recommended on these grounds, with a statement included as to why any particularly early consent is needed. If the architect still delays, it would appear to become a case of unreasonably withholding consent, as no reason has been given or at least substantiated. The contractor cannot proceed without consent and must revert to the remedies mentioned in the preceding paragraph.

Such early action may also be needed to secure a suitable subcontractor who would otherwise withdraw his tender in favour of taking other work. Provided the architect has been properly informed, the simple fact that the contractor has lost a more favourably priced subcontract is a cause of extra expense to him. In any of these instances the contractor must base his claim upon the general principle of a breach of duty by the employer's architect and not upon a specific clause. Recompense here can be included in the final account (as distinct from being sought in proceedings) only with the employer's express agreement, uncomfortable though this may be for the architect. Conversely, however, the contractor cannot delay finding a subcontractor just because he is still looking for one to match the price of one lost.

When major cases of subletting are involved and when the contractor knows whom he is choosing, it is highly desirable to clear them during precontract discussions. This gives certainty to the contractor and avoids later disputes. It does not debar him from second thoughts, leading to a fresh consent, so long as he has not committed himself to a subcontractor.

Two cases demonstrate that it is necessary for a subcontractor to take care over the terms on which he enters into a subcontract. In *Martin Grant & Co. Ltd* v. *Sir Lindsay Parkinson & Co. Ltd* (1984) a subcontractor had carried out work under a non-standard domestic subcontract which contained a provision 'at such time or times and in such manner as the Contractor shall direct . . . the Sub-Contractor shall proceed with the said works expeditiously and punctually to the requirements of the Contractor so as not to hinder hamper or delay the work'. It was held that the contractor was not obliged, as the subcontractor put it, to 'make sufficient work available to the [subcontractors] to enable them to maintain reasonable progress and to execute their work in an efficient and economic manner'. The terms were sufficiently clear and express to preclude an implied term, such as the subcontractor sought.

In *Piggott Foundations* v. *Shepherd Construction Ltd* (1994) it was held that the subcontractor was not required to comply with the main contractor's programme, on the interpretation of the DOM/1 subcontract. The words 'the progress of the works' in the opinion of the judge meant that the subcontractor had to carry out the works in a manner as would not unreasonably interfere with the actual carrying out of any other works which could conveniently be carried out at the same time. There was no contractual obligation upon the subcontractor to carry out his work in any particular order or at any specified rate of progress. This appears to mean that a contractor may have some difficulty in maintaining control over his programme and that of the subcontractor.

Employer's or architect's initiative

The other way of prefabricating a subletting is contained in JCT clause 19.3, but is not available under the IFC contract, where the use of a named subcontract is more restrictive. This provides for 'certain work measured or otherwise described . . . and priced by the Contractor' to be sublet to one of a given list of three or more persons given in the documents. The person chosen by the contractor 'at [his] sole discretion' then becomes in all respects like any other domestic subcontractor. The clause goes on to allow either employer, architect or contractor to add further persons to the list post-contractually, subject to consent from the others. This still does not force the contractor to accept any particular name.

If the available list drops below three names sometime before there is a binding subcontract, again names may be added, this time by agreement, which has the same effect as before. Alternatively, at this stage the contractor may choose to carry out the work himself or sublet it in the way first described. Again the contractor is not forced to accept any one person and it is not stated which option is to prevail. It appears to be whichever is implemented first. If so, the contractor may secure the unassailable right to do the work himself, even if two persons remain on the list. If he wishes to sublet in this case, it is effectively the same as if he were to seek to add names to the displaced list, as consent is still required. The clause is not very helpful, however good in intent.

There is still no contractual provision in this instance for the contractor to be granted extension of time or to be reimbursed for loss and expense if the procedure runs into trouble, despite the greater activity from the employer's side, whose remedies are just as troublesome.

The initiative of the employer and/or his agents in influencing the letting of 'domestic' subcontracts must take into account the context of the contract and any financial effects that may follow from actions taken by the employer. It is necessary for an architect or an employer to be careful over the manner in which he lets subcontract work if it is not clearly by nomination or naming procedures in JCT 80 or IFC. An example referred to in Chapter 6 is *St Modwen Developments* v. *Bowmer & Kirkland Ltd* (1996), where it was decided that the employer had let work as domestic subcontracts. However, the quantity surveyor did not value attendance items and allow preliminaries and supervision in adjusting provisional sums included in the contract for work eventually executed by the domestic subcontractors. The bills of quantities in the contract documents included provisional sums for this work and it was held that attendant items, not valued by the employer, were due. This was held because the method of dealing with the subcontract work in the bills of quantities departed from the Standard Method of Measurement for placing what, in effect, were nominated subcontractors and did not use the procedures contained in JCT 80 for nomination. The case illustrates that departures from standard documentation and procedures, done in this case in order to control the placing of the subcontracts, should be done with great care and caution and may well be taken against the party arguing for it, under the contra proferentem rule.

Subcontract terms

The standard form available for optional use for a domestic subcontract under the JCT contract is the DOM/1 subcontract, whereas that under the IFC contract is the IN/SC subcontract. In all respects relevant to present concerns, they respectively follow closely the wording of the forms for nominated and for named sub-contracts looked at hereafter, except there is no involvement of the architect in actions and agreements. The various forms and procedures related to the JCT and IFC main contracts are considered in *Building Contracts: A Practical Guide*.

NOMINATED SUBCONTRACTS

Nominated subcontracts lie at the far end of the spectrum from their domestic cousins, there being greater interaction with them from the employer's side, both precontractually and later. Named subcontracts lie somewhere between, but are taken last to ease comparison. Some broad differences between the nominated and named varieties may be discerned from Figure 7.1.

Nominated subcontracts are subject to more control by the architect over all the factors of quality, programme, design and price described above. Quality is usually the overriding factor, with design either required or not. The other two tend to come in because they are difficult to leave out, although this is managed with named subcontracts.

JCT clause 35 regulates the position in some detail and relies on a number of subcontract-related documents, as named in it. It limits or even removes several ways in which the contractor can act on his own initiative, including:

(a) *The contractor does not price nominated work in his tender*: He simply includes a given prime cost sum and prices any profit and attendance stipulated.
(b) *Choice of the subcontractor rests with the architect*: He obtains tenders and instructs the contractor which one to accept, subject to some right to demur and the obligation to agree subsidiary detail. There is the possibility of the contractor himself tendering for nominated work in a suitable case.
(c) *Payments at all progress stages, including retention*: These are controlled by the architect and related to amounts agreed by the quantity surveyor. This extends to the possibility of direct payment of the subcontractor, if the contractor defaults, subject to the counter-possibility of the contractor setting off charges which he may have.
(d) *Final sum payable*: This is effectively agreed between the subcontractor and the quantity surveyor, although the contractor is involved procedurally over the whole account and directly over any items of payment which are in dispute with him from either direction.
(e) *The architect is crucially involved in particular decisions*: These include responsibility for delay and extension of time and when practical completion occurs, even though the contractor is responsible for keeping the subcontractor to programme and otherwise organising how he fits in.

(f) *Determination of the subcontractor's employment prematurely*: This is closely monitored by the architect, both over reasons for its occurrence and the terms resulting.
(g) *Replacement of the subcontractor by renomination*: This is undertaken by the architect, although the contractor usually receives reimbursement of the difference in amounts payable.

Despite these elements, the clause maintains that there is no privity of contract between employer and contractor due to the basic position established or to actions taken within that framework. Contractors often state that 'the nominateds' are not 'their subbies', but the legal position is sound. Even so, employers and architects need to watch their actions carefully, to avoid creating some implied relationship leading to effective privity and responsibility. There is a direct relationship over design by the subcontractor, but this is created and moderated by a special agreement outside the subcontract arrangement.

Authority for nominated subcontracts

JCT clause 35.1 deals with the right of the architect to introduce a nominated subcontract. The ways of doing this are as follows:

(a) The use of a prime cost sum included in the contract bills or specification. This has been mentioned and is the main method, in itself straightforward as it gives the maximum warning of intentions.
(b) A variation instruction or the expenditure of a provisional sum, both of them less predictable by the contractor.
(c) Agreement between the architect and the contractor, by definition unexceptionable once the agreement has been reached. No rules are given over this, to allow complete flexibility and not to attempt to force anything.

The second option is the one needing care. In the case of a variation, the right of the architect to select a subcontractor, which is what is in question, is limited by two characteristics, *both* of which must apply. The first is that the work must be additional to the work in the contract documents, which by inference is to preclude the architect from taking work priced by the contractor himself and giving it to a subcontractor. The second requires the additional work to be of a similar kind to the work already in the contract and allocated to nominated persons.

These two characteristics go with clause 13.1.3, which specifically excludes from the definition of a variation the nomination of a subcontractor to perform 'work of which the measured quantities have been set out and priced by the Contractor' in the contract documents. Clause 13.1.3 forbids the architect taking out work originally included and giving it to another, but in itself strictly does not negate him giving additional work of the same kind to another, even though this might be contrary to the spirit of things and to practicality. The present clause does not positively allow him to do this, by virtue of its imitations, and he has only such powers as the conditions give him.

The architect is thus limited to 'such additional work . . . of a similar kind' and 'not further or otherwise'. This must be interpreted as referring to work of the same basic type, such as more suspended ceilings or mechanical engineering installations, when such work is already the subject of prime cost sums. In such a case, it could not reasonably be stretched to mean terrazzo paving (not included as a prime cost item) as being further internal finishings, in the same broad category as suspended ceilings. Given mechanical engineering installations in the first instance, but no plumbing, it would be not unreasonable to bring at least comparable plumbing work within the scope of a nomination. Highly distinctive sanitary work might push matters to the limits. There is some room for doubt here and need for tolerant interpretation.

In practical working situations, the distinction between a variation instruction extending an existing subcontract and one introducing a new one might constitute the watershed over kinds of work. But, in principle, the clause does not exclude the introduction of a new subcontractor, rather it exists to facilitate this happening.

With all this recorded about variations, the stark fact is that no restrictions are placed upon the architect when instructing the expenditure of a provisional sum. This sum may have been quite an inexplicit insertion in the documents, such as a sum for 'Tank Room complete', so that the contractor had little warning of the precise type of work required. This would rank as a provisional sum for undefined work if SMM7 is in use with bills of quantities (Chapter 6), but in the present situation and with any type of provisional sum this does not prevent the architect, when instructing over expending the provisional sum, from introducing nominated subcontractors for work not so covered in the original contract. This includes work for which he does not have the authority when dealing with a variation, anomalous though this may be. Clause 13.1.3 has no bearing on provisional sums.

Given work of a different kind from any originally envisaged, of a specialised nature and arising in the context of a variation, it is best for the architect and the contractor to agree under the last option that it be dealt with by nomination. The alternative is for the contractor to perform it himself or to seek consent to him subletting it. Both agreement and consent are not to be unreasonably withheld under the respective provisions, so that a first round deadlock on a quid pro quo is possible, assuming the contractor is not equipped to do the work himself. With the provision in clause 35.4.1 for the contractor to object (again reasonably) to a particular subcontractor, it is unlikely that a deadlock will exist except over money, and here an agreement on a nomination could take account of the contractor's margin in a special case. Responsibilities over delay or disturbance here could vary, according to the detailed pattern of events.

JCT clause 35.2 allows the alternative of the contractor himself tendering for work included in the contract as the subject of nomination. Although it hedges procedures about rather fussily, it does not produce problems in interpretation or practice.

Nomination procedure

The procedure for nomination of a subcontractor starts within the contract at the point at which the architect has obtained tenders for the subcontract work and has decided which person he wishes to nominate and has also set up the warranty agreement between employer and subcontractor which will run collaterally to the subcontract itself. This is so even if tenders have been obtained ahead of awarding the main contract, and even if then the contractor knew the name of the firm when he himself tendered. The procedure is closely linked to the various documents listed in clause 35.4, all of which are obligatory for each nominated subcontract. These documents and their place in the detailed procedure are reviewed here in relation to problems which may arise and are discussed more widely and closely in *Building Contracts: A Practical Guide*. Here are the key elements in the procedure:

(a) The contractor has an initial right of reasonable objection under clause 35.5 to entering into a subcontract with any person nominated as such, rather than objection to any terms put forward, this being more likely to succeed on such grounds as that the proposed subcontractor has a very bad record of performance as would be of particular interest to the contractor in organising the works overall.
(b) The architect issues a nomination instruction under clause 35.6 to the contractor (with a copy to the subcontractor), accompanied by various documents, including those parts of the tender documents which bear upon such matters as any modifications afforded to the subcontractor over obligations and restrictions imposed by the employer in the main contract.
(c) All being in order, the contractor and subcontractor enter into the subcontract within ten days under clause 35.7, having been through the documents to see that they are to their individual satisfactions.
(d) If there is delay expected beyond the ten days, the contractor is to notify the architect under clause 35.8 of when he expects to enter into the subcontract or of any problem encountered.
(e) The architect is to respond under clause 35.9 by fixing a later date for entering into the subcontract, or by informing the contractor whether or not he considers him justified in not having complied with the nomination instruction in time or at all.
(f) If the architect considers the contractor to be justified, then again under clause 35.9 he has three options: instructions to remove the problem (e.g. over attendance matters) and usually therefore extra cost within the contract, omission of the work from the contract, or a fresh nomination.

This process is often criticised on the grounds of the complexity of its detail within the several documents, quite apart from the inherent desirability of nominated work. But it is essentially normal good practice codified, a necessity arising out of frequent lapses in its application in the past. The documents optionally available for domestic subcontracts are close in all essentials, giving comparability in operation.

The procedure is always at some risk in that the architect obtains a tender which leaves some elements with financial implications to be agreed between the contractor and the subcontractor after a tender sum has been given to the architect. This risk is minimised by the requirement for the architect to fill in quite a bit of detail about the contract and the terms under which the subcontractor will be operating, before he sends information to tenderers. It is therefore essential for the equivalent information to have been included in the main contract, so that it forms the basis of the contract sum.

A special case is that of the subcontract programme duration. This usually arises out of the contractor's own programme, with the possibility that he may require the subcontractor to operate within an uneconomically short span, so leading to a higher tender or haggling when it is received. The higher price is not always easily detected. The best safeguard is to consider main contract and subcontract tenders together, so that programme effects can be sorted out before any commitment. If subcontracts are not available then, the alternative is to examine the contractor's programme to see if it is adequate over subcontracts. Even though the programme is not a contract document, the times for significant subcontracts could be embodied in the contract.

In *M. Harrison & Co. (Leeds) Ltd* v. *Leeds City Council* (1980) the contractor's tender was based upon his programme which showed fairly late execution of the structural frame, this programme not forming part of the tender proper or of the contract. The latter was on the 1963 JCT form. A quotation from the nominated subcontractor for the frame had been obtained in advance and, in a special condition, showed erection at an earlier stage with a cleared site. This quotation was not known to the contractor when he tendered. Nomination was made by a 'variation order' accompanied by the quotation and special condition. In the end, a compromise programme was agreed. It was held, on the detailed facts, that there was not a variation, that the contract bills should not be corrected (so leading to a 'deemed' variation) and that the nomination was not a breach. The 'variation order' amounted to an instruction of postponement leading to payment for loss and expense. Documents like Tender NSC/T do not prevent this type of problem, but should identify it in time for avoiding action.

Because each subcontract may be different in, for instance, its call upon temporary facilities on-site, there will be some margin for uncertainty, which is the reason for allowing the interchange between contractor and subcontractor. Although the architect will not wish to concede extra reimbursement for elements which should be included already in either the main or the subcontract tender, he should remain alive to the possibility that there may be some genuine gap. Even more, there may be a situation in which it is good commercial sense to accept a lower subcontract tender, even though this is at the price of some extra payment to the contractor, perhaps for a facility or some hoisting. The standard framework may need to give occasionally and it is at these points that the architect may be giving 'further instructions' within the ten-day period allowed. He may be asked for these earlier and should comply sooner if at all possible to avoid delay.

Equally, if the contractor validly objects to a proposed subcontractor, or if the tenderer will not settle terms or withdraws, the architect has to act. This may simply mean starting on another tenderer, but it may allow some attempt to lure the earlier tenderer back by a concession, according to why he has withdrawn.

The importance of all this wheeling and dealing is not just that some money is being negotiated, but that time is running out on the parties. Nomination is treated as an instruction so that, as with other instructions, the contractor may be entitled to extension of time or reimbursement for loss and expense, if the instruction is late. The subcontractor stands outside this delay problem, as he has not yet entered into a binding commitment. On the other hand, he has an incentive to obtain the work and to look to future relationships with the other players. The critical question is, What constitutes lateness on his part or the architect's, in these circumstances when three persons are interacting?

Here nomination instructions are no different from any others, in that the contractor has to request them 'in due time' under clauses 25.4.6 and 26.2.1 (Chapter 15). This means that he is responsible for assessing dates when he needs subcontractors on-site within his programme and, more difficult perhaps, how long they need from nomination to get there. It is up to the architect to decide the earlier dates for obtaining tenders to suit. In principle, all of this is straightforward. It can only be helped to work by close cooperation between architect and contractor, allied to preliminary enquiries of likely subcontractors, if unusual timescales are involved.

All of this is prior to when clause 35 itself starts and is quite direct in application. After tendering, it results in a nomination instruction, but this is still conditional upon successfully treating with the proposed subcontractor before there is a subcontract. This depends on the outcome of the negotiations, which may be affected in several ways implicit in what has been set out above:

(a) *The tender may be unsatisfactory to the contractor*: This may be because the architect has invited it on an inadequate basis or has failed to tie up the details which are properly his concern before passing it to the contractor. If this leads to delay, a late nomination instruction may in turn lead to extension of time or loss and expense reimbursement.
(b) *The contractor may be awkward over elements of agreement or simply take too long*: Here he has no cause for favourable treatment. In an extended case at a critical juncture, he could eventually run into liquidated damages.
(c) *The subcontractor may cause delay*: Although the architect has selected the person, he probably cannot be held to guarantee that he will not delay at this stage, although he should take whatever action is needed if the ten-day period is in jeopardy. The case of *Rhuddlan* v. *Fairclough* (discussed below) may be relevant here. This was decided in the different circumstances of renomination, so that the architect was held not to have guaranteed there would not be even a reasonable delay in obtaining a replacement subcontractor.

There is plenty of room in these instances for questions of fact to be introduced, while the clause is silent about any inefficiencies in agreement, as distinct from complete failure to agree. The ten days do serve to keep people on their toes and to trigger remedial actions, but it is highly desirable to allow as much extra margin as possible in the programme of activities. For the architect to be faced with accepting a poor deal because time has run out is unfortunate at the least.

Also within the contract provisions, there is mention several times of the warranty agreement between the employer and the subcontractor. This has to be entered into alongside the rest of the activities. Its effects are generally more important at later stages, although they do give the facility for the architect to order fabricated work and design direct from a potential subcontractor ahead of nomination, so gaining programme time. Should the nomination not go ahead, there are provisions for payment. Some related sorting out in the main contract will then be needed over such effects as discounts and other margins of payment to the contractor related to amounts paid direct and to handling etc of fabricated items.

Renomination of subcontractors

During the nomination process, a proposed subcontractor may drop out for some reason and this may cause quite a disturbance or delay, as noted. If a nominated subcontractor drops out *after* nomination and particularly during the execution of his work, the effects are more complex for both progress and the contractual aspect. Several reasons are listed in JCT clause 35.24:

(a) *Default of the subcontractor*: The reasons given in subcontract clause 7.1 and 7.3 leading to a determination by the contractor are
    (i) Unreasonable suspension of the subcontract works
    (ii) Unreasonable failure to proceed with the subcontract works
    (iii) Serious failure to remove defective work or materials
    (iv) Subletting without consent
    (v) Corruption
(b) *Insolvency of the subcontractor.*
(c) *Default of the contractor*: The reasons given in subcontract clauses 7.6 leading to a determination by the subcontractor are
    (i) Unreasonable suspension of the main works
    (ii) Unreasonable failure to proceed with the main works
(d) *Remedial work not due to the subcontractor's default*: Where there are special circumstances, such as damage caused consequent upon architect's instructions and the subcontractor is not reasonably available, he is relieved of performance.

The wider questions of determination itself, of these reasons and of the financial and other consequences are considered in Chapter 18. In outline, in the first case, the contractor has to carry the architect with him over whether the default apparently justifies determining the employment of the subcontractor, before he effects a determination. In the other cases, the architect is presented with a *fait accompli* in his dealings with the subcontractor. In all cases he has to act to obtain another subcontractor, that is he must renominate.

It is enlightening in this instance to go back as far as the JCT 1963 contract, under which there was doubt for some years over whether the architect had any obligation or even right to renominate when a subcontractor had been lost. The 1963 contract provided nothing more than in clause 11(3) that 'The Architect shall issue instructions in regard to the expenditure of prime cost and provisional sums' and in clause 27(a) that 'prime cost sums . . . shall be expended in favour of such persons as the Architect shall direct', leaving the question of what happened to find successors open. There were three principal areas of uncertainty:

(a) Who was to select a replacement subcontractor.
(b) Who bore the difference in cost between old and new subcontractors.
(c) Who bore any delay costs or loss and expense flowing from the disturbance.

A test case was brought and reached the House of Lords on clause 11(3), to clarify the position over a relatively small amount as the additional price of a replacement subcontractor. In *North West Metropolitan Regional Hospital Board* v. *T. A. Bickerton & Son Ltd* [1970] it was argued by the employer that the architect had discharged his responsibilities once and for all by the original nomination instruction, and that the contractor was responsible for completing the works upon the failure of *his* subcontractor at no extra charge to the employer. The case was decided against the employer on both points. From this it appeared that the architect must renominate as a duty, whereas the contractor need not and must not find a successor, as this would infringe the architect's right of nomination, which remained alive. The question of resultant costs of delay or loss and expense was not in issue and was not decided, although later cases have dealt with it, as stated below.

As a result of this decision, the JCT introduced its present provisions. The Tribunal (and the RIBA, as copyright holders) had received quite an amount of criticism judicially and otherwise at the time, although in fairness the then editions of the ICE and GC/Works/1 contracts were equally deficient in these areas and were quietly revised ahead of their present editions to deal with the problems, in somewhat different ways.

Subsequent cases have also been related to the 1963 JCT form, but have a bearing on the 1980 form. In *Percy Bilton Ltd* v. *Greater London Council* [1982], another House of Lords case, a subcontractor had become insolvent and the contractor had acquiesced in his repudiation, but brought his action over the delay element, which had three strands:

(a) Reasonable and inevitable delay in obtaining and nominating a new subcontractor.
(b) Further delay due to the employer's dilatoriness in this last matter.
(c) Yet further delay due to the extended time required by the new subcontractor to complete.

It was held that the subcontractor's default was not a 'delay on his part' within the meaning of what is now JCT clause 25.4.7; see *Westminster* v. *Jarvis* [1970] (Chapter 15). Furthermore, the employer was not by any express term responsible for the subcontractor's repudiation or obliged to maintain a subcontractor continuously available to the contractor. Loss of a properly accepted subcontractor was one of the risks of subcontracting work.

It was commented that the contractor had not raised any 'reasonable objection' to the nomination of a replacement subcontractor who required a longer period to complete than the original. It appears that his case might have been helped if he had, and if he had not acquiesced in the repudiation initially (see *Rhuddlan*

v. *Fairclough* below). In the circumstances of the *Bilton* case, the contractor won extension of time and relief from liquidated damages only in respect of the second cause of delay listed. Under the JCT 1980 form, again the contractor would receive an extension of time for the second cause of delay, as this would be the direct result of an instruction of the architect, producing (as it happens) a renomination. The first cause would still remain his risk, as the next case indicates.

In *Rhuddlan Borough Council* v. *Fairclough Building Ltd* (1983) the position went further. Here a nominated subcontractor defaulted and there was a repudiation, leaving work partly done and defective. The architect spent several months in trying to get as good a deal as possible for the employer in terms of the new subcontract price and its basis. He nominated a subcontractor whom the contractor declined to accept. Two key reasons were that the work in the renomination excluded the remedial work, so the contractor would be left paying for it, and that the new subcontractor's period ran to beyond the main completion date. The contractor accepted that he would have to concede to the employer the amounts already paid through him to the first subcontractor for work now defective, but the cost of remedial work was several times this amount. In the end, after further delay, another subcontractor was nominated.

It was held that the architect was reasonable in seeking an advantageous price for the employer and taking a reasonable time in doing this, even though it was causing delay to the contractor. But the contractor was not obliged to perform remedial work when the subcontractor had not completed his work and it remained the subcontractor's responsibility. The contractor was therefore entitled to a renomination which included the remedial work as part of the price. (JCT clause 30.6.2.1 allows the deduction in the final account of the amount already paid to the subcontractor for defective work as such.) Furthermore, because of the difficulties, the contractor was entitled to an extension of time to cover the additional subcontract time over and above that due to the original delay in renomination.

Debate over whether the architect (or in the actual case, the supervising officer) had actually to instruct the contractor to determine led to *James Longley & Co. Ltd* v. *Reigate and Banstead Borough Council* (1982), which related to the previous Green Form subcontract. It was held that he was not, so the employer was not liable for delay and disturbance flowing from his inaction. The case is described in Chapter 16 in relation to determination, and its impact is modified in the current contracts. The effect of JCT clause 35.24 is therefore that the architect is to issue an instruction to the contractor when the contractor notifies a default by a continuing subcontractor, over whether to determine or not. The architect may also instruct the contractor to come back to him for a further instruction. This postpones matters, while the subcontractor may mend his ways. Whenever the architect agrees with the contractor, the architect must instruct a determination, after which the contractor has an option as to whether he actually determines.

As all instructions under clause 35 qualify potentially for extension of time, the time before renomination would appear to be eligible for inclusion in the calculations, contrary to the foregoing cases related to the JCT 1963 form. However, there is no entitlement to loss and expense on this count. This leads to the need for a careful balancing of the various factors in a complex case. When determination occurs due to the subcontractor's insolvency, there is no architect's instruction prior to that of renomination, so an extension is not available. When it is the subcontractor who determines for the contractor's default, the position is similar, as might be expected, but the payment provision is different (Chapter 18). Following *Rhuddlan*, JCT clause 35.24.10 provides for the architect to have a reasonable time within which to renominate.

NAMED SUBCONTRACTS

Just as nominated subcontracts are peculiar to the JCT 1980 form, so named subcontracts are virtually peculiar to the IFC 1984 form. There is in fact just one reference to 'naming' in JCT clause 35.1, as a means of identifying a nomination as an alternative to using a prime cost sum. In practical terms, this can refer only to a nomination arising out of a provisional sum, and has been ignored under earlier headings.

The process of naming is akin to nomination, in that the architect puts a specific person to the contractor for acceptance. Also, if the subcontractor drops out during progress, the architect is involved in selecting and naming a substitute and the amount payable by the employer may change as a result. The likeness goes little further. The subcontractor is otherwise like a domestic subcontractor, with the contractor pricing the work competitively in his own tender. As Table 15.1 indicates, there is no special treatment of the subcontractor himself over dealing with such matters as interim and final payments, extension of time, loss and expense and completion or failure to complete on time. He *is* the contractor for these purposes. Only when he provides design to the architect is his position distinct, and this is over a matter properly outside the present contract itself (Chapter 5).

Naming subcontractors

IFC clause 3.3 gives two methods for introducing a named subcontractor: a detailed inclusion in the contract documents, or a provisional sum expended wholly or partly in favour of such a person. Either way, only one person is actually put to the contractor, the architect having carried out the earlier work of inviting tenders and making the ultimate selection.

The more likely method is the naming of a single person at the stage of the main contract tendering. The architect has to use defined documents, including a tender form, the most intriguing section of which is that for listing the tenderers for the main contract. Although this is there so that subcontract tenderers are warned as to who may be their main contractor, and so can register any objections early and save time and complication later, the added complications of broadcasting the list of main tenderers do not need elaborating! There would appear to be no fundamental objection to leaving the space blank and clearing with the subcontractor whether the main contractor is not desired, or only desired on different terms, as soon as the main contractor becomes known. This, after all, is what happens in reverse with nominations under the sister form.

The resulting tender in all its detail about price and stipulations is to be made available to the main contract tenderers, who include the price with their own additions for other factors. What the contractor prices will depend upon the basis of the contract: it may be a bill of quantities, a specification or some other description of works. It is essential that it is drawn up so as not to conflict with what the subcontractor priced or to be discrepant with it. This is important, even though the subcontract tender documentation has to accompany the main invitation.

Within twenty-one days of entering into his own contract, the contractor under clause 3.3.1 is to enter into a subcontract with the named person. This may mean some quick action, especially if 'entering into' covers some implied action, such as a letter or just starting on site. There is the possibility that the two parties to the proposed subcontract may run into an area of disagreement over the particulars given in the subcontract tender, as the subcontractor is not under a fixed obligation to contract. At the extreme, the subcontractor may not be willing to enter into the subcontract at all, if his own tender has been lying with the architect for some considerable time and other work is to hand. The moral is that the subcontract tenders should be obtained just before the main contract tenders, with a requirement that they be kept open for a sufficient time to complete the process. If they are 'dated', confirmation should be obtained that they still apply, with any amendments.

The clause recognises that the best laid plans may suffer in the usual way. If, therefore, the contractor is unable to subcontract on the terms provided by the subcontractor, and on which the main tender was based, he is to inform the architect, who in turn has three options:

(a) *To change the particulars to remove 'the impediment'*: By implication, this will mean a financial change, and usually in one direction only. It may be sensed that this is another area where undesirable collusion could occur, given the wrong participants.
(b) *To omit the work*: This leaves the possibility of placing it as a direct contract, which may temper the tendency to operate unfairly under (a) but is counterbalanced by the contractor's right of reasonable objection under clause 3.11. If suitable, the work may simply be left out of the project.
(c) *To substitute a provisional sum for the work designated for the named subcontractor*: This postpones the final decision and means that the work may not go to a named subcontractor at all, under the rules for provisional sums.

The first two options result in a variation, to be valued accordingly, and possibly leading to extension of time and loss and expense, as the clause states directly. As the action all occurs quite early in the programme, it is to be hoped that the effects will be minimal. If the architect seeks fresh subcontract tenders, it is possible that a reasonable time spent doing this would not qualify for either of these redresses, on the basis of the decision in *Bilton* v. *GLC* discussed in relation to nominated subcontractors. Although its broad effects may be seen as quite possibly transferring to named subcontractors, there is the distinction that the *Bilton* case related to failure of a subcontract already in being (where it was observed that the employer could not be responsible for constantly maintaining a subcontractor in being at all costs) and resulting delay, not to delay in the setting up of one initially, with the architect still in a primary position in the action.

Furthermore, the possibility of seeking fresh tenders is not specifically mentioned in the clause, which refers only to an instruction to 'change the particulars', something rather less than one to 'change the sub-contractor', it might be thought. The only clear way of introducing a different subcontractor is under the third option. The architect still has to act to instruct the expenditure of the provisional sum, where a delay may also produce these effects, owing to a late instruction. The likely applicability of *Bilton* v. *GLC* remains the same in this situation. The position with each option will be governed by when the subcontractor work is needed, and this may be late rather than early.

Nothing is said about what happens if the twenty-one-day period is overrun without a valid subcontract being entered into and without any delay to progress as a whole resulting. Clearly, there is no case for the contractor receiving extra time or payment, and it would appear that there are no consequences upon which the employer may act over the technical breach which exists.

Most often, a procedure like this will run without hitches. If they do occur, they are likely to be due to the triangular pattern of activities between architect, contractor and subcontractor. The time when most is known, but distributed among all concerned, while least is committed, is the main contract tendering period. Desirably for the employer, the tenderers should endeavour to remove any problems during tendering, but are unlikely to do so when other more pressing aspects concern them and later adjustment is available. It would be possible to make some system of resolution a condition of tendering, as effectively happens when domestic subcontractors are named under JCT clause 19.3. This is not likely to commend itself to the one subcontractor in the present case, when he already has the job virtually in his pocket. It would eliminate some problems but retain others, such as allowing the subcontractor to call the tune and allowing the identity of all tenderers to be known to a third party.

When the named subcontractor is identified by way of a provisional sum, it becomes necessary to value the financial effects under clause 3.7. This should be reasonably straightforward, if the 'control' figures have been made available in the contract bills or other analysis, set against the original subcontract tender.

Naming another subcontractor

Although 'renomination' is an accepted term, even if slightly inaccurate in its implications, 'renaming' does suggest something unintended – perhaps a conversion from heathendom! At any rate, the topic falls within clauses 3.3.3 to 3.3.6, which also deal with determination (taken in Chapter 18, as also for nominated subcontractors).

The reasons justifying a determination, or at least appearing to do so, are substantially those occurring under the arrangements of the JCT form. These have been listed above in relation to nominated subcontractors and, again, they are discussed in Chapter 18. In this contract there is the specific debarring of the contractor from himself accepting a repudiation of the subcontract.

Although the JCT clauses envisage only the possibility of a renomination, the present clause 3.3.3 has three avenues available:

(a) Naming another person to perform the work or such part as remains.
(b) Instructing the contractor to perform the work (or presumably any remainder), with the possibility of him subletting.
(c) Omitting the balance (or presumably the whole) from the contract, although perhaps having it executed by a direct contractor.

Whether the presumptions indicated are justified, is a matter for doubt, as the precise wording supports the distinctions shown without the bracketed portions above. There is little reason which can be put forward for tying up affairs in such knots, especially when it runs against what has been done over a hiatus in the original appointment of a subcontractor. If the distinctions are upheld, it restricts the alternatives open to the architect to (a) or (b), if subcontract work has not been started on or off site, or to (a) or (c), if it has been started.

In relation to the cases discussed under the nominated position above, the present arrangement generally follows the same pattern. However, several extra elements are introduced, presumably in the light of the decisions which were handed down in the cases after JCT 80 was published, or of a review of that contract. Thus, in option (a), clause 3.3.4 designates the instruction as leading to extension of time when this is justified, but specifically excludes reimbursement of loss and expense. In options (b) and (c) he is entitled in principle to both these recompenses (see the case study in Chapter 33).

These specific provisions cut across the concept of the architect having a reasonable time in which to secure a new subcontractor or otherwise act, perhaps by delaying an instruction to the contractor to make his own arrangements until possible subcontractors have been explored and eliminated. Extension of time, for instance, starts to run from any qualifying delay occurring. This is still a matter which has to be established in the particular circumstances, so that some specific effect on critical activities is needed. By contrast with the nominated situation, delay on the part of an outgoing named subcontractor over his own subcontract period does not lead to extension of time.

If the architect has been unable to find someone to name, the contractor is not obliged to sublet. The architect should consider carefully whether he would wish the contractor to perform the work himself, having once used the named subcontractor system presumably to secure control. On the other hand, there is at least the possibility that the contractor will not be able to find someone either, or not someone acceptable at any rate, but he may not be able to perform the work directly. This is an acute problem when the work involves special components obtainable from the original source only, but it is not unique to the naming system. The result may be redesign and the delay which goes with it or omission of the work, if this is practicable, or perhaps both. In the worst instance, existing work may have to be removed and replaced.

CHAPTER 8

CERTIFICATES AND PAYMENTS

- Status of certificates
- Dissatisfaction over certificates
- The Housing Grants, Construction and Regeneration Act 1996
- Deductions from certified amounts
- Ownership of materials and goods
- Off-site materials and goods

The topics of this chapter are linked, because all payments to the contractor and many to subcontractors depend upon the issue of an architect's certificate. Not all certificates deal with payment, however, as some deal with approvals or the deciding of contractual issues between the parties. The cases in which the architect may or must issue a certificate are strictly limited and a complete list of certificates of all types under the JCT 1980 and IFC 1984 contracts is given in Table 5.2. They may be set beside the architect's instructions, as his other major way of communication during progress, in addition to drawings. Only certificates covering payment are treated here.

Here are the main areas of dispute over certificates, particularly those governing payment:

(a) Whether they are properly authorised, correctly issued and on time.
(b) Whether they have been subject to any undue influence from the employer.
(c) Whether the contents are entirely in order, including the principles and amounts certified.
(d) Whether the employer can modify the effects, such as by making a reduced payment.

These matters and others concerning certificates are dealt with more widely and in detail in various works, such as *Building Contracts: A Practical Guide*, as they affect the employer, the contractor and subcontractors. This section is limited to the more contentious elements, as they bear on the purpose of this book, by being incidental to other disputes or perhaps bringing them on.

STATUS OF CERTIFICATES

The authority for issuing certificates is covered by the clauses in Table 5.2, and any stipulations about their contents are within the clauses. Those relating to payment, in JCT clause 30 and IFC clause 4, are particularly detailed and should be consulted where there is uncertainty.

It would appear that the contractor is entitled to have work certified for payment as soon as it has been performed and a certificate is due. This is so, even if this is well ahead of what evenly distributed progress would indicate, just as he is entitled to finish early and be paid off. Equally, he may run late with his programme and receive a greater reimbursement for fluctuations. The programming aspects of this are mentioned in Chapter 4 under 'Payment'.

What an interim certificate does not do is to give final sanction for inclusion of any item within it in the final certificate, something demonstrated as far back as *Tharsis Sulphur & Copper Company* v. *McElroy & Sons* (1878). There, the interim inclusion for payment of some work did not formally establish a right to final payment, in the absence of a proper architect's instruction for the work to be done. The architect in question appears to have been negligent in ever including the work, so far as can be told at this remove. A similar lack of finality will exist when amounts are included on account for loss and expense due to disturbance of regular progress, for which no architect's instruction is issued – understandably! It would be imprudent, at the least, for an architect to include amounts when he had not decided in principle that loss and expense was due, but it does mean that he may include approximate amounts and later revise them downwards, if needs be. The broad issue of including such amounts and the contract authority is set out in Chapter 15, and *Croudace* v. *Lambeth* in Chapter 17 should be noted, as should the next case, which was decided in mid 1998.

In *Beaufort House Developments (NI)* v. *Gilbert Ash NI and Others* (1998) the House of Lords ruled that an architect's interim certificates are not binding and have only 'provisional validity'. They can now be opened up and reviewed by a court (courts now have similar powers to those of an arbitrator, reversing the *North Regional Health Authority* v. *Crouch* decision referred to later in this chapter). The certificates that are reviewable are interim certificates, not the final certificate that remains final and binding; see *Mowlem* v. *Eagle Star* (1995). Now interim certificates have a 'provisional validity' in contracts and are both subject to review by an arbitrator and by a court.

The other side of this consideration is that, when the architect has issued a proper certificate, the employer usually must honour it under JCT clause 30.1.1.1 or IFC clause 4.2, even if unhappy over its apparently high amount. This was the issue in *Killby & Gayford Ltd* v. *Selincourt Ltd* (1973), where the Court of Appeal held that the JCT 1963 contract was mandatory over the time margin for payment (as is the current version), that there was no suggestion of defects in the extra work to give a counter-claim, and that the defendants had produced no evidence to show that the work was not properly instructed. The need for the employer to raise his query direct with the architect was evident.

However, in *C. M. Pillings & Co. Ltd* v. *Kent Investments Ltd* (1985) the employer was judged as entitled to challenge the accuracy of a certificate in arbitration, the contract being in the JCT prime cost form. The estimated final cost exceeded the original estimate by over 100%, as a result of hundreds of variation instructions. The contractor claimed that he should be paid and then the accuracy checked. The employer's case was accepted in view of the large discrepancies and the apparently casual attitude of the architect, alongside the possibility of arbitration over whether a 'certificate is not in accordance with this Contract'. This is rather parallel to a sensibly quantified set-off under a subcontract (see below). The Court of Appeal referred to its judgment in *Ellis Mechanical Services Ltd* v. *Wates Construction Ltd* (1978), where the employer had 'on balance . . . raised a bona fide arguable contention'. This is not helpful to contractors, any more than the decision in *Lubenham* v. *South Pembrokeshire* mentioned below.

Cases of this type point up the importance of the architect keeping the employer in the picture over the way that events are moving financially, to avoid clashes or worse. This is especially so when there are matters of the nature of loss and expense, which the employer may not see coming, as he should see the cost of variations, at least when they are substantial and then track back to his initiation of what has happened.

DISSATISFACTION OVER CERTIFICATES

Contractor's remedies

The remedies of the contractor vary according to the cause, if he is dissatisfied with matters relating to a certificate. One of the least satisfactory aspects is over exactly what is included in a certificate. Although it may be argued that contract provisions should be quite definite about this and although limited approximation is to be expected in valuation and certification, the contractor has no specific remedy to promote realism, other than adjucation and arbitration, if the valuation for a certificate is grossly low, or indeed if this keeps on occurring. On this matter adjucation arbitration may take place immediately under the JCT form, as it may on any matter under the IFC form. Unless he decides on nuisance tactics, the contractor is unlikely to use arbitration, or even threaten it, except for a very large discrepancy or if there is a suggestion of improper valuation, in the sense of incompetence or wilful undervaluation.

In the early stages of a contract where there is a substantial amount due for variations, or especially if sums are due for loss and expense, the possibility of underinclusion of a provisional allowance in certificates is high. When there is considerable uncertainty, the contractor has no realistic remedy other than to provide sufficient details as soon as he can of his valuation of the amount due by making his own 'application'. This is recommended practice in JCT Practice Note 26 and will be emphasised with the 1998 amendments to JCT valuation/certification that allow interest to be paid if procedures are not followed. There is no excuse for a quantity surveyor or an architect not trying to make reasonable inclusion, bearing in mind the possibility of reducing the amount later, as indicated by *Tharsis* v. *McElroy* above, and balancing this against the risks of overinclusion and overcertification.

Decisions from the courts have shown that it is not certain that interest will be payable to a contractor for what he regards as undercertification. Unless it can be shown that an architect or engineer acted in bad faith, it is unlikely that interest will be due on 'late payment' arising from what is viewed as undercertification in interim certificates. The ruling in *Morgan Grenfell Limited and Sunderland Borough Council* v. *Seven Seas Dredging Limited* (1990) held that (in the judge's words) 'if the arbitrator revises his (the Engineer's) certificate so as to increase the amount, it follows that the Engineer has failed to certify the right amount'. But the ruling in *The Secretary of State for Transport* v. *Birse Farr Joint Venture* (1993) was that unless the engineer undercertifies due to some mistaken principle or some error of law, that is his certificate can be seen to be a bona fide assessment, then no interest is due. This has been upheld in *Kingston Upon Thames* v. *AMEC Civil Engineering Limited* (1993) and in *BP Chemicals Limited* v. *Kingdom Engineering (Fife) Limited* (1994), where it was held that only the sum due under the certificate was the sum certified, and the larger sum due under an arbitration award was not payable until that award was made. It is interesting that the sixth edition of ICE has expanded the wording of clause 60(6) from that the fifth edition to make it clear that interest is payable on undercertification, to take account of the *Morgan Grenfell* v. *Seven Seas* decision. Under JCT contracts it appears that undercertification will not allow interest to flow unless it can be shown that the undercertification was a breach, arising from bad faith or lack of bona fide assessment. It all depends on what is meant by 'undercertification'.

Even if there is an improper deduction from a certificate, the contractor can do little more than seek adjudication, arbitration or to have the position rectified in the next certificate, although this has been modified under the Housing Grants, Construction and Regeneration Act 1996 (discussed in more detail below) by allowing suspension of obligations of the contractor in JCT contracts and a contractual right to simple interest at 5% over the Bank of England base rate. But if there has been a 'deduction', meaning a lower certification than the contractor believes is correct, then the position is and remains more complicated. An example of this was stated in *Lubenham Fidelities & Investment Co.* v. *South Pembrokeshire District Council and Wigley Fox Partnership* (1986), where the contractor (actually a bondsman completing through subcontractors) had determined his employment because the architect had purported to deduct for defective work (which should not have been included, presumably) and for liquidated damages when the completion date had not been passed. Although the architect had been negligent, the contractor had no right to determine, so he was in breach and responsible for the employer's losses.

Although the Latham Report recommended changes in practices to improve interim payment disputes and although some of this has been incorporated into standard contracts to reflect the provisions of statute (see the requirements of the Housing Grants, Construction and Regeneration Act 1996, referred to as 'the Act' and discussed below), it does not seem likely the new provisions and procedures will avoid all the instances that are now reviewed. At root is the continuing possibility of disputes over 'the value' of the work carried out, perhaps in contrast to what a contractor may view as 'the cost' of the work he has carried out, that is the amount he contends he is due from calculations using the pricing structure in the contract. This seems to remain a problem when two quantity surveyors may approach valuation from different points of view. Interim valuation is just that, so the context of the whole contract must be remembered when considering the 'value' to the employer of work which may have cost £2*x* but which may not be worth more than £*x* at the stage the valuation is done.

A number of cases concerning certification continue to illustrate that there may, even now with the Act, be no easy remedy for a contractor who disputes, for instance, the amount included in a certificate by an engineer. In *Balfour Beatty Civil Engineering Ltd* v. *Docklands Light Railway* (1996) the Court of Appeal confirmed that it has no general power to open up and/or review decisions of an employer (in most contracts it will be the agent of the employer, say the engineer). In this case the employer's representative took the place of the engineer. Also, the clause under the ICE contract dealing with settlements of disputes had been deleted, which would normally have allowed reference to arbitration. The contractor appealed against the decision of the Official Referee, who held that there was no power for him to open up a certificate. It was also stated that even if there is no express obligation, there is an implied term that the employer (and his agents under a construction contract) must act honestly, fairly and reasonably in reaching his decision. If the contractor can establish that the employer breached his duty to act in good faith, the court may chose to remedy that; but if there is no breach, the court will not act.

Until *Beaufort* this was unlike the power of an arbitrator, where it was stated (again) in *Oram Builders Ltd* v. *M. J. Pemberton and C. Pemberton* (1985) that certificates could be opened up for review/amendment. *Oram* followed the ruling in *Northern Regional Health Authority* v. *Derek Crouch Construction Co. Ltd* (1984), where it was established that a court has no power to intervene unless the contract specifically provides for intervention or the facts show that an architect has patently 'got it wrong'; see above for a reversal of that decision in *Beaufort* v. *Gilbert Ash* (1998). And remember that the Courts and Legal Services Act 1990 already gave a discretion to the court to accept such an issue coming before the courts when the parties agree to this being done.

Each case must be determined on its facts, and a 1996 case ended with the court substituting its decision on certification for that reached by the architect. In *John Barker Construction* v. *London Portman Hotel Ltd* (1996) it was again confirmed that there is an implied term that the architect will act lawfully and fairly. The arbitration clause had been deleted and replaced by a provision to the effect that the proper law of the agreement was to be English law and the English court was to have jurisdiction. The court held that this was not sufficient to vest in it a general power to open up, review and revise certificates, but see *Beaufort House* (1998) above. However, it held there to be an implied term that certificates were to be given honestly, fairly and reasonably. Failure by an architect to act honestly, fairly and reasonably may enable the court to declare the architect's decision invalid. The court may not substitute its decision solely on the basis that it would have reached a different decision. If this were to happen, it would usurp the role of the architect and the court had, until 1998, resolutely refrained from doing this. However, where the contractual machinery has broken down, the court will make its own assessment, and the Official Referee took on that duty in *John Barker*. This was pointing to the decision that came in *Beaufort and Gilbert Ash* in mid 1998.

In the unreported case of *Tarmac Construction* v. *Esso Petroleum Co. Ltd* (1996) the contractor disputed the amount of additional payment for adverse ground conditions and went to court. The employer maintained that the court had no power to replace the engineer's decisions with its own, but the court found that 'clause 66 of the ICE contract expresses an intention that the engineer's decision is not final and may be revised by the court. . . . The court will be enforcing the parties' rights and will not be doing anything for which it does not have jurisdiction or which it does not regularly do.' These cases reinforce the idea that the circumstances of each case must be taken into account and the full report, where possible, should be read to obtain the context and facts.

If a dispute on certification, arising from claimed 'undervaluation' due to whatever difference between the parties, has occurred and eventually has been settled in the courts, then interest has often been claimed on any 'extra' amount awarded by an arbitrator or a court. This is because some considerable time may have passed on a considerable sum and, other things being equal, one might think that interest should be payable. It may be that a contract specifically deals with this, but unless it does, there is not much precedent to allow a successful claim for interest on late payment. The courts have an overriding principle and it appears to be that provided an architect or an engineer has acted in good faith in valuing, there should be no interest awarded on the increased amount which may come from a tribunal or courts award.

In *Morgan Grenfell Limited and Sunderland Borough Council* v. *Seven Seas Dredging Limited* (1990) there was temporary comfort for English contractors obtaining interest. In the Scottish cases of *Nash Dredging* v. *Kestrel Marine* (1987) and *Hall and Tawse* v. *Strathclyde Regional Council* (1990) and then in *BP Chemicals Ltd* v. *Kingdom Engineering (Fife) Ltd* (1994) it was held that there was no entitlement to interest under an ICE fifth edition contract, where clause 60 (6) had been deleted. In *A. E. Farr Ltd* v. *Ministry of Transport* (1977) and *The Secretary of State for Transport* v. *Birse Farr Joint Venture* (1993) it was held that a certificate which bona fide assesses the value of work done at a lower figure than was later found, and which does not involve a contractual error or misconduct by an engineer, will not rank for interest under ICE clause 60(6). Under JCT, in *Lubenham Fidelities and Investments Co. Limited* v. *South Pembrokeshire District Council* (1986), although there was undercertification by the architect deducting liquidated damages from the face of the certificate, it was held that this did not constitute a breach of contract and therefore no interest was due.

Equally unsatisfactory is late issue of certificates: no specific remedy is available, other than repeated complaints to the employer. The absence of a certificate was, however, no bar to payment in the special circumstances of *Croudace Ltd* v. *London Borough of Lambeth* (1986); see Chapter 17. Over late payment by the employer, the provisions are clear: the contractor may determine his employment under the contract, subject to observing the procedure over notices. This is a most drastic remedy, and so unlikely to be used except in the most drastic situation. It is also no use when late certification follows practical completion, the final certificate being the ultimate situation and a common one for delay in settlement. In JCT 80 and IFC 84 contracts there is now provision for interest on late payment of certified amounts, without due notification of the intention to deduct by an employer or main contractor. With this goes the right to suspend contractual obligations to proceed with the works, to obtain an extension of time and probably loss and expense for the effects of the contractor 'downing tools' while he waits for an employer to honour the amount certified.

In a loss and expense situation, as distinct from a variation situation, the contracts being quoted do allow the possibility of the contractor recovering financing charges incurred during the period of delay between notification and certification, in the form of direct loss and expense. This is distinct from the general position that interest cannot usually be applied to business debts, without special contract terms, a position which applies to building contracts as much as any others. There is no certain legal authority for recovering financing charges incurred through late or inadequate certification in general. It may be that the attitude of the courts is changing here, in that there is more readiness to allow at least simple interest on some classes of late payment. Cases discussed under 'Interest and financing charges' in Chapter 12 may be consulted, with the mention of the courts 'distinguishing' cases. An extension of their principles to inadequate or late certification is at least possible, including situations that arise from late preparation of the final account.

The remaining aspect to note is the employer trying to interfere with a certificate or to influence the architect over it, a serious act because the architect is supposed to exercise his functions impartially between the parties. Again, the remedy available is determination, present mainly as a deterrent, but more likely to be used in this situation. The contractor needs to steer a careful path here between what is miscertification and what is the result of interference. In *R. B. Burden Ltd* v. *Swansea Corporation* (1957) the contractor had determined on interference, whereas the House of Lords found that it was only miscertification (after a history of prior mistakes), so the contractor was incorrect in determining.

Dissatisfaction also arises in certificates that concern valuation for parts of work carried out as variations. Generally these items must be valued according to rules which are discussed in Chapter 6. If the works have not been authorised, it may be that they fall to be valued under quantum meruit (Chapter 9). In *Costain Construction Ltd and Tarmac Construction Ltd* v. *Zanen Dredging and Contracting Co. Ltd* (1996), on appeal from an arbitrator's award, an Official Referee held that, under a subcontract for work in a casting basin for a marina, the instructions were not authorised variations under the subcontract and that a quantum meruit was an appropriate method for some of the valuation of the works.

Subcontractor's remedies

Subcontractors have no direct remedies over certificates, as they are not parties to the main contract. Only nominated subcontractors receive any special treatment in certificates; and when it comes to lack of inclusion of their amounts in whole or part, they are in the same basic position as the contractor himself, the position outlined above.

The contractor has an obligation under a clause of each subcontract, for example clause 1.13 in the nominated subcontracts, to 'obtain for [the subcontractor] any rights or benefits of the Main Contract' where they apply to the subcontract. This all-embracing but nothing-specifying provision perhaps requires him to press the employer over matters of variation instructions etc affecting subcontractors, when there are irregularities. In the case of domestic or named subcontractors, the relationship between main and subcontract payments is not fixed by what the architect has certified, and the contractor is due to pay irrespective of whether he is being paid some related sum. He is not entitled under the standard forms to operate 'pay when paid'. These subcontractors therefore have to press the contractor in isolation from the main contract position.

Subcontractors of any variety have a right to suspend their work after due notice, for example under nominated clause 4.21 if the contractor has not paid them, and to continue suspension until they are paid. In the case of nominated subcontractors, this may also mean when they have not received direct payment from the employer, which is discussed below. This may become effectively a weapon against the employer in their case, as the contractor will be entitled to an extension of time if the suspension is due to the employer failing to operate the direct payment provisions properly, when these are obligatory upon him. Named and domestic subcontractors also have a right to determine if they are not paid, this being a weapon against the contractor, as he decides what to pay them, even if the dates are related to the main contract dates. In summary, this section illustrates an area of potential for disputes in construction contracts that will remain whatever statutory and procedural good practice measures are put in place. The potential for disagreement over valuation of work in progress is large and will remain so. It is imperative that an employer, through his architect or engineer, exercises his powers reasonably. An architect must use his discretion fairly and without improper influence from an employer. As stated in *John Mowlem & Co. plc* v. *Eagle Star Insurance Co. Ltd and Others* (1992), a term will be implied that the certifier will not be exposed to improper influence and will be left free to exercise independent judgement, either by reference to his own standards or by reference to the contract requirements. It was stated that an architect who does submit to pressure to issue falsely undervalued certificates may commit the tort of wrongfully interfering with the performance of contractual relations between employer and contractor. This may leave him exposed to damages to the contractor. An architect's mere submission to such a pressure does not normally constitute a tort of procuring or facilitating a breach of contract by the employer.

THE HOUSING GRANTS, CONSTRUCTION AND REGENERATION ACT 1996

The Housing Grants, Construction and Regeneration Act 1996, referred to in this chapter as 'the Act', is discussed in several sections of this book. It has been introduced to put into effect some of the proposals that the Latham Report recommended, namely the right to adjudication (Chapter 27 and other chapters), an entitlement to stage payments in all construction contracts that last over forty-five days, the right to suspend work if monies due are not paid, the outlawing of 'pay when paid' clauses and the restriction of the right to 'set-off' payments. In this chapter it is only necessary to discuss matters concerning certification and how some of them have been incorporated into JCT contracts.

The Act has now required that some of the common law rights be modified in construction contracts. Since 1 May 1998 it has caused standard forms of contracts to have to be amended to take account of sections 110, 111 and 112 of the Act concerning payment rights.

Section 110 of the Act requires that every construction contract (exceptions to what is a construction contract do not matter for the purpose of this volume) shall

(1a) provide an adequate mechanism for determining what payments become due under the contract, and when, and

(1b) provide for a final date for payment in relation to any sum which becomes due.

(2a, 2b) provide for the giving of a notice concerning any set-off or abatement and for specifying the amount of the deduction and the basis on which it is calculated (this is provided no set-off or abatement was permitted by reference to any sum claimed to be due under one or more other contracts).

(3) if a contract does not contain such provisions, the Scheme for Construction Contracts shall apply.

The parties are free to agree how long the period is to be between the date on which a sum becomes due and the final date for payment.

Section 111 requires that a party to a construction contract may not withhold payment after the final date for that payment unless he has given an effective notice of intention to withhold payment.

Section 112 gives a right (without prejudice to any other right or remedy) to a person to whom a sum is due but unpaid in full by the final date and for which no effective notice to withhold has been given then to suspend performance of his obligation to the party that has withheld payment.

The JCT issued amendment 18 to JCT 80 and amendment 12 to IFC 84 to comply with the Act and the intention and effect of these is now examined.

JCT 80 provisions

Clause 30 has been redrafted in amendment 18 to require that an interim certificate must state what the payment relates to in the terms of the Act section 102(2).

Clause 30.1.1.1 provides that the certificate shall have a statement specifying what the payment relates to and the basis on which the amount was calculated, the final date by which it should be paid and provisions for paying simple interest at 5% over the base rate of the Bank of England on the amount that is late in being paid. This is to be treated as a debt due from the employer to the contractor.

Clause 30.1.1.3 provides that the employer not later than five days after the date of issue of an interim certificate shall give a written notice in respect of the amount stated as due.

Clause 30.1.1.4 provides that not later than five days before the final date for payment of the amount due, the employer shall give a written notice of the amount, if any, that the employer proposes to withhold and the reasons why.

Clause 30.1.1.5 provides that if the employer does not give such a notice then he shall pay the contractor the amount due in interest under clause 30.1.1.1.

IFC 84 provisions

Section 4 of the IFC 84 contract has been redrafted to provide for the Act by incorporating requirements within the redrafted form.

Clause 4.2(a) provides that the architect shall certify the amount of interim payments, to state the amount to which it relates and the basis on which it has been calculated, as well as giving a final date for payment of the certificate; and it obligates the employer to pay simple interest on the amounts unpaid for the period during which they remain unpaid.

Clause 4.2.3 provides that if the employer intends to withhold or deduct from any amount pursuant to clause 4.2(a), he must notify this by the necessary written notice to the contractor.

Clause 4.2.3(b) provides that no later than five days before the final date for payment, the employer may give a written notice to the contractor specifying any amount proposed to be deducted from the certified amount and the grounds on which it is proposed to deduct each amount.

Clause 4.2(c) permits the contractor to submit details to the quantity surveyor of what he considers to be a fair valuation. If this is done then the quantity surveyor shall submit to the contractor a similar amount of detail if he disagrees with the application of the contractor. This was recommended to be done in JCT Practice Note 26 *Valuation and certification for interim payments including variations*, but perhaps in practice was not often done.

Clause 4.4A provides a right to a contractor to suspend the works if, without written notice, the contractor fails to pay the contractor in full by the final date the amount due in a certificate. If seven days have passed of non-payment following a written notice to the employer and the architect, the contractor may suspend performance of his obligations under the contract. Such suspension shall not be suspension to which clause 7.2.1(a) applies (default without reasonable cause by the contractor that could lead to determination) or clause 7.2.1(b) (failure to proceed regularly and diligently with the works). It will affect, for example, obligations such as insurance and indemnity provisions and the employer and contractor will have to look into these matters carefully.

The suspension will count as a reason for extension of time under new clause 2.4.18, as an event referred to in clause 2.3. A new clause 4.12.10 provides that suspension for non-payment shall be a 'matter to be referred to' for disturbance of regular progress, under clause 4.11.

As a result of these and other changes coming out of the Act, there have been amendments to interim payment on practical completion and to the final certificate under JCT contracts (clauses 4.3 and 4.6) if deductions are to be made by an employer from the amount stated as due in such a certificate.

These measures are intended to make payments not subject to deductions and to require simple interest to be due on overdue final certificates (once they have been issued, under JCT 80 clause 30.1.1.1 and IFC 84 clause 4.6), but it remains to be seen how practice and the law will determine application of this and other unresolved matters.

DEDUCTIONS FROM CERTIFIED AMOUNTS

Employer's rights

The right of the employer to make deductions, or set-off, from certified amounts for counter-charges, mentioned in JCT clause 30.1.1.2 and IFC clause 4.4, is founded in the common law right. Counter-charges, by definition, are in respect of something other than what is in the certificate: they are deductions from its total and not adjustments of it. In fact, the architect may not deduct these amounts when arriving at the amount stated as due in a certificate. They remain with the employer under the contract wording, even though the architect may well advise him of his rights.

As such, the deductions exist unless the terms of a contract expressly forbid them, or there is a necessary implication to that effect. They may extend, for example, to amounts for remedial work which the employer has performed when the contractor will not do the work, work that is not itself within the certificate. This example and its terms are made specific by the contracts. So too is the right to deduct liquidated damages when, and only when, the architect has properly certified that completion is overdue. Proper certification was emphasised in *Token Construction Co. Ltd* v. *Charlton Estates Ltd* (1973). These instances follow the common law right, but there is the implied qualification that the employer cannot make frivolous unsupported deductions. He should show reasonable quantification of what he contends is his due, as is expressly provided in the subcontract arrangements mentioned below.

Contractor's rights

Similar provisions for regulating payments between contractors to subcontractors have now to be incorporated in their contracts, as a result of the Act.

There are two basic situations in which the contractor may deduct from subcontract payments: where he himself has suffered a corresponding deduction from a certified amount and where he has some contra-account on which to recover. In dealing with a named or domestic subcontractor, he acts entirely on his own, in that the subcontractor has no recourse to the architect. When a nominated subcontractor is involved, the position is more complex.

This is because of the obligation of the contractor under nominated subcontract clause 4.16.1 in general to pay on to a nominated subcontractor precisely what he has been paid, as the settlement of the subcontractor's account. This amount is set against the prime cost sum in the contract, so the decision on the amount payable effectively rests with the architect and the quantity surveyor. If the contractor fails to pay in this way, he may face action by the architect implementing direct payment of the subcontractor. When the employer has exercised his right to deduct from a main contract amount because of a subcontractor's item, there should be no difficulty, unless the employer has overlooked letting the architect know about his action.

When the contractor has made a deduction for some reason of his own, he should ensure that the architect is aware of the reasons, so that the possibility of direct payment (discussed below) is minimised. Among these reasons may be unliquidated damages for delay, of which the architect will be aware, as he has to be involved in the granting or otherwise of any extension to a nominated subcontractor. If the architect is not convinced of the adequacy of the contractor's reasons, and especially if the subcontractor complains to him and he checks the position, he may still implement direct payment.

The other reasons why a contractor may deduct from a subcontract amount are several, and are common whatever the type of subcontractor. The contractor may have rendered some service by way of attendance etc, which is reimbursable. Only the amount, rather than the principle, may be in dispute. Alternatively, there may be a question of loss and expense (such as provided for in nominated subcontract clause 4.27.1 and not such as to be passed up to the employer) or damage to other work by the subcontractor, with the reinstatement by the contractor or another subcontractor. Here both principle and the contingent amount may be in dispute. In all of these instances and provided there is something more than a frivolous deduction, the architect should make haste slowly over any direct payment action. He certainly should not be drawn into the rights and wrongs of the issue. At most, he or the quantity surveyor may wish to check that the deduction is not at an absurd level, suggesting that the issue is being used as a pretext for delaying payment.

The question of quantification mentioned above was highlighted by the legal cases mentioned under the next heading, and has become part of the provisions of the JCT nominated subcontracts in clause 4.27.2. It is also present in the corresponding domestic subcontract and in the IFC named subcontract and domestic subcontract. These documents all make it a precondition of the right of the contractor to deduct. This remains true even when the deduction originates with the employer, although the details should then be readily available. A perusal by or for the architect may be useful for the reasons just advanced. To resolve any gross injustice, it is usually sufficient to leave the matter to the adjudication procedure provided in the several subcontracts.

And subcontractors, when faced with contra-accounts which relate to other subcontractors or the employer, should continue to act and pay through the contractor, as they have no privity of contract with these other persons. The exception occurs when they have made a direct arrangement with another subcontractor over a private issue.

Set-off in subcontracts

The position over set-off or counter-charging subcontractors is in essence as given above, so that separate discussion is not needed. However, there have been a number of legal cases on the theme affecting subcontractors directly, of which two may be mentioned as their vibrations are still sometimes felt.

Dawnays Ltd v. *F. G. Minter Ltd and Trollope and Colls Ltd* (1971) was a successful appeal by a subcontractor for legal action to be allowed to proceed over a matter of money withheld by the contractor for delay. The defendants contended that arbitration under the Green Form of subcontract (predecessor of the JCT nominated subcontract) should be followed, which would have led to long delay until after practical completion. It was held that cash flow was vital to the industry (almost, it appeared, more than to most). More important, it was held that counterclaims for sums, relatively vague and unagreed in amounts, could not be deducted from sums due. These had already been paid to the contractor for paying on to the subcontractor. Specifically, it was held that the words 'any sum or sums which the Sub-Contractor is liable to pay to the Contractor' in clause 13 of the existing form meant any sum or sums ascertained by proceedings or otherwise agreed, rather than those where the legal liability was not disputed, but the amount was vague.

Several cases followed from this, and the expression 'Dawnay principle' gained currency. Some doubt arose over elements of the decision, illustrated by the next example, although they did not formally reverse it.

Gilbert Ash (Northern) Ltd v. *Modern Engineering (Bristol) Ltd* (1974) related to a deduction by the contractor from an interim payment to a nominated subcontractor as set-off for breach. The sum was slightly less than what was calculated by the contractor, but was not agreed by the subcontractor. It related to bad workmanship and delay which were not disputed in principle. The subcontract form was not the Green Form, but specially produced, and allowed the contractor 'to deduct . . . or otherwise to recover the amount of any bona fide contra account and/or other claim'. It was held that the contractor was entitled to make the deduction under the wording, even though not agreed, and that the then main JCT form did not remove this right as between employer and contractor. Several collateral points emerged as obiter dicta during the judgment of the Lords:

(a) A right to set-off at common law or in equity existed but could be contracted away by a party, although it had not been done in this case.
(b) The terms of the main contract did not exclude a set-off there by the employer, although one was not in question in the case.
(c) Although the *Dawnay* case was not being reviewed in the judgment, it would appear that were a case following it to come before the House in the future, it would be held by a majority (if not all) of the present judges to have been misdecided by the Court of Appeal in disallowing a set-off. This was despite the strong view taken in that case of the words 'is liable to pay'.

Direct payment of nominated subcontractors

Alone among subcontractors, the nominated variety may be paid direct by the employer in some circumstances. One instance is for work etc performed before nomination, which is not likely to lead to dispute, except over accounting detail. The other arises after nomination in the limited case of the contractor failing to discharge an amount already included in a certificate, so that it remains unpaid, or more precisely cannot be shown to have been paid, by the time that the next certificate is due to be issued. There is neither right nor duty to pay subcontractors direct, instead of paying them through the contractor, however strong the fear that the contractor will not pass monies on. Payment can be made only after default by the contractor. Chapter 7 deals with subcontracts, and the main stipulations of JCT clause 35.13 may be summarised as follows:

(a) The employer has an option whether to exercise this power except that, when there is an employer/subcontractor agreement in force, he must exercise it as one of the terms of the agreement.
(b) An architect's certificate stating the amount of the non-payment is required before the employer may act.
(c) The amount is to be paid to the subcontractor at the same time as the corresponding certificate is honoured. It is to be deducted by the employer from 'any future payment otherwise due', so that it falls into the category of set-off, rather than being something which the architect can use to reduce the amount to be certified. Certificates will therefore continue to show the unreduced amounts.
(d) The inference of 'any future payment' is that the employer may make the deduction in more than one bite, if there is not enough due at one time, or if he is seeking to support an ailing contractor, when deduction is optional. When there is not enough available as 'otherwise due' to allow a complete deduction without the employer being temporarily out of pocket overall, he is never obliged to pay more than is available – and should not.
(e) Within these main provisions there are subsidiary rules about retention and about apportioning the available funds when more than one subcontractor is owed money.
(f) If insolvency of the contractor is coming about, the whole apparatus of the clause ceases to have any effect, as it might well transgress the legislation by giving one or more creditors a privileged position as against others.

The provisions exist almost entirely for the benefit of nominated subcontractors, as part of their VIP treatment, as it appears to be. As indicated, neither named nor domestic subcontractors have any such safeguard. The only obvious benefit to the employer is that he may be able to avoid a desirable nominated subcontractor suspending work, or even repudiating his subcontract if driven too far. It is not in the interests of the employer to face disruption of the programme, even when he has assured remedies at hand. A smooth run is better, so long as the price is not too high.

Again, the employer should be wary of operating these provisions when there are negotiations going on between contractor and subcontractor. There may be contra-accounts for extra facilities provided or for damage to other work or materials, with the principle or amount in dispute. This is why the clause requires an architect's certificate as the trigger to action by the employer. Provided this is available, the employer may make a deduction from the main contract payments without fear that he is transgressing under his contract, even though the architect may be open to criticism if he has not taken enough care. In calculating the amount, the employer should be advised to deduct retention from the original gross inclusion, as this is part of the fund for nominated subcontractors and is still held back from the main payments. He should not deduct cash discount, as this is due to the subcontractor as recompense for late payment.

For all other types of subcontractor, the employer should not make direct payments at all. Again, he may be entering areas of dispute about which he knows nothing. He certainly risks paying twice, as the contractor is always entitled to be paid for work performed by a subcontractor in the absence of an unusual provision like that for nominated subcontractors. Under the named and domestic subcontract forms, the subcontractor has the extra weapon of determination available in the provisions.

OWNERSHIP OF MATERIALS AND GOODS

Ownership can be a particularly difficult area when dealing with payments during progress. All that can be done within the present limited scope is to warn of the more problematic issues which arise, centring around the question of transferring ownership from one party to another. Detailed information may require consultation of specialist legal sources.

All standard contracts and subcontracts provide for payments to be made for materials and goods which have been delivered for incorporation into the works. The contractor or subcontractor may rely on these provisions as giving them a right to be paid, provided he complies with requirements about timing of deliveries and storage etc. Usually, there are also provisions for payment for similar items off-site, these being available for use by the architect entirely at his discretion, so the contractor or subcontractor cannot rely on them. They are most commonly used when there are severe problems over space on-site, or when advanced, extensive prefabrication is occurring. Not all of these provisions are legally watertight in all the situations which arise.

The basic problem over ownership and payment is that no one can confer a better title than he himself possesses. If, therefore, the employer pays the contractor for materials properly on-site and supplied by another, he usually obtains a good title as and when the contractor pays for them, but not necessarily before. In particular, if they are a subcontractor's consignment, property does not pass until the subcontractor has been paid by the contractor, as there is an agreement for works and materials and not for sale of goods. The period of risk is while the payment chain is being completed, if it starts from the wrong end, as it often does when merchants are offering credit which exceeds the time for honouring certificates. The most common risk is that of the contractor becoming insolvent after himself being paid and before paying a subcontractor. The issue has been around for a long time, but was brought to a head for the construction industry comparatively recently.

The earlier form of the issue centred on the position between employer and the liquidator over materials for which the employer has paid. It was held in *Re Fox ex parte Oundle and Thrapston Rural District Council* v. *The Trustee* (1948) that the trustee in bankruptcy had no claim to materials on-site which had been paid for in interim certificates under a predecessor of the JCT form, containing similar provisions to the present clause 30.2. These materials had not been in the reputed ownership of the contractor, that is there was a presumption of ownership by the employer. On the other hand, materials paid for but at the contractor's yard were in the reputed ownership of the contractor, in the absence of proof otherwise. This doctrine of reputed ownership does not apply when liquidation of a company occurs. The predecessor of JCT clause 30.3 was introduced to give protection here, although it has weaknesses over subcontractors.

A development of the situation came in *Aluminium Industrie Vaassen BV* v. *Romalpa Aluminium Ltd* (1976). A supplier incorporated a 'retention of title until payment' in a contract for sale of goods, which admittedly suffered from complications in translation from the Dutch and otherwise. It was held that the clause was valid against the buyer, who had not only taken delivery, but had converted some of the goods by manufacture and sold some on to others. The goods or proceeds were in trust until the seller had received payment in full. Without the clause, the legal position would have been reversed. This indicates one area of severe difficulty. The bibliography may be consulted for further information.

In 1979 there was the case of *Humberside County Council* v. *Dawber Williamson Roofing Ltd.* This was brought under the old Blue Form subcontract, which preceded the present range of standard domestic subcontracts. The subcontractor delivered roofing slates, the employer paid the contractor for them, the contractor went into liquidation, and the subcontractor had not been paid. The equivalent of JCT clause 16.1 was in the main contract. The employer prevented the subcontractor from removing the slates as his own property and used them for completing the works, arguing that he had paid for them and they were his property for this reason and by virtue of the clause. It was held that the employer had to pay the subcontractor for the materials and other damages. Title did not pass from subcontractor to contractor under the subcontract by any provision in it, as it was a subcontract for works and materials. The notice which the subcontractor had of the main contract terms did not import the ownership provision into the subcontract. The contractor had no sufficient title to transfer: he had not paid for the materials, nor had they been fixed.

Terms now exist in the JCT main and subcontract forms to overcome these problems for the employer and protect him. These appear broadly effective, although not free from weaknesses. When in force, they do reduce the protection available to subcontractors.

Debate has been inconclusive over the position of the quantity surveyor in preparing valuations and the position of the architect in issuing certificates. Strictly, it would appear that, before inclusion, they should check the status of every brick or pipe lying on-site. When the chain of payment may extend back through three or four persons, the matter becomes quite impracticable. The implicit attitude seems to aggregate as 'press on and hope for the best'. The situation is not entirely unique to the construction industry. To protect themselves, some consultants have suggested warning employers that they are implementing the requirements of the contracts, but these requirements may put them at some risk. But it remains to be established whether, in any action over professional negligence, employers would then or otherwise be held to be aware of this fact.

It certainly is necessary for the terms of contracts and subcontracts to be observed carefully, especially over payments for materials off-site, and despite flaws in those clauses. In the case of Scottish contracts, these clauses do not satisfy the distinct legal position in Scotland, and the alternative contracts of purchase available there must be used.

OFF-SITE MATERIALS AND GOODS

Another result of changes introduced in April 1998 to JCT 80 and IFC 84, at the time of those required by the Act, is provision for off-site materials and goods to be paid for by an employer according to conditions that include the use of a schedule and advance payment bond in respect of the materials and/or goods. It is hoped that this will enable another area of potential dispute to be lessened by making the payment of money to a contractor by an employer more secure, by bonding in the event of the contractor defaulting in some way and the employer having paid for materials which have not arrived on-site and will be essential for completion of the works.

PART 3

DELAY AND DISTURBANCE: INTRODUCTORY CONSIDERATIONS

CHAPTER 9

REMEDIES FOR BREACH OF CONTRACT

- Damages for breach of contract at common law
- Standard provisions equivalent to damages in building contracts
- Liquidated damages
- Costs of preparing claims for extension of time and loss and expense

This chapter is included primarily to outline a number of principles of the background law. These affect the application of the remedies available beyond the contract terms, but also give substance to the provisions of the contracts themselves. They relate particularly to damages and so to the theme of loss and expense taken in later chapters, but also include the question of liquidated damages in this chapter. Note that these principles are full of complexities and they are treated at length in legal works, because of the cases which have led to their elucidation and continuing development. The present treatment is given to set out some pointers, which should be followed up elsewhere if the need arises.

This chapter also indicates that there are more available remedies when one party is in breach than those set out specifically in standard contract forms, or indeed in any contract as written. What is included over the two major areas of extension of time and loss and expense for disturbance of regular progress exists to regulate affairs by providing a first point of reference and action.

In the case of extension of time, the provisions given will usually delineate the boundaries of action fairly closely, as may be seen from discussion in Chapters 10 and 16. Provisions over disturbance of regular progress are also usually the major consideration and exist to be used when no other provisions within the contract provide reimbursement, as the clauses state; Chapters 10, 11 and 12 cover the general principles and Chapter 17 explains the JCT provisions. These contract provisions are also by the same token without prejudice to, and so do not exclude other remedies, meaning arbitration and legal proceedings.

It is thus open to an aggrieved party to go to other remedies if dissatisfied with his treatment. Most commonly, this is likely to be the contractor reacting against decisions of the architect or the quantity surveyor, or against some entrenchment of the employer, but it could be the employer doing the reacting, as both architect and quantity surveyor are there under the contract to act impartially between the parties.

The expression used in JCT clause 26.1 and IFC clause 4.11 in relation to recompense for disturbance of regular progress is 'direct loss and/or expense'. It is generally held, for reasons set out below, that this gives amounts similar to those which would arise by the award of damages in the courts. It has just been indicated that the existence of clauses like these does not remove the possibility of the contractor seeking damages as an alternative to payment for loss and expense under them, or even when he is not satisfied with what he has already obtained under them. There is also the possibility of issues arising that are common law breaches which do not fall within the ambit of the clauses. These must be pursued at law if the employer is unwilling to meet them by negotiation, so as to include the resultant settlement within the final account by specially authorising the architect and quantity surveyor. It is also possible that not all of the matters in the clauses would constitute breaches in all circumstances.

DAMAGES FOR BREACH OF CONTRACT AT COMMON LAW

Nature of breach of contract

A breach of condition at law is a major breach, to be distinguished from a breach of warranty, as discussed in Chapter 3. What is termed a 'condition' within standard or other written contracts is not necessarily a condition in the immediate sense, which is restricted to legal nature and importance. There is a growing tendency for the courts to distinguish conditions and warranties, not by abstract definition of their statuses in particular instances, but by the extent of the effects of their breach. However, every breach of condition or warranty leads to the possibility of an action for damages.

In addition, when the breach is of a condition and so fundamental in its effect, it may entitle the aggrieved party to treat the contract during or before progress as repudiated by the other, so as to bring it to an end, or to treat the precise and apparently binding terms of the contract as overruled. The courts are careful to restrict recognition of fundamental breach. Thus, in a case where such breach was alleged over the non-availability of a hired ship due to the defendant's fault for twenty weeks out of a total hire period of twenty-four months, it was held that there was not fundamental breach, as the plaintiff could still obtain a large part of the hire benefit; see *Hong Kong Fir Shipping Co. Ltd* v. *Kawasaki Kisen Kaisha Ltd* (1962). On the other hand, the installation of defective pipework which led to the complete destruction of a mill by fire was held to be fundamental, as in *Harbutt's Plasticine Co. Ltd* v. *Wayne Tank & Pump Co. Ltd* (1970). Here, in fact, work had been completed, but the decision overruled an exemption clause in the contract, put forward originally by the defendant.

It is also possible in these circumstances for the party treating the contract as repudiated to select the alternative of suing on a quantum meruit basis instead of damages, that is usually for expenditure incurred. This may be more advantageous in the right circumstances, but can only be pursued in extreme circumstances. Thus, in *Davis Contractors Ltd* v. *Fareham Urban District Council* (1956) it was held that an extension of the contract period from eight to twenty-two months did not frustrate it and allow a quantum meruit, which is discussed in more detail later in this section, under its own heading. The contract to erect a number of houses as envisaged could still be performed and was not frustrated in its nature, whatever the problems and costs introduced by the extended period, as long as the potential hazards which had come into reality were not beyond the contemplation of the parties. This of course did not deny the contractor whatever reimbursement was open to him under the contract.

In contract, it is generally possible for an injured party to seek specific performance or an injunction as an alternative to damages over breach. Specific performance is unlikely to be granted, as it would be difficult to oversee to make it effective, whereas an injunction is seldom relevant, especially against the employer.

Some principles of damages

Principles of damages are similar, but not identical, whether the action is in contract or tort. The purpose of damages under contract is to put the injured party back into the position he held before breach, so far as it is possible to do this by pecuniary means. The aim therefore is to compensate that party for loss, and not to inflict some penalty on the other. It is what the one party has suffered, and not what the other has gained, which is relevant. This applies whether there is non-performance in part or whole, or misperformance of some part, although the relevant ingredients of the damages will vary. The broad parallel may be noted with liquidated and ascertained damages, discussed below, over compensation rather than penalisation.

Direct and indirect (or consequential) damage

The compensation afforded by *damages* is in respect of *damage* sustained. The test over damage is its remoteness or otherwise, where the position is that it must not be so remote, or indirect/consequential, as not to be direct, in the terms of the contracts being considered. A direct loss is one which flows from the cause without the interposition of some other cause. This is a complex matter affected by the actual sequence, the nature and weight of the respective causes and the questions of 'Contemplation of the parties' (see next heading) and 'Concurrence of causes' (Chapter 10). Several broad variations may be suggested as a minimum:

(a) Cause A leads to loss W, without any intervening cause. The loss is direct.
(b) Cause B leads to cause C, and cause C to loss X. The loss may be direct or indirect, depending on its proximity.
(c) Cause D and cause E occur independently in parallel, leading to loss Y with some degree of contribution from each.
(d) Cause F is followed by, but does not lead to cause G. Loss Z is caused primarily by cause F, but cause G aggravates it.

These stark examples are minimal in their illustrative power. Within construction works, the situation often arises when there is a loss by delay due to one cause and further loss by defects due to another. If only the delay can be laid at the employer's door in terms of reimbursement, its costs should not be inflated by the extra costs due to the presence of the defects, which have arisen independently of the prime cause of delay. Their conjunction is fortuitous and has led to excess costs which should be settled separately where they lay.

As one judge has remarked, 'The word "consequential" is not very illuminating, as all damage is in a sense consequential.' However, in *Millar's Machinery Co. Ltd* v. *David Way & Son* (1934) it was stated that the word had come to mean 'not direct'. This was quoted in *Saint Line Ltd* v. *Richardsons, Westgarth & Co. Ltd* (1940), where a clause excluding 'any indirect or consequential damages' was under review. It was held that loss of profit on the use of a ship, wages and stores and fees paid for superintendence were all direct and recoverable. Similarly, in *Croudace Construction Ltd* v. *Cawoods Concrete Products Ltd* (1978), loss of productivity, delay costs and meeting indemnity costs to a subcontractor were all held to result directly and naturally from failure to deliver on time and from defects in the items delivered.

The meaning of 'consequential' continues to cause confusion and 'consequential loss' appears not to have a clear legal meaning. In *British Sugar plc* v. *NEI Power Projects Limited* (1997), an unreported case because it was a judgement on a preliminary issue, a number of cases were reviewed. The judge held:

> I have come to the conclusion that 'consequential loss' must be construed as meaning such loss as the plaintiffs prove *over and above* that which arose as a direct result of such breaches as the plaintiffs may prove in accordance with the rules laid down in *Hadley* v. *Baxendale*. The limitation or 'cap' does not apply to such damages that arose within those well known rules which have stood the test of time for nearly a century and a half. I reject as inconsistent with those rules the dichotomy of 'normal' and 'consequential' loss contended for [by the defendant]. (Emphasis added)

The judge quoted from *Croudace* that 'the judge was right to hold that the word "consequential" does not cover any loss which directly and naturally results in the ordinary course of events from late delivery.'

To avoid uncertainty, some lawyers advise against using the term 'consequential' and recommend an explicit statement of the categories of loss or damage that are accepted without limit. Here is an example: 'The contractor shall be liable for the reasonable costs of repair, renewal and/or reinstatement of any part or parts of the works. . . . The contractor shall not be liable for any other losses incurred.'

Contemplation of the parties

This principle of directness of loss is then moderated according to the knowledge held or which reasonably should have been held by the parties about the loss likely to flow from a breach, especially the defendant in an action. A narrower view of what might be assumed to have been in the contemplation of the parties is taken in contract than in tort, so that reasonable information is required to be available about the possible nature and extent of loss flowing from breach, in the absence of crystal clear knowledge. How serious the amount of the loss may be is a distinct matter, and there is room for much wider variation in information.

The basic case on contemplation is *Hadley* v. *Baxendale* (1854), concerning failure to deliver on time a replacement shaft for a mill, which in consequence was out of production, so causing loss to the owner. It was held that the plaintiff could not succeed: the defendant was not reasonably to have known that no spare shaft was held by the mill, as might have been thought in accordance with general practice. There was no indication to the contrary in the order for the shaft. The critical section of the judgment reads:

> Where two parties have made a contract which one of them has broken, the damages which the other party ought to receive in respect of such breach of contract should be such as may fairly and reasonably be considered *either arising naturally*, i.e. according to the usual course of things, from such breach itself or such as may *reasonably be supposed* to have been *in the contemplation* of both parties, at the time they made the contract, as the probable result of the breach of it. (Emphasis added)

The two branches of the decision relate to damage which arises in the normal course of affairs, and to damage which arises in special circumstances. Damage in special circumstances is only admissible during assessment if the defendant was specifically aware of the circumstances when contracting. In a later case, *Victoria Laundry (Windsor) Ltd* v. *Newman Industries Ltd* (1949), the two branches were rolled into one, forming the first of the three propositions given in the judgment:

(a) Recovery is limited to 'such part of the loss actually resulting as was . . . reasonably foreseeable'.
(b) What was then 'reasonably so foreseeable depends on the knowledge then possessed by the parties . . . [especially] the party who commits the breach'.
(c) Two classes of knowledge may be possessed 'one imputed, the other actual', the former relating to what is reasonably foreseeable, the latter to what must be specially made known when contracting.

The point at issue in the case was distinct from that in *Hadley* v. *Baxendale*, in that there were two levels of profit involved. One level was profit attainable on normal business, but lost by late delivery of a boiler. This profit was recoverable, as the plaintiffs had made it plain that they were 'most anxious' to put the boiler into use 'in the shortest possible time'. It was held that the defendants must have known the implications of these statements in relation to taking on work, by their own experience in engineering. The second level was profit in prospect from some 'highly lucrative' contracts. This, it was held, the defendants could not have foreseen unaided, so that damages were to be limited to an assessment of normal profit for the type of contracts concerned.

Later still, in *The Heron II* (1969), the House of Lords modified this position by stating that the criterion was not whether the damage as such should have been foreseen, but whether both parties should have had the probability of its occurrence within their contemplation. At this point, contract is narrower than tort, where a position more akin to possibility rather than probability applies. The liability is not, however, to be limited by whether the actual consequences, as distinct from their probability, are more serious than could have been reasonably contemplated. It may well be that damages are out of all proportion to the contract amount, as happened in *Harbutt's Plasticine* v. *Wayne Tank*, where a fairly small defective installation led to destruction of the whole building into which it had been fitted.

Extent of damages

It is the duty at law of a person sustaining damages to do all that is reasonable to mitigate his loss, that is to reduce its impact by suitable countermeasures. He must not allow loss to build up unheeded over a period, if he could stop it or reduce it early on. It may be that his act of mitigation itself introduces extra costs which he may have reimbursed, again subject to them being reasonable. A net saving should therefore be in view, although what is done may turn out with hindsight to be more expensive. The key issue is whether proper judgment was exercised over matters as they appeared in the circumstances prevailing at the time. Several illustrations of the problems are given in Chapter 18 and in case studies in later chapters.

There are many examples of cases in the courts where damages have been awarded for breach of contract in building disputes. Two recent cases illustrate that the courts are reaching rulings that do not slavishly follow restitution for breach, in all cases.

In *Ruxley Electronics Ltd* v. *Forsyth* (1996), which went all the way to the House of Lords on a claim of £21,560, it was decided that damages for loss of amenity of £2,500 were due. The dispute over the construction of a swimming-pool, which was built to a lesser depth than specified, took two years to resolve. A lower court awarded in favour of the contractor, the Court of Appeal held that it was reasonable to allow restitution to the householder of £21,560 as the cost of rebuilding the pool, but the House of Lords reversed this, holding that 'reasonableness' is not confined to the issue of mitigation. Where it is 'unreasonable' for a plaintiff to insist on reinstatement (e.g. where reinstatement would be out of all proportion) then damages should be confined to the difference in value. Contracts for a personal preference are in the same category. As obiter dictum (a side remark of the judge, not crucial to the finding) one of the judges stated that the court has no concern as to how a party uses an award of damages. Once liability for loss is established, any intention as to subsequent use of the amount is not relevant. Failure to achieve a contractual objective does not necessarily mean there is a total failure of the contract. Damages should reflect that the pool was perfectly usable and therefore reinstatement cost was not the measure of damages, but damages for loss of amenity were due.

Earl Freeman v. *Mohammed Niroomand* (1996) followed the *Ruxley* case by stating:

> Damages are not limited to diminution in value of a property or the cost of reinstatement, but may be awarded in an intermediate figure to reflect the loss of amenity or convenience suffered by the householder, or simply the personal

and intangible loss which arises from the fact that he has not obtained what he wanted and what was contracted for, the so-called 'consumer surplus'.

Quantum meruit

Quantum meruit is important in the area of construction disputes because it is often claimed as a method of payment in the context either that a contract does not exist or that the terms for payment in the contract are not appropriate for the conditions under which work was carried out. Quantum meruit means 'as much as he deserves – a reasonable sum'. It is often used as a synonym for quantum valebant 'as much as it is worth'. It is a measure of payment (i) where a contract has not fixed a price or (ii) where, for one reason or another, the contract price is no longer applicable.

There are four conditions when quantum meruit may be applicable:

(a) Where work has been done under a contract without any express agreement as to price.
(b) Where there is an express agreement to pay a 'reasonable price' or a 'reasonable sum'.
(c) Where work is done under a contract which both parties believed to be valid at the time but which is in fact void (i.e. the contract has no legal effect, or it is a nullity).
(d) Where work is done at the request of one party but without an express contract, for instance on the basis of a letter of intent. In these circumstances the claim is for payment under a quasi-contract or one for restitution.

In *British Steel Corporation* v. *Cleveland Bridge & Engineering Co. Ltd* (1981) it was stated:

> In most cases where work is done pursuant to a request contained in a letter of intent, it will not matter whether a contract did or did not come into existence; because if a party who has acted is simply claiming payment, his claim will usually be based upon quantum meruit, and it will make no difference that the claim is contractual or quasi-contractual. A quantum meruit claim straddles the boundaries of what we now call contract and restitution.

Here 'restitution' means an obligation of one party to restore goods, property or money to another and is intended to avert possible injustice. In *McAlpine Humberoak* v. *McDermott International Inc.* (1992) it was held by the lower court that the contract had been frustrated because of the issue of variations in circumstances that frustrated the contract virtually from the start. The Court of Appeal reversed this decision, holding inter alia that where a contract provides a mechanism for the ordering and valuing of variations, it must be impossible to say it is that very act which sets the contract aside. Therefore, the valuation of work should be determined by using the mechanism provided in the contract. There was no provision for liquidated damages in this contract but the court held, in what must be regarded as obiter dictum, that any variation, however trivial, which is issued during a period of culpable delay does not of itself put 'time at large'; see also *Chestermount Properties* for further consideration of this point.

STANDARD PROVISIONS EQUIVALENT TO DAMAGES IN BUILDING CONTRACTS

Liquidated damages

Liquidated damages provisions wear their communal heart on their sleeve. They are a prior arrangement to cope with one form of damage – damage flowing from delay in completion by a contractor, but not usually delay by a subcontractor. By virtue of their special character, they present a number of distinctive features and are dealt with separately later in this chapter. Under the NEC they may be provided to cope with non-performance as well as delay in completion.

Normally in construction contracts they are used only as a means of recompensing an employer over delay by a contractor, whereas the next form of provision equivalent to damages is used solely to recompense a contractor or a subcontractor.

Direct loss and expense or damage

Within the particular sphere of construction contracts, an important distinction is determined between a contract performed, albeit perhaps with difficulty, and a contract broken off. Taking the contractor's aspect, when a contract is performed but with some element of breach from the employer's side, the measure is loss and expense incurred by the contractor in performing his obligations. When he is unable to perform, as in a determination, the measure is what he might have expected to obtain from the contract and has been unable to realise. In later chapters, this leads to a distinction over profit foregone and overheads laid out in each situation. The question of financing charges to cover delayed payment is also discussed.

The term in standard contracts for the former case of breach within the continuing performance of a contract is 'direct loss and/or expense', or something similar. This echoes, and means, the concept of direct rather than consequential, indirect loss when damages are under consideration. It means that care has to be taken to segregate those matters, as the clauses term them, which are or are not the responsibility of the employer at main contract level, or those matters which are or are not the responsibility of contractor or subcontractor at subcontract level. Only these, or a share of the total situation, are the responsibility of the party concerned. 'Loss' is usually related to some return on resources which the contractor is unable to realise, and 'expense' to additional outlay incurred. Fortunately, no segregation of these elements is required, as the practical distinctions are not always easy to sustain. For instance, when productivity of plant drops – a loss of return on existing resources – there is often also the need to spend more on additional plant – an added expense.

The term in standard contracts for the latter case of breach, breach at determination, is 'direct loss and/or damage'. This carries the same essential meaning, although it is used within the provisions in such a way as to distinguish some elements of loss from others. These distinctions are brought out in the respective chapters.

LIQUIDATED DAMAGES

Damages are by their nature generally unliquidated, that is they will turn out to be the amount that a contract-breaker may be due from the other party as a result of that party's action or inaction. It is not possible to foresee the amount of damage that will be done by a breach without knowing its circumstances and extent. A large body of legal decisions and discussion underlies the use of liquidated damages, a device by which the damages payable upon a defined breach of contract are calculated in advance and made a term of the contract. The parties have thus agreed to their inclusion at the level stated. Upon the breach occurring, it is then a case of the party in breach paying the sum in question, without the need for legal proceedings to discover the amount of the liability and therefore how much is due. Needless to say, this simplicity is an understatement of reality. As the cases below illustrate, parties can contest that the liquidated damages which appeared to be deductible in their contract may not in fact be liquidated damages or, if they are, that they should not be deducted in the circumstances obtaining.

Purpose and level

The usual purpose of liquidated damages in main building contracts is to provide against late completion of the works by the contractor (a note on subcontractors is given at the end of this section). Some contracts, for instance the NEC (Chapter 22), have provisions for liquidated damages for non-performance, as well as for failure to complete on time.

In standard contracts, this is their sole purpose and they are usually stated as so much per week or other period. It is therefore assumed that any such period within a total overrun has the same effect on the employer in terms of his loss, any inaccuracy here being outweighed by the simplicity of the system and the ease and promptitude of collection. Inaccuracy may relate to the need to forecast or to accommodate different disturbance effects, according to just when they occur and for how long. The figure inserted also gives the contractor a clear measure of what he has to forego for a given delay on his part. He may even decide whether or not it is worth his while to catch up at a particular stage of the programme!

It is not possible to include just any figure in the contract, at least not one that is too high. If the sum is demonstrably high, it will be treated by the courts as a penalty, a sum put in terrorem, so that it will be set aside, even though included with the agreement of the contractor and whether or not he knew at the time that it was unreasonably high. This distinction was established in *Re Newman, ex parte Capper* (1876), whereas *Law* v. *Redditch Local Board* (1892) established that a single sum in respect of a breach of any scale or character was unenforceable because of its very uncertainty. It remains, however, a question of fact in the actual situation under any contract whether or not an amount is a penalty. The test is whether it is a reasonable pre-estimate of anticipated loss at the time of entering into the contract, as distinct from the actual loss which flows from the breach when it occurs. If it is held to be so, and the matter may be tested, then the courts will not interfere to set the amount aside. If it is set aside, unliquidated damages will probably be substituted by the courts; the employer does not lose his right to damages, only the unduly high level that he sought.

It may be that the effect of reasonable prior calculation, based on the evidence available at the time of entering into the contract, may work in the interests or otherwise of either party if conditions change. For instance, a contract to build a filling-station may have the level of liquidated damages related to a given volume of business that is likely to be lost if the station is late opening. Upgrading of the road by which it stands may dramatically increase the traffic, hence the takings. Conversely, construction of another road may syphon off much of the existing traffic and cut takings. In either case the level of liquidated damages is not affected by hindsight, but remains as contract, if this was reasonable at the time.

The amount of liquidated damages must be a reasonable pre-estimate of anticipated loss, but actual loss need not be demonstrated; see *BFI Group of Companies Limited* v. *DCB Integration Systems Limited* (1978). The damages can still be enforced even if the employer suffers no loss, as in *Clydebank Engineering and Shipbuilding Co.* v. *Don Jose and Castenada* (1905). If the sum is 'extravagant and unconscionable in amount in comparison with the greatest loss that could conceivably be proved to have followed the breach' then it may be held as a penalty and become unenforceable, as in *Dunlop Pneumatic Tyre Co.* v. *New Garage and Motor Co.* (1915). Having said this, the courts have been reluctant to hold that liquidated damages are a penalty, principally because the parties have agreed to them, and it is up to the contractor to prove that they are a penalty, *Finnegan (J. F.) Limited* v. *Community Housing Association* (1993). In this important case it was found that the amount was not a penalty, there was a requirement for a certificate to be issued under clause 24.1 and for this notice to be in writing under clause 24.2. It was stated that the notice should be before the deduction of damages but could be contemporaneous with the deduction. The notice should say whether the employer was claiming payment from the contractor or making a deduction of part or the whole of the damages. These were held to be conditions precedent to the employer deducting liquidated damages under JCT 80. It is not helpful that in *Jarvis Brent Limited* v. *Rowlinson Construction* (1990) the lack of a written notice was held not to affect the employer's right to deduct liquidated damages. Again, the circumstances of each case should be noted. JCT 80 and IFC 84 now have requirements for an employer to state whether he is deducting liquidated damages from a certified amount in a certificate, following the Housing Grants, Construction and Regeneration Act 1996 (Chapter 16).

No case has settled how liquidated damages should be calculated but calculation by means of a formula is possible, as the *Finnegan* v. *Community Housing* case has shown. In that case it was decided the employer had to issue a notice requiring the payment of allowance of damages, and the notice should include the period of overrun and the method of calculation upon which the employer relied for liquidated damages.

JCT contracts have a procedure for deduction of liquidated damages, and this can work to the disadvantage of the employer if the procedure is held not to have been followed. Sometimes compliance with the extension of time clause is not made, as in *Token Construction Co. Limited* v. *Charlton Estates Limited* (1976); in this case deduction was held to be void. When an architect failed to issue a non-completion certificate in *A. Bell & Son (Paddington) Limited* v. *CBF Residential Care and Housing Association* (1989), liquidated damages were not enforceable. The circumstances of each case are important and a full report should, as always, be consulted. If the employer causes delay, except where properly provided for by the extension of time clause, then damages may not be due, as in *Peak Construction Limited* v. *McKinney Foundations Limited* (1976).

In *Bramall and Ogden* v. *Sheffield City Council* (1985) the amount of damages was not properly stated, and this meant the right was lost. Where the amount of damages was stated in the contract to be 'nil', in *Temloc Limited* v. *Errill Properties Limited* (1988), it was held that liquidated damages were not applicable – unliquidated damages probably were. In *Baese Pty Limited* v. *R. A. Bracken Building Pty Limited* (1990), an Australian case, it was held that unliquidated damages could be claimed instead.

These cases emphasise that a provision for liquidated damages needs to operate from one specific date to another, and if there is any uncertainty the right may be lost.

Validity and counter-defences

As so often, the test is reasonableness not precision. Indeed it may not be possible to attain precision, given all the imponderables of a venture. It is no defence against the application of a provision for liquidated damages that the calculation was beset with difficulty and necessarily approximate. The principle is still, Was it done with reasonable care and judgment? If so, it stands. Because this is a calculation of damages, it must be related to the employer's loss and not to what, if anything, the contractor may save by not finishing on time. The employer's loss also does not bear a fixed relation to the contract sum, as the purpose of developments varies so much, although some broad correlation is to be expected on general economic principles of value, cost and return.

It is important that the contractor should be fully aware of the purpose of the project, and so of how reasonable is the level of damages, more particularly when this level is apparently high. Although he agrees the sum in that he knows what is going into the contract, he also knows that the sum may be set aside as a penalty, if it is too high. So if he has not been adequately warned of the purpose, hence indirectly about the likely return on the project, he may assume that his liability is not as drastic as it appears. According to the precise facts, he may be held to be excused from liability, if his ignorance warrants it, following the principles in *Hadley* v. *Baxendale* and other cases discussed above.

Apart from the matter of penalty, several defences against enforcement of liquidated damages exist, especially over the procedural niceties of granting an extension of time. These are discussed in the context of the standard contract provisions in Chapter 11. There is also the general question of the employer failing to meet his obligations under the contract and so causing delay. If he does, in the absence of any modifying contract provisions, he then places time for completion at large, so the contractor may finish within a reasonable but not excessive time, and so the liquidated damages provision becomes ineffective. For this reason, standard contracts regularly include a number of events which are in essence a lapse or even a positive act on the part of the employer, and which lead to an extension of time being granted. Without these provisions, the employer would lose his right to damages (even unliquidated damages) because he has at least contributed to the delay. An extension of time clause is as important to the employer in these circumstances as it is to the contractor, who also needs some of its other, more neutral provisions.

Even this outline suggests that clauses for liquidated damages and extension of time are inserted for the benefit of both employer and contractor, at least in the form in which they occur in standard contracts. In non-standard contracts, if they are inserted for the benefit of one party, they risk being construed contra proferentem, that is against the party putting them forward, as happened in *Peak Construction Ltd* v. *McKinney Foundations Ltd*, referred to above. Here the employer was dilatory in settling problems which had arisen out of defective work by a nominated subcontractor. The employer sought to rely on an event leading to extension of time given as ‘other unavoidable circumstances’. It was held that the employer’s delay was not unavoidable, whereas the form was described as containing ‘the most one-sided, obscurely and ineptly-drafted clauses in the United Kingdom’, which is probably an accolade without equal and not highly to be prized at that!

Provided a standard contract is a consensus document, that is one agreed by a group representing both sides to contracts, this type of risk is not present. Although the JCT forms are safe on this basis, those emanating from one body, such as the GC/Works/1 form (Chapter 20) and the ACA forms (not considered in this book) may well not be.

However, amendments or incautiously inserted provisions may remove the employer’s protection. In *Bramall & Ogden Ltd* v. *Sheffield City Council* (1983), the liquidated damages provisions were construed contra proferentem and held to be unenforceable, even though the employer was recognised to be operating the provisions in a reasonably fair manner. He was declared not to lose his right to seek common law damages. In the JCT appendix, a sum had been inserted at a rate per house for each week of late completion. There was a single completion date and no sectional completion supplement. To circumvent the problem, the employer advanced the argument that a total sum for liquidated damages over the whole contract should be calculated from the available data then apportioned under the routine partial possession provisions, even if those optionally available for sectional completion were not present. This argument was not accepted, on the basis that it gave a single sum which would have become a penalty if the employer had not taken possession progressively.

Subcontracts

In building subcontracts, it is not usual to provide for liquidated damages, but for an amount payable equivalent to loss or damage and to be calculated after the event. This is because the effects of a delay per week may be quite varied in the context of interacting site operations, according to the total length of delay and just what stage the contractor and other subcontractors have reached. It may render the contractor liable to the employer for liquidated damages or to other subcontractors for disturbance costs, or both, as well as causing disturbance costs for the contractor himself. Usually affairs are rather more clear-cut for the employer at main contract level, so liquidated damages are considered appropriate there.

COSTS OF PREPARING CLAIMS FOR EXTENSION OF TIME AND LOSS AND EXPENSE

A contractor (or a subcontractor) may choose to expend some amount of effort in drawing up a claim over one of the major contingencies which the standard forms recognise. Alternatively, through either lack of capacity or a particular need of expertise, he may engage a specialist to draw up a formal claim document or to give advice on how to present a claim in negotiations. Whichever way this particular enterprise is undertaken, it used to be thought that he is not entitled to receive payment in the settlement for doing it.

The reasoning here is that the contracts do not require a contractor to prepare such a document or to engage outside assistance. Contracts and subcontracts generally require the contractor (or subcontractor) to provide the architect or, in the case of loss and expense, the quantity surveyor with limited information in support of a notice or application from which the person acting may calculate the contractual entitlement. The principle is similar in subcontracts, although the subcontractor (except when nominated) will be dealing with the contractor alone, which may colour his approach to settlement. As an example, here is the information under the JCT main form:

(a) *For extension of time*: Particulars of expected effects, an estimate of the expected delay and further information from time to time to keep the architect up to date.
(b) *For loss and expense*: Information to enable the architect to form an opinion in principle on the application, and details of loss and expense as are requested and necessary for the architect or the quantity surveyor to ascertain the amount.

It is then the responsibility of the person dealing with the application to perform the calculations and grant the extension or settle the amount. It is true this gives a rather simplified picture of what the contractor has to do in presenting even what is necessary as a bare minimum, and often he may consider that his case will be better served by presenting as cogent a document as possible. But, just as with the actual final account (of which a loss and expense amount is but part), so with these other exercises, he is due to make an allowance somewhere to comply with his contractual obligations. Although he anticipates the final account, he perhaps does not anticipate these other contingencies, at least not in any precision, but he should prudently make an average allowance in tenders at large.

The matter has been highlighted in *James Longley & Co. Ltd* v. *South West Thames Regional Health Authority* (1983), where the contractor was generally successful but was not allowed the original fees incurred in drawing up a claim. He had, however, incurred subsequent fees from the same source for preparation of expert witness submissions for arbitration, and these were admitted. This follows the usual pattern that costs in presentation in arbitration or court are admissible and follow the award or judgment.

Some slight modification of the position may have come about as a result of *Tate & Lyle Food and Distribution Ltd* v. *Greater London Council* (1982). There it was indicated that additional 'managerial and supervisory resources' expended as a result of the actions of the defendant 'can properly be claimed'. In principle it would appear that these may include costs of assessing loss and expense, if not of preparing a full-scale claim. The actual case failed on this point because the plaintiff put forward a flat percentage, rather than an analysis of expense incurred, but not on the principle involved (Chapter 12). It has also been argued how it is implicit in the reasoning of *Merton* v. *Leach* and *Croudace* v. *Lambeth* that where there is clear authority for the view that the architect's failure to ascertain, or instruct the quantity surveyor to ascertain, the amount of direct loss/expense suffered or incurred by a contractor is a breach of contract for which the employer may be liable in damages if the contractor can establish he has suffered damages as a result of the breach. In the *Croudace* case the judge said:

> Unless it can be successfully maintained by Lambeth that there are no matters in respect of which Croudace are entitled to claim for loss and expense under [what is now JCT clause 26], it necessarily follows that Croudace must have suffered some damage as a result of there being no one to ascertain the amount of their claim – the employer in that case having failed to appoint a successor architect when the named architect retired.

The decision in *Babcock Energy Limited* v. *Lodge Sturtevant Limited* (1994) found in favour of reimbursement of managerial time involved at head office, including the cost of preparing a claim. If a claim is referred to arbitration, an arbitrator must exercise his discretion on costs connected with or leading up to the arbitration. Normally cost will only be awarded as incurred after service of an arbitration notice. But in the circumstances of a particular case it may be that any parts of the costs of preparing a claim document, which ultimately form part of pleadings but are spent before an arbitration notice is issued, may fall into 'costs in contemplation of arbitration'. It is important for a claimant to consider recording these to try to establish recovery at a later date.

The position of the contractor in seeking reimbursement is made rather tougher by the procedures in the current JCT documents. Compared with earlier forms, they make a more explicit allowance for progressive notification and preparation of claims for extension of time and loss and expense. The emphasis is upon the contractor giving notification then either the architect or the quantity surveyor performing the actual calculations, as discussed in later chapters.

CHAPTER 10

DELAY AND DISTURBANCE OF PROGRESS

- Nature of delay and disturbance to programme and progress
- Concurrence of causes of delay and disturbance to programme
- Nature of disturbance
- Some factors related to disturbance
- Patterns of disturbance

This chapter is concerned with the key reasons why the contractor may become entitled to extension of time, may also be entitled to payment for delay if it is for compensable reasons and may also be entitled to payment for expense or, under the JCT arrangements, for loss and expense. For simplicity the JCT terminology, including the use of 'architect', has been employed throughout.

How entitlements are to be arrived at in practice and the underlying provisions of specific forms of contract over the scope and content of those entitlements form the subject of much of the rest of this book. Here the two topics of this chapter are taken in turn before some of their interactions and distinctions are considered.

NATURE OF DELAY AND DISTURBANCE TO PROGRAMME AND PROGRESS

Because the two main topics of this chapter are related and yet need discussing separately, their outline relationship is set down here, at risk of anticipating later comments aimed at refining their definitions and going into subsidiary areas. Both delay and disturbance are the result of causes, active or passive, which impact upon the process of construction in such a way as to alter its intended flow through to completion. The main distinction between them is patent:

(a) Delay is the putting back and possible attenuation of performance of part or all of the contract work by some supervening cause or causes. Delay may also be such as to put back the finishing date for the contract, assuming that no rectifying action is taken. It is possible for it to occur with adequate warning, so that it produces no disturbance, but is absorbed within an orderly rearrangement of activities over a longer period than planned.

(b) Disturbance is the upsetting of the orderly performance of work, again by some supervening cause or causes. Disturbance may occur in such a way that rectifying action to maintain orderly performance is made difficult and to a degree impossible, so that extra cost is incurred. It may exhibit one or both of two characteristics: disruption of activity or prolongation such that the sequence of related activities is upset. Whether one or both occur, it does not necessarily follow that there will be delay sufficient to lead to the finishing date for the whole contract works being put back.

Delay and disturbance are expanded upon later in this chapter.

From these descriptions, it may be inferred that demonstrating either delay or disturbance so exists as to activate the contract mechanisms does not automatically demonstrate the other also exists to do so. They need to be demonstrated separately, frequent though it may be that they both so exist.

It is the underlying position when there is an entire contract (Chapter 2), that the responsibility to complete on time and without extra payment lies prima facie upon the contractor, this being modified only by the express terms of the contract, as usually happens to quite a degree. In this chapter it is relevant to the nature of the causes of either delay or disturbance. These may be placed into one of three groups when allocating responsibility: as stemming variously from the employer, from the contractor and from other sources which are to be regarded as 'neutral' because they are neither of the others, not because they are harmless. According to whether delay or disturbance is in view, some causes are routed differently by particular forms of contract. The tendency is for a greater number of active causes to accrue to delay than to disturbance.

There is clearly a stronger connection of delay with prolongation alone than with disruption alone. Thus, demonstrating that one exists will tend to help towards or even be concurrent with demonstrating that the other exists. Nevertheless, their separation remains an important principle and point of practice.

Delay

Delay as defined above may be caused by the contractor, by the employer or those acting on his behalf, or by others; or it may result from external events, that is external to the contract operation. Delay is valid as leading to extension of time only when it leads to failure to achieve the completion date and not simply when it affects some part of the programme before completion only, that is when delay is caused to activities not upon a critical path for the project and when they themselves do not then become part of a new critical path.

When the contractor is not to be responsible for delay, the contract provides for him to receive an extension of time, but still no reimbursement for the delay in isolation from any other disturbance effect, and even then only for some causes (Chapter 15).

There are stipulations to various extents in contracts over the timing of granting extensions and over related procedures, to lessen the possibility of either party suffering through the laxity of the other, or having events manipulated against his interests. Lists of reasons for granting extensions of time are given in standard contracts and the architect cannot step outside this list when exercising his power to grant extensions (Chapter 15). Little consideration is given in contracts to possible overlaps between causes of delay, when at least one of them may lead to extension, although they must be taken into account in calculating extensions.

Quite a number of issues lie dormant within these basic points and fall to be examined in this and other chapters. In anticipation it may be said that the tidy contractual positions given by contracts are often disturbed in practice by such things as the interaction of several causes of delay – to say nothing of the failures of the participants to carry through the procedures as laid down. It is therefore useful to remember the basic legal positions set out in Chapters 9 and 16, while moving into some of the practical problems, which are what disputes and claims so often concern.

The significance of delay is that it leads either to the contractor being liable to concede liquidated damages to the employer, or to the employer having to concede extension of time for completing the works to the contractor (with or without compensating payment).

When an extension is conceded, or granted, it is usually defined as to be one that is 'reasonable', but little else is stated, owing to the difficulty of drafting such detail. No expression corresponding to 'ascertainment' is used to indicate what should be done, just as for loss and expense. In so far as progress can be seen to occur or not, matters are more easily observed. However, what cannot be done is simply to note the lateness of completion and then grant the whole of this length of time as an extension; there are several reasons for this:

(a) There may be delay due to the contractor's fault, with (b) following as a special case of this, or due to neutral causes not entitling him to an extension.
(b) There may be an overlap between qualifying and non-qualifying causes, needing some examination of responsibility.
(c) Although the contractor's programme for the works is prima facie evidence for what should have happened, it may have been inaccurately prepared so that, for instance, there may have been too much time allowed for earlier activities, leading to an effective delay by the time that a cause for extension crops up.

These are parallels to the difficulties arising with disturbance assessments, where there is also the equivalent problem of not being able to take the contractor's tender or final cost figures as adequate starting-points for ascertainment of loss and expense (Chapters 11, 12 and 17).

Disturbance of regular progress

Disturbance of regular progress is a convenient common term for the concept dealt with by the precise terminology of individual contracts. For instance, some of the contracts taken in this book include as broadly equivalent the following expressions, considered in detail in Chapters 16 and 17:

(a) *JCT clause 26.1*: 'Direct loss and/or expense . . . because the regular progress of the Works or any part thereof . . . [is] materially affected.'
(b) *IFC clause 4.11*: 'Direct loss and/or expense . . . due to . . . the regular progress of the Works or part of the Works being materially affected.'

(c) *GC/Works/1 condition 46(1)*: 'Directly incurs any expense . . . which unavoidably results in the regular progress . . . being materially disrupted or prolonged.'
(d) *ICE clause 31(2)*: 'Delay or cost beyond that reasonably to be foreseen by an experienced contractor.'

Comparable but varying expressions are used in other clauses.

In the case of the NEC contract, note that there is no provision for retrospective dealing with delay and/or disturbance under the comparable form dealing with a contract let on bills of quantities or an activity schedule. There is only provision for the contractor to obtain the project manager's agreement to circumstances being present that will lead to a 'compensation event'. The contractor will then price for this, including all delay, disruption etc, and agree this with the project manager before the work is carried out. The new quotation for the work will therefore not relate to the prices in the contract (or adjustments, say as under JCT 80 clause 13) but will take into account all foreseeable requirements to catch up lost time and 'loss and expense' that may be involved. If the contractor and project manager do not agree, the project manager can instruct the contractor to carry out his instruction and then the adjudicator may be called in.

The term 'materially affected', as an example here, is discussed in Chapter 17 as meaning some substantial impact, greater than the comparatively incidental fluctuations due to ordinary external influences and internal inefficiencies which are experienced in all manner of projects, and especially those which occur easily in the uncertain circumstances of a construction site.

These instances restrict rigorously the influences affecting progress which will qualify as leading to reimbursement of loss and expense. Standard construction contracts differ in the matters which they permit, although the JCT and IFC forms give the same list of matters. All of them are discussed in their respective chapters. They constitute just some of the larger, corresponding list of items qualifying as 'relevant events' (JCT terminology) and so leading to extension of time. In particular, they exclude those neutral causes which secure extension of time. There may also be other factors which entitle the contractor to recompense, but these must be pursued by common law remedies (Chapter 9) if there is resistance to payment. The contract clauses and their procedures cannot then be invoked. These limits to the ideas in this chapter should be carefully noted.

CONCURRENCE OF CAUSES OF DELAY AND DISTURBANCE TO PROGRAMME

When it comes to the assessment, calculation or ascertainment of extension of time or loss and expense, leaving aside any distinctions in these methods of reaching an answer, the exercise is complicated when there is overlap in causes leading to them. This refers not to a single cause leading to both results, although this may happen as well, but to more than one cause occurring within the same time span and also contributing to the same result. This is usually known in this context as concurrency. There is some distinction between delay and disturbance here.

Delay to progress

In the case of delay, the underlying legal position means the employer cannot insist that the contractor shall complete on time if the employer has prevented him from so completing. This is then modified by the presence of express terms for the contractor to receive extension of time for (perhaps among other causes) acts or omissions of the employer or those acting on his behalf, such as the architect. The contractor therefore has the original contract period plus the length of the delay caused to him in which to complete (and, as a corollary to this, the employer has not lost his right to a continuing liquidated damages provision contained in the contract). If there is also delay caused by the contractor or delay in other ways that do not entitle the contractor to extension, this is then to be borne by the contractor. In legal decisions this has been worked out to the effect that the contractor will receive his full due relief from liability for liquidated damages and then bear liability for any excess, which may be only part of the length due to the non-qualifying cause because of the overlap; see *Balfour Beatty Building Ltd* v. *Chestermount Properties Ltd* in Chapter 15. This situation is among those illustrated in Figure 10.1.

This broad position should be compared with the equally broad position in the case of disturbance considered below, where liability runs more according to the circumstances rather than to a single prevailing rule.

Two further considerations have been developed. One of them is work already delayed by the contractor's default, which when performed is subject to further delay that is not his responsibility and is greater due to its actual timing. In *Walter Lawrence & Son Ltd* v. *Commercial Union Properties (UK) Ltd* (1984), delay due to exceptional weather occurred after other delay that was the contractor's responsibility. In consequence, the weather delay was more prolonged than might have been expected. It was held that the weather delay was to be assessed as it occurred and not as some putative shorter delay, as though the other delay had not preceded it. Wider application of this principle tends to support the larger actual extension for a contractor, rather than a reduced one, in a range of appropriate circumstances.

The other consideration is granting an extension of time to the contractor when he is in culpable delay, that is when he has already overrun the original or amended date for completion, especially if the extension is due to some cause like extra work that is now instructed for him to perform. There was some debate over whether the contractor should have the date flowing from his own delay substituted for the due date already fixed and then have this extended by the now pending extension, or whether he should have a comparable, reasonable date substituted because the provisions over liquidated damages are now ineffective. It has now been held quite firmly that the extension should be granted from the amended date last fixed, even though this action is retrospective. This means that any excess due to the contractor's prior default should remain his responsibility, even though the extra work now instructed may well move the contract date for completion back to a date before the actual date when the extra work was instructed; see *Balfour* v. *Chestermount* in Chapter 15.

In this latter instance the courts lean, as elsewhere, towards what may be seen as a commonsense solution. The contractor is still protected from damages in accordance with causes given in the contract, but does not receive what is sometimes termed the 'gross' extension of time, but rather the 'net' extension due solely to those qualifying causes.

When several causes combine to produce delay, some being qualifying causes and some not, the question of the allowable extension of time arises. For this purpose, the causes may be grouped into four categories:

(a) Employer's responsibility leading to extension, such as due to architect's instructions (relevant subcontractors may be included here for present purposes) – qualifying causes that will result in payment to a contractor, sometimes called 'compensable' reasons.
(b) Contractor's responsibility not leading to extension, such as poor programming or organisation – non-qualifying causes, sometimes called 'non-compensable' reasons, which may also result in an employer being able to deduct liquidated damages.
(c) 'Neutral' causes leading to extension, such as strikes or exceptionally adverse weather conditons – qualifying reasons in most standard construction contracts but not leading to compensation or to deduction of liquidated damages.
(d) 'Neutral' causes, such as failure of a domestic supplier, not leading to extension and possibly entitling deduction of liquidated damages.

Of these, (a) and (c) as 'qualifying' causes may be 'relevant events' for an extension of time under JCT contracts, whereas (b) and (d) as 'non-qualifying' causes may not be. The guiding principles for calculating the duration of any extension when more than one cause occurs, are related to these two condensed groupings: qualifying and non-qualifying, sometimes known as compensable and non-compensable reasons or events. Several of the possible combinations are shown in Figure 10.1. They all relate to delay actually caused and making for delayed completion, and are based upon these propositions:

(a) A qualifying cause would have happened independently of a concurrent non-qualifying cause, so the delay which it causes is held to be primary in effect and to be the measure of the extension.
(b) Two or more overlapping causes in the same grouping, qualifying or non-qualifying, may lead to a different overall delay than that calculated as their sum, and this different overall delay will usually be less than the sum, but perhaps more than the length of the greater single delay, leading to the longer extension being granted.
(c) A qualifying delay occurring entirely after a non-qualifying delay, but affected in duration by the 'amended' time at which it occurs, is to be allowed at its actual length, longer or shorter.
(d) A non-qualifying delay occurring entirely after a qualifying delay, and affected as in (c), does not attract even a part extension due its timing.

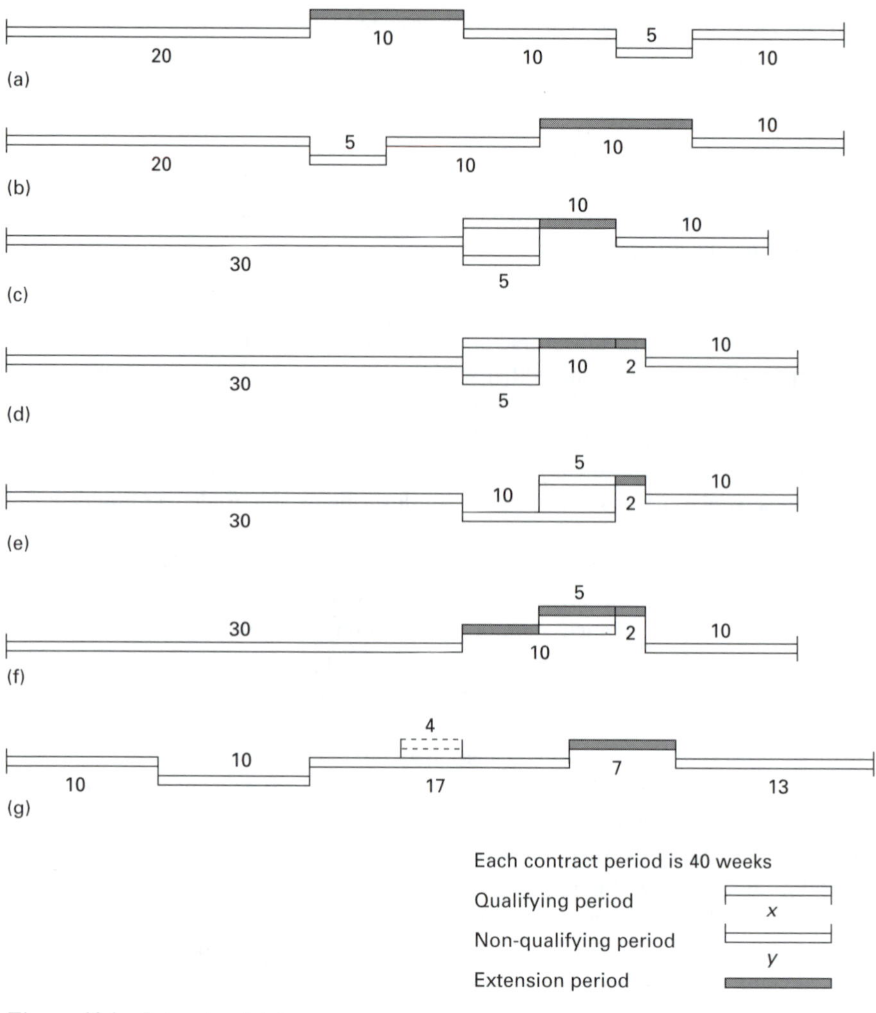

Figure 10.1 Interrelated delays

For greater completeness, two versions of non-overlapping delays are shown first. There is also one illustrating the principle in the legal decision mentioned below. The order in which qualifying and non-qualifying causes occur has no effect on the extension granted, unless the qualifying cause actually falls after the due date for completion has passed, as discussed under *Balfour Beatty Building* v. *Chestermount Properties*. The calculation of a time extension is difficult to show with any realism in a book, because of the limitations on data, but its relationship to a network analysis is shown in Chapter 30. Figure 10.1 shows the effects of delay in a programme of 40 weeks original contract duration following the notation in the figure:

(a) A qualifying delay of 10 weeks and a non-qualifying delay of 5 weeks: *10 weeks extension.*
(b) The same delays as in (a), but in reverse order: *10 weeks extension.*
(c) A qualifying delay of 10 weeks entirely overlapping with a non-qualifying delay of 5 weeks: *10 weeks extension.*
(d) The same delays as in (c), but creating a further 2 weeks delay by their interaction: *12 weeks extension.*
(e) Delays as in (d), but with the qualifying and non-qualifying values reversed: *7 weeks extension.*
(f) Qualifying delays of 10 and 5 weeks respectively, overlapping but creating a further 2 weeks delay by their interaction: *12 weeks extension.*
(g) A qualifying delay which alone would be 4 weeks, preceded by a non-qualifying delay of 10 weeks, and so displaced into a period in the programme when it becomes 7 weeks: *7 weeks extension.*

For (g) see the case of *Walter Lawrence & Son Ltd* v. *Commercial Union Properties (UK) Ltd* (1984), where delay due to exceptional weather occurred after a delay that was the contractor's responsibility. As a result, the weather delay was more prolonged than might have been expected. It was held that the weather delay was to be assessed as it occurred and not as some putative, shorter delay that may have been the delay due to weather if the contractor-caused delay had not preceded it. Wider application of this principle tends to support a larger actual extension for a contractor, rather than a reduced one, in a range of appropriate circumstances. Also, in *Balfour Beatty Building Ltd* v. *Chestermount Properties Ltd* (1993) it was held that an instruction, issued in a period when the contractor was already in delay due to non-excusable, non-qualifying reasons, would entitle a contractor to an extension of time for the period that it would take to carry out the instruction issued. In this the contractor was around 40 weeks in delay and an instruction was then issued to carry out something that would take around 4 weeks. The contractor claimed that, because he could not possibly complete the work now instructed until 4 weeks time, it was illogical he should therefore not receive an extension for the whole contract until 4 weeks hence. It was held that his contract should only be extended by 4 weeks and that the contract would be viewed as having been due for completion say 40 weeks earlier than the date on which the instruction was in fact issued. This became known as the 'net' method of calculation of extensions of time.

Disturbance of progress

Any disturbance that may be reimbursable to a contractor may be complicated by the presence of other, overlapping, disturbance that, while causing loss and expense to the contractor, does not qualify for reimbursement by the employer. Essentially, this may be of two types: (i) items at the contractor's risk (including those which are 'neutral' items, set out in the contract as qualifying for extension of time only) and (ii) items which arise out of the contractor's own inefficiencies. The concept of contractor's inefficiency as related to 'unavoidable' wastage is developed in Chapter 11.

A definitive list of the non-qualifying disturbance items would need to be open at both ends, as the candidates for inclusion are legion. Fortunately, the criterion for meeting the contractor's loss and expense is that an item should appear on the list in one of the contract clauses as something the employer has admitted in advance as an item he will recompense to a contractor if 'the event' or 'matter' occurs. Nothing else qualifies for reimbursement of its effects as contractual loss and expense, except the effect of antiquities when under JCT clause 34, which may be taken as within the present pattern, despite less detail in the clause. It is suggested that it should be similarly treated under other contracts, desirably by the issue of instructions requiring a postponement or a variation.

The question of the overlap of causes in disturbance, or concurrency, is thus more complex than in delay. With delay, the contractor is interested in avoiding paying liquidated damages by obtaining extra time for completion to counter the employer's prevention of due completion, and that single thing is what is on offer, regardless of the existence of other delays, otherwise caused. The essence is identification of the qualifying cause of delay and giving it full weight. With disruption, a contractor wishes to be reimbursed amounts that are entangled with other amounts: at least the amounts which he expected to pay in the absence of disturbance, but possibly also (and significantly) amounts which he did not expect to pay in the sense of planning for them, such as for his own failures in operation. With delay, the issue is the amount of that delay the employer has caused, irrespective of other delays whether or not concurrent. With disturbance, the view moves to the possibility of apportioning liability and then trying to quantify it.

Several possible solutions have been floated. One is to take the first cause of disturbance that occurs and to give it pre-eminence. This may often give an illogical effect in causation, especially when the following cause would have occurred anyway and where its effect is overwhelming in the effect produced. A second solution, when two disruptive causes appear to be approximately equal in effect, is to apportion the effects accordingly. This may be fair at times, but leaves aside the question of culpability: an essentially neutral cause may reduce the impact of one plainly due to the default of the employer.

The third solution, most widely accepted but still not always suitable, is that of the dominant cause. Here the attempt has to be made to identify the cause which overrides any others as leading to the disturbance. This may mean that it has some precedence, not necessarily in time, so conditioning by some form of leverage what happens 'later', that is the effect of other causes. In this case it should be taken in its entirety for its own effects and also to the extent that it modifies the following effects. Alternatively, or even as well, it may be clearly the major cause of disturbance, so that other causes are swapped and become insignificant.

These three options having been set out, it will still be apparent that in a particular situation some overlap of their cogency may occur, so more than one may be in play. In the chapters concerned with ascertainment hereafter, it is implicitly assumed that a dominant cause is always at work, as nothing is to be gained here by introducing a scenario of 'ifs and buts'.

None of this means that there cannot be other issues when the employer is alleged to be liable; but if they do occur, they do not fall to be dealt with by the present machinery. They may be common law breaches, needing appropriate action if the employer does not acknowledge liability and instruct the architect to authorise the quantity surveyor in turn to agree and include the amount in the final account. The prime issue under the present heading is the need to separate out proper causes of loss and expense, as the contractor does not stand to recoup amounts which are his proper risk and responsibility.

It is useful also to refer to the section 'Background use of tender and cost figures' in Chapter 11.

NATURE OF DISTURBANCE

Throughout the JCT and IFC contracts mentioned as the datum for discussion in this book, the noun 'disturbance' is not used as corresponding to 'materially affected', any more than the two other nouns that commonly come into the present reckoning: 'disruption' and 'prolongation', as used specifically in the GC/Works/1 form of contract.

Of these, 'disruption' has the meaning of a more violent and thoroughgoing disturbance, but may be taken as synonymous with a disturbance of orderliness which reaches significant proportions. The other noun 'prolongation' clearly refers to something distinguishable, so for contract purposes, the case must be made out for including it also within 'disturbance of regular progress'. All of the terms are used in the following discussion within the broad bands of meaning that they hold.

Disruption and prolongation

Disruption can take many forms, as the art of claimsmanship demonstrates. It is possible to discern three broad, basic conditions that lead to disruption:

(a) A change in the sequence and pattern of the intended work, without a significant change in what is to be produced eventually, perhaps due to delay in or modification of the information needed to carry out work. (Examples are discovery of the ancient fort in Chapter 30 and discovery of the power cable in Chapter 32.)

(b) The introduction of some significant element of additional work, which may be seen in the finished project, in such a way or at such a time as to cause a change, complication or delay in the sequence and pattern of the original work. It is extremely unlikely that significant omission of work will produce disruption, however else it may lead to financial recompense. (Examples are the annexes, the infusion plant and work due to the floor damage in Chapter 30.)

(c) The introduction of numerous, individually insignificant additions or omissions or other changes of mind, in ways or at times which cumulatively upset the smooth running of the project. (Examples are difficult to give in the nature of things, although numerous interruptions corresponding to the delays under 'Amendment option 1' in the network example on p. 179, would relate.)

The essence of each of these conditions is that the contractor has to absorb the effects into his programme of work when elements of that work are in progress, so he cannot plan in advance how to fit them in smoothly and in the most economical way. Even if he had adequate warning, he might still incur additional costs in performing the original work because of changed conditions, but these costs would be reflected in the rates for variations under JCT clause 13.5 or IFC clause 3.7. They would be due to rearrangement, not disruption, and would fall to be adjusted to allow any additional payment under the valuation provisions of the contract.

By distinction, prolongation occurs without any change in sequence and it changes the pattern only to the extent that it may lengthen the contract period. Disruption may or may not extend the overall completion date, that is lead to prolongation on its own account. Conceivably, disruption need not extend the programme period for even a subsection of the works. It is also quite common to have subsidiary prolongation of a non-critical activity without extension of the overall programme.

The essence of prolongation is a straight move backward of the programme by enforced cessation of some or all work, or alternatively by attenuation of activity due to some stifling effect, such as shortage of enough information to proceed at the usual speed. In its pure form, prolongation does not affect the sequence of work at all, but simply when it happens along the time line. As a cessation of work, it may be known about in advance or it may crop up suddenly, the main difference being the practical one of how resources may be harboured according to the opportunity to redeploy them. As attenuation, it tends to creep up gradually, otherwise a halt might be called by the contractor for economy's sake. In practice prolongation is hardly likely to appear in its completely pure form, but when it does it must be emphasised that it can lead to payment for loss and expense only when the appropriate 'matters' provided in a contract are the cause. For other reasons, given for instance in JCT contracts as 'relevant events' only, there can at best be extension of time only, which is usually seen as a fair apportionment of risk.

Whichever way it comes about on its own, and particularly when compounded with disruption, prolongation does have the effect of disturbing regular progress. Such progress is not to be seen simply as planned and orderly, but also as steady and even. It materially affects regular progress and costs the contractor money, which may be classed as loss and expense in suitable cases. Notably, both clauses allow an instructed 'postponement of any work' (which could be the whole) as a matter; whereas delayed possession of the site clearly does not cause disruption, unless it concerns the ordering process or other concurrent contracts.

There is the possibility of substantial amounts of extra work, fed in as variations instructed well in advance, extending the contract period. Although they should lead to extension of time, they will necessarily not lead to disturbance as considered here, since they have appeared in an orderly manner. All their financial effects should be taken up under the valuation of variations (Chapter 30), including any inflation effects in a fixed price contract.

In general, the single word 'disturbance' is used in subsequent discussion to cover both disruption and prolongation, according to the context.

Some basic examples of disturbance

By way of simplified illustration, two examples are taken here to isolate principles from the above. They raise two points.

One point is that the contractor is under no express, contractual obligation to mitigate loss and expense, as he is required to do over delay leading to extension of time. He is, however, under a general common law duty to do so and not to allow the employer's costs to mount unnecessarily. This does mean, say, that the architect or the quantity surveyor cannot sit back after the event and pare down the loss and expense amount to some theoretical minimum. It just means that the contractor is to use reasonable discretion over the measures which he takes. In general, he is under an obligation 'regularly and diligently [to] proceed with the Works', even if this is more than he needs to do to complete by the due date. It may suggest a bias to make progress, even when this is expensive. Cases on the question of mitigation show that the 'innocent' party suffering damage has a fairly wide discretion in special circumstances.

The second point is that the architect in turn has no power to instruct the contractor positively over what action to take, that is to say he cannot tell him how to organise the construction of the works when things are going wrong, any more than he can when things are going right. There is not even express permission for him to require the contractor to do what is reasonable to achieve some economy (as there is over maintaining progress when there is delay), although the contractor still has the general legal responsibility on his own initiative to mitigate his loss. All that the architect has is the drastic option of holding up work completely by postponement, if this is likely to reduce the excess expenditure.

The point of the contractor putting in an 'application' over loss and expense under the standard contract clauses is to warn an architect or engineer of what is happening, so that he can request any data and keep his own records. It also allows an architect or engineer to take such action as is open to him. This may well be restricted to putting right the things that are causing the loss and expense, if they are within his control, and to instructing postponement. Postponement is desirable only if the benefits are seen to outweigh the disadvantages, as illustrated below.

Structure example

Figure 10.2 shows an egg-crate structure consisting of a slab foundation, with alternate lifts of in situ concrete walls and slabs (they could almost as well be precast) rising from it. The obvious sequence is to put up each lift in turn for the full width of this fairly narrow structure. For some reason, the detailed information for the left-hand ground-floor lift of wall is not available in time. If the contractor proceeds with as much work as he possibly can before he stops due to this shortage, he will erect everything below the heavy line stepping up to the right. When eventually he carries on to completion, he will have performed in all six wall lifts and five slab lifts, and all but the last slab lift will be smaller than the three of each sort which he would originally have performed. This introduces piecemeal working

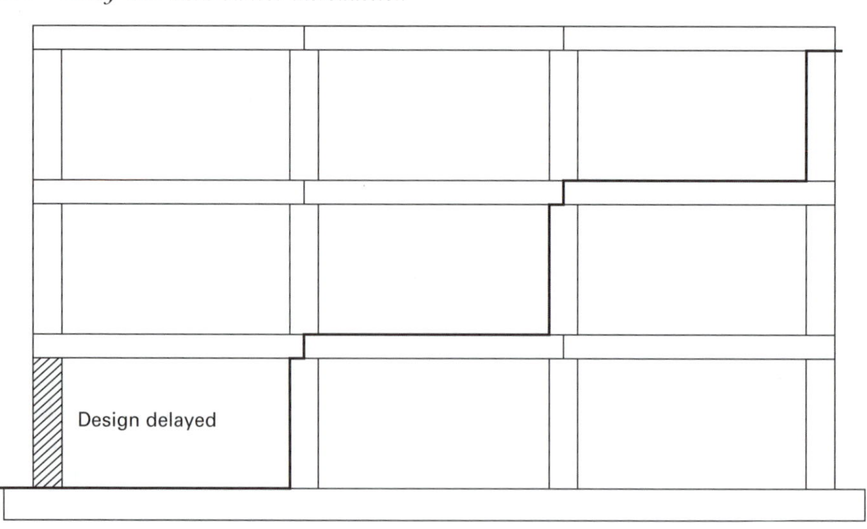

Figure 10.2 Structure example

for labour and plant and, in a more complex example, would carry the risk of the gangs getting in each other's way.

In such a simple case, it may be questioned whether the contractor would not be better advised to wait for the missing information at some intermediate point in his work, perhaps even before performing any walls. To put up the one second-floor wall in particular may be seen as really pushing progress at the expense of cost.

Network example

Figure 10.3 shows an initial network analysis for a project and two options that each allow for a single lapse in information. This is delay in providing the design for operation C.

The first option for the contractor is simply to wait until the information is provided at event 2A, giving a waiting or dummy operation (B). If operations E and H are on the critical path, this may create dummy operations (F) and possibly (H) to absorb delay. The extents will depend upon the float within operation F. Operations J and L are not affected in duration. The finishing event 10 is delayed, so the project is completed late.

On this basis, operations before the dummies are not affected, whereas those following the dummies are simply performed later, with possible inflation and reorganisation costs. There is a general prolongation effect and the dummies represent a disturbance of regular progress by interruption.

The second option for the contractor is to attempt some reorganisation to limit delay. He might find it possible to perform parts of operations J and L (which are not on the critical path) and H (which is), before achieving events 4, 7 and 8. This creates operations J1, J2, L1 and L2, and also H1 and H2, in place of the original

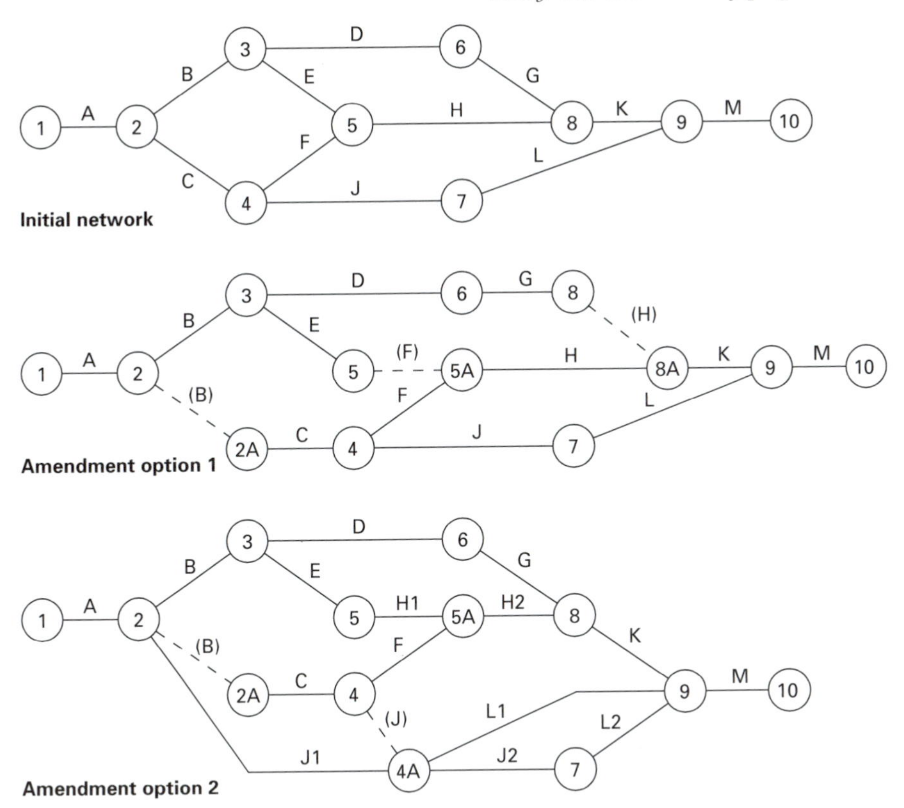

Figure 10.3 Network example: horizontal time intervals are equal

three, with the possibility that extra costs are incurred through smaller-scale working and perhaps otherwise. The programme shows complete recovery of total time, but this may not happen. Even if it does, it gives disturbance without extension of time.

In both options the separation of operations B and C from running in parallel will provide a further extra cost beyond prolongation costs, if they were intended to use common facilities, say batching plant or scaffolding. This is despite them not being closely interrelated at the point of application on-site or in their effect on following operations, as the network logic indicates and as a distinction from the effects on operations J and L.

Equally valid here are comments made on the previous example, the egg-crate structure, concerning what the contractor is obliged to do. The possibility of splitting operations highlights a little more the dilemma between mitigating the employer's extra cost and the contract responsibility to maintain progress, unless an instruction of postponement is given. On balance, the first option suggests a mitigation approach, whereas the second suggests maintenance of progress, within the simplified illustration provided. In practice this conceptual division often cannot be identified with confidence. At most, the architect can suggest action by the contractor, and even here he needs to act with care to avoid running unwittingly into some excessively expensive situation. The question of the contractor's inefficiencies is discussed later in this chapter.

SOME FACTORS RELATED TO DISTURBANCE

Variations of physical work

The effect of variation instructions is to cause a replanning of work in the area concerned. If the replanning is then implemented in an orderly manner, all the financial effects are to be covered in the valuation of variations, including any undue costs of the replanning and the effects on other work associated with that varied, but which remains unvaried. Only direct disturbance costs, or the excess costs of performing work in disturbed conditions are to be allowed as loss and expense.

The distinction is not always easy to maintain, although it is easier than for variations of obligations and restrictions (see next section). There is a special need to deal clearly with such elements as overheads, as some extra for these elements will be included in the valuation of additions variations themselves, a more likely source of disturbance than omissions. This is illustrated in the case study calculations in Chapter 31.

Variations of obligations and restrictions

Under all standard constructions, for instance JCT 80 clause 13.1.2 and IFC 84 clause 3.6.2, the architect may instruct variations of 'any obligations and restrictions imposed . . . in the Contract Bills'. It is argued in Chapter 6 that these clauses limit his power to varying by addition or omission those categories of obligations and restrictions alone which are already given in the contract bills (although this is not the point at issue here).

Just as with variations of physical work, so with this variety the intention is for the immediate valuation to embrace all the financial effects, and it is provided for the related value of unvaried work to be adjusted to take account of the repercussions of the variations. This means valuation which is at the price level of the contract bills, whether by a pro rata method or one of fair valuation, taking care of overheads, profit and other elements in the contract sum. In particular, there may be adjustment of items in the preliminaries and change of allowances for working conditions as originally embodied in measured and other prices. All of this is to take into account the effect of variations instructed early enough for the contractor to reorganise his work so that it is performed regularly and without disturbance of progress. The valuation will include for any reorganisation to this end.

It is worth repeating the list of obligations and restrictions discussed in Chapter 6:

(a) Access to the site or use of any specific parts of the site.
(b) Limitations of working space.
(c) Limitations of working hours.
(d) The execution or completion of the work in any specific order.

In a number of cases the instruction of a variation of one of these items at short notice will have effects on how effectively and efficiently the contractor can go about reorganising his activities, more particularly if there is an intensification of the obligation or restriction. If it is eased, he should be facing lower costs and be able to rearrange his work without disturbance, although much will depend on the length of warning given.

If there is a disturbance effect, it falls to be reimbursed as loss and expense, because it is due to a variation instruction. The distinction between the two categories, variation valuation and ascertainment of loss and expense is important for the reasons given in Chapters 6 and 12. It will be seen, from the nature of the items listed, that segregation of the categories is easier in concept than in practice. If an important criterion of disturbance of regular progress is that work is performed out of sequence, then where is the distinction to be made in the case of item (d) above? The point is illustrated in Chapter 32, where the programme is severely modified without clear reference to obligations and restrictions.

Acceleration of the works

There is no contractual provision for acceleration anywhere in either the JCT or the IFC form. Indeed only postponement and extension of time are envisaged: things either stay the same or get worse! In the NEC form there is provision for acceleration to be requested and the intention of the form seems to be that the contractor is expected to accede to the request and give in effect an all-embracing quotation. ICE and GC/Works/1 contracts do also require a contractor to accelerate, but it could be discussed to seek agreement so to do. Normally this must be very carefully agreed so that payment for acceleration is seen strictly against controllable circumstance, otherwise disputes may easily arise if acceleration is not achieved, despite extra payments having been made for it.

It therefore follows that no acceleration-related effect appears among the matters leading to loss and expense in standard construction contracts. If the employer wishes to advance the date for completion, or simply to regain lost time, he can only approach the contractor as to whether he is prepared to modify a term of the contract on some negotiated financial basis. The contractor does not have to agree to such a thing at all.

A similar position exists over any element of acceleration cost put forward by the contractor within other loss and expense items. There is no need for him to work overtime beyond what is standard on the site to catch up, and any argument that he had to work overtime for some other reason should be investigated with care. Suggestions that it was or is necessary to attract or retain labour are a regular instance. One area where some sensible margin may exist is presented by an attempt to mitigate a loss and so the employer's payment, but even here the contractor is well advised to secure agreement to what he proposes before he implements any policy. In the network example given earlier the choice of options over whether to rephase work could involve some incidental overtime which might well be justified as part of an amount.

Elements of cost such as additional supervision may sometimes edge into this grey area, but are usually less clearly straight cases of acceleration expense. They should be considered equally carefully just the same.

PATTERNS OF DISTURBANCE

Although disturbance may be seen as disruption or prolongation, separately or together, it occurs in various patterns. Again, these patterns are by no means entirely discrete occurrences, but frequently they are blurred combinations of the matters and other causes.

Distinguishing these patterns may be unnecessary, but often the key to a solution of a loss and expense amount may be in the way that the causal matters are linked or have occurred. This is particularly true when responsibility over causes is divided.

For many, these patterns may be so self-evident as to make their setting out rather trite. However, with respect, it is suggested that the ways in which the contractor's working system may be disrupted are not always considered by some who may be concerned with the evaluation of loss and expense. Someone who is aware by education and experience of the systems within which effective briefing, evaluation and design proceed, will know how such activities can be nullified by random inputs from outside the system, even when actual construction proceeds later in an orderly manner. Equally, someone who understands the need for a disciplined methodology in cost control on behalf of the client within the design process and thereafter, knows the ways in which runaway decisions made in isolation from other team members can wreak havoc, even when construction is not disturbed. The former gives an inefficient design, the latter weak cost control. But neither need mean excessive payment for what is produced.

But just as these areas of activity differ from one another and have their own problems, so is the construction activity different from either. It is also the most inflexible, in that it is no longer a straightforward case of jettisoning and substituting ideas, but of altering physical work or halting and rearranging a complex of people and machines. Following operations are due at dates which mean that deliveries of materials etc are already in hand. To add to the potential for confusion, this organisation on and off site consists of a whole array of firms, each with a different perception of success in the operation. They have different goals and means of making profit, they have other work in different places and their timescales of achievement vary. They are not all committed to the success of the project as a whole. Therefore, those out of the direct field of building production need to sit back and visualise or, better, analyse closely what is happening on-site.

Two ways of classifying disturbance are given here. The first shows in graphic form some of the overall effects which result from particular disturbances, whereas the second describes some of the ways in which matters can crop up to produce these effects. The two ways are complementary and the latter may be fed into the former to achieve one of the many syntheses that are possible.

The first is set out in Figure 10.4 in a succession of diagrams, each representing the programme of a project in block form:

(a) This shows the original programme as originally anticipated and as achieved without any problems, as occur in each of the instances following. This is perhaps the most unexpected version of all.
(b) Here are two patches of pure prolongation, without disruption. In the earlier patch there is postponement which brings all work to a halt at once and then it starts again. According to how much warning there is and what is available elsewhere, the effects could be appropriately minimised by taking resources away from the site to other useful work. In the later patch, one or more matters lead to work being slowed down. The absence of disruption here is somewhat unusual, but conceivable. Extension of time would flow if the causes lay within the contract terms.
(c) This is the other extreme of pure disruption without prolongation, so that the contract period is not extended. This situation is perhaps even more unusual than in diagram (b), but again conceivable, the more so if the disruption affects only elements not on the critical path of the project. An attenuation not on the critical path is also possible and could lead to reimbursement for loss and expense.
(d) A further possibility is shown here in its simplest form. This is the introduction of major extra work in one piece, so that is displaces by its own duration all outstanding work and also causes disruption to the earlier parts of that work. A far more likely (but graphically more complex) version of this would be for original contract work to go on partly alongside the extra work, but still to be disrupted by it. There might well be a prolongation effect as well. At any rate, the effects here would be fairly clear to identify.
(e) This shows a related pattern to the more likely version of diagram (d), that is the effect of a number of minor matters (of whatever type) which cause both prolongation and disruption spread at large across an extensive section of the programme.
(f) This gives a variation on pattern (e) by having two breaks in progress, so that part of the disturbance is quite obvious. Other effects either flow from these breaks or are caused by other matters of a more minor nature; these more minor matters are intermingled, so they are more difficult to identify.

Of these diagrams. (e) and (f) may be seen as illustrating situations where considerable care is needed in keeping records and segregating costs clearly and cogently. Or where justified, they may lead towards the global claim situation (Chapter 14).

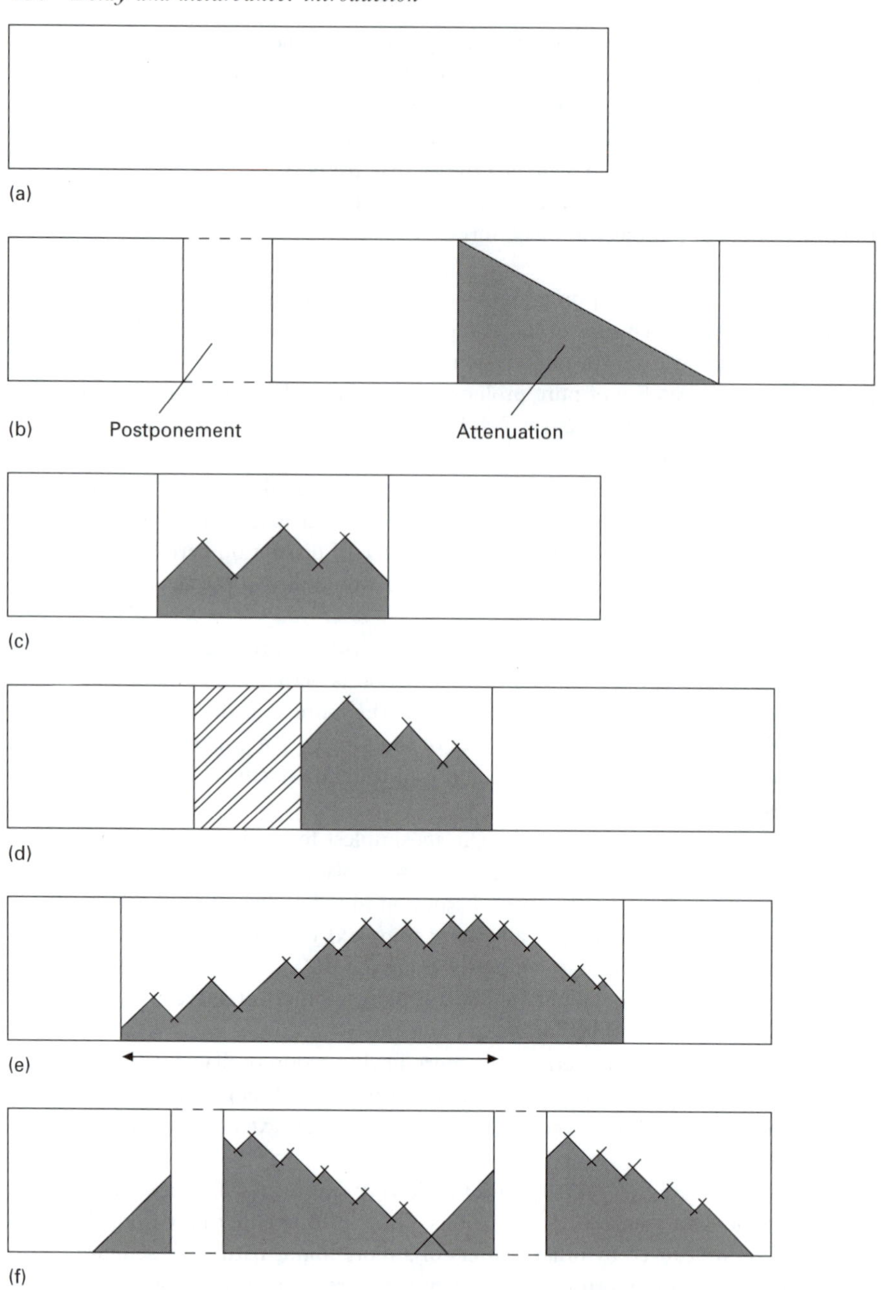

Figure 10.4 Some patterns of prolongation and disruption resulting from postponement and disturbance: (a) no delay, no disturbance; (b) prolongation; (c) disruption of original work but without prolongation; (d) introduction of major extra work; (e) cumulative minor disturbances with prolongation; (f) prolongation with stop-go effect

The rest of this chapter examines the second and descriptive way to classify patterns of disturbance.

Single self-contained matters

The structure example and the network example earlier in this chapter both fall into the category of single self-contained matters. One mishap occurs and its effects run their course without interaction with any other problem. As the examples show, the effects may be fairly complex with ripple sub-effects; but once the original matter comes about, the results are predictable within close limits. If the effects work themselves out fully and the contractor reorganises his programme in their wake, any subsequent single disturbance will also have quite separate effects.

Series of self-contained matters

This category may be defined as where two or more matters occur quite separately in time, but close enough for their effects to overlap, so that there is some compounding of them. A later matter starts off 'on the wrong foot' in that its effects are injected into an already disturbed situation. This means that the countermeasures which the contractor is taking to restore regularity to the programme are upset before they have become fully effective.

This pattern is likely to become most troublesome when the matters are individually below the contract threshold of 'materially affecting regular progress', but together build up into a significant effect. For instance, the contractor receives a relatively minor variation instruction, followed in turn by late receipt of information and by an instruction to postpone some work. They are in three distinct physical areas of the work, as well as distinct in time, but they lead to quite considerable disturbance in a fourth sector.

An insidious version of this pattern occurs when an almost interminable series of such matters crops up through a project, causing major disturbance incrementally. The problem for the contractor is to know when affairs are building up to the 'material' stage, particularly when delayed effects are being produced. He needs to rely on such words as 'has been' if he is a bit late, or 'likely to be' if he is a bit early. On the whole, he is better to be early, but without running into the alternative trap of crying wolf by overreacting to every little pinprick.

In view of the practical sequence of events here, the contractor may be advised to approach the architect to agree some rationalisation of a potential maze of separate applications over, for instance, every late piece of information. It could be established that a near-constant shift into lateness of requested information is to constitute one matter for purposes of the procedures. This fits with the wording of the clauses which refers to matters as groups, such as delayed instructions, as well as the reference to making an application relating to 'one or more of the matters'. For any one matter or group, once established, it is then necessary to make only one application to cover any effects, past and future. When the same category of matter recurs at widely spaced intervals, the system of separate applications should still be followed.

Group or hierarchy of matters

Another conceptual option is for matters to occur at about the same time, so that more than just their effects overlap. Several separate matters may act together to produce a greater effect together than they would have done separately. On the principle of Murphy's law, they never manage to cancel each other out! Such matters may just be of equal status, or there may be one or more governing matters, which control the others and affect their severity.

The remarks just made about establishing some relationship among the applications are valid here also, although the opportunity to present them as one will be more apparent, assuming the contractor retains his sanity!

Total blanket of matters

There comes a point at which any formal categorisation breaks down. In the present context, this happens when it becomes practically impossible to separate out the effects of matters in any sure or meaningful way. A barrage of matters may simply swamp the project, with interactive effects which cannot be logged. It may be described as any combination of the foregoing classes which appears to fit, but description does not aid analysis or, ultimately, ascertainment of the loss and expense incurred.

Although this case may come, every attempt must be made to keep heads above water, in the interests of progress and of retaining a basis for settlement. At some point a contract becomes subverted, so that time and money almost become at large: this needs to be avoided if possible. If it does come about, the architect must seek the agreement of the employer to any change of basis, however inevitable it may be.

PART 4

DELAY AND DISTURBANCE: PRACTICAL APPROACHES

CHAPTER 11

ASCERTAINMENT OF PRIMARY LOSS AND EXPENSE

- Some basic considerations
- Labour and plant
- Materials
- Site overheads

It is convenient to divide loss and expense into two categories: primary and secondary. Although secondary expense is reserved for Chapter 12, there are a number of considerations introduced here which spill over into that chapter, so the two should be considered together. Specific discussion of JCT provisions appears in Chapter 17.

SOME BASIC CONSIDERATIONS

Elements of cost affected by loss and expense

In discussing the concept that loss and expense are to be ascertained, it is necessary to identify the main elements which most commonly enter into their composition and which are discussed later on:

(1) Primary
  (a) Labour
  (b) Plant and equipment
  (c) Materials
  (d) Site overheads, such as supervision and temporary structures
(2) Secondary
  (e) General or head office overheads
  (f) Profit relating to loss of other business
  (g) Financing and interest charges in particular situations of delayed ascertainment and payment of amounts
  (h) Extra inflation or fluctuations costs

In principle each amount for loss and expense should be arrived at by isolating and allocating some part of (a) to (h) that has been increased or incurred in the circumstances.

It is plain that only the first four can possibly be observed on-site whatever the circumstances. The others regularly need to be demonstrated by less direct means and are likely to involve more intricate argument over entitlement. This problem is treated at the beginning of Chapter 12 in relation to the more 'hidden' or 'global' nature of the elements considered in that chapter.

Items (a) to (d) are also the ones where the temptation to resort to the original tender levels is the most seductive, or where some practical use of them creeps in most easily without always being observed. This is because, as direct costs, they can often be calculated on some quantitative basis for loss and expense purposes, and they are also known to exist for the build-up of prices in the contract bills. They thus offer themselves more obviously for analytical comparison. This is somewhat illusory, as the division between them and general overheads and profit remains the problematic element in any analysis of prices. The issue is expanded below.

Three starting-points

From what is said elsewhere (Chapters 10, 12 and 17) about the broad question of loss and expense, it may be seen that this concept relates to the actual amount by which the contractor is out of pocket, compared with where he would have been had the element of disturbance not occurred. It is the equivalent of common law damages and not a payment for extra work or expenditure, as is reimbursement for a variation. It should therefore leave him with the same total profit as he would have had, again if disturbance had not occurred. It is also distinct from allowing a quantum meruit to either party, as explained in Chapter 9.

The problem with ascertaining an amount for loss and expense is that it is not waiting 'out there' to be found in the same way that the amount of a variation may be. In that case the contractor is paid for a measured variation on the basis of the product of quantity and price, with the quantity objectively present on the site and the price in the contract bills or deduced from them at least fairly objectively. Whether the result reflects the contractor's actual expenditure is another question, but this is the accepted method within the contract system for most standard construction contracts. (Chapter 22 discusses the NEC contract and its departure from this method.) By distinction, loss and expense occur in the cracks between the various events which should have taken place, in the hiatuses when nothing is happening, or when something is happening unevenly, out of the normal control assumed when the contract was made. It is a form of waste and so is not fully visible and open to detection as an entity. It certainly is not measurable: one cannot say, 'Here it is – and there, get hold of it and add it up in isolation from the productive bits.' At completion of a contract, extra work issued as variations can generally be seen, but by definition, loss and expense will not be seen – it is inefficiency.

Although the courts are charged with the task of awarding damages which reflect the loss and expense concept (even if they describe it slightly differently), there should be no illusions about judges having some judicial crystal ball which renders discharge of the task accurate to the point of infallibility. Judges are mere mortals too, who have to use whatever guidelines are available, as outlined in Chapter 9. At the end, in a case of divided responsibility, they may still be found saying, 'Sixty per cent damages are awarded to the plaintiff', with even the total open to some uncertainty.

Having said that, there are three datum levels which might be considered when ascertaining loss and expense in a building contract:

(a) The contract sum, which is analysed by the contract bills or similar documents into a particular distribution or loading of the figures, and which is adjusted by the final account to allow for variations etc, even without any loss and expense amounts. The original distribution may have been done for ease of tendering or with an eye to cash flow or the pricing of variations (Chapter 3). There may even have been a claims situation in mind. At any rate, this represents in its totality what the contractor hoped would be his costs plus profit, but not the detailed reality of events in terms of costs.
(b) The contractor's actual costs as they would have been had no disturbance of progress occurred. These include any elements of extra efficiency compared with the tendered expectations, but also any elements of greater inefficiency. The extent of the disturbance will govern how far these costs are confused within the total body of costs that have occurred, or how far they have been changed into other costs in substitution for what would have been incurred.
(c) The contractor's actual costs as incurred, inclusive of the elements in (b) and also of loss and expense.

Each level has its problems. The first is not reality, in its parts or necessarily its total, but a hypothetical construct for commercial purposes. As such, logically it cannot enter into a precise calculation of a damages-equivalent figure: real 'out there' loss and expense. To take the difference between it and the actual costs as loss and expense would be to reduce the reimbursement formula in the contract to one of prime cost. The result simplifies to

Total payment = (a) + [(c) – (a)] = (c)

The second level is a reality which did not actually materialise. As such, it is forever beyond reach. How far it differs from the third, which is the amalgam of everything actual, will depend on the impact of the disturbances that have occurred. If it were accessible via the crystal ball, then the difference between it and the third level would represent one approach to loss and expense, subject to some qualifications. One of the more important of these, discussed hereafter, is the problem of how to allocate overheads and how far they should differ between the two levels. The other is that the levels of efficiency and inefficiency (see next section) may vary between the two, although it is probably fair comment in most cases that any divergence should be attributed to the disturbance itself leading to extra chaos or greater discipline. But alas, the second level is missing by definition, so this approach is impossible in any precise form.

It appears highly unlikely that tenderers could make any reasonable allowances in their tenders for the effects of disturbance, were they told of them in advance, although this is what the GC/Works/1 contract does over perhaps smaller-scale expense due to variation instructions. By 'reasonable allowances' here is meant amounts which would be comparable within the normal tolerances of estimating. This unlikelihood exists, even allowing for the possibility of anticipating and damping down some of the consequences. This is another indication of the problem of sorting out affairs in the circumstances of actual, but (worse) unexpected disturbance.

Inefficiency

Before following the main theme through, the question of inefficiency may be pursued a little further. In itself, it is usually viewed as something which happens on-site due to indifferent or poor management, but it may better be seen as representative of a number of factors, broadly falling under the heading of inaccuracy and imperfect operation.

Like many other matters, inefficiency is a relative affair. Hindsight will show numerous times and avenues for greater efficiency than was achieved, whereas any sensible tender and programme (which is the reverse of hindsight) make allowances for 'contingencies', 'float' and other titles which cover inefficiency, remedial work, overoptimism, error or whatever the causes may be termed. In any assessment which depends upon looking at what has actually happened, there is therefore a strong argument for allowing some margin for these elements. It must be assumed that the contractor, in tendering and in operating on-site, will have made some allowance for the inevitable aberrations. If these then are the cost to the employer of progress unimpeded by him, it is unreasonable to extract them from the reckoning when he has impeded progress. Following this line of argument, it is suggested that some normal level of wastage should be made in calculations where appropriate, as much for time-related elements like labour and plant, as is routine for materials.

As before, this suggestion is more easily made than implemented, as inefficiency, or for that matter efficiency, is not readily to be isolated or quantified. The step should not be made uncritically from the productivity assumed in tendering, to what occurs in conditions of disturbance, because the former is dependent upon or masked by all sorts of factors. These include:

(a) Inefficiency assumed.
(b) Bonus payments against hourly rates.
(c) Keenness of tender, affecting the trade-off between basic costs and profit.
(d) Loading of areas of pricing for reasons other than strict 'scientific' estimating.
(e) Estimating errors.
(f) How certain any analysis of tendered rates may be, which is another way of stating the foregoing points as a group, but also in terms of data for performing the analysis, such as apportioning between labour, plant, materials and overheads.

With these reservations, it must be admitted that there are occasions when practitioners use some degree of analysis of what was intended in order to achieve one side of the balance in a settlement of loss and expense. This may be ignorance of the principles, but equally well it may be the need to use some base rather than none. Sometimes this is done without it being made explicit, as explained in later discussion of detailed calculations.

Any approach which uses 'adjusted average' times etc for operations performed in particular situations is effectively leaning upon distilled experience derived from many projects. This may be done when loss and expense is being assessed, just as much as when tendering. In neither case is 'reality' being invoked directly, as the principles underlying ascertainment properly require. It is one step further away to use the narrower database of the particular contract currently operating, and open to strong objections for the reasons listed above. But there are occasions when it may form a useful cross-check on what is being done by other means (Chapter 25).

If any concession to the inevitability of using such information is made, the effect of 'inaccuracies' or 'inefficiencies' must be included or excluded on both sides of the balance. Two crude, but strictly theoretical, examples illustrate the position:

(a) Original figures from the tender are used for an element of loss and expense, and a total cost of what happened is also used. How directly one is set against the other is not an issue here. If inefficiencies are detected in the cost, they should be allowed if their level is reasonable, which it may be assumed is also allowed in the tender. But they should be deducted if they are excessive, even when there is disturbance of regular progress.
(b) An amount for loss and expense is calculated by taking some original figure and allowing 30% of it. If the original figure includes an inefficiency allowance of 3%, the loss and expense percentage should have been abated proportionately to avoid increasing the intrinsic amount for inefficiency, unless the conditions of disturbance somehow warrant an increase. Inefficiency within loss and expense is a suspect concept.

The crudity of these examples is obvious, while the theoretical aspect arises because of the assumption that inefficiency can be readily segregated from disturbance.

What is not being argued for here is the approach which says, 'Let's see what has been lost, and then dress up the results to look like a genuine ascertainment of loss and expense.' This is just the prime cost method, rejected above. It comes about when the whole contract is so subverted that the parties agree on the need to introduce a different basis. This means the substitution of a new contract for the old, and so is outside the bounds of this book, which deals with matters inside a given and continuing contract. The argument here is for a limited allowance being made for normal working slacknesses, which the employer pays for in the give and take of tender pricing based originally upon feedback from site.

Background use of tender and cost figures

Background figures are also considered in Chapter 25, but from a different viewpoint. Both initial and final figures have a value as regulative of the bounds of possibility over loss and expense. This is useful for the architect or the quantity surveyor, whoever happens to be ascertaining amounts. It may also help the contractor, if he is thinking of flying a kite!

Provided the tender figures are reasonably distributed and provided the keenness of the tender has been assessed, they can be used in conjunction with the final account, which is derived from them, to give a guiding framework within or around which the total costs should have been incurred. The detail will become progressively less reliable for this purpose, as it is chopped finer. The more the amount relates to the contract as a whole, because there has been widespread disturbance, the more use the figures will be, on the 'swings and roundabouts' principle. This is likely to apply to a rolled-up claim (Chapters 17 and 14) or one where there are just one or two major and closely related parts. In any instance it is necessary to analyse out sub-assessments for labour and plant, as the major elements subject to stress on-site. These should usually be taken together, as the original intentions for using one or the other is unknown or may have been changed. The ebb and flow between the two is one of the uncertainties in dealing with loss and expense.

The approach may still be of use for assessing the feasibility of amounts relating to sections of work, provided they can be isolated quite rigidly either on the site layout or in the construction programme, that is in space or time, but preferably both. Considerable care is needed to ensure that common facilities etc are allowed in a rational manner, as they may not show separately in the contract bills. Once there are site costs which run on from one period to another, between several sections of work in a straggly manner, the corresponding initial figures will become much less useful. Within such a framework, for instance, it may be possible to demonstrate that a suggested drop in productivity for some part of the works represents an unduly large increase in costs in percentage terms, when set against what presumably has been achieved on the rest of the works.

This approach becomes more valuable when used in conjunction with a review of the total costs actually incurred. If it is instituted by the quantity surveyor, say, it means asking the contractor for quite an amount of cost information which would not normally be forthcoming, but which falls within the terms of what the contract clauses permit. In fact, the decision about what is needed rests with the one who is carrying out the ascertainment, subject to the requested information being reasonable. In the case of major disturbance, the contractor could be asked for all of his cost framework, to be supported later by any detail required.

This information usually gives a different distribution from what is found in the bills and final account. The bills (and to some degree the final account) split the works into finished products, such as buildings and external works, themselves divided into work sections or similar; whereas the cost detail splits the costs into types of inputs, each of which may have gone into various parts of the works.

Input information comes into play in considering whether particular suggested excess costs fall within the right range. It is most valuable when discrete and identifiable parcels such as single categories of plant are involved, which are used widely over both the disturbed and the undisturbed areas of the works.

Suppose the analysed bill and final account information shows that these two areas (both taken as undisturbed) should have absorbed about one-third and two-thirds respectively of the plant category. If these figures in practice show a moderate swing, matters may be feasible, given strong disturbance and remembering that what comes off one side in a swing goes on the other, as a gearing effect. If they are reversed, whatever their intrinsic values and relationship to those extracted from the contract accounts, there is a factor of 4 change and something catastrophic must have occurred in practice – or in drawing up the figures! At least, a broad control has been established, before too much effort has been expended by going off in the wrong direction. A slightly different illustration, based solely upon costs, is given in Chapter 25.

The value of anything beyond the global figures depends on how efficient and reliable the contractor's costing system is in presenting feedback. He certainly has no incentive to hold anything back, as far as his total expenditure is concerned – the more the merrier. Only if some element reveals the wrong thing can he wonder whether to keep it to himself: what aids loss and expense may have repercussions elsewhere! It is possible for a whole sweep of information of this type to be fabricated, but it becomes very difficult to do this convincingly in advance of events and not knowing in which direction they may take off. Even leaving something out tends to show. Commonly these records, at the micro level of what individual workpeople and cost clerks produce, present many idiosyncrasies (as a convenient euphemism) which attest their reality.

Again, this approach becomes less useful as the section under review becomes smaller. It tends to be least useful when the section exists alongside others being performed at the same time, so that misallocation by laxity or a philosophy not concerned with loss and expense may occur. It is also most useful when it is performed by looking retrospectively at data, so that it has probably been prepared without thought to its use in this way, or alternatively has been prepared by the contractor and the quantity surveyor acting closely together to ensure its suitability and reliability.

But given the problems of both tender and cost data, they may be seen as cutting the cake in different directions. This is useful, if it is remembered that the dissection does not involve precisely the same cake, or at least not the same tier in the layer cake of price and cost. To modify the analogy, at least some grid lines have been ruled on the map, even if they are somewhat blurred and wavering.

It bears repeating that neither of these sets can be used to give a final and definitive amount of loss and expense, but simply to establish guidelines for closer work.

A 'what if' example of background figures

Here is a simplified example of the data which may become available during a survey such as has been suggested above, along with some of the questions which may be asked. An outline of what to look for next is sketched, but no attempt has been made to arrive at answers, as they will require a deeper examination, using the regular means of ascertainment with the present figures suggesting the best directions in which to look.

(a) The measured final account for the contract, including other elements such as daywork, gives a total of £300,000.

(b) The claim under scrutiny contains total amounts of £20,000 for extra labour and £10,000 for extra plant, not impossible at one-tenth of the whole account, or even of the net labour and plant content.
(c) However, the affected elements of the final account are isolated at a total of £100,000.
(d) Within these elements, the net labour content is £25,000 and the plant is £15,000; both are based upon an approximate analysis of the rates, largely those used in the tender or derived from them by regular means. They are referred to below as 'tender inclusions', because of their level.
(e) How reasonable these rates are may be viewed against the level of the tender overall and the consistency of the rates from section to section, as hopefully has been noted when the tender was examined. Any comparison now performed should be based upon what would have been included in the tender at a competitive but profitable level.
(f) An analysis of costs incurred, taken from the contractor's records, gives a net labour amount of £35,000 and a net plant amount of £20,000.

These figures may be compared as follows.

	Labour	Plant
Tender inclusions	25,000	15,000
Extra amounts in claim	20,000	10,000
Percentage increases	80%	67%
Totals of tender and claims	45,000	25,000
Costs incurred	35,000	20,000
Claim as percentage of last	57%	50%

The costs incurred happen to be precisely halfway between the tender inclusions and the tender and claims together, not that this affects the principle of what should be done. Some of the effects of varied levels may be noted:

(a) To hold to the claim figures for present purposes (i.e. without further enquiry into the merits of the claim as such), it would be necessary for the tender inclusions to be re-evaluated and drop to £15,000 and £5,000, but this would increase the percentage increases against the earlier figures to 133% for labour and 200% for plant. The previous figures look high, but these begin to look dramatic in the extreme and so much easier to evaluate in the round against what has happened on the works, and the reasons for the percentage changes not being the same may be sought, possible though they may be.
(b) To hold to the tender and cost figures, it would be necessary for the claim figures to drop to £10,000 and £5,000 without further enquiry.
(c) Had the cost figures been much closer to the tender plus claim figures, it would have suggested that the claim was at least possible in figures, whatever its strengths otherwise.

(d) Had the whole area of the final account been involved for the claim, while the claim had remained at the same level, a similar but more feasible effect would have obtained, as a function of the smaller proportionate amounts of the claims figures, so making the present line of enquiry less useful.

This illustration is by no means profound, but the substitution of alternative 'what if' figures helps to set markers for the next stage. This type of survey should relate either to a large discrete area of the project where the final account and costs permit overall comparison, or to a smaller section where distinctive types of labour or plant are involved for that section and at that time, but not elsewhere in the project. These two are similar in principle.

The discussion of comprehensive claim presentation in Chapter 26 may be read alongside this section.

LABOUR AND PLANT

Two elements of the contractor's costs are linked here because they are interchangeable within limits, on-site and elsewhere, as has been suggested in passing in the last example. Thus, the contractor may suffer because he is forced to exchange one input for the other and at greater cost in the circumstances in which he has to perform the works. More probably, this would be labour used instead of plant, because of such problems as piecemeal working or restricted access or space.

Labour

The distinguishing features of labour for present purposes are that it is engaged by the period of time, that it cannot be 'stored' when its use is temporarily halted and, above all, that it is human. (For convenience, labour is termed 'it' here, because to the extent that it costs money it may be narrowed to a commodity.) Plant is similar on the first count, although it may actually be owned, and it can be semi-stored when not in use, while still costing money. It misses out on the last count.

Any period of engaging labour, assuming it is not in the permanent employ of the contractor, is preceded by some effort and time to secure it. Equally, it needs time to disengage it. In the meantime there is a period of building up the workforce, numerically and in its familiarity with the precise task or tasks on the particular site. (Again for convenience, 'site' is used to cover all workplaces.) Within 'task or tasks' is concealed the facts peculiar to the project about how far there is repetition and how intricate the work happens to be. It is the intention of building contracts cast in the standard forms that, once the contractor has assessed these elements and tendered upon them, he should be allowed to perform what he has priced without disturbance and without change of the contract sum, other than by defined occurrences, such as instructed variations and allowable market price changes.

The loss and expense provisions are included to cope with disturbance of regular progress, which upsets the contractor's rhythm of working, be it otherwise smooth or bitty, as he properly anticipated it and provided the disturbance is due to a proper 'matter' (Chapter 9), as set out in the contract. Here are some of the most common and important labour items put forward as loss and expense:

(a) Drops in productivity due to disorganisation, frustration etc.
(b) Loss of bonuses due to inability to meet targets anticipated in the contract prices, and similarly of overtime payments.
(c) Standing time when nothing can be done.
(d) Extra activities, not covered by the original contract or by variations etc, including reorganising activities to meet change.
(e) Familiarisation time for picking up fresh aspects, including time for engaging fresh squads after a gap.
(f) Extra supervision by leading hands and forepersons.
(g) Aggravation of costs of labour-related items, such as transport and messrooms.

It is obvious that there are potential overlaps between several of these items, leading to practical problems of accounting and calculation. At present, these problems are left aside: the principles are in view rather than the practicalities of segregation – always an author's alibi for what he leaves out! Some illustration is implicit in the case study calculations in Chapter 31. The first item is the core item in many cases and introduces the greatest number of issues.

For now, it is possible to ignore the effect of changes (usually increases) in labour costs when there is a delay; it is a distinguishable issue that does not require explanation. But its calculation is treated under 'Inflation on fixed price contract' in Chapter 31 pages 519–522.

Drops in labour productivity: smaller cases

Drops in labour productivity may occur when work is proceeding continuously, if not economically, on tasks in conditions of disturbance. It is related to directly employed workpeople paid on a time basis, rather than on piece rates. The effect on piece rates may be deduced, if it is remembered that they are rather like overgrown bonuses (see below).

The argument has already been advanced that loss and expense cannot readily be seen 'out there' in many instances, and this is particularly true of labour, in view of its characteristics. Furthermore, calculations are not to proceed on the basis of tender figures and actual costs alone, but are to include the elusive middle figures of what would have been expended (Chapter 30) to meet the criterion of reimbursing loss and expense as the equivalent of damages. Purity of approach is seldom possible in practice, and it may be noted that a combination of figures is suggested in what follows.

The simplest case, and therefore the least likely, is a straightforward prolongation affecting only tasks which also occur during the programme in conditions of normality. Provided there is no suspension of activity, so that work proceeds fairly continuously, but with reduced vigour, the first approach here must be to look at an increase in labour costs proportionate to the increase in time.

It may be possible to draw upon evidence from extension of time granted, but this has its limits. It relates to the project as a whole, whereas the prolongation may affect only some part, which may or may not have contributed to some part of the extension of time. More precisely, an extension of time is granted in respect of delays affecting activities which are on the critical path of a network (or its equivalent) or which come on to the critical path because of those delays, and so extend the completion date. Subcritical activities are not taken into account in granting an extension, but may feature in a loss and expense item, as happens over the ancient fort affecting the laboratory in Chapter 30. It is also the case that an extension of time is granted in advance of its full run, or usually should be, and then confirmed or adjusted after practical completion, whereas loss and expense is calculated as or after it is incurred, making for complications of record keeping.

Furthermore, it does not follow that labour hours have been extended directly with the contractually granted extension of time, which is given for quite distinct reasons. Although this may be a starting-point and useful regulator for calculation, it needs to be evaluated against such other evidence as may be obtainable. Such actions as spreading labour across other work, so it is engaged more thinly on the affected work, may reduce the impact of the delay in terms of cost, if not in terms of programme time. Other activities, such as dealing with non-qualifying delays and remedial work, may need to be excluded; this is discussed in Chapter 15. On balance, the extension of time approach must be seen as a maximum assessment method (as it is for overhead costs), to be trimmed down by closer analysis to reflect the realities of the situation.

If this approach is suitable, it should be seen as absorbing all such effects as occasional stopping times. No attempt should then be made to evaluate them separately in the way mentioned below. However, any major break is best taken out of the reckoning and dealt with quite distinctly. It becomes more difficult to apply the method if any task is finished during a slowed-down period or if the nominal pace would have been changed in either direction during such a period. It is then necessary to make proportionate adjustments to allow for such 'wedges' of work. It definitely requires close records of what is being performed during any period of delay, and whether all parts are affected by the delay. If they were under pressure in the unaffected programme, they may even be helped to greater economy on occasions.

The clear alternative to prolongation is a straightforward case of disruption alone within the same overall timescale, with only the same tasks involved before and during the disruption, and possibly after it has ceased. This is something of a laboratory condition, likely to be encountered only when there is a very localised dislocation. Here the aim should be to secure data on the productivity being achieved before, and perhaps after, the disruption and to compare this with what happens during it. There is a potential weakness in using data from after the disruption, unless it is the major share of the work, as there is the possibility of a special acceleration of work affecting the results.

To operate this alternative, it is necessary to obtain output rates per unit of work in both conditions and to calculate the excess in the disturbance situation, after allowing for any other variables applying to the 'same' tasks. In the case of the disturbed work, this is coming close to a daywork assessment, and the usual precautions should be taken of checking that due efficiency was observed in deploying resources. It is, however, the substitution of one set of costs for another and not for measured valuation. If it is done, it must be done for the whole period of disturbance: use of 'the first week or so', for instance, is suspect. Apart from deliberate excessive slowing of work while it is under observation, there is the possibility that work will inevitably be slower then, as the contractor is adapting his organisation to changed conditions. Disruption may also produce effects which vary from day to day all through the period.

More likely, and almost always when there is an extensive and extended disruption, the set of tasks will not be entirely the same in both conditions. Some will start and some will stop, whereas others would have changed momentum even without the disruption. When a number of matters conspire to create large-scale confusion, perhaps pulsating in intensity, the practicability of 'doing neat little sums' is not there. Nevertheless, the general principles of the two approaches for prolongation alone and disruption alone should be held to, so far as possible. At some point up the line, the smaller cases become larger cases (see next section) or there may even be a global claim situation (Chapters 14 and 12).

Drops in labour productivity: larger cases

It is at the point of pious platitude just reached that advice tends to run a little thin and the parties are left to their combined ingenuities, although several general guidelines may be given.

The two approaches given may be combined, to delineate some limits to the amount of loss and expense over labour. When disturbance is extensive, it is most likely to include both elements and the question may be looked at from each direction in turn, to see what emerges. The difficulty is that a large disturbance does not take place inside a neat and tidy box of space and time, but has ragged ends and edges, so that it starts and stops irregularly.

Another line of action is to break the disturbed area into subsidiary areas, but not just into pieces which are manageable in size while bearing no particular relation to one another. Dividing a problem into parts sometimes eases its solution, but it may also obscure the possibility that the whole happens to be larger than the sum of the parts. For this method to be fruitful, it should represent a systematic analysis of how the components affect each other. The contract programme offers an obvious starting-point here, as in the use of the network in Chapters 30 and 31. Each subsystem can then be looked at on its own, to see how it is affected by the ripple effects of preceding or parallel subsystems and also to see what effects it creates itself. The resultant pieces can then be analysed on their own, using whatever tools are best. This method has the advantage of maintaining the relationship between the several matters of disturbance which have led up to this complex disturbance.

A further way into the problem is to look unashamedly at some form of comparison between the contractual valuation emerging from all elements other than loss and expense on the one hand, and the contractor's costs on the other. As indicated earlier in this chapter and in Chapter 25, this may show up some area of discrepancy so intense that it cannot be shrugged off as entirely due to errors in pricing or site inefficiencies, or the like. If the contractor has incurred genuine loss and expense because of inadequacies from the employer's side, he may need to be given the benefit of the doubt. This is quite distinct from cases in which he is pumping up a claim to make what he can out of a situation which has fallen to be exploited.

Lastly, within the realm of ascertainment proper, allowance can be made for the impact of the other causes of labour expense which are considered under the headings following. If there are clear areas which can be taken out of the relatively vague sphere of general productivity, all well and good. Although this area has been taken first because of the issues it raises, it may be the last part of a claim to be settled.

Beyond ascertainment, there are finally instances where the only recourse open is to employ straight judgment of the effects. This is actually present in all the other approaches already suggested, as none of them can be used uncritically, and various asides indicate this. The difference now is that there are situations in which nothing but judgment is available. This may be dressed up under all sorts of guises, but that is what it often is. Affairs sometimes come back full circle to the judicial gadgetry mentioned earlier in this chapter. The result is negotiation (Chapter 25) as an alternative to seeking the judicial solution, which depends so much on case and counter-case. So does negotiation, but it also involves considerations of ultimate expediency for the parties, such as speed of settlement and avoidance of the extra risks and costs in going to arbitration or court.

Loss of bonuses and overtime

A common point advanced in a loss and expense situation is that disorganisation of work has meant it has not been possible for personnel to achieve bonus targets or have overtime available, for reasons beyond their control. As a result, it has been necessary to make enhancement payments as compensation and to avoid demoralisation causing further loss of output, but without any positive benefit accruing to the contractor and in addition to payment for extra hours worked. Some examination of this contention is needed.

When the contractor formulates a rate for an item of measured work, for example, he does so on the basis of a certain labour content. This consists of some combination of the rate per hour payable, inclusive of allowances for overtime and other such directly variable costs, and of the expected bonus payments which will be earned. This takes account of the saving in actual hours due to the incentive of the bonuses available, so arriving at a global labour cost for the unit item. Here is an illustration:

100 hr (without incentive) × £5.00 hr = £500

[80 hr (with incentive) × £5.00 hr] + £100 (bonus) = £500

80 hr × £6.25 hr (consolidated rate) = £500

In essence the contractor prices for a given quantity of production for a given payment to labour. In doing so, he allows for the expected split of the benefits of increased productivity built into the level of bonus targets (be it money or programme time saving) and for the rate of production likely, so that faster working saves payment of the one type, wages, but increases payment of the other, bonuses. The actual levels of targets are negotiated with the labour force, gang by gang and week by week or as necessary, according to the work immediately in prospect. These levels may be intended to give the contractor some benefit in lower total payments to labour. There are also economies of supervision and other overheads to be gained by the contractor from more rapid working.

If a disturbance situation occurs, the contractor is faced with lower output, as discussed under the preceding heading. From the above figures, it may be seen that a drop of say 20% in productivity has the effect of increasing the contractor's costs by 25%, whether he was previously paying any bonus or not – always assuming that he has set his figures at the correct level when tendering. So if, in lieu of bonus, he arranges to pay his workforce at rates per hour which compensate them for the loss of actual bonuses, he should remain with a shortfall in recovery of the same margin of 25%, as his costs will be

100 hr × £6.25 hr = £625

It is therefore important to be clear whether or not the hourly productivities and rates used at each stage are based upon incentives.

The figures assume there is a precise lack of the extra production which bonuses are intended to achieve. If the result is an intermediate position of some extra productivity, hence payment of bonuses, then his extra costs will also fall into an intermediate position. A greater drop will lead to an increased loss. Whatever the position achieved, the argument also assumes the contractor does his sums correctly in arriving at payments in lieu of bonus and in supervising work in the possibly confused situation of disturbance. There is therefore some room for manoeuvre, but it should be the case that any plea is unfounded for complete reimbursement of lost bonus amounts in addition to other labour elements.

There are also several reasons to be cautious over an argument for the payment of additional bonuses to make up time when it has become possible to resume normal progress. One reason is that extra speed should reduce total basic hours, just as less speed has increased them. Another reason is that there may be the possibility of engaging more people or working overtime, rather than working faster, although both of these measures have their costs. A third reason is that behind this argument is a hint of less efficient working, which may affect quality as well as productivity, as the original working must be assumed to have been at the optimum pace. But most cogently, there is the contractual position that the contractor is not required to regain lost time, which would be an acceleration, but only to lose no more than is inevitable. He therefore should not be reimbursed for extra costs which he has incurred without authority. It is difficult to separate this in all circumstances entirely from the contractor's general duty to mitigate his loss, discussed in Chapter 10, as sometimes an extra expense in one place may avoid greater expense, or loss, somewhere else. The message for the contractor is clear: seek authority, obtain agreement in principle in advance.

To the extent that overtime is worked and paid as an incentive to attract and retain labour, it falls into the same category as bonus payments. This may mean that it becomes necessary to pay enhancements to compensate for lack of opportunity to work longer hours, but either the enhancements or the non-productive time should be seen as giving a similar cost per unit of work produced. If so, the same argument applies as for bonuses.

Other elements stand in the background of these payments made directly to labour. They include supervision, transport and insurances, which are by way of on-costs. To an extent, they may be seen as costs which offset against hourly wage payments, rather as do bonus and overtime, so that a claim for them to be reimbursed may not always be valid. But this is only to an extent, and they are discussed separately hereafter.

Standing time

There may be a complete stoppage of significant duration, as distinct from a minor hiatus during a period of reduced productivity, as already mentioned. Here the main position is clear: the contractor is due to be reimbursed for payments made without any return, so that hourly amounts and compensation payments for incentive amounts all become eligible, without arguments over which elements offset which. A distinct set of records can usually be kept.

It will be necessary to take account of how much of a rundown period there is before cessation occurs, as it is not always a straight case of 'stop on Friday'. There may be a phasing out of work and activities to bring affairs to a tidy position without much progress. Similarly, there will usually be a phasing-in period as work recommences after the break. This means time in refamiliarising the workforce with what is to be done.

When a break is obviously going to be prolonged, there is a case for the contractor reducing the level of personnel standing unoccupied. He should therefore be expected to look for alternative work, which most likely means on another site, unless he is able to put them on relatively unsuitable (and so uneconomic) work on the same site. The alternative is that he must lay off workpeople altogether, which is reasonable only if the delay is really extended. Otherwise the costs of taking on and familiarising a fresh set of labour will outweigh the saving.

The break-even point in this latter instance does not simply fall out by entering a few values in a formula. Worse, the decision is easier in retrospect than when things are actually still going sour. An architect's instruction to empty the site for six weeks is susceptible of advance assessment, whereas a creeping paralysis may pass the break point before it is clear and then leave too short a time for any laying-off to be of value. The various situations may be discerned in the case study chapters hereafter, although not in strict isolation, which is symptomatic of the problem.

This illustrates the difficulty in using with any rigidity the principle which applies to all loss and expense, namely, that the contractor should mitigate his loss, and hence mitigate the loss of the employer. The contract confers no power on the architect to direct the contractor as to how he should act during a period of disturbance. There is power to issue instructions on matters specified in the contract and so relieve the situation indirectly, if this is practicable, entirely at the architect's discretion. All that there is beyond this, by implication and on common law rules, is a right to reject elements of cost which stem from failure to mitigate. In situations where the parties are proceeding affably in the face of adversities, there is much to be said for consultation over how to proceed for everyone's benefit.

Extra activities of reorganisation etc

Extra activities of reorganisation etc are items left over when variations have been valued, so as to cover as many aspects of cost as possible, which should be the primary approach, allowing payment which includes a profit margin, among other things. Many matters of site reorganisation, for instance, fall under the variations provisions over obligations and restrictions. Whenever the variations provisions do not fit and in other 'untidy' situations, it is necessary to resort to the loss and expense provisions, with all their stricter bounds. Then any activities should be dealt with as entities, allowing records to be kept or a separate assessment to be made.

Familiarisation time

Familiarisation time has been mentioned earlier, as part of the resumption costs after a break. It may also occur when sequences are changed, so that certain personnel are engaged upon tasks previously assigned to others. It is therefore likely to be a feature of piecemeal working. Whether it needs to be separated from a general productivity drop will depend upon how records are best kept.

Extra supervision

Supervision breaks into two levels: overall supervision of the project or major sections of it and close supervision by leading hands and forepersons of the detailed production of quite limited parts of the work by their own specialisms.

Overall supervision is taken hereafter as it is affected differently and structurally by conditions of disturbance, and because its cost is not usually related so directly to the total labour content of a project. Close supervision is meant here, and it may be said that in many instances its cost is proportional to the cost of the labour being supervised. This is because the span of control of a supervisor of this type is fixed fairly closely, so that more workpeople means more supervisors on a proportional basis.

But often the direct supervision of detailed operations needs to be intensified, so that there occurs a structural change among the persons concerned. If this happens, greater costs will arise which need to be assessed on their merits. It is difficult to argue the case for any saving of supervisory costs during disturbance, except by suggesting that there has been a failure in the duty to provide people to keep adequate control, hence incurring labour costs.

Labour-related items

There are numerous costs incurred during a project, at or near the site, which are related to the cost of labour, but not necessarily in a direct, straight-line way. Two examples may be given.

Transport for labour, such as buses, tends to rise fairly directly with the numbers employed, so the costs may be expected to rise with labour costs. This pattern is affected particularly by the indivisibility factor, that is buses cannot be chopped in pieces to allow for changes in the workforce. So if the workforce is attenuated, while the transport provision must remain constant, the cost of transport will rise more rapidly than the cost of labour itself. This is especially the case if labour is being brought in from several locations, or (as a special case of this) if a number of subcontractors are involved. Alternatively, an intensification of labour in conditions of disturbance may lead to better use of transport, hence to no proportional increase of cost against the cost of labour itself; it could even lead to a decrease. Whether there is still an intrinsic increase will depend upon the detailed relationships.

Messrooms and other facilities for labour, such as toilets, are in a different category. Usually a disturbance of regular progress will not change the basic pattern of their use, as they just stand there. It is not practicable to reschedule their availability, but if there is a significant drop in the rate of use, it may be possible to divert operating staff to other tasks or sites. Unless there is a huge increase in the workforce, more facilities are not likely to be needed, as they are somewhat elastic items. An exception could come about if delayed work led to bunching of labour demand, when put alongside later work. More commonly, these items are likely to be extended in their period of use, so falling into the same category as other site overheads considered later.

Plant

Plant has some characteristics partly in common with labour; they have already been introduced and are expanded here. Other characteristics are new and are introduced now. The labour associated with plant is to be treated with other labour, as discussed above. Because of the differences between plant and labour, the next section treats some general factors in plant costs which apply even in the absence of disturbance.

Special cost features at all times

Like labour, plant is engaged by the period of time. This is strictly so when plant is hired and effectively so when it is owned, as it is realistic to account for its cost on a time basis, by taking in its special features of cost of capital and depreciation, as well as its immediate operating costs. It can, however, be set aside and so 'stored' when not in use, as labour cannot. The practical effect of not incurring some costs when plant is idle is that there are two hourly rates for plant: working and standing. This difference is more significant when plant is owned by the contractor himself.

This is because a hire firm will charge at a rate, by the day or week, which does not vary to take account of how intensive the contractor's use of the plant is going to be. An average rate of use is assumed, which any one contractor may exceed or not. When the plant is not actually working, the contractor stands to save fuel costs and labour costs, in so far as it is his own labour, but there is no change in wear and tear costs to him. Capital charges, depreciation and maintenance are also covered in his hire payments. When the contractor owns the plant, wear and tear and so maintenance of most parts is cut down when plant is not being used, along with fuel etc. The total effect here is to give a greater difference between running and standing costs when plant is owned, so the status of plant is important when loss and expense is in prospect. The distinction has been observed in judicial circles in *Bernard Sunley Ltd* v. *Cunard White Star Ltd* (1940), when a reduction was made for standing time of the contractor's own plant. The case is also interesting for the refusal to cover loss of other output in the absence of reasonable proof that this had occurred.

But the status of plant is important for two other reasons. One is that the hourly rates for hired plant are generally higher than for owned plant, partly because it is hired at overall rates, but also because it is usually hired for shortish periods. The contractor's own plant, by contrast, is with him somewhere for a long time, relieved only by when he can hire it out to others. He hires plant in at relatively high rates when he does not have enough use for particular items over a matter of years to cover their initial purchase. The policy followed is also complicated by features like taxation arrangements, resale options and other matters which cannot be pursued here.

The other reason why the status distinction matters is simply because hired plant is not permanent – it can be sent back if necessary, so ending the hire charges completely. This is another reason why hirers charge more. The result is more flexibility, important for economy when things are going wrong on-site. For his own plant, the contractor is dependent upon having another use elsewhere, if he is to save costs. For neither category of plant should the rates be as high as those for plant used in daywork. This is usually work incidental to the main work, and daywork schedules recognise this by relatively high rates for small parcels of work. They also enhance the rates, which are for working hours only, to allow for a related share of standing time.

But just as there are costs associated with sending labour packing, so there are with plant. These are costs of installation, removal and transport, which tend to inhibit how much it pays to send plant away, even when there is nothing for it to do. There is no advantage to the contractor in sending his own plant away if there is no work elsewhere, unless it is just in the way of whatever work he can do.

These basic costs of plant may be summarised as follows:

(a) Bringing to site and commissioning.
(b) Working or productive time.
(c) Standing or idle time.
(d) Dismantling and removing from site.

Some special features over loss and expense

Plant is less flexible than labour, in that it can only perform a few tasks (perhaps only one) per item. It also occurs in units which are dearer per hour and so less divisible in terms of cost. The combination of these factors means that when plant is subject to disruption effects in particular, it is more of a problem. Against this, it does not suffer the psychological effects of disturbance, except to the extent that its operators are affected and pass it on!

Related to these factors is the substitution characteristic. If a piece of plant is rendered sub-optimal by disturbance, there are four basic options:

(a) *To carry on as best may be*: This is the only feasible choice when matters are running in a stop-go phase.
(b) *To substitute another piece of plant*: This is possible when the length of the problem can be assessed fairly closely. It usually means a similar smaller piece of plant, but may mean something different in function.
(c) *To substitute labour*: This is necessary when either the amount of work or its positioning renders anything else impracticable.
(d) *To suspend work*: This applies when all else fails.

These options are in some order of feasibility, but much depends on what work is involved. Two conditioning factors are the nature of the work and the practicability of carrying on with it, without running completely dry. Another factor is cost, but this must be seen as the total cost of work and not just the cost of the plant or the immediate operation on which it is engaged. There are some large items of plant, such as concrete batching plant, where the cost of substitution is very high compared with the gain to be experienced.

It is thus true, even more of plant than of labour, that the decisions about how to reorganise it in conditions of disturbance depend upon the whole structure of production which the contractor has set up and which is now under threat. Because they have no powers, even less can the architect or the quantity surveyor, whoever is dealing with the ascertainment, dictate what steps are to be taken, although again the contractor may consider it expedient to consult over major dispositions of plant. Apart from anything else, he may need guidance over how long the delay is going to be. Something fundamental to many site operations, such as a tower crane, can be whisked away only if the contractor is convinced of the rationality of the action.

One point coming out of this is that any attempt to question whether the contractor has done his best to mitigate the loss and expense must be based upon very sure ground. It is even harder than with labour to make statements based upon hindsight, and this is all there may be to go on. When things have gone wrong in any major way, the contractor must be given the benefit of any doubt – a recurring theme!

Relationship of plant and labour

The immediate effect of inadequate use of plant is to incur extra costs of the plant and usually directly proportional extra costs of the labour that operates it. Occasionally, it may be possible to use less labour in the immediate task of keeping plant going, such as banksmen and other peripheral persons. How far this reduces expense will depend upon what those released are then able to do. Sometimes use of plant in restricted or difficult areas may actually increase attendant costs of labour.

The other side of this is that plant working at reduced output may mean that less labour can be kept going at the next stage of production. This is especially true of plant which hoists and distributes materials, although this type of excess cost may show for one or more of the primary reasons covered under labour. It is usually easier to look at labour as demanding the output of plant, than plant as conditioning the output of labour. Any side-effects should be looked at and assessed carefully over the logic of their supposed occurrence.

As noted in option (c), substitution of labour for plant may be expected to mean higher costs. Otherwise, it is reasonable to expect that labour would have been employed in the first instance. There are likely to be reorganisation costs and perhaps extra supervision, with again the question of what happens to specialised plant operators. This sort of activity is a likely one for introducing ripple disorganisation effects.

These are all cases in which the effects are difficult to trace or calculate. More critically, unless care is exercised, there may be double-counting of costs.

MATERIALS

It is usually the case that any unanticipated costs of materials can be fully covered in valuing variations. Measurement and daywork are both related to quantities of materials, so the actual amounts can be assessed; whereas pricing can take account of small consignments and factors like undue wastage due to special cutting and similar causes. An instruction to remove materials because a change in design has rendered them superfluous is classed as a variation under the JCT clauses, so it should be valued under their rules, allowing for any credit value.

Loss and expense may occur when materials deteriorate in conditions of delay or disorganisation, through exposure to the elements or through excessive handling. As with labour, there is the question of inflation costs, if a section of work is moved back in the programme. Even here the provision for work performed under dissimilar conditions is likely to prove more convenient, because of the measurement aspect. It should also be used to take the costs into a variation, because it is the prior contractual option.

Materials stored off-site for a protracted period may lead to additional storage costs or disturbance to the contractor's yard or a supplier's facilities. Unless the available contractual option of paying for such items while still off-site is exercised, the contractor will face a constriction of his cash flow, which may well warrant recompense (see 'Financing charges' in Chapter 12).

It may be argued that the contractor should not obtain materials far in advance. Provided he has a firm contractual indication that the items are to be supplied, by virtue of quantities or detailed drawings, it is quite reasonable for him to obtain them early and make any saving in cost that results. He is not obliged to wait the good pleasure of whoever may wish to initiate an unheralded variation and sweep firmness away, although he perhaps cannot demand that he should be given advance information which just does not materialise early of its own accord. This is an inference which may possibly be drawn from the limitations on early information in the relevant events and matters clauses. If the contractor does choose to purchase materials unduly early, he must obviously be prepared for the possibility of having to fund the purchase himself for the excess period, so reducing the overall financial benefit which he secures. He also takes a risk if he orders materials from inadequate information at any time.

SITE OVERHEADS

The term 'site overheads' is used to cover elements of cost which are commonly and desirably priced in the preliminaries section of bills of quantities or other such documents. SMM6 lists them in its Section B and SMM7 in its Section A, where it also includes a number of provisions which cannot be distinctly and separately priced. For now, the more important elements of loss and expense may be grouped as follows:

(a) General supervision and administration of the project, covering staff and their related costs.
(b) Plant items which are not closely related to the quantities in individual work sections.
(c) Temporary facilities, such as accommodation, roads, fencing and services.
(d) Setting up and running down the project.

There may be considerable room for overlap between loss and expense costs and amounts for variations here. As usual, the primary rule is that valuation within the variations framework is to be used where possible. This is discussed briefly in Chapter 6. In view of the complexity of this area of cost, the present discussion is limited as well, so the main discussion is taken in connection with the case studies in Chapter 31 to unite the various aspects.

The main overlap is with variation valuation. Both there and in ascertaining loss and expense, any amounts are not related in any necessarily direct or pro rata way to those other costs, such as labour, which come about in the immediate production of physical work. A considerable amount of additional work there may cause little in the way of additional supervision, for instance; whereas a small amount of such work may require comparatively large efforts on the part of staff.

The actual segregation of supervision costs between a variation and consequential loss and expense can become a task needing the wisdom of Solomon. Especially when variation of obligations and restrictions is involved, it may be questioned whether this wisdom over the division is really needed, so long as everything is covered once and only once (see example in Chapter 32). The problem becomes essentially conceptual – as Solomon found!

Usually, temporary facilities are not drastically changed in a physical sense. If they are, the adjustment may fall within the variation orbit. Here again, obligations and restrictions can be particularly awkward. Plant, on the other hand, may often figure in loss and expense calculations. Here the greatest care is needed over defining which items of plant are to be dealt with separately in a global way and which are to be treated within smaller parcels of calculation, as already discussed; the problem is illustrated in Chapter 31 over plant already allowed in Chapter 30. The 'end costs' of the project, in setting up and closing down, are not so likely to figure in loss and expense calculations, although a fragmented end to work may give an exception (see the various case study chapters).

Because these elements of cost are not related directly to the quantity of work performed, they tend to need quite distinct treatment when changed by variation or otherwise. For variation valuation, the figures in the contract bills may be unrepresentative, or even non-existent, as mentioned in Chapter 31. As a result, calculation from first principles may be needed. This is as true of loss and expense, for reasons of legal principle, as well as practical calculation.

Two main types of assessment are common. One is the special concentration of extra expense when there is some relatively localised disturbance, especially relevant to supervision and plant. The other is the result of prolongation, sometimes rather inaccurately termed 'extended preliminaries'. This may affect supervision and plant, and is the most likely way that temporary facilities will be affected. It is also more likely to occur when there is an extended disturbance due to multiple, interacting causes. This head of claim may be pressed when there is extension without disturbance, and should then not be met, although extra preliminaries due to extra work extending the time may be allowable (Chapter 31). On the other hand, it may come about when there is a prolongation at some point within the programme, not on the critical path and not leading to extension of time. Plant is especially vulnerable here.

As site overheads build up during the earlier stages of a project, run at a comparatively constant level and then tail off, the period at which any prolongation occurs is important. It may mean extending overheads at whatever level applies in the adjacent period. Alternatively, it may mean a relatively accelerated build-up or a deferred tail-off. Some possibilities are shown diagrammatically; Figure 11.1(a) shows a build-up period and Figure 11.1(b) a peak period, which could also represent a rundown period, if inclined to the right. The lines relate to the following levels of overheads:

- A is the original, unaffected programme time and levels of expenditure.
- B is the extent of prolongation.

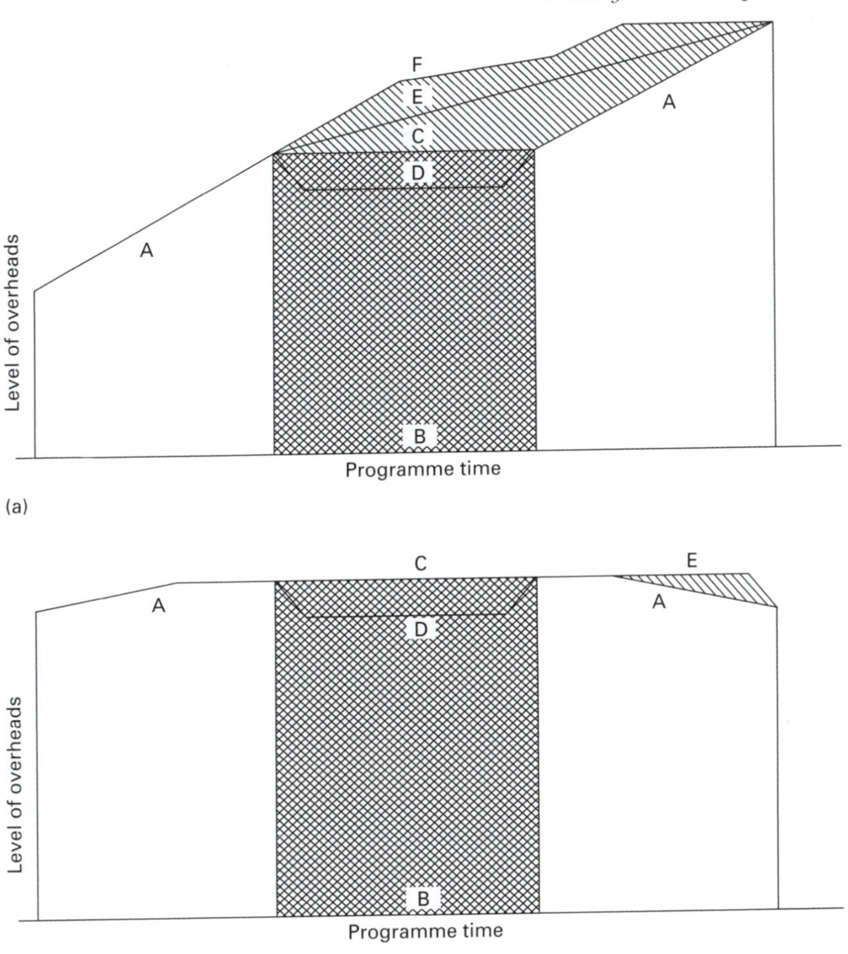

Figure 11.1 Prolongation of site overheads

- C is a simple extension of the level of expenditure at the beginning throughout the prolongation.
- D is the effect of being able to reduce the expenditure at the beginning and reinstate it before the end, both depending on some forecast of duration.
- E is extra expenditure incurred fairly evenly and deliberately during (in the first case only, it is assumed) and following prolongation, to overcome some of its effects.
- F is extra expenditure incurred during and following prolongation, but partly due to inability to avoid further build-up and partly due to extra effort to overcome effects.

The principle outlined here is similar to that given under 'Client's risk analysis' in Chapter 1 in respect of initial and ripple effects.

It is important to distinguish 'site overheads' (often known as 'preliminaries' items because they appear to have been priced in the 'preliminaries' section of contract bills of quantities) from the more widespread overheads in the accepted sense of the expression. The more widespread overheads are incurred at the level of the enterprise as a whole, and so cannot be allocated directly to the individual project, whereas site overheads are incurred quite specifically because the individual project is being carried out. However much they may need care in allocation, site overheads are capable of separate itemisation and evaluation. The other class of loss and expense (secondary) is considered in Chapter 12, with quite a different approach.

CHAPTER 12

ASCERTAINMENT OF SECONDARY LOSS AND EXPENSE

- Head office overheads
- Loss of profit
- Interest and financing charges

The preceding chapter considered loss and expense which, with minor exceptions, is incurred entirely at the site of the works or off-site for inputs to those contract works alone. It is then possible in principle to deal with loss and expense on a contained basis, as being the sum of items related to the one contract only. Difficult questions of apportionment between various projects under the contractor's control are thus unusual. In moving from primary to what is here termed the secondary categories of loss and expense, matters are very different, in that apportionment or other formulaic ascertainment may be appropriate.

Secondary categories are those areas of cost etc that are not incurred immediately on the site or in relation to discrete elements of work performed elsewhere for it, but are related to the conduct of the enterprise as a whole. The major categories are head office or general overheads, profit and financing or interest charges. Although the individual project causes or contributes to them in its own measure, they are diffuse across the organisation, rather than contained within the project. The three categories require somewhat different treatment, despite this measure of similarity.

In this respect, it may be seen there is a shift in methodology between much of what comes into primary loss and expense and much of what comes into secondary loss and expense. For the primary items, the temptation, as observed at the beginning of the preceding chapter, is to try to relate calculations to original tender levels as well as to what has happened on-site. It is also the area where the expectation is that the global claim approach will be eschewed except as a last resort, when segregation of effects cannot be pursued any further. The global approach tends to come in most insistently for site overheads, where attenuation and other diffuse effects occur.

By comparison, secondary elements are often such as to have effects which are global: they may impact upon the operation of the head office, for example, in ways that do not flow identifiably from single events on-site to single costs at head office, but from a build-up of events to a gradual accumulation of costs. The nature of these effects is usually taken for granted in matters concerning claims. It contributes to the difficulty of ascertaining quantum, so that sometimes the argument becomes: There cannot be any effect, because it cannot be isolated. This is fallacious, as comes out in discussion in this chapter.

The remarks about the nature of damages for breach of contract at the beginning of Chapter 10 should be borne in mind here. The purpose of such damages and of reimbursement of direct loss and expense under a JCT contract is to put the party suffering because of the breach back into the position financially in which he would have been, but for the breach. It is not to treat and evaluate the elements of loss and expense as extra turnover or business. Equally, the elements are not to be priced at so high a level as to act as a punishment to the defaulting party. This is especially relevant to profit, as discussed later. This principle underlines the value to the contractor in most cases of being reimbursed by fully valued variations for preference, so that he secures a sure profit margin (Chapter 6).

The loss and expense considered here is still the amount that is 'direct' and so equivalent to damages in its scope. It so happens that the type of cost considered in this chapter is largely the type that accountants class as indirect, rather than direct, but this is quite a distinct way of looking at affairs. Their concern with indirect costs is broadly with costs incurred other than at the point of production. An indirect cost in that sense can always qualify as a direct expense for present contractual purposes.

The most comprehensive set of considerations arises within the overheads category, so that is considered first. Although 'overheads and profit' are often dealt with as a single percentage for pricing purposes, just because they are both usually available as percentages, they are not strictly linked even there. They are even less linked for present purposes. With overheads there is the prospect of expense, but with profit there is the prospect of a loss – that is, loss on future business not obtained because of the present contract.

HEAD OFFICE OVERHEADS

Identification of overheads

The term 'overheads' is used to cover expenditure relating to the business as a whole, such as staffing, premises and insurances not project-specific, at least for most of the time, although they may be concentrated during tendering, main ordering and similar periods. These costs are usually allowed for when tendering by a percentage spread across prices for the contractor's own direct work, that is excluding the work of subcontractors. Sometimes a percentage, perhaps smaller, is allowed on subcontractors as well, so the first percentage is then reduced. How all this has been done in the tender detail is immaterial to the question of reimbursement in the event of loss and expense, just as it is with site overheads and how they have been included in the contract sum. Originally the contractor must have assessed the cost of these items in relation to his turnover, i.e. historically, before arriving at the percentage or percentages that will include any adjustment for changes in levels of cost which he foresees.

The same problem arises with these costs in loss and expense as with those incurred on-site, e.g. labour: the extra amounts are not directly visible 'out there' as showing separately from the costs which would have been incurred without the disturbance occurring. The problem is 'merely' intensified by the fact that the costs cannot be related in any simple way to quantities of work performed on-site.

The principle of extra overhead expense

There is, however, a further aspect to the problem, perhaps almost a new problem. This is the more philosophical question about whether extra expense ever occurs at all, or at least very often. It is sometimes argued that particular events on a site just do not affect the contractor's costs at head office at all. He will maintain the same organisation, but the attention which it gives to this project or that will switch according to what is needed at any time. When there is a small case of disturbance on a site, this is usually true. The contractor takes it in his stride, as his organisation is not so finely tuned or hypersensitive as to be thrown out by a distant ripple on the pond. The exception may occur when there is need for highly specialised attention to deal with a highly specialised disturbance. Even here it may just mean diversion of resources and not any extra cost.

The counter-argument is that there often is extra cost, because the disturbance is significant in scale or complexity. It is just not feasible that no extra cost ever arises. On first principles, that the contractor has tendered for and planned the project on the basis of a straight run through, this argument logically must have the upper hand. The other argument appears to be based upon the difficulty of being able to detect just what the effect has been, and so declares it non-existent. The two arguments are not entirely opposed, as both tend to use such qualifications as suggest there is some threshold below which no effect is produced because any organisation has some slack or resilience with which to deal with minor problems during progress.

The extreme form of the second argument is embodied in uncritical use of one of the formulae for calculating extra overheads (and sometimes profit), which are considered below. Formulae assume, inter alia, that there is a direct and inevitable link of a mathematical nature between certain quantifiable elements (usually money and time) in the contract and the cost of disturbance of regular progress. All the assumptions made are open to question, but the formulae do at least illustrate the ways in which loss and expense may occur. The burden of proof remains with the contractor to justify that he has lost or incurred the expense of extra overheads and this is referred to in decided cases below, after exploration of formulae for overhead calculation and when it may be appropriate to use them.

It must be suspected that most contractors price work by taking their total past turnover, including loss and expense amounts, and applying their total past head office overheads to obtain a percentage for future use, with any relevant adjustment for changed conditions. In theory they might extract all past loss and expense reimbursement from their turnover, and so arrive at somewhat higher percentages for overheads in tenders, but this is theory. The difference may be shown by a simple, if exaggerated, illustration.

Method 1	
Total turnover, excluding overheads	£10,000,000
Total overheads	£1,000,000
Overheads as percentage of rest of turnover	10%
Method 2	
Total turnover, excluding overheads and loss and expense reimbursement	£8,000,000
Total loss and expense reimbursement	£2,000,000
Total overheads	£1,000,000
Overheads as percentage of reduced turnover	12.5%

If method 1 is used, and if there has been significant loss and expense in the past (usually higher on sites than at head office, as will be suggested below), then contractors will be adding somewhat lower percentages than would otherwise be the case in future tenders, which of course exclude loss and expense allowances. The recovery of the balance of their overhead costs will then depend upon enough loss and expense claims arising to bring in the difference. Given very bad experiences one year and very good the next, contractors could be out of pocket in the good year! They will equally be out of pocket if the ascertainments of loss and expense granted them exclude overhead allowances. Although the figures shown are a slight parody of reality as it usually is, they do point up the strength of the argument for at least some reimbursement of overheads, remembering that what contractors can actually add for overheads in future tenders is conditioned by how competitive the market then is, rather than by whether they would like to recover any past underreimbursement in loss and expense payments.

This example suggests merit in paying the contractor some reasonably assessed amount for extra overheads, and no doubt this often occurs. The attitude of the courts is not sympathetic to excessive use of such an approach, to put it mildly, as is shown hereafter. Any contractor faced with some offer in this direction, but wishing he had more, should weigh up whether to go to court or whether it might not be better to settle with his adversary!

The two discrete methods of ascertaining loss and expense in this area may be contrasted as by use of a formula or by isolation of individual items of cost. Whether they may be segregated quite so rigorously in practice is another point, and so is the approach of the courts. It is convenient to start to answer these points by looking at the formula system of calculation. A review of law cases on overhead recovery, particularly involving the use of formulae, comes at the end of the next section.

Formula calculation of head office overheads

There are several formulae widely known within the industry, and three of them are considered here. They aim to take account in various ways of factors of delay time and of extra money payments for primary costs arising from disturbance, by making fairly broad assumptions about their effects on secondary or overhead costs. It is the broad assumption aspect which is disliked by the courts, as discussed below. By definition, a fixed formula cannot be ascertainment in a variable situation.

Two of the formulae also include profit in their orbit. This may be noted, but otherwise ignored at this juncture, as profit must receive separate treatment and cannot be put in with overheads. The formulae are first described critically and then compared. An example of their effect is given in relation to the case study in Chapter 31.

The Hudson formula

The Hudson formula was included by the editor in the tenth edition of *Hudson's Building and Engineering Contracts.* It includes an allowance for profit, which is shown here, but which has a different significance from what is apparent at first sight, as is explained later. The editor himself enters his own caveat about the profit allowance. The formula has three main sections, and is as follows:

$$\frac{\text{Head office and profit (\%)}}{100} \times \text{Contract sum} \times \frac{\text{Period of delay (weeks)}}{\text{Contract period (weeks)}}$$

The first section needs little comment. It assumes that overheads and profit are to be linked in one composite percentage. Each may have been included at different levels in various parts of the tender, perhaps in respect of direct and subcontract work, with the further distinction between domestic and nominated subcontracts. If profit is treated separately, this point becomes less significant, but does not disappear.

For arithmetical correctness here, and assuming a single flat rate percentage, the percentage used must be that of overheads (and profit if included) as a percentage of the contract sum and not as a percentage of the balance of the contract sum, which is how it will have been calculated initially. The effect of misapplying the formula may be illustrated as follows.

Contract sum, including overheads and profit as 10% of itself	£100,000
Resulting amount of overheads and profit	£10,000
Same overheads and profit as percentage of remainder, i.e. as calculated when tendering	
$\frac{10,000}{90,000} \times 100\% =$	11.11%
Substitute 11.11% into the formula, to obtain an incorrect base amount for overheads and profit	
$\frac{11.11}{100} \times 100,000 =$	£11,110

What this erroneous calculation has done is to allow overheads and profit on the original overheads and profit. It will then be applied compound fashion to any increase due to disturbance.

The second section needs comment because it consists of just the one item of the contract sum. This also ignores the distinction between direct and subcontract work, mentioned above, but it also ignores the difference between the contract sum and the final account. Many elements in the account will include something for head office overheads, because they are calculated by a pro rata or similar method based upon the original level of pricing; Chapter 6 covers variations and there is a case study in Chapter 31. Whether therefore the account is higher or lower than the contract sum, there will already be some automatic adjustment of overheads in the arithmetic of quantities and prices to be set against that provided by the present formula used raw. The third section of the formula is based upon time change alone, so that any adjustment here will lead to some overlap of reimbursement, given a higher account due to the effect of variations.

This is not a simple relationship, and certainly not one of using either financial values or time periods alone. It is pursued further in the case study of Chapter 31.

The third section has the effect of reimbursing overheads at the percentage derived in strict proportion to the time overrun on the project. This is in itself a suspect concept, and one of the central areas underlying the antipathy of the courts to such formulae, as it assumes a straight-line increase of cost directly with time and without proof. But it also ignores the distinction between those delays resulting from events which are also matters leading to loss and expense payments, those resulting from 'neutral' relevant events and those which are the contractor's responsibility. The effects of these neutral events are shared between the parties, with the employer bearing the delay and the contractor the extra expense.

The consequence of this section of the formula is also that there is no reimbursement of overheads if there is no time overrun. Although this is quite possible if a disturbance is not on the critical path of the project, it by no means follows rigidly.

The Emden formula

Predictably, the Emden formula also takes its name from the book in which it appears, *Emden's Building Contracts and Practice*. It is very similar to the Hudson formula, differing only in the calculation underlying the percentage for head office and profit. Whereas the Hudson formula uses the value included in the tender, the Emden formula calculates the percentage:

$$\frac{\text{Total overheads and profit}}{\text{Total turnover}} \times 100\%$$

This becomes the numerator in the first section of the formula. Here 'total' means the total in each case for the whole of the business for a year.

It would appear best to use the percentage resulting from the year or total contract period (as in the Eichleay formula below) in which the loss and expense occurred, according to how prolonged the disturbance is, although this may mean a delay before figures are to hand. The alternatives in order of preference are those for the immediately preceding year or those applying when the tender was made, each with some allowance for any expected change in the current period. The reason for this order is to comply as closely as possible with the criterion that the contractor's *actual* loss is what is being sought. To this extent, by relying on actual values rather than amounts hoped for when tendering, the present formula has some advantage over the Hudson version.

Otherwise, the formula attracts the same set of comments as the Hudson formula. The use of total annual figures means it does not matter in quite the same way that percentages used in tendering may have been different for various elements of work. A less refined method is used on both sides of the calculation, so removing any particular bias. The need to ensure there is not a calculation of compound overheads and profit remains to be watched just the same.

The Eichleay formula

The Eichleay formula draws its name from a case in the United States in which it was applied. It has been used to quite an extent across the Atlantic, although reservations appear to exist even there. It is usually set out in three stages, which as ever, it is possible to conflate into a single formula.

Stage 1

$$\left(\frac{\text{Contract billings}}{\begin{matrix}\text{Total contract billings}\\ \text{for contract period}\end{matrix}}\right) \times \left(\begin{matrix}\text{Total head office}\\ \text{overheads for}\\ \text{contract period}\end{matrix}\right) = \text{Allowable overhead}$$

Here 'contract billings' means the amount of the project final account and the amount of all final accounts, or portions of them falling to be considered. This is the equivalent of annual turnover in British practice. As with the other formulae, it is necessary to compare like with like, so that all billings should either include or exclude overheads (and if appropriate, profit). As they appear above and below the line of the fraction, it does not matter which in this instance. Inclusive amounts are easier, as these form the basis of billings, or payments.

Stage 2

$$\frac{\text{Allowable overhead}}{\text{Days of performance}} = \left(\begin{matrix}\text{Daily contract}\\ \text{head office overhead}\end{matrix}\right)$$

Here 'days of performance' are the total days spent on the project. It is arguable that their use will give an attenuated spread of overheads in particular patterns of disturbance, but the reverse is possible.

Stage 3

$$\left(\begin{matrix}\text{Daily contract}\\ \text{overhead}\end{matrix}\right) \times \left(\begin{matrix}\text{Days of}\\ \text{compensated delay}\end{matrix}\right) = \left(\begin{matrix}\text{Amount of}\\ \text{recovery}\end{matrix}\right)$$

The use of a rate per day or per week in the various formulae is not of the essence, and the periods could be interchanged.

Combining all three stages Running the formula into one and using British terminology, gives the following expression for the recovery:

$$\left(\begin{matrix}\text{Total overheads for}\\ \text{contract period}\end{matrix}\right) \times \left(\frac{\text{Final account}}{\text{Total accounts}}\right) \times \left(\frac{\text{Days of delay}}{\text{Days of performance}}\right)$$

This rearrangement makes for simplicity in overall presentation. It also shows that the formula amounts to taking the contractor's total overheads during the relevant site work period only (in most cases), apportioning from them by other amounts payable to the single project and then apportioning from them to the period of delay. It is thus the only formula of the three to take some notice of the final account, although care is needed to arrive at comparable figures throughout.

This method apportions overheads without using a percentage as such, because it uses total real amounts, but the effect is the same. It also introduces overall amounts payable, rather than just the original contract sum. This is an improvement in that it comes closer to actual amounts, but still needs care in the detailed treatment about loss and expense amounts taken in across the contractor's whole activity, as well as on the contract being adjusted. All the questions remain about whether the method represents reality. It also means more investigation of the contractor's business at large, but this is also implicit in any non-formula method which the courts regard with any measure of favour.

Limitations of formula calculations

The three formulae considered may be set down in brief form for comparison (HO = head office):

Hudson

$$\left(\frac{\text{HO\% tendered}}{100}\right) \times \left(\begin{matrix}\text{Contract}\\ \text{sum}\end{matrix}\right) \times \left(\frac{\text{Delay}}{\text{Contract period}}\right)$$

Emden

$$\left(\frac{\text{HO\% actual}}{100}\right) \times \left(\begin{matrix}\text{Contract}\\ \text{sum}\end{matrix}\right) \times \left(\frac{\text{Delay}}{\text{Contract period}}\right)$$

Eichleay

$$\left(\begin{matrix}\text{HO total for}\\ \text{period}\end{matrix}\right) \times \left(\frac{\text{Final account}}{\text{All accounts}}\right) \times \left(\frac{\text{Delay}}{\text{Actual period}}\right)$$

Hudson differs from the other two by its reliance upon the tendered overhead percentage (the question of profit being ignored now). This is a weakness in that the calculation is based on intentions or hopes, not actual overhead, to calculate loss and expense. (The caveats expressed in Hudson are mentioned in a later section on profit.) The other two agree by taking actual figures in the same way, even if the arithmetical approaches differ superficially.

Eichleay differs from the other two by taking the final sum (preferably inclusive of primary loss and expense amounts) not initial sums, and by taking the delay in relation to the actual total period, not in relation to the initial period intended. Use of the final sum is a definite move nearer to accuracy. Use of the actual total period simply rearranges the time figures by reference to a different base time; provided calculations are done carefully, the results should not differ.

With this comparison to suggest that the formulae rank in ascending order of reflecting reality, their limitations should not be underestimated:

(a) They do not ascertain actual loss and expense, which is the primary reason why the courts are loath to use them without great caution and in the absence of proof of what they suggest. Arbitrators, and more particularly those negotiating settlements, are sometimes not quite so constrained, but do well to take note of judicial views.
(b) They need care in applying the arithmetical detail to avoid inaccurate results, as has been explained above.
(c) They do not allow for differences within the structure of the contractor's pricing, such as different weighting of overheads applied to his own and subcontractors' work.
(d) They ignore the uneven rate at which work proceeds on-site, whether subcontracted or not, by allocating overheads evenly over time. There is a build-up to a peak and then a rundown on most projects, commonly known as the S-curve (Figure 12.1(a)), and the timing and effect of any delay due to disturbance is not reflected by calculating overheads pro rata to time alone.
(e) They do not allow for some disturbance effects causing higher or lower concentrations of head office costs than normal, needing special assessment. Minor disturbances may be absorbed from this point of view, although when they lead to extension of time this may be countered by the general principle that overheads should be spread according to attenuation of work.
(f) They are strongly time-related, so they do not cater for disturbance without delay, which may occur when activities affected do not lie on the critical path of the programme network or its equivalent. There may still be additional head office costs in some instances.
(g) They take no account of the difference in overheads recovery which may arise due to amounts contained within payments for varied work, fluctuations etc.
(h) They take no account of the general common law principle that the party suffering under a contract must take such measures as are possible to mitigate his loss, such as by redeployment of resources. No formula can do this, as what is possible depends on circumstances, but an adjustment must be made when appropriate.

When the first two formulae are used to deal also with loss of profit (which they link with overheads, but which is considered separately below), some of these points recur, perhaps with a different reference, and should be noted.

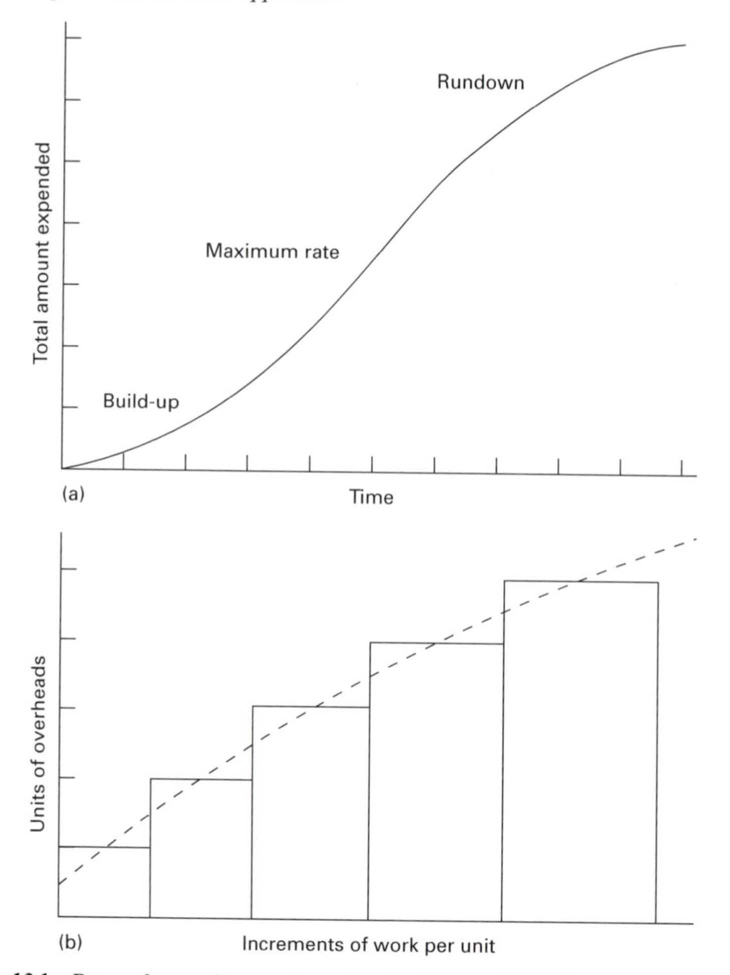

Figure 12.1 Rates of expenditure and overheads: (a) expenditure S-curve and (b) overheads on work

With these limitations noted, it remains that formulae of some sort must enter into some overhead calculations, if only because overheads are measured when accounting and allowed when tendering on a percentage basis, that is by formula. The main burden is the question of proof that the percentage and how it is applied are both reasonable. This is indicated by *Tate and Lyle* v. *Greater London Council* (1982), already referred to in Chapter 9, where the judge said, 'I am satisfied that this head of damage can properly be claimed, [but] I am not prepared to advance into an area of pure speculation when it comes to quantum.' This was with reference to an addition of 2½% for managerial expenses as a customary amount, but without any supporting calculation. It is therefore clear that such overheads are reimbursable in principle, but equally that they must be properly quantified, even if then expressed as a percentage.

In a Canadian case, *Ellis-Don Ltd* v. *The Parking Authority of Toronto* (1978), the contractor demonstrated that a tender allowance of 3.87% for overheads and profit was matched by an actual earning of 4%. As a result, he recovered 3.87% (oddly) on unproductive staff because they 'would otherwise be earning such an amount for the contractor'. There was 'no reason why the plaintiff should not recover' in this way in these circumstances, as distinct from recovering profit lost from some other and particular contract.

The most suitable occasion for use of a percentage is when there is a series of possibly small disturbances adding up to major delay, occurring without the opportunity to reschedule work and such that the several delays and their effects run into one another. Large isolated delays are more likely to lend themselves to isolated assessment without a formula.

Acceptance of a formula method of assessment of contractor's overheads was made in *J. F. Finnegan Limited* v. *Sheffield City Council* (1988). The judge referred to an unreported case, *Whittall Builders Co. Ltd* v. *Chester-le-Street District Council*, where a Recorder had applied a formula, quoting a simple formula in *Hudson's Building Contracts.* In *Finnegan* v. *Sheffield* the judge said he preferred the Hudson formula and that it should apply in the case. In fact, the judge slightly misunderstood the formula and applied something more akin to the Emden formula. Following this decision, formula methods were beginning to become accepted but doubt was raised by the *Tate and Lyle* case where the judge accepted that overhead was acceptable as a head of claim but that the method of calculation was not accepted. The judge remarked that it was up to managers to keep time records of their activities to justify additional overheads. It has been submitted that *Tate and Lyle* casts doubt on formula methods only where a contractor is unable to demonstrate that he has suffered a loss of opportunity to use his resources on other sites. This may be, for instance, in times of recession in the industry. If a contractor was reducing his organisation to 'scale down', it may not be easy for him to show that he has suffered 'loss' of overhead recovery.

In assessment of overheads, allowance should also be made for additional recovery that has happened by any increase in the value of the final account, where resources have been significantly reduced or deployed on other contracts. Naturally allowance should also be made for any staff recovered in any sections of contract recovery, such as 'preliminaries' allowed for in the contract.

In *Babcock Energy Limited* v. *Lodge Sturtevant Limited* (1994), the court awarded head office overheads and other costs, including those incurred in preparation of a claim based on accurate records. This seems reasonable, in that it was held:

> If, however, the magnitude of the problem (calculating/preparing the claim) is such that an untoward degree of time is being spent on it, then their costs should be recoverable. . . . If they (the plaintiffs) can satisfactorily demonstrate that the costs to them of time unnecessarily spent and therefore lost . . . it is for the defendants to show that the losses prima facie incurred are not the correct measure of damages and this [the defendant] failed to do.

In this case payment was made for managerial time involved at head office, including the cost of preparing the claim.

In *Alfred McAlpine Homes North Ltd* v. *Property and Land Contractors Ltd* (1996) recovery of overhead and profits based on the Emden formula was determined. The contract was for housing carried out under JCT 80 and McAlpine postponed the works. Property and Land submitted a claim for loss and expense; this claim was disputed and referred to arbitration. The arbitrator awarded a sum for 'additional head office overheads incurred and not recovered elsewhere'. The arbitrator further found that recovery was not permissible on the Emden formula but that recovery was recoverable on the basis of overhead actually expended and not recovered elsewhere during the period of postponement. There were unusual circumstances in that Property was a subsidiary company of another company and only contracted, in essence, for a limited amount of work at any one time – it was a 'single contract company'. McAlpine appealed and the Official Referee confirmed that the award should stand and Property was entitled to recovery of overhead. The judgment categorised overheads as 'unabsorbed overheads', arising from reduced activity when the contract no longer pays its way towards the overheads and 'additional overheads' where in a period of prolongation of the contract period the contractor has had to allocate more overheads than contemplated at the contract date. The arbitrator had agreed figures given to him and these were not disputed by the parties. The Official Referee also held:

> In ascertaining direct loss or expense under clause 26 of the JCT conditions in respect of plant owned by the contractor the actual loss or expense incurred by the contractor must be ascertained and not any hypothetical loss or expense that might have been incurred whether by way of assumed or typical hire charges or otherwise.

Property and Land owned plant and had to show that they had incurred actual losses or expense in order to be entitled to recovery. In this sense they had to show 'cost' and were not allowed hypothetical charges or 'costs' that they would have incurred had they hired plant.

In *St Modwen Developments Ltd* v. *Bowmer & Kirkland Ltd* (1996) an Official Referee upheld the award of an arbitrator that a shortfall in recovery of overheads arising from a period of prolongation was admissible. The award was made using a formula method and the Official Referee stated:

> It was specifically accepted . . . that the construction industry was buoyant/booming at the material time. These are precisely the conditions in which a formula approach may be acceptable. It must not be forgotten either that the arbitrator, clothed as he expressly was to draw on his own experience, was entitled to conclude from the evidence that was before him in the manner that he did.

The cases of *McAlpine* v. *Property and Land* and *St Modwen* v. *Bowmer* illustrate that formula use is applicable in some instances but not in others. First and foremost, liability must be established and then the method of calculation will follow from the circumstances, the degree of records that are judged appropriate and the information made available to the tribunal.

Northwest Holst Limited v. *CWS* (1998) and *City Axis Limited* v. *Daniel P. Jackson* (1998) have reinstated five tests to be applied before a contractor's entitlement to payment for loss and/or expense is sustainable. Only after this is use of a formula sustainable. The GC/Works/1 contract refers to recovery of expense (not loss) and this is generally held to exclude any loss of contribution to overheads based on a loss of opportunity formula. A contractor will be left with proving by records the actual cost of extra resources employed in an office to justify recompense.

The wider question of overhead ascertainment

Although formulae have their limitations on practical and legal grounds, if used alone, the approach of direct and detailed ascertainment labours under the problems of practicalities.

When it is used, it requires the eliciting of individual costs, such as staff and equipment time. It may be practicable to isolate time for key personnel, such as directors, departmental heads and senior assistants, but the lower down the hierarchy, the less easy it is to make enquiries. This is because higher-level staff will perform tasks related to the disturbance which are unusual in nature and in keeping with the disturbance itself, but other staff will often be performing comparatively routine tasks. These latter tasks may be out of sequence, but they may also be so mixed in with expected tasks as to render clear accounting impossible.

The practical answer may be to assess such staff, and often equipment for similar reasons, as a percentage of those staff who can be separated out. This can be seen as a reversion to a formula approach at a different level within the organisation. It also becomes the only sensible way to deal with premises and similar large, indivisible items of overhead; more detail can be found in the case study of Chapter 31.

It follows that direct identification of head office overhead costs, when used alone in any but the simplest cases, is likely to result in an inadequate reimbursement of costs, rather than an excessive reimbursement. So it appears there is some merit in using both this approach and a formula, modified for any of the reasons given above, to attempt to reach the truth out of what each has to offer. This is partly the method used in the case study of Chapter 31.

Overheads are calculated in the normal course of events for future application as a percentage by reference to total overhead and non-overhead expenditure in the immediate past. This assumes implicitly that the curve for the percentage rate of change, expressed graphically, will be smooth in any part of its length, rather than jump vertically at points, so as to be virtually discontinuous. This, however, is a practical simplification for such purposes as accountancy and estimating. In reality, many elements of overheads present the vertical step effect, as when an extra member of staff is taken on or extra accommodation is acquired. The usual method of representing overheads smoothes out these steps, as indicated in Figure 12.1(b), and this is quite acceptable for most purposes in any enterprise large enough for the inaccuracies not to warrant greater finesse.

It is usually stated that uncritical adoption of a formula of the type discussed will always lead to an excessive reimbursement of overheads, that is a disturbance is bound to occur on a horizontal or slightly sloping part of the actual cost curve. Although this is frequently the case, there is the possibility that it will lead to underreimbursement if a step is caused solely by a particular disturbance. This is likely to be difficult to demonstrate conclusively, as there is seldom a single cause, unless the extra overhead is highly specialised and can be attributed to the one project and its disturbance. From previous discussion, it is clear there still needs to be care in using the available sources of data, as it cannot be argued that the approach should be a formula calculation plus extra cost of jump.

This leads into the area of contributory causes, as with extension of time, and the possibility of allocating costs between the parties. The problems of responsibility for overlapping events are mentioned in Chapter 10; they are as relevant here as the principles for resolution.

These slightly digressive points may be concluded with the comment that the objection to using a formula for head office overheads is not usually carried into assessment of site costs. All-in labour rates, for instance, carry allowances expressed as percentages, and the underlying amounts do not necessarily increase absolutely pro rata to the actual hours expended. Although allowance should be made, as mentioned, when dealing with labour loss and expense for the effect of changes in overtime working etc, a formula can be a relatively crude method of doing this. However, any other approach would be impractical to use. There appears to be room for rather greater tolerance in the courts over head office matters too.

LOSS OF PROFIT

It may also appear that there is room for greater tolerance over the element of profit in loss and expense calculations. But this section merely sets out the strict position which follows from the philosophy that direct loss and expense is to be calculated as equivalent to damages at law (Chapter 9).

Entitlement to reimbursement

The intention of either basis of reimbursement is to put the aggrieved party (in this case the contractor) in the financial position he would have occupied but for the occurrence of disturbance of his regular progress. There is not scope for including extra profit on the loss and expense itself, because this would be equivalent to allowing profit on damages. It may be argued that loss and expense, or more particularly expense, amounts to additional turnover which has been forced upon the contractor under the terms of the contract, and so is rather like variations where additions (and for that matter omissions) include a profit margin. With this view much sympathy may be felt, but it does not find legal favour. The position should be distinguished from that occurring when the contract is determined by the contractor, when he may be entitled to payment of the profit on work which he is unable to perform because of the employer's breach; see *Wraight Ltd* v. *P. H. & T.*

(Holdings) Ltd (1968) in Chapter 18. Even here it is profit on what has not been done which is in view.

The root distinction may be made that, although the contract gives the employer a clear mandate to introduce variations within limits via the architect (Chapter 6), it does not give any mandate to disturb regular progress. The basis of the contract is that the contractor is to be afforded complete and uninterrupted possession of the site and full information as and when he reasonably requires it, among other things. If the employer does not afford these things, he is in breach of contract. The provisions over ascertainment of loss and expense are present to ease matters by giving a procedure for settling which may save recourse to the courts, but without spelling out the formula for settling, as do the variations provisions.

What the contractor may be entitled to, in lieu of profit on the additional 'turnover' which he unwillingly acquires, is reimbursement of loss of profit on other work which he might have been performing with the capacity which he has had tied up against his will. Even here it is necessary for him to be able to demonstrate a reasonable likelihood that he could have been so engaged, and not just to allege that this would have followed without question, a point established in *Bernard Sunley* v. *Cunard* (Chapter 11).

This does not amount to showing that he has been unable to take on a further contract or so, simply because of the overrun on the present one. Except in extreme situations, this is most unlikely to occur as a single event which can be traced so inexorably within his records. What he needs to do is to show that he is running with a reasonably full order book across the contract period and beyond, so that his prime resources are adequately committed without the disturbance element now in question. His prime resources are those which condition how much work he can undertake, such as his capital, permanent organisation and workforce, provided the elements concerned cannot reasonably and economically be supplemented to carry him over the hump period.

In *Ellis-Don* v. *The Parking Authority of Toronto* (1978) it was said, 'If a contractor is entitled to damages for loss of income to cover head office overheads, why should he not also be entitled to damages for loss of income that would result in normal profit.' The point was made in *Peak Construction (Liverpool) Ltd* v. *McKinney Foundations Ltd* (1970), which dealt with a contract in which a year's suspension had occurred in conditions of uncertainty about resumption. Although the plaintiffs appeared to hesitate in some respects over the basis of their case, the Court of Appeal came to the view that they were arguing for loss of opportunity to earn profit on other work which they could not take on. This was the case as presented earlier to the Official Referee. On this basis, the Court of Appeal rejected the defendants' submission that no loss of profit had been established, this being 'probably untenable' on the facts presented. However, the Court was quite strong over the sort of evidence which the Official Referee might need in a new trial:

> Every issue under this head should be open for the official referee (sic) to consider. It might be of some help to him not to be left only with the evidence of the plaintiffs' auditor on this point; possibly some evidence as to what the site organisation consisted of, what part of the head office is being referred to,

> and what they were doing at the material times, could be of help. Moreover, it is possible, I suppose, that an official referee might think it useful to have an analysis of the yearly turnover from, say, 1962 right up to say, 1969, so that if the case is put before him on the basis that work was lost during 1966 and 1967 by reason of the plaintiffs not being free to take on any other work, he would be helped in forming an assessment of any loss of profit sustained by the plaintiffs.

Notably, this suggests a long period spanning the actual time of suspension and considerable data made available for assessment of the likely position, although not to prove cast-iron loss, which is impossible in most instances. This is not therefore a matter of absolute proof, but of showing a sensible balance of probability that the disturbance has imposed an extra load on the contractor beyond his expectations. Rather like overheads, it is a case in which the capacity of the organisation has some flexibility to cope with a single contract subject to disturbance, but where any instance can be argued as causing some theoretical effect that cannot be absorbed. If the contractor is subjected to disturbances on several contracts then, according to his scale of business, the result can be assumed to be that his general capacity is under some strain.

This overload is a situation in which a time overrun, as well as the financial overrun, may lead more easily to a loss of profit being considered. On the other hand, a plain intensification of working within the same overall period may mean little more than taking on extra labour or plant, to be absorbed within the main organisation of the project. There is no categorical answer to dealing with profit, although it is clear that allowing this element becomes the more credible, the larger the rest of the amount for loss and expense.

The Hudson formula discussed above is given by the editor of Hudson as subject to the caveat that the element of profit, as distinct from overheads, included in the standard formula ranks for inclusion only if it would be reasonably possible for it to be earned elsewhere (as *Peak* v. *McKinney* shows), in the absence of the disturbing events. This lessens objections to the Hudson formula over profit, but not overheads, provided the profit corresponds to the whole of the disturbance. The formula actually shows it as profit on the balance of the loss and expense itself, so that it would need to be stated formally in the ascertainment as being rather profit lost elsewhere. However, the broader criticisms of the formula remain.

But it would no longer seem to be in doubt that, however it is calculated, there is an entitlement of loss of profit. In *Obagi* v. *Stanborough (Developments) and Others* (1993) it was held that a party could recover damages for loss of profit which he had a substantial chance and not a mere speculative possibility of making were it not to have been for a breach of contract by the other party. He does not need to prove on the balance of probabilities that he would definitely have made such a profit.

Calculation of loss of profit

The amount of profit to be included consists of two factors: its extent and its level. Once the inclusion of profit has been established in principle and as related to all elements of disturbance, the extent may be a matter of taking the amount of loss and expense otherwise ascertained as equivalent to work displaced elsewhere which the contractor has been unable to perform. Here this factor of the profit calculation is the same as if profit had been allowed upon the additional 'turnover' within the contract generated by the disturbance.

It is not always suitable to work like this. There may, for example, have been a small disturbance early in the programme which alone does not substantiate loss of profit on other work, followed much later by a significant disturbance which does. If these two are quite distinct, there would appear little justification for allowing loss of profit due to both. On the other hand, a series of individually small disturbances may add up to a combined case for loss of profit. Between these two extremes, the gradation of possibilities is considerable, with room for difference of opinion over what is allowable.

The second factor is the profit level to be applied to the first factor. Here, as indicated by the legal cases under 'Some principles of damages' in Chapter 9, the guide is the level of profit which, before the disturbance, the employer may have expected that the contractor would reasonably have earned on the alternative work foregone. Usually this is to be equated with the normal level of profit available to a comparable contractor operating in similar market conditions and with a similar range of work. Only in peculiar circumstances made known to the employer in advance of the contract will the contractor be able to claim for an unduly high rate of profit, on the principles enunciated in the *Victoria Laundry* case in Chapter 9.

At this point, the distinction between allowing profit on additional turnover and allowing loss of profit on work foregone becomes important again. This is because the measure of profit allowed does not depend directly upon the profit level included in the immediate contract, which need not be demonstrated here, and whether or not this profit has been achieved or even exceeded. It is the level of profit foregone which is relevant (see the example in Chapter 31). Given even conditions of business for the contractor, this distinction may not matter. If conditions are fluctuating within limits, such as might be anticipated even by a moderately businesslike employer, then it is the general level which counts. Even if the contractor obtained the contract at cost by adding no profit margin, this in itself will not debar him from being reimbursed at a positive level. On the other hand, he may then need to demonstrate that he left off his profit margin by mistake, rather than of deliberate policy to obtain work in a slack period. Otherwise he will be needing to show that work prospects have improved since, so that his order book has cheered up!

INTEREST AND FINANCING CHARGES

Interest and financing charges have a number of precise but differing definitions, according to when they are being used and by whom. For present purposes, they are taken as having limited meanings:

(a) *Interest*: The charge on monies owed by one person to another, most particularly the amount that may be due from the employer to the contractor on outstanding sums.
(b) *Financing charges*: The cost to one person, and here the contractor in particular, of servicing his working capital or maintaining an overdraft to keep solvent.

It may be seen that one man's interest is another man's financing charges. Normal recurring amounts are covered by contractors within their head office overheads, whether taking the form of payments against borrowing or dividends to shareholders, with dividends becoming elements of profit. For loss and expense leading to extra amounts, the situation is similar to profit: the contractor is not entitled to charge interest, but he may possibly be able to recover financing charges broadly as charged to him, with the actual amounts in practice not too dissimilar. These mainly relate to the delay in performing work and receiving payments generally, due to disturbance, and so leading to an impaired cash flow for the enterprise. They are higher to him for his own costs than for those of subcontractors, where they may not arise for him, although the subcontractors may have their own expenses. An illustration is given in Chapter 31.

Interest

It remains the general legal position that interest is not payable as damages simply because a sum legally due is paid late. This appears to go back to the ancient dislike of usury, backed perhaps more recently by fear of the possibility of a creditor not desperately needing his money and failing to press for it, while welcome interest mounted up. The point has been maintained in *London, Chatham & Dover Railway Co.* v. *South Eastern Railway Co.* (1893).

In Chapter 8 this subject was referred to under 'certification' and it is now time to consider it, along with a contractual right under JCT contracts for deductions made by an employer from certified amounts. It is necessary to consider interest on what may collectively be described as 'undercertification', although that is not in many instances the correct position. More neutrally it is the position where a larger sum of money is paid at a later date, either in a claim for loss/expense or in an award by an arbitrator or a court.

Although the courts do not wholeheartedly embrace the *London Chatham* position today, they are not completely free to avoid it without proper legislation, but have 'distinguished' it in some cases. What legislation there is, does not give complete freedom. JCT contracts now have the right for a contractor to receive interest on amounts certified but not paid but that is not the matter being considered here. It is also the case that interest in the courts can be awarded as simple interest only, not as compound. An arbitrator may, however, award interest on an amount still outstanding and included within an award, by virtue of the Arbitration Act 1996 and there is discretion for him to award simple or compound interest. It would appear that there may even be compound interest in calculating financing charges up to the date of an application over loss and expense under JCT clauses, although only simple interest thereafter (see *Minter* v. *Welsh Health* below).

The line taken in distinguishing other cases from the *London Chatham* case is to take *London Chatham* as relating to 'general damages' for delayed payment. By distinction, 'special damages' may arise when the serious possibility of loss by the other can be supposed to be in the contemplation of the defaulting party. This is the line of reasoning flowing from *Hadley* v. *Baxendale* and other cases taken in Chapter 9. In *Wadsworth* v. *Lydall* (1981), the Court of Appeal awarded special damages for interest due to late settlement of provisions in a dissolution of partnership. The House of Lords has approved this decision in *President of India* v. *La Pintada Compania Navigacion SA* (1982), a shipping case, stating that the *Chatham* decision 'did not extend to claims for special damages'.

Thus, in a straight loss and expense application, at present the contractor has no absolute assurance that all inclusions of interest as such will be met with favour. Things are different if, but only if, the contract expressly provides for interest on overdue amounts, as the Institution of Civil Engineers' contract provides over late honouring of certificates. In *Lubenham Fideliteis* v. *South Pembrokeshire District Council* (1986), mentioned in Chapter 8, it was held that even when an architect deducted liquidated damages on the face of his certificate (the employer should have been the party to make such a deduction, assuming it was correct), and as such it was held to be 'undercertification' the judge found that it was not a breach by the employer, hence the contractor had no entitlement to interest as a result of the architect's undercertification. In *BP Chemicals Limited* v. *Kingdom Engineering (Fife) Limited* (1994), mentioned in Chapter 8, it was held under an amended ICE form (fifth edition) that the only sum due under the certificate was the sum certified and that a larger sum due under an arbitrator's award was not therefore payable until the award was made, hence interest was not awardable from before a date when the cause of action began (i.e. the date of commencement by notice of arbitration).

Financing charges

The essential distinction with financing charges is that they are costs incurred by the contractor in operating his business, so that an excess may rank as 'expense' when due to disturbance. They may arise in two ways: the contractor may use his own capital to run his business, and so be deprived of a return which he might gain on his money invested elsewhere, or he may need to borrow by way of overdraft or otherwise, and so incur interest charges himself. As this interest is an expense to the contractor, it does not rank as interest charged directly to the employer. It is thus admissible as loss and expense in suitable circumstances. It needs to be proved reasonably (like loss of profit), and could not often be supported in a small claim, when no extra financing would really arise. This is considered in Chapter 33, where the comments on small contractors should be noted.

The old view, perhaps helpful to smaller contractors and set out by the House of Lords in *The Edison* (1933), was that financing charges must arise out of the claimant's 'impecuniosity as a separate and concurrent issue' distinct from the cause of direct loss, so that they rank as indirect. The House of Lords and other courts have tended not to follow the decision here, or at least to 'distinguish' the position in other cases, as the argument is highly suspect in the light of modern financing arrangements throughout industry. So, for instance, in *Dodd Properties (Kent) & Others Ltd* v. *The City of Canterbury & Others* (1979) allowance was made for inflation relating to an indemnity amount.

Some leading cases

In *Sunley* v. *Cunard* (see above under 'Loss of profit' and Chapter 11 under 'Plant'), the contractor was awarded, among other things, both depreciation of and interest on the cost of plant tied up in a delay, as wasted costs incurred. Somewhat more recently in this field, it is worth going back to the position obtaining under the 1963 JCT contract, as related legal cases lead into the distinctions introduced by the 1980 JCT contracts and the 1984 IFC contract.

Salient in this respect is the Court of Appeal case of *F. G. Minter Ltd* v. *Welsh Health Technical Services Organisation* (1980). This upheld that direct loss was the equivalent of damages for breach, so following the distinction in *Saint Line Ltd* v. *Richardsons, Westgarth & Co.* (1940) between direct and indirect or consequential loss. This means that all loss flowing without intervening cause from the primary cause is direct, so that financing charges are in principle recoverable. In *Minter* v. *WHTSO* the contract under examination was JCT 1963, and in particular its clause 24(1), predecessor of the 1980 clause 26.1, and also the then clause 11(6) which dealt separately with loss and expense due to the instruction of variations. These clauses contained wording (emphasis added) as follows:

11(6) If the said application [over loss and expense] is made within a reasonable time of the loss or expense *having been incurred* . . .

24(1) If upon written application being made . . . the Architect is of the opinion that the Contractor *has been involved in direct loss and/ or expense* . . .

It highlights the critical wording that was held to mean an application had to relate to loss and expense which was entirely past and incurred at the date of application. Minters had applied after the loss and expense had been incurred, mainly on-site, but in the action sought to recover financing charges from the time of loss and expense on-site up to the date of payment following the architect's certificate.

It was held that Minters could recover only the portion up to their application, as only this was past by then. It would appear from the judgment that they would have been able to recover up to the last application, had they given a series of later notices at the appropriate intervals stating that they had incurred further loss and expense by way of financing charges. The Court of Appeal definitely disagreed with the High Court's view that such later charges were indirect expense rather than direct. They thus confirmed that financing charges were in principle admissible in loss and expense claims.

In that case, the Court of Appeal rejected the defence that Minters were not claiming direct losses at all, but were making 'a naked claim for interest'. This is an instance of the courts tending to undermine the long-standing principle about interest-related damages noted above.

The JCT 1980 form amended the critical wording by giving, in clause 26.1 (which absorbs both the earlier clauses), the words 'written application . . . that he has incurred *or is likely to incur*' (emphasis added). The effect of this is thought to be that one application made 'as soon as it has become . . . apparent' will suffice in principle to cover all aspects of loss and expense, including financing charges (see the next case mentioned below). It is, however, prudent for the contractor to mention this element in his application when it applies, as it is not one which will always apply and not one that may be obvious to an employer. It may also be a very large item. Otherwise, he may face a rebuttal on the grounds that the urgency of settling had not been made clear, so that charges have been mounting up at least partly avoidably.

More recently, but still in relation to the 1963 JCT wording, there is *Rees and Kirby Ltd* v. *City of Swansea* (1985). Here the Court of Appeal considered a situation in which the contractor had duly given notice of claim. This was not pressed with maximum vigour, because the council (City of Swansea) and the contractor entered into prolonged alternative negotiations about whether to turn the fixed price contract into a fluctuating one or even to settle for an ex gratia payment in a time of high inflation. This was seen by the contractor as a way of absorbing the loss and expense. Unfortunately, the negotiations broke down, leaving an extended period to be covered for financing charges. The contractor then lodged his claim for these charges, which was rejected by the council, so leading to the action. The judgment ranged over several points:

(a) The correct timing of the application was contested, but upheld by the court as it had been made after the loss concerned, even if delayed by the intervening negotiations. In the *Minter* case, there had been no such following application.
(b) The notice given had referred sufficiently, if in a 'most general' way, to the charges being incurred. The need to make successive applications, so that they were within a reasonable time of the suffering of loss and expense, was endorsed. This had been done. It was indicated in passing that this specific mention might not be necessary under the JCT 1980 form.
(c) The council's defence was not accepted that there was a cut-off in principle for any amounts to accrue at practical completion. Charges could continue to build up until the last of the periodic applications which the contractor might make before a certificate is issued, including the amount ascertained. This effectively means until agreement of the amount and would appear as relevant under the JCT 1980 contract.
(d) The council's argument was accepted that a part of the delay during which charges accrued was not due to the disturbance. This was on the basis that the negotiations, as a collateral activity, were not caused by the disturbance and so were not a direct contribution to the expense, but only indirect by being an intervening cause. This is based on the special factors in the case and is not of general applicability.

(e) The contractor was allowed the charges at compound interest, rather than simple interest, which is normal when a court allows interest as such. In this instance the basis was that financing charges as such are calculated on a compound basis, so the contractor should recover on the same basis as that on which he had lost.

Rees and Kirby Ltd thus received a substantial part of what they had claimed, but not all. The special element here was the delay to which they were party while the abortive negotiations went on. But for this delay, it would appear that they would have been entirely successful in the action as it would have been. In particular, they would have been entitled to a resumption of entitlement after the invalid period, provided notices had been resumed under the JCT 1963 requirements.

The question of such delays is an interesting one in view of the way that negotiations often proceed over a long period, even when they relate directly and solely to ascertaining loss and expense, and are allied to the possibility of either party unduly protracting discussions. If the contractor drags things out, the employer may be on good ground in refusing to meet the whole amount. If the employer is the tardy party, the contractor should watch carefully that he does not lose his right to charges by default. He should keep up his notices if under a JCT 1963 contract or equivalent, and advise the employer of his continuing claim for financing charges under a 1980 form, where a notice may look ahead. This is against the background of the desirability of making it clear, whenever these charges arise, that they will be included in the claim, as this last case emphasises.

Advanced warning of these charges under JCT 1980 should lead to a desire to settle early and avoid a build-up of charges. There is also a clear argument for the employer to pay amounts progressively, and perhaps on a provisional basis, in interim certificates and so reduce the level of charges accruing up to the point of settlement.

Although the contractor may be able to keep his right alive up to the point of inclusion in a certificate, or virtually so, by his last notice of loss or by the date of agreement of the amount ascertained, he cannot secure any extra reimbursement of financing charges when once a certificate has been issued and payment is then not forthcoming. Amounts for loss and expense are no different from any other when certified payment is delayed, and what would be straight interest can now be claimed; Chapter 8 looks at the effects of the Housing Grants, Construction and Regeneration Act 1996, which has meant there is now a statutory and a contractual right to simple interest on late payment of certified sums, whatever they represent. Although it did not relate to financing charges, the case of *Croudace* v. *Lambeth*, considered in Chapter 17, shows that the employer cannot hide behind the absence of a certificate to put off payment, when it is missing due to his own dilatoriness.

Calculation of financing charges

Several points arise over calculation of amounts, without going into technical detail here. It has been noted that compound interest may be applicable. Whether it is actually appropriate will depend on whether it was chargeable to the contractor himself. Equally, the rate of interest used should reflect the rate chargeable, not necessarily the actual rate which the contractor paid, but rather the general commercial rate obtaining in the market-place during the period concerned.

This reflects the possibility that the contractor may have been charged a specially high rate due to his own circumstances and standing. As the employer cannot be held reasonably to have anticipated this as the likely level of loss, unless he had explicit warning of the fact, he cannot be expected to pay the difference (see the discussion on damages at the beginning of Chapter 9). This should be distinguished from the situation when the contractor incurs charges at all because he is, for instance, in overdraft because he is 'impecunious', that is has inadequate capital. Here he is not debarred from the appropriate normal level of recovery, as discussed above regarding *The Edison.*

Whatever happens, it is necessary for whoever is ascertaining loss and expense to check what actually is being paid by the contractor to his regular funding body, modified perhaps in the light of published general lending rates plus the usual 1% or 2% applicable in the circumstances. The rates are likely to fluctuate frequently over the lengthy sort of period which warrants the inclusion of financing charges as a head of claim. Within limits, some sensible averaging can be used to avoid what could become daily calculations. Compounding of the amounts should be applied at the intervals applicable to the contractor himself in any routine case.

Taxation and the abatement of financing charges

As a tailpiece, it is worth mentioning taxation and the abatement of financing charges. In *Tate & Lyle* v. *Greater London Council* (see 'Formula calculations') it was decided that, when calculating interest, account should be taken of the effect of setting off the sums still due against profits for the year in question, when making returns for corporation tax. This has the effect of deferring the payment of some tax, so that money remains with the contractor or other person claiming reimbursement of financing charges and reduces the charges effectively. It appears that contractors dealing with central government departments are meeting pressure for this sort of abatement to be made in all relevant situations.

The calculation of amounts may be quite complex, owing to changes in the rate of tax, the period over which it arises to be set off against, and the particular tax status of the contractor and whether he can defer payment, quite apart from the immediate issue of financing charges as an element of loss and expense. The existence of several contemporary contracts with this element further complicates affairs, over whether the abatement should be shared and precisely how. This is a case where the contractor is responsible for producing figures, as always, so that loss may be ascertained, but where the onus of proof of precise quantum perhaps lies with those acting for the employer.

It appears that the argument for such an adjustment under any form of contract and with any type of employer is sound, and this type of calculation may be expected to increase, given commercial life as it is. It is an area where some reasonable averaging of amounts should be acceptable to avoid effort reaching the near absurd in minor instances at least.

Costs of preparing a claim for extension of time and loss and expense

This has been covered in some detail in Chapter 9. For the completeness of this chapter, it is emphasised again that the circumstances of each 'claim' will need to be examined but that in principle a contractor will be able to recover these costs. It is arguable that preparation of details for an extension of time should be those that will be valuable within the provisions of the contract, unless they are very extensive. Costs for preparing claims for loss and expense may also fall into this area, but two matters should always be reviewed.

Firstly, the cost of producing submission documents in support of a claim that arises from provisions in a contract for the contractor to carry out such submissions in order to obtain an entitlement to payment under the contract, it is unlikely they are going to be recoverable, unless this is expressly stated in the contract. Although it is extra work, beyond what a contractor could have foreseen, a contractor (or subcontractor) is complying with the requirements and provisions of the contract. This can produce a circular argument in that it can become a matter of degree over how much 'a reasonably competent and experienced contractor' could be deemed to have assumed that he should include for dealing with 'claims' – in what after all is often meant to be a 'fully designed contract', hence something that should not be subject to disruption and delay through bad information flow, many instructions and variations etc.

Secondly, where the conditions of contract require an architect or engineer to receive a notice and details from a contractor (or a contractor to receive details from one of his subcontractors) and then to ascertain loss and expense, if there is a failure by the architect or engineer to ascertain, the result will be a breach of contract. This may well lead to additional costs being incurred by a contractor that would fall within the *Hadley* v. *Baxendale* rules. Chapter 9 refers to *James Longley* v. *South West Thames Regional Health Authority* (1983), *Tate & Lyle Food and Distribution Ltd* v. *Greater London Council* (1982), *London Borough of Merton* v. *Stanley Hugh Leach Ltd* (1985), *Croudace Ltd* v. *London Borough of Lambeth* (1986) and *Babcock Energy Limited* v. *Lodge Sturtevant Limited* (1994).

CHAPTER 13

CLAIMS AND MULTIPLE CONTRACTS

- Scope of claims in multiple contracts
- Broad nature and selective features of multiple-contract projects
- Special aspects of claims in multiple-contract patterns

This chapter does not deal with any standard form of contract or even with a clearly identifiable contract arrangement, but partly with any pattern of multiple contracts (Chapter 2) and partly with subcontract relations (Chapter 7) and cognate ones like those of works contracts in management contracts. To ease discussion, it is mostly related to multiple *direct* contracts, as described in Chapter 2, with a few direct mentions of subcontractors, as generally reference to subcontractors and others may be inferred fairly easily. Aspects specifically relating to loss and expense of all categories of subcontractors under JCT contracts are discussed in the last section of Chapter 17, with a mention of the three-way case.

Here is an overall survey of some considerations affecting claims approach and then some governing points which may be observed in dealing with claims, for which the internal principles will often follow those considered at length throughout this book. More precise discussion has been left aside, as there will often be special written or drastically adapted contract conditions and other documents to suit the nature of the project.

SCOPE OF CLAIMS IN MULTIPLE CONTRACTS

As throughout the rest of this book, the term 'claim' (not actually used in most contracts) has a particular reference to categories leading to 'loss and expense', as for instance in the JCT documents, or to cognate expressions and results under other contracts such as the GC/Works/1 and ICE forms, although distinctions in effect must be noted (Chapters 20 and 21). The term may also reasonably be extended here to cover such relief as comes by way of extension of time or other contractual provisions such as enhanced rates for measured work.

But it may also cover reimbursement for the extra costs of contractors due to the various effects of integration of the complete project works under the directions of the client's project manager, such as those listed below. These may be covered by instructions under the terms of the individual contracts, or may be of the nature of loss and expense. Either way, they represent outgoings or loss not to be anticipated when working in the single main contractor role. What is therefore significant is that the causes can be much wider than the lists in the JCT contracts, whereas the results may sometimes be incurred in a more orderly manner that does not lead to disturbance of regular progress, even though there may be loss and expense in any of the categories described in Chapters 11 and 12. By way of summary, the following is an indicative, if incomplete categorisation:

(a) Matters affecting regular progress or otherwise causing prolongation or disruption, so leading to loss and expense.
(b) Instructions to perform additional work, perhaps of a permanent nature, needed to fill some gap in what has been anticipated, or in substitution for similar work by another contractor who may, for instance, already have left the site.
(c) The non-availability of common site facilities, operating space or other features which a contractor would reasonably expect to be available, leading to some form of loss and expense and possibly additional work to be paid for under other provisions of the contract.
(d) 'Any act, omission or default of the client or any person or other contractor for whom he is responsible', which may be of the nature of breach of contract (Chapter 9) so far as the present contractor is concerned.

Category (a) is straightforward claim territory, depending upon the provisions of the contract concerned. Categories (b) and particularly (c) may well be a mixture of measured or whatever work the contract contains and of loss and expense, but with the causes lying at least in part beyond the scope of JCT disturbance.

The quotation in (d) is cast in similar terms to the wording of provisions in the JCT and JCT related subcontracts, with the substitution of 'client' and 'contractor' and this is intended to show the similarity of effect. The whole phrase is wide enough to cover elements otherwise within the scope of breach of contract, as are some of the JCT 'matters' for loss and expense, but stretching further by the word 'any'. How a contract will express this may vary, but the essence is that it becomes necessary to cope with the degree of irregular working which may come about with the interaction of several independent contractors and of a client who may himself be active. In the case of process and other engineering work, this may occur because the client is changing his mind along the way, at least partly because technological or market developments are making what was intended obsolete in part even before it comes on stream. This may be justified in terms of the overall trade-off or the client's business, although it must be suspected that such necessities also breed an atmosphere in which changes are then spawned almost as of habit. So far as advice to the client goes, others have to seek to discern the differences and when to advise what.

BROAD NATURE AND SELECTIVE FEATURES OF MULTIPLE-CONTRACT PROJECTS

Nature

The pattern covered for present purposes by the term 'multiple contracts' is that in which there are several firms working on-site, all in direct relationship with the client and not in contract with one another, although as a minimum with some obligation to the client not to interfere with one another's operations. Some of the key features are set out here to prepare the way, as they include the types of issue which come into play when disputes or claims particular to this operating pattern come about. They are not unique here and many, with some different nuances, crop up in main contractor and subcontractor relationships and have done throughout this book.

These features are also largely applicable to the situation existing when there is a joint venture or other consortium project, such that the client is usually in contract with a single specially constituted body, which exists primarily to hold the separate firms of substance together. This almost man-of-straw body is then in contract with its individual working member firms on a basis such as is now being reviewed. This introduces an additional layer at which there may be a formally distinct layer of claims, but not effectively any more procedural complication than that – which is as well! The constituent firms will each claim against the body of which they are all a part, and that body in turn against the client.

Features

As has been suggested, the multiple-contract situation is likely to arise in response to a project with a band of several of the following characteristics:

(a) Extremely large-scale works.
(b) A resultant need to compress the total time span to avoid undue attenuation.
(c) Several diverse types of work, going on in parallel and with broadly similar scale of working, so that no one contractor emerges as suitable to coordinate the whole project, even if essentially qualified for such a role in wider circumstances.
(d) Substantial design by various of the contractors, who may be in large measure consultants for the project.
(e) Overlap of design and construction for a contractor and between contractors.
(f) Uncertainty over just what is required in terms of work and perhaps over other priorities, such as time and timing, emanating from the mix of the other characteristics.

The contractual arrangements used for individual contractors may vary considerably, but should have a common strand running through them to regulate practical working on-site. A major part is played in this by the client's project manager, who is responsible among other matters for seeing that the several contractors operate together in accordance with their contracts and to secure the optimum realisation of the project. These two objectives may clash if preparation for them has not been thought through adequately, as often it cannot be. It thus may be that the project manager needs to issue instructions which lead to payments for extra work performed by some contractors involved and to claims by others. These may be in the following areas:

(a) Harmonising working areas (including temporary use of areas ultimately to be occupied by the work of others), access, storage etc within physical site constraints.
(b) Harmonising working within programme constraints, affecting who is where and when, but also what a contractor needs to have done ahead of his own work, what he needs to perform in stages with the work of others coming between, and how he will leave his work in respect of following contractors. These are not things to be left between contractors to sort out, as they are not responsible for the overall programme.
(c) Harmonising working methods, to the extent that they may cause undue disturbance to other contractors in respect of discomfort and disturbance.
(d) Ensuring that protection of finished work of preceding contractors is effected, and cognate but not so obvious matters like maintaining humidity or dryness.
(e) Commissioning work operationally which needs to be integrated with that of others and where the commissioning may occur well after the main work of installation.
(f) Pursuing charges against defaulting contractors for infringement of the foregoing matters, where another contractor has necessarily had to incur expense to rectify matters.
(g) Ordering, paying for, receiving and storing materials etc which, because of long lead times or comparable pressures, are to be installed by a contractor yet to be appointed.
(h) Reallocating parcels of work between contractors, if special factors intervene.
(i) Allocating the availability of restricted common facilities when there is competing demand.
(j) Some mediation between disputing contractors, at least to seek amicable ways to resolve disputes when their origin lies in project manager's instructions.

Some of these areas will be subject to the 'common strand' conditions, such as that contractors give proper notice of requirements, but a disturbance element or other excess cost frequently enters in. The provision of temporary facilities needs special care in view of common use: there may be a clash over timing or interference with others in and by the very use. Stipulations and related procedures about what is to be provided by the contractor and what is to be provided to him, with the all-important question of what is free of charge and what is chargeable, are needed over the following salient elements:

(a) *Accommodation for staff and workpeople*: Any part of this is best used by a single contractor at a time, except for canteens and sanitation and welfare, which should be managed centrally.
(b) *Workshops*: These are best used by one contractor at a time, even if the space itself is provided and allocated centrally.
(c) *Services and removal of waste*: These present problems because they are threaded through the working areas of a project and workshop areas and they need frequent adaptation as the project develops. Such items as main distribution boards and chutes are often best provided centrally.

(d) *Scaffolding and staging*: These are similar to item (c) and may actually be shared by several sets of workpeople at one time. The backbone provision is quite often a central responsibility.
(e) *Handling and hoisting equipment*: This tends to be concentrated at particular locations, but it also tends to be needed by everyone. The need for unified control over all major items (especially cranes with their wide sweeps) is paramount, unless the site is quite open.

Figure 13.1 gives a basic picture of what may occur. This is *not* a site layout, but a diagrammatic illustration of interrelationships to bring the foregoing list together. The hexagon shows six contractors, each of whom interfaces with some of the others. Some of them need scaffolding, sharing this with another at their interface in each case. A crane is provided for use by all of them, when the question of queuing is likely to present itself on a regular basis. Off the construction area, accommodation divides into the common, shared items and items particular to each contractor. Services and other facilities are available to all the foregoing within the construction area and outside and will need adaptation quite frequently to various extents.

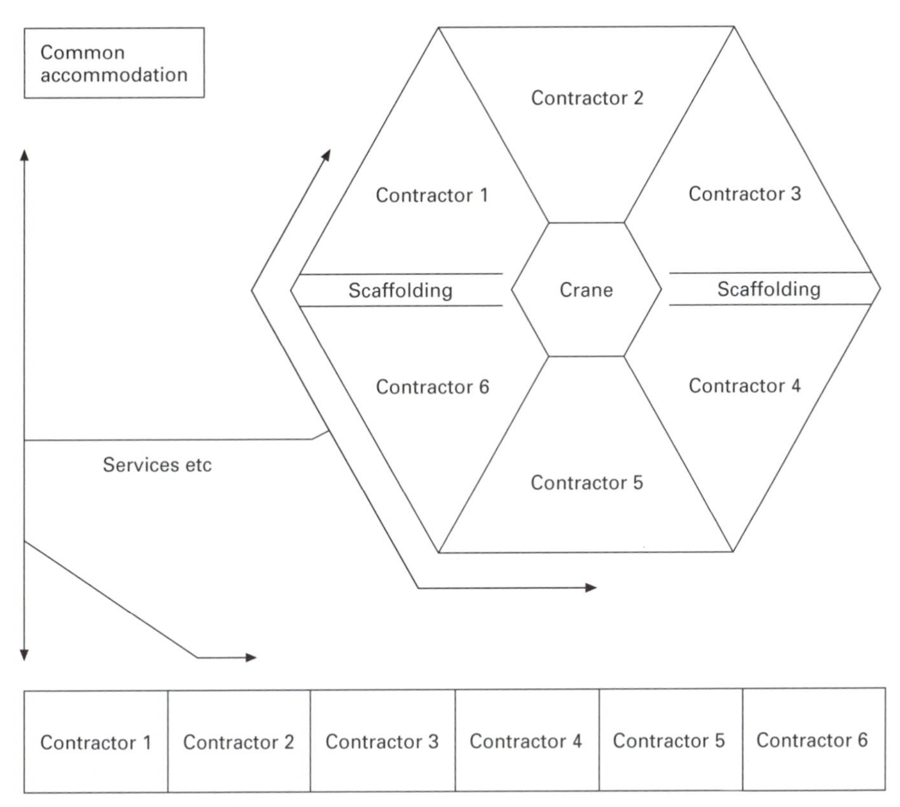

Figure 13.1 Temporary site facilities for construction: interrelationships

Again, what is described above is in principle and detail largely what exists when a main contractor engages subcontractors, or within a management contract. Perhaps the main distinction is that it is assumed in the descriptions above that the client will make facilities available rather more generously than often happens with at least the main and subcontract situation.

At first glance, the advantage of charging for any facilities provided to contractors is clear: the incentive to be economical in their use, particularly with such items as power supplies which are relatively expensive and scaffolding which may be shared. However, this has to be set against the problems of efficient metering and billing, when several contractors are working in the same area and off the same supplies and when disputes may be difficult to resolve. A uniform solution may not be the best for all projects.

SPECIAL ASPECTS OF CLAIMS IN MULTIPLE-CONTRACT PATTERNS

Many of the claims which can arise within a multiple-contractor setting may be dealt with on the same principles and with similar detail to what is needed in the single-contractor setting. Essentially any one contractor has any claim entitlement arising under contract against the client (with whom alone he has a contract), even if the root is in the actions or inactions of another contractor, whereas any redress of the client will lie against that other contractor. This pattern is comparable to that existing for subcontractors, as outlined in Chapter 17. Indeed that very pattern may also exist within the present setting as the various contractors will have their own subcontractors, and for some of whom there may even be nomination arrangements. The main lines of liability may be set out:

(a) Those between the client and contractor, possibly passed up in whole or part by a subcontractor.
(b) Those under (a) but where the client then has a cause of redress (effectively a counter-claim) against another contractor, which that contractor may be entitled to pass down in whole or part to a subcontractor.
(c) Those where the client has such a cause of redress against a contractor, arising out of some contingency that does not track back to another contractor.
(d) Those between contractors or subcontractors and not involving the client, which arise out of special agreements into which they may have entered on their own initiative over some parcel of work. Here, in effect, there is a subletting between the firms, which the client may or may not know about, according to the requirements of the head contracts which he has with his contractors.

In instances (a) and (b) the client faces clear cases where, frequently, the matter turns into a claim in the sense regularly considered in this book, although other cases may crop up. Instance (a) is a linear situation, wherein the claim travels straight up the line. So is instance (b), with the claim travelling round a bent path, and the temptation to the client to treat it as a three-way claim by seeking to pass it across from one contractor to the other must be resisted. The recipient contractor has no contractual relation with the other and so is entitled to shrug the matter off while, as between two independent contractors, it is less likely that the aggrieved one will receive his entitlement. The client has created the scenario and should bear the complications.

The problem for the client here is that he has two sets of negotiations over the same issue, but this is how it is, even though he is due (if correct in his contentions) to receive his direct loss and expense, and so pass the same amount through from the other contractor. He also faces the problem of settling too high with the first contractor and meeting resistance from the second. In this case he is as well to go so far with the first and then sound out the second, acting as a form of postman or even arbiter, but still not trying to withdraw from negotiations.

In instance (c) the cause for claiming (in a different sense from that generally used in this book) may be something like liquidated damages where the amount has to be settled on the terms in the relevant contract. Nevertheless, follow-on claims by another contractor or more coming in later in the programme may have to be settled on their own terms without reference to the settlement in the earlier contract.

Instance (d) is one of which the client should generally keep clear, unless it embraces a subletting situation, which he must take into account for that reason alone.

Although it is easy to segregate these cases in the placid pages of a book, the overarching problem is often that the various varieties entangle with one another, and then often in parts which vary between contractors and subcontractors, so that negotiations do overlap where they hopefully might not and then reiterate several times. It may be necessary to hold a meeting between the main players in what appears to be the ball of wool in question and try to separate out who is dealing with whom over the various strands, even if this results in redefining the ball of wool.

A small illustration of the type of problem which can arise, even in the more restricted circumstances of a modest building project – a problem between architect, contractor and subcontractor – is given under 'Subsidiary problem with the steelwork' in Chapter 32.

CHAPTER 14

GLOBAL CLAIMS

- Introduction
- Development of case law on global claims

INTRODUCTION

Chapters 16 and 17 discuss the provisions in JCT 80 and 84 contracts for recovery of primary and secondary loss when the scope of the works and/or the circumstances described in the contract documents do not occur. Considerations for recovery of secondary loss have been examined in Chapter 12, in particular the 'hidden' nature of costs that may result when a programme is disrupted and delayed. In Chapter 11 the effects that changes to the content and/or circumstances of a contract have on primary matters such as labour, materials, plant and site overheads ('preliminaries', in quantity surveyor's and contractor's jargon) have been discussed. On the majority of contracts, that should have been preplanned in content and context, if variations and changes in circumstances do occur, there should be little difficulty in reaching agreement of final figures, for instance, using valuation provisions in JCT 80 clause 13 and IFC 84 clause 3.6 alongside loss and expense provisions in the JCT contracts.

The correct and still the usual way in the majority of circumstances found in construction contracts (assuming that negotiation has not enabled a settlement between the parties) is for a 'claim' to be presented in a final account or before an arbitrator or a court by linking each cause with each effect. A basic tenet of any negotiation and particularly of any adversarial proceedings is that a party should be able to know the material facts on which the other party seeks to rely for its claim or for its defence. A simple example would be that if a contractor has been delayed for a week by exceptionally adverse weather, his records should show that no work could be carried out in critical areas of the works. Meteorological reports, with hot or cold temperatures or rainfall and snow records on the site would be produced to show the effect this had on critical construction work, and to confirm these records supported the argument of exceptionally adverse weather. It should then be a relatively simple matter to show cause and effect and, in this example, if the bad weather occurred at a critical time in the construction of the works, to show this and to have an extension of time granted by the contract administrator, under a JCT contract. The extension of time may or may not be for a week, less than or more than a week, depending on the circumstances.

Although circumstances will often become more complicated than in the previous simple example, the accepted manner of presenting a claim remains in essence to link each claimed cause with each claimed effect on, for instance, delay and/or disturbance.

It is when circumstances become more complicated and events compound on one another, intertwining into an apparent spiral of delay and/or disturbance, that theory and practice sometimes part company. With one event, such as exceptionally bad weather, it should be possible to record and prove the effect of that single isolated event; but when circumstances intertwine, recognition of individual events may not be so easy. Even less easy may be identification of individual effects from individual causes, and keeping records of individual costs arising from individual causes may seem almost impossible.

For example, if an architect issued information late in a construction period, it may have an effect on the completion date of the works. If this followed a period of delay due to weather, postponement of part of the works due to finding antiquities, delay by a nominated subcontractor and the issue of a large number of instructions that significantly varied the works, then keeping records on each of these effects may be very difficult, if not apparently academic and arbitrary. Normally, a contractor will seek to show that the late issue of instructions, say about groundworks, had an effect on completion of the groundworks. Because these works were, as often, on a critical path for the contract, the whole of the works were therefore claimed to be delayed. In this instance it could then be argued that the effect of the late issue of information by an architect was clearly a delay to part of the works and to completion of the contract as a whole. Here cause and effect have clearly been linked.

In the case of delays and/or disturbance to a contract arising from late supply of information, variations to the works, deferred or suspended work, it is the intention of the contract, backed by practice in litigation and arbitration, that liability for each event caused by an employer is established and then the physical and monetary effects (quantum) are proved by evidence separated into discrete effects, each backed by separate records. Perhaps because of the willingness of contractors to attempt to bypass the need to show cause and effect and perhaps because of the difficulties of sifting the wheat from the chaff and what may appear sometimes to be an academic exercise, 'global claims' remain attractive (and maybe appropriate) in some circumstances.

DEVELOPMENT OF CASE LAW ON GLOBAL CLAIMS

Global claims remains a subject where the exact circumstances of each dispute must be examined very carefully and comparison made with the circumstances of decided cases before a decision is made to present or reject a claim founded in this manner on any particular contract.

Since the late 1970s contractors began to argue that, in the circumstances of some contracts, it was not possible to link each item of cause to each item of effect.

If it was accepted that in some circumstances it may not be possible to demonstrate the effect of each event in isolation, then perhaps a series of causes which in total resulted in delay and/or disturbance to the works, could not be clearly demonstrated to show each cause and effect as separate events. In these circumstances, under a 'global claims' approach, all causes of delay and/or disturbance were lumped together and one overall delay and/or disturbance (with associated costs) was therefore claimed to have been the result. Previous requirements to link each cause of delay or disturbance with each, separately identified, result of that delay and/or additional cost were therefore ignored. Two leading cases gave support for this concept.

In *J. Crosby and Sons Limited* v. *Portland Urban District Council* (1967) a case under the ICE fourth edition, the contract for laying a water main overran considerably. An arbitrator ruled that the contractor was due thirty-one weeks out of forty-six weeks claimed as the period of delay. He awarded the delay as a lump sum of delay, rather than giving individual periods of delay against nine matters that had been argued as causes of delay. The arbitrator justified his award by saying:

> The result, in terms of delay and disorganisation, of each of the matters referred to above was a continuing one. As each matter occurred its consequences were added to the cumulative consequences of the matters which had preceded it. The delay and disorganisation which ultimately resulted was cumulative and attributable to the combined effect of all these matters. It is therefore impracticable, if not impossible, to assess the additional expense caused by delay and disorganisation due to any one of these matters in isolation from the other matters.

The respondent contested the arbitrator's award, in particular the findings of providing a lump sum delay, without giving individual amounts for each head of claimed cause and effect. Rejecting his appeal, the judge agreed with the arbitrator and said:

> I can see no reason why he (the arbitrator) should not recognise the realities of the situation and make individual awards in respect of those parts of individual items of claim which can be dealt with in isolation and a supplementary award in respect of the remainder of these claims as a composite whole.

A similar case was *London Borough of Merton* v. *Stanley Hugh Leach* (1985), concerning a JCT 1963 contract for constructing dwellings. The contractor claimed that all the delay to the contract was as a result of instructions issued by the architect and sent a single letter, lacking in detail, in making an application for loss and expense covering both clauses 11(6) and 24(1) of the JCT 1963 form. It was held by the judge (again considering an appeal from the award of an arbitrator) that 'the loss and expense attributable to each head of claim cannot in reality be separated'. This case established a difference between liability and determining quantum. The judge held:

> Implicit in the reasoning of Donaldson J. (the judge in *Crosby*) was first that a rolled up award can only be made in a case where the loss and expense attributable to each head of claim cannot in reality be separated, and, secondly, that a rolled up award can only be made where apart from that practical impossibility the conditions which had to be satisfied before an award can be made have been satisfied in relation to each head of claim.

Although *Crosby* and then *Merton* established circumstances where a global or rolled-up claim may be successful, it should be remembered that they are only authority for the financial consequences of delay, they are not authority for the pleading of a claim for rolled-up liability for the delay itself. Neither case was concerned with determining whether or not there had been any *excusable* delay (i.e. an event, such as weather, that allows a contractor time for the delay, relief from liquidated damages but no extra payment) or with the existence of *compensable* delay (i.e. an event which delays progress or completion and entitles a contractor to be recompensed by valuation and/or loss and expense for any financial effects of the event). These aspects were not in contention.

The 'global claims' approach seemed to remain acceptable, in some circumstances, for a time but was then thrown into question by *Wharf Properties Limited and The Wharf (Holdings) Limited* v. *Eric Cumine Associates, Architects, Engineers and Surveyors, (a Firm)* (1991). In this case the plaintiff made no attempt to link each cause of delay with each effect. The plaintiff was the employer of an architectural practice and made a claim against the practice for, among other things, failing to manage the work of contractors and that because of this the project was considerably delayed. Six periods of delay occurred but the statement of claim made no attempt to show how individual delays were the result of individual breaches by the architect. Wharf pleaded that, due to the complexity of the project, to the very large number of factors and to the interaction of disruptive factors, knock-on effects, and so on, it was not possible to identify individual delays in the manner that the defendant architect required, at the pleadings stage. Wharf stated that individual reasons would not be known until the trial itself and that its pleadings could not be and need not be specific before then. Eric Cumine, the defendant architect, was successful in his appeal about lack of particularity when the Court of Appeal of Hong Kong decided that the pleadings were so 'unparticularised' (i.e. so lacking in detail of each event, cause and its claimed effect) that they should not be allowed to stand. Because of the cases of *Crosby* and *Merton*, the *Cumine* case was referred to the Privy Council of England, which held that although it was very unlikely that Wharf would be able to succeed in establishing a claim on the general basis pleaded, the pleading did nevertheless establish a cause of action. The Privy Council upheld the Court of Appeal on the alternative basis that the 'unparticularised pleadings' were an abuse of the process and that 'the approach is enough to make this (that is, the action) speculative litigation, which would throw an enormous and quite unfair burden on the defendants which should not be allowed to continue (that is to trial)'. In the judgment, the Privy Council also held among other things:

> It may be proper for an arbitrator to make individual financial awards in respect of claims which can conveniently be dealt with in isolation and a supplementary award in respect of the financial consequences of the remainder a composite whole. . . . The failure even to attempt to specify any discernible nexus between the wrong alleged and the consequent delay provides, to use [counsel's] phrase 'no agenda' for the trial.

But remember that *Wharf* v. *Cumine* was a very unusual case with special circumstances, the events of which took place over a seven-year timescale.

At that stage in the development of global claims, in 1991, the editors of *Building Law Reports* made the following observation, in Volume 52, page 6:

> *Crosby* and *Merton* are to be confined to matters of *quantum* and then only where it is impossible and impracticable to trace the loss back to the event. The two cases are not authority for the proposition that a claimant can avoid providing a proper factual description of the consequences of the various events upon which reliance is placed before attempting to quantify what those consequences were to him. (Emphasis added)

So it is now held that Crosby and Leach are cases about money and not about liability. They now provide no shield to a pleader who either cannot or will not plead his case on liability with specificity.

Global claims progressed further in definition, though not in new rulings, with *Mid Glamorgan County Council* v. *J. Devonald Williams and Partners* (1991). This was another application to the court to strike out an action on the grounds that the pleadings were not particular and were an abuse of the court (i.e. in taking up time that would not lead to prosecution of a case). Extracts from the judgement in *Mid Glamorgan* summarised matters from *Crosby*, *Leach* and *Eric Cumine* as follows (from the *Building Law Report*):

(a) The pleading of complicated cases requires a proper cause of action to be pleaded.
(b) Where specific events are relied upon as giving rise to a claim for moneys under the contract then any preconditions which are made applicable to such claims by the terms of the relevant contract will have to be satisfied, and satisfied in respect of each of the causative events relied upon.
(c) When it comes to quantum, whether time-based or not, and whether claimed under the contract or by way of damages, then a proper nexus should be pleaded which relates each event relied upon to the money claimed.
(d) Where, however, a claim is made for extra costs incurred through delay as a result of various events whose consequences have a complex interaction that renders specific relation between event and time/money consequence impossible or impracticable, it is permissible to maintain a composite claim.

The area of global claims then moved on with *Imperial Chemical Industries* v. *Bovis Construction and Others* (1992), which was decided in the Official Referees Court. This case arose out of Bovis renovation and enlargement of ICI's headquarters in London, which took much longer and cost much more to complete than expected. In this case *Wharf* (which had resulted in the plaintiff losing its right to proceed with its case because of lack of particularity in its pleadings – very special circumstances lead to this decision), was referred to and the use and detail that may be put into a Scott schedule was commented on. *Imperial Chemical Industries* shows that pleading is ultimately a matter of degree and the amount of detail should match accordingly. It also demonstrated that a defendant may be successful in having a plaintiff's pleading struck out, and the amount that an Official Referee used Scott schedules and revised schedules/documents, rather than amended documents. It showed the degree to which a party could be ordered (or not be ordered) by a court to present its case or defence in a certain way. Among other things, the judge said:

> There must be an 'agenda' for the trial; there must be discernible connection between the wrong and, where delays are relied on, the consequent delay. It seems clear that rolled up claims will be acceptable in financial claims subject to the parameters laid down in the aforementioned cases [*Crosby*, *Merton*, *Wharf* and *Mid Glamorgan*]. However, cause and effect will have to be linked in respect of delay claims. There is going to be a great deal of argument in the future concerning global claims. Contractors and sub-contractors are therefore advised to get it right first time and link the cause and effect of every event giving rise to a claim.

Against this background, a practice direction was issued by the Lord Chief Justice in January 1995. This required that judges sitting at the first instance 'assert greater control over the preparation for and conduct of hearings than has hitherto been customary'. It meant to ensure that 'every pleading must contain, and only contain, a statement in a summary form of the material facts on which the pleading relies, but not the evidence by which those facts are to be proved, and the statement must be as brief as the nature of the case admits'. This was issued to put into context the habit of making 'forest' pleadings, that is a pleading where the reader is left to wander through a 'forest' of enormous supporting schedules in order to find the real nature of the plaintiff's case.

Then in *GMTC Tools and Equipment Limited* v. *Yuasa Warwick Machinery Limited* (1995) the Court of Appeal found that the Official Referee in the lower court was wrong to try to enforce that the plaintiffs had to particularise their pleadings in a particular way (in fact he considered that the plaintiffs, after two unsuccessful attempts to be more specific, should not be allowed to proceed to trial). At the appeal, the judge was overruled – the Court of Appeal held that a plaintiff should be allowed to formulate his claim for damages as he wishes and not be forced into a way chosen by the judge or by his opponent. This suggested that the degree of specificity is a matter of degree, arising as always from the circumstances of the case. Provided there is seen to be sufficient specificity, it seems that 'global' may suffice, at least during the early stages.

In *British Airways Pension Trustees Limited* v. *Sir Robert McAlpine and Sons Limited and Others* (1995) the Court of Appeal continued with the theme that specificity was a matter of degree. It reinforced that pleadings have the basic purpose of enabling an opponent to know the case against it in order to be able to answer it, not to plead in detail matters that were not apposite at the time. Specificity or particularisation only needed to be sufficient to meet the objectives of pleading. In the light of *GMTC Tools* and *British Airways*, it appears that global claims may not fall at the pleadings stage but, during a hearing, failure to particularise may well mean that a claim will fail.

Inserco Ltd v. *Honeywell Control Systems Ltd* (1996), a case involving printing machinery and interruption to its use, demonstrated that where there is an extremely complex interaction between consequences of various breaches, variations and additional works and where it is impossible to make an accurate apportionment of the total extra cost between the several causative events, it is legitimate to make a global award as to claims arising. The circumstances of the *Inserco* case, given in a 405-page judgement, were unusual but again they demonstrated that if it is impossible to make an accurate apportionment between issues then a 'total' claim is admissible.

Detail required in pleadings must always be differentiated from detail required in a statement of case. In *Bernhard's Rugby Landscapes* v. *Stockley Park Consortium* (1997) the judge emphasised the importance of considering the contractual claim submissions and extracting only the material that ought properly to be in the statement of claim, leaving schedules to cover any details of the specific allegations and, where appropriate, embodying the relevant parts of the original claim. More about this is given in Chapter 26 on preparation of a comprehensive claim document.

In summary, case law has now indicated that, provided effort has been made to identify the breaches and the resultant effect and damages within a statement of claim, applications are unlikely to succeed in having pleadings struck out on the grounds that they are not specific and therefore global. This is because pleadings are a means to an end and specificity at that stage is not required, provided sufficient material is produced to give each party a fair hearing (at a hearing or other subsequent presentation of case and defence). Proving the case at a tribunal will still need to take into account the requirements given in *Mid Glamorgan* v. *Devonald Williams* and *Inserco*. Although there has been much activity trying to have global claims struck out in the pleadings stage, as a matter of policy the courts have allowed rolled-up claims to proceed provided the claimant has made every effort to be specific to explain any link between a claimed effect and a claimed cause.

Provided the fundamental distinction is drawn between pleading and proof of facts that give rise to liability on the one hand, and on the other the pleading and proof of the financial consequences of that liability once established, the global approach may be permissible in some circumstances. The problem remains that, at trial, a failure to link cause and effect is likely to prove crucial to success or failure of the claim. Global assessment of quantum may still be appropriate if, once linkage of individual causes to individual effects has been accepted, circumstances as existed in *Crosby* and *Merton* pertained. It appears there will be little success for a pleader who either will not or cannot plead his case with specificity, but how to interpret specificity is an art not a science and remains fairly obscure at that.

PART 5

DELAY AND DISTURBANCE: JCT CONTRACT PROVISIONS

CHAPTER 15

CAUSES JUSTIFYING EXTENSION OF TIME AND/OR REIMBURSEMENT OF LOSS AND EXPENSE

- Some principles
- Causes leading to extension of time only
- Causes leading to extension of time and/or loss and expense

This chapter sets out the various causes or reasons which may lead to extension of time or loss and expense under the JCT and IFC contracts and their related subcontracts. This is done here before considering the wider aspects of the law and contract clauses and their various effects in later chapters, to bring together the causes which are given in each contract. Most of these causes are the same, even though there are differences of emphasis or effect due to the contract in which they occur. They may thus be compared across the range of contracts most conveniently, and may also be referred to from comparisons in later chapters.

The contracts and subcontracts use the term 'event' or 'relevant event' to mean something which leads to extension of time and the term 'matter' or 'relevant matter' to mean what leads to reimbursement of loss and expense. The relevant events in the following forms are taken into account:

- JCT 80 main contract forms (in any edition)
- IFC 84 main contract form
- JCT 91 nominated subcontract form NSC/C
- IFC 84 named subcontract form NAM/SC
- Domestic subcontract form DOM/1 (used optionally with JCT 80)
- Domestic subcontract form IN/SC (used optionally with IFC 84)

The clauses concerned in the first four of these are tabulated for outline comparison in Table 15.1. Clause numbers are not given in commenting upon the various events or matters, but may be traced by reference to this table, which also gives clause numbers relating to determination as discussed in Chapter 18.

SOME PRINCIPLES

Causes and effects

The expression deliberately used above is 'may lead to extension of time or loss and expense'. It is still necessary to demonstrate that an actual effect of delay or loss and

Table 15.1 Comparison of relevant events and matters[a]

Event or matter		Extension of time				Loss and expense			
		JCT 25.4	IFC 2.4	NSC 2.6	NAM 12.7	JCT 26.2	IFC 4.12	NSC 4.38.2	NAM 14.2
Force majeure	[D]	1	1	1	1	–	–	–	–
Weather conditions		2	2	2	2	–	–	–	–
Loss by perils	[D]	3	3	3	3	–	–	–	–
Civil commotion etc	[D]	4	4	4	4	–	–	–	–
Delay nominated firms		7	–	7	–	–	–	–	–
Statutory powers		9	–	9	–	–	–	–	–
Shortage of labour		10.1	10[b]	10	9[b]	–	–	–	–
Shortage of materials		10.2	11[b]	10	10[b]	–	–	–	–
Statutory bodies		11	13	11	12	–	–	–	–
Deferred possession		13[b]	14[b]	14[b]	13[b]	–	–	–	7
Suspension of work		–	–	13	14	–	–	–	8
Performance-specified work		15	–	–	–	–	–	–	
Terrorism/terrorist threat	[D]	16	16	16	16	–	–	–	–
Instructions regarding									
Discrepancies etc	[D]	5	5	5	5	3	7	3	6
Variations etc	[D]	5	5	5	5	7	7	7	6
Postponement	[D]	5	5	5	5	5	5	5	5
Antiquities		5	–	5	–	(34)	–	–	–
Nominated firms		5	–	7	–	–	–	–	–
Named subcontractors		–	5	–	5	–	7	–	6
Opening and testing	[D]	5	6	5	5	2	2	2	2
Information schedule		6.1	7.1	6.1	6.1	1.1	1.1	1.1	1.1
Other instructions		6.2	7.2	6.2	6.2	1.2	1.2	1.2	1.2
Work not in contract	[D]	8.1	8	8.1	7	4.1	3	4	3
Materials from employer	[D]	8.2	9	8.2	8	4.2	4	4	4
Ingress and egress	[D]	12	12	12	1	6	6	6	5
Forecast quantities		14	15	15	15	8	8	8	9
Compliance with CDM	[D]	17	17	17	15	9	9	9	10
Suspension after non-payment		18	18	18	18	10	10	–	11

[a] The JCT and IFC main contracts and the related nominated and named forms are given. Slight differences between the clauses could lead to different effects on occasions. Causes are listed in the order in which they are discussed in the text.

[b] These causes are optional and must be specifically included to be active.

[D] These cause may also lead to determination under the main contract forms; there are other causes for determination.

(34) This cause is given in clause 34.

expense has been caused by the item identified. The simple occurrence of a specific item cannot alone lead to a concession to the contractor or subcontractor: it must produce effects of a suitable nature and sufficient magnitude at the right time and place, so that some adjustment of time or money is needed under provisions which serve to lighten the contractor's or subcontractor's responsibilities under an entire contract.

Equally, an item may lead to either extension of time or loss and expense, without the other: they are not inexorably linked in such a way that both must be conceded when one is proved to be justified. Most obviously, this is shown by the longer list of items leading to extension of time, compared with the list of items leading to loss and expense reimbursement in all the forms. It is also demonstrated by the general proposition that time can be extended *without* disturbance that costs money. But also there can be such a disturbance within a programme which is not extended overall. However, most commonly delay leading to extension and disturbance go together. Often the two effects are linked, but the proof that one has occurred is not enough to show that the other follows. Certainly, there is no unbreakable legal link, as the existence of two clauses in each document suggests. The distinction was rehearsed particularly clearly in *Henry Boot* v. *Central Lancashire Corporation* (1980), discussed for other specific reasons below. The point is important enough to be repeated in several later chapters.

Events and matters

It has been noted that only events which result in an overrun of the final completion date, not those causing some delay contained within the total programme, can be used to justify an extension of time. Similarly, matters must cause actual loss and expense to lead to reimbursement. The point is self-evident when put so baldly, but much energy can be wasted if a clear view is not taken in a given instance. Furthermore, it is only in relation to the events and matters actually specified that the architect can act at all, any others being either completely the contractor's responsibility or those constituting breach on which he should approach the employer himself, even if he chooses to do so through the architect.

The clauses covering extension of time in JCT 80 and IFC 84 contain almost identical sets of events, which divide into two conceptual categories:

(a) Events which are the responsibility of the employer or the architect on his behalf. These may be lapses or deliberate acts taken in recognition that they carry potential repercussions. As they have this status, the employer is protected by their inclusion from losing a defined, if extended, completion date.
(b) Events which are not usually the responsibility of either party, so that they are effectively 'neutral'. As these events have this status, it is the contractor who is protected from what he would otherwise bear under an entire contract. Occasionally, it is not entirely the case that the contractor does not have a margin of responsibility for some of these events.

The first category contains those items which are also matters possibly leading to loss and expense. The third possible category of event, covering items for which the contractor alone is responsible, is not included within the ambit of the clauses, as might be expected. Questions of drafting and sequence aside, here are the salient differences between the lists in the two groups of contracts and subcontracts, JCT and IFC:

(a) Nominated subcontractors and suppliers occur only in the JCT group, and named subcontractors occur only in the IFC group.
(b) Antiquities and the exercise of statutory powers occur only in the JCT group.
(c) Inability to obtain labour or materials occurs in both groups, but is an optional event only in the IFC group.

Subject to these considerations, the various events and matters from the contracts and subcontracts may be taken together, arranged under the two conceptual categories mentioned above and otherwise following the order of JCT 80 where there is any difference. The interpretation of each cause is in view, as the possible interaction of more than one is reserved until Chapter 11.

If any clauses within one of the forms are likely to be considered for special amendment in a particular contract, the present are strong candidates for change, when they do not appeal to one party or the other at some point. This may be understandable, and even justifiable in some circumstances, but it causes problems over the fluctuations provisions when there is an overrun of the completion date, in view of JCT clause 38.4.8 and its equivalents. An ill-considered change of wording is especially likely to fall foul of the contra proferentem rule in these sensitive areas of the contracts.

The contractor and subcontractors

The comparable clauses in the main contracts and the subcontracts for events and matters are very similar, because they are dealing with the same basic causes. They can therefore be discussed once, whichever level of the contractual relationship is in view, over their main effects and subject to some overriding comments here. Usually the reference is to the contractor for simplicity.

Some events or matters are mentioned in exactly the same terms in each set of clauses, such as adverse weather and discrepancies in documents. Others are qualified in subcontract clauses as happening to either contractor or subcontractor, such as late architect's instructions, because what happens specifically to the contractor may affect subcontractors in their progress by way of a follow-on. Sometimes subcontract clauses refer also to the contractor's directions. These nuances are not mentioned in each instance, but should be borne in mind. Even when the point is not made in a clause, a follow-on effect may occur, so that a subcontractor is affected because the contractor has been thrown out by an event or matter. Such a subcontractor is entitled to put forward a notice in these cases under the clauses concerned.

Within the subcontract clauses, there are also extra provisions to any in the main contract clauses. One is the additional question of suspension of work by subcontractors, which still counts as an event, but which does not run up to main contract level to allow the contractor the possibility of an extension of time.

The other provisions are concerned with 'act, omission or default' of either contractor or subcontractor. In the case of extension of time, they can only refer to the contractor; whereas in the case of loss and expense, either party may cause the problem. This is not because a subcontractor cannot delay the contractor, but because the appropriate remedy must be entirely in loss and expense, as the subcontractor cannot give the contractor an extension. If the contractor takes proper action to progress a subcontractor, he may gain an extension from the architect when the subcontractor is nominated, but this is outside the terms of the subcontract (see JCT clause 25.4.7 hereafter).

The default situations are not spelt out in the subcontract clauses in the same detail as those governed by events or matters; this is because they are always breaches, potentially more variable and not to be restricted to a few causes which modify the contractual position of entirety in a defined way. They are not discussed separately for this reason, but can be of importance. The procedures embodied in the clauses are as relevant over them as over the other affairs, although the detail given varies between nominated and other subcontracts.

Although more strictly a concern of later chapters, it may be stressed here that it is up to subcontractors to remember they are responsible for taking their own action under their subcontracts, by way of notices and so forth to secure extensions of time and reimbursements for themselves. Even though these may be affairs to be pursued with the architect on their behalf by the contractor, it is not the responsibility of the contractor to remind subcontractors to act, or even to put forward his own ideas of what is due. It may be he does not know with any precision that there is a case or, in the extreme, perhaps he does not know at all.

In the case of named and domestic subcontracts, the amount agreed in time or money does not relate directly to what the architect and the contractor agree, which may not isolate it, even though a money amount is loss and expense and strictly should be the 'direct' amount. The contractor has to come to an agreement which he should ensure includes subcontract elements, and then deal with his subcontractors afterwards if necessary. If possible, he may find it useful to settle with subcontractors first, so that he knows where his baseline for settlement with the architect lies. Alternatively, he may want to see how the main settlement is going before actually finalising with subcontractors. This is the standard dilemma over agreeing rates and all such features of final accounts.

In the case of nominated subcontracts, the architect is to be involved directly in agreements and effectively to make them. This in practice gives such subcontractors a more obvious spur to act initially, in that the architect has to be introduced and will wish to look at all aspects together to form a balanced assessment. They may thus 'miss the boat' if they delay. Other subcontractors may feel that they have a little more leeway, because only the contractor is involved in the settlement. Often it works like this in practice, but subcontractors should remember that the timetable set out in subcontracts can run against them, if they delay unduly and the contractor settles without taking account of their representations. He will not take kindly to granting what he cannot recover, and the provisions are on his side.

CAUSES LEADING TO EXTENSION OF TIME ONLY

To emphasise the distinction in the heading, the term 'event' is used throughout this section.

Force majeure

Force majeure is the nearest thing to 'etc' in the list. It refers to an uncertain area of exceptionally severe events. It is a difficult peg on which to hang a request for an extension, since in its extreme form it is likely to amount to frustration, whereas in less than the extreme form it may be impossible to demonstrate at all.

Weather conditions

The conditioning term in this event is in the other two words of the clause 'exceptionally adverse'. The thrust of this is that the contractor must allow in drawing up his programme for the routine run of adverse weather which all but permanent underground dwellers must be expected to know about in these climes. He must know about rain, frost, snow, ice, continuous sun and wind running through long periods of any year and in certain positions of exposure and so forth abounding above what is otherwise normal. He must also anticipate a limited measure of any of these somewhat outside the strict limits.

What he cannot allow for is an exceptionally severe incidence or duration of adverse weather, or weather which is completely misplaced. He has quite a difficult task here to prove his point, as anyone acquainted with the weather patterns of the United Kingdom will be aware! To establish whether there has been a distinct aberration from the norm, it is necessary to go to the meteorological records for the region covering a prolonged period, at least a decade, and to establish a statistically significant deviation, not just something rather severe for the time of year.

It was held in *Walter Lawrence & Son Ltd* v. *Commercial Union Properties (UK) Ltd* (1984) that the applicable test was the exceptional nature of the weather and not how exceptional was the delay produced. Furthermore, it was held that the contractor was not obliged to programme 'in the strict sense', so that reasonable allowance for expected conditions was what was to be made. Interaction with the contractor's own delays in this case is mentioned in Chapter 16.

It will of course depend on the stage of works, such as whether there is a lot of earthworks, wet trades, cladding, roofing or work under cover going on, as to the effect produced. This is also a good example of an event where the responsibility of the contractor to mitigate the extent of any delay is high. He is responsible for protective work of various sorts, often by express provisions of the specification or preamble clauses, although this does not mean that he must necessarily press on in all weathers irrespective of the cost involved. Equally, the instructions of the architect to suspend work which the same clauses may confer, do not necessarily give a right to extension of time, although here the position is entwined with the provision over extension due to postponement. Such clauses should be so worded that the architect is acting to stop the contractor ignoring weather conditions and the specification, but is not introducing a definite postponement for other reasons. The instruction should then require the contractor to suspend work to comply with the specification.

Loss or damage by perils under JCT clause 22 or IFC clause 6.3

The loss or damage here relates to the works themselves and to unfixed materials on or adjacent to the works. It does not cover damage to adjoining property, through which the contractor may have a negotiated access, although here the employer may be liable for an extension of time under the 'ingress or egress' provision below. Perhaps more pressing for the contractor, it does not cover loss or damage to materials off-site, even when they have been paid for by the employer, or to his plant, accommodation or other temporary items. If any of these contingencies also lose him time, the risk is entirely with him, so he cannot secure an extension.

The perils leading to the possibility of an extension are the same under the JCT and IFC contracts; JCT contracts use the term 'Specified Perils' (Chapter 18). The full range covered by 'all risks', as in the insurance clauses themselves, therefore does not apply here. If perils are to lead to a determination, they must not be due to the negligence of the contractor, subcontractors etc.

As with a number of the events, there is no corresponding matter available in this instance in the clauses covering loss and expense. Both parties are advised to consider their insurance cover against the categories of risks and results which are not within the contract scope, and for which cover is possible over each party's own delay costs, as it is not over some other events under these clauses. The contractor may find himself running into liquidated damages here. Alternatively, the employer may lose his right to recover liquidated damages.

Civil commotion etc

The various elements given here are effective at any stage in the production process leading up to the delivery of materials etc at the works. There is therefore considerable scope for this event to come into play, provided its effect can be demonstrated, which is the more difficult side of it. Damage due to civil commotion is covered by the preceding event.

Delay on the part of nominated subcontractors or nominated suppliers

The primary nature of this event is delay in the execution of work by one of these persons. The clause affects the usual responsibility of a contractor towards an employer for whatever subcontractors or suppliers do or fail to do. It is initially discussed in relation to subcontractors alone, as they present more facets; suppliers are taken by comparison in closing.

Nominated subcontractors

The clause allows the contractor to secure an extension and avoid damages when a subcontractor has defaulted in the manner stated, provided the contractor has done everything reasonable in the circumstances to mitigate the effects. If therefore the contractor has taken appropriate action, he is not liable to the employer. Further, the subcontractor has no liability to the contractor, as the contractor is released. He is already without liability to the employer directly, as there is no privity of contract. This is a most odd provision when left like this, without the extra arrangements mentioned below. It is made more so by the uncertainty attaching to the expression 'delay on the part of'.

The provision received consideration by the House of Lords in the case of *Westminster Corporation* v. *J. Jarvis & Sons Ltd and Another* (1970), which was brought on the same wording in the JCT 1963 form. A nominated subcontractor completed piling work and left the site. The piling was found to have been carried out with defective concrete, and remedial work caused substantial delay. The contractor sought extension of time, arguing that there had been delay 'on the part of' his subcontractor, which 'he had taken all practicable steps to avoid or reduce' (meaning in this instance that he could not have done anything about it). The main argument was around whether such delay could arise once the subcontractor had finished work, apparently satisfactorily. The defects came to light when an excavator nudged a pile cap, which crumbled away. It was held that this did not constitute delay within the meaning of the clause. The case was fought effectively between the employer and the subcontractor, with the contractor hardly represented in the action, as he had little at stake.

From the judgment, several deductions appear to follow about the naked clause itself:

(a) The term 'on the part of' is not to be interpreted as equivalent to 'caused by'. It must be limited to delay in performance by the subcontractor himself and not be extended to include delay which is the consequence of any default, such as poor work. In the circumstances of the actual case (and these are always critical), the resultant delay came after the subcontractor had completed. Specifically, it was held that the term did not apply to any breach of contract or to complete failure by the subcontractor to perform his obligations because of repudiation, insolvency or other determination of the subcontract.

(b) It would appear likely that delay caused during the subcontractor's activity, owing to his defective work or some cognate matter leading to him being late would qualify as being 'on the part of', provided it delayed him and the contractor in turn. If it delayed the contractor without any interposed delay to the subcontractor, say by affecting large areas of the contractor's work and little of the subcontractor's, it would appear to be 'caused by' and so not likely to qualify.

(c) Provided the contractor operates a procedure suitable in all the circumstances to 'take all practicable steps' to limit delay, the clause removes the employer's protection against the subcontractor's default in being late in completion. This may recognise a degree of special relationship existing in nomination, but exposes the employer totally.
(d) The corollary of this is that the subcontractor can protect himself against liability for damages by proving two things: that he was in default and that the contractor took 'all practicable steps', which he then ignored or could not match. This is obviously a fall-back position from proving that he was not in default: either 'I am innocent' or 'I am doubly guilty'.
(e) The contractor can remain fairly passive, so long as he chases the subcontractor in the first place. Either the subcontractor responds and delay does not occur, or he does not respond and the contractor is covered against damages, although he still has to act against the subcontractor over the cost to himself of the delay.

This catalogue presents a quite unsatisfactory position for the employer. One answer would be to delete the clause entirely, and this would not lead to problems of cross-dependence of other clauses. It is presumably included at all to give some reassurance to contractors when accepting nominations of performers of uncertain quality, and so may not be lightly removed, apart from the reactions of tendering subcontractors.

The answer afforded in the JCT system is the obligatory employer/nominated subcontractor agreement, NSC/W (Chapter 7), which has a number of virtues. For present purposes, they give the employer the right to secure recompense from the subcontractor for the liquidated damages which he has had to forego when the contractor receives an extension of time. Like all such rights, it is worth as much as the substance of the subcontractor or his insurer in meeting his liability. Further discussion and legal cases are given in Chapter 7, under renomination procedures.

A nominated subcontractor who does not delay, but who is affected by the delay of a second such subcontractor, is in a position similar to the contractor. If the contractor receives an extension of time, so does the affected subcontractor, to whatever degree is appropriate. Conceivably, he might receive one when the contractor does not. The architect is to be implicated in the granting of any extension for a nominated subcontractor, with this covering even the situations in which the contractor is at fault and not receiving an extension himself (Chapter 16). Among them would be an extension where the contractor has not taken all steps to avoid or reduce the other subcontractor's delay.

In view of the multi-level dealings between architect, contractor and subcontractor, the contractor needs to pay particular attention to the procedural aspects considered in relation to subcontracts in Chapters 16 and 17 and in relation to multiple contracts and their claims in Chapter 13.

Nominated suppliers

In principle the effects of delay on the part of nominated suppliers are the same as those of nominated subcontractors, so far as the contractor is concerned. The related subsidiary document is the Standard Form of Tender by Nominated

Supplier, which is strictly optional, although something close to it is always necessary to fit with the requirements of JCT clause 36.4.

It is for the architect to set up matters by obtaining tenders on this form to protect the employer. A critical provision is that delivery is to be 'in accordance with any delivery programme agreed'. Beyond this, the document is silent in its original version. It does not involve the architect in the post-contract phase in giving his opinion over programme matters, or provide for how the supplier receives any equivalent of extension of time. It does, however, provide an equivalent to agreement NSC/W, which inter alia covers an indemnity by the supplier to the employer over loss of liquidated damages and over loss and expense, when the supplier is at fault. JCT 80 gives a fairly limited selection of causes for varying the programme, which are carried into the revised form of tender.

As between contractor and supplier, the position is regulated more widely by the general law. Much will depend on the precision with which the delivery programme is laid down and any extra clauses which the parties to the supply contract are able to introduce or exclude.

Exercise of statutory powers

Exercise of statutory powers is included only in the JCT contract, intended as it is for larger projects. Its scope is both wide and restricted.

Its width arises from the unbridled reference to 'any statutory power'. This is, however, tempered in several ways. It is an 'exercise after the Date of Tender', so the contractor has to foresee and make allowances when tendering, as usual. It is an exercise 'by the United Kingdom Government' and not by some nationally owned body, which may of course introduce policies that affect progress in some way. There may be fine points of distinction here, when the government influences policy by its own pressures fairly directly applied. Furthermore, it must be an exercise which 'directly affects the execution of the Works', rather than some less direct effect, such as an order applying to builders' merchants. These indirect effects may be covered by the events in the next clause.

Whatever the action, one effect is that it restricts 'the availability or use' of labour, so the contractor may not be able to obtain it at all or may not have free disposition of what he can obtain, say by an imposed limitation on working hours. The other effect is 'preventing or delaying' the contractor's securing of essential materials, fuel or energy. Delaying here is the equivalent of restricting use in the case of labour, as both stretch out the supply for the contractor.

In the case of materials, and to a smaller degree labour, there is some possibility of the architect instructing variations to overcome the problem of delay by introducing different materials or techniques. This he does not have to do, but realistically he will consider the option when some serious delay is in prospect. Often there is little that can be done. This is an instance in which the contractor's obligation to give warning of problems under the procedures for extension of time has very direct point. If he fails, it is more than a lapse in procedures when avoiding action could have been taken. However, the onus of proof that there was a reasonable way round the delay will be on the architect.

The materials etc must be those which are 'essential to the proper carrying out of the Works'. Usually any materials which the contractor plans to use, and which are likely to hold up progress by their absence, will be those which have been specified sufficiently precisely to prevent him making any substitution on his own initiative. If he has any latitude, he must use it, as he should not receive any extension when the possibility of ameliorating action is open to him. This will be so whether or not he can obtain a substitute at equivalent cost. If these comments are true of permanent materials, they are even more likely to be true of fuel or energy, where the option of switching to an alternative is often more readily present. This might mean using a dearer type of concrete batching or placing system, or a change from work on-site to yard or the reverse.

In all of these cases there is some reasonable limit to what the contractor should be expected to do, although this will vary with the seriousness of the delay or the level of expense involved. It would be ludicrous to expect a change to hand-mixing of concrete for a project of any scale at all, whereas an interchange between site and ready-mixed concrete may be reasonable. If a site does not permit a large batching plant, this will affect what is reasonable in that direction, whereas a change from fossil fuels to electricity will be affected, but not necessarily ruled out, by whether complex plant has already been set up on-site.

Inability to obtain labour, goods or materials

The two clauses relating to this group of events are given as optional in the IFC contract and its subcontracts, and an entry must be made in the related appendix to render them effective. It would be possible to include one and not the other, or to have them applying in one document and not the other. The latter would lead to unevenness of treatment, and should be resisted by the party adversely affected.

When the clauses apply, the criterion for an event to qualify is the same as under the immediately preceding clauses about government action, that is the elements concerned must be 'essential to the proper carrying out of the Works'. All that applies there is true here also, although the absence here of any mention of fuels and energy is important. For example, shortage of petrol without statutory action causing it, does not qualify for an extension of time.

The immediate event is the contractor's 'inability . . . to secure' labour or materials. Again, 'inability' does not mean 'inability at more or less the same cost', so the possibility may have to be faced of inducement or importation of labour with attendant higher costs, if the local supply is taken up by some other contract. There is no power in the contracts for the architect to agree to extra payment in these circumstances, and no obligation on the employer to meet such amounts. For his part, the contractor has to weigh the additional costs against his possible liability to liquidated damages, quite apart from goodwill.

Although this may be correct in principle, it may be wondered whether quite such a hard line holds in the case of a smallish local contractor for whom the operation of engaging and importing labour from any distance would fall into the category of being beyond his ability. The same might apply to obtaining materials from national markets. Perhaps for such a firm it is reasonable for the architect to lower his sights somewhat in applying the provision. This is again a matter of degree and one which may lead to consideration in drawing up a tender list according to the urgency of the project.

There are two qualifications to the contractor's inability, which underline these points. One is that there must be 'reasons beyond his control', which would be a likely term for the courts to imply, even if it were not express, to give 'inability' substance. The result is perhaps tautological, but avoids uncertainty where it is often felt. The other is that the inability must be such that he 'could not reasonably have foreseen at the Date of Tender'. He should not contract to perform the impossible, but should warn of anything in this field. Otherwise, he will be held to have anticipated the problem and to have made allowance in some way for it.

Work in pursuance of a body's statutory obligations

There are situations in which the contractor may engage a local authority or statutory undertaker to perform work under a subcontract, when this clause does not apply. If then the subcontract is domestic, no extension of time is due whatever happens; whereas if it is nominated or named, the position is affected accordingly and is discussed in Chapter 7. These positions hold even if the contractor has no choice over whether to have the body as his subcontractor.

There are other cases for which these clauses are intended in which the body is acting, not as subcontractor, but in the role given it by statute. When this occurs, the contractor may find himself impeded by the simple performance of the work concerned, or by some lapse in its performance. The work may or may not be on his site, although it may be the provision of some service to the site. It must be 'in relation to the Works', and not just causing some hindrance while completely extraneous.

There are areas of uncertainty here, as the position of a statutory body in relation to the contractor is not always clear. In *Henry Boot Construction Ltd* v. *Central Lancashire New Town Development Corporation* (1980) work was allowed as provisional sums in the contract bills for statutory undertakers to perform, and delay occurred. It was held, despite the inclusion, that the undertakers were employed directly by the employer as 'artists and tradesmen' under the 1963 JCT contract, so the employer was liable to suffer both extension of time and loss and expense. Had they then been statutory undertakers performing their statutory duties under what is now clause 6.3, only extension of time would have been available to the contractor, as the present clause provides. This decision has been criticised on several counts, but remains the law at present.

Deferment of possession of the site

This is a relevant event only when JCT clause 23.1.2 or IFC clause 2.2 is stated in the related appendix to apply. It is similarly included in the sub-contracts to the IFC contract, but apparently in error, as it is superfluous.

The clauses allow the employer to defer giving possession by a period as stated there, but with the intention (unless modified) that this is by not more than six weeks. Whether the effect on the contractor's programme will be a straight deferment of the same amount, will depend on how closely before the date of possession he actually receives the notice of deferment and how advanced his own preparations are. He may, for instance, have orders placed which are fitted into a supplier's own schedule and which will be displaced by an undue amount if once they lose their place. Alternatively, he may welcome the extra breathing space.

If the deferment is agreed with the employer on the signing of the contract, often the completion date will be moved at the same time and usually by the same amount. This case does not fall within the clause, which can relate only to post-contract events when the architect is charged with fixing a revised date for completion.

Under a contract without a provision like the present, failure to give possession on time is a breach entitling the contractor to the usual remedies. In practice, he may well be happy to negotiate in such a way as to give a result similar to that under the above contracts, although he does have the additional lever available.

Suspension of work by subcontractors

Suspension of work by subcontractors occurs only in the subcontract forms and is applicable to all sorts of subcontractors over extension of time, provided the appropriate standard subcontract form is in use in the case of domestic subcontracts. It is given in clause 14.2 of the subcontract forms to the IFC contract as a matter leading also to reimbursement of loss and expense, but there is no equivalent in the other subcontract forms. It could be a difficult provision to have, as it extends the power of the subcontractor in a situation in which the default is hard to place, by giving him recompense over a matter which he triggers off.

In respect of extension of time then, the event entitles only the subcontractor to an extension, as might be expected. He is the person aggrieved by non-payment, which is why he may institute suspension. This he may do when the contractor has failed to pay him or, alternatively, when the employer fails to pay him direct on the contractor's default. A number of considerations arise over the procedures for suspension, as are discussed in Chapter 8.

Even though this event lies entirely between the contractor and the subcontractor, the action of the architect under JCT clause 35.14 is still needed for a nominated subcontractor to obtain a valid extension. The architect is not implicated under any of the other subcontract arrangements.

Terrorism or threat of terrorism

The terrorism provision covers the results of actual or threatened terrorism and also of official activities that may become necessary to take in response to either.

CAUSES LEADING TO EXTENSION OF TIME AND/OR LOSS AND EXPENSE

As already indicated, extension of time and reimbursement of loss and expense are not necessarily linked, although the cause may be common. To distinguish what is said in this section from what has been said earlier, the term 'event or matter' is used throughout.

Compliance with architect's instructions

Compliance with architect's instructions is to be distinguished from that under 'Late instructions etc' below, where delay in the issue of instructions is in question, for instance by non-provision of information as required under Information Release Schedules. Under this provision of the contract, the instructions are issued in proper time, but their nature leads to prolongation of the contract period and/or disturbance. Alternatively, variations of design or of obligations and restrictions may lead to reduction of an extension in suitable cases.

Design questions

Discrepancies, divergences or inconsistencies (treated slightly differently as between events and matters) may not be discovered until a late stage just before the work concerned is to be performed, with the result that corrections take some time to work out before being instructed, or to put into effect after that. Even worse, they may not be found until some part of the troublesome work has been done, so the correcting instruction inevitably leads to some disturbance of sequence. This is of course subject to the contractor giving warning on such issues as soon as he can, which includes him checking over information as he receives it for the more obvious points.

There is room for difference of opinion here as to when the contractor should find any mistake, not only difference between architect and contractor on a particular project, but between any two architects or contractors when faced with a similar inconsistency and a delay in bringing it to the architect's notice. Although the contractor is held to be a person skilled in the process and technicalities of building, the intricate and sometimes innovative nature of much construction does mean that not everything can reasonably be discerned in time. The problem is accentuated by the preparation of some drawings by subcontractors, so the contractor is receiving information in different formats. If the architect has not fully coordinated this information, it is not always fair to hold the contractor responsible in his position of longstop. The subcontractor's responsibility is a separate issue, as mentioned in Chapter 5.

Variations are not something which the contractor should be requesting – and very few really look for them! On the other hand, he does have a duty, by virtue of the wording for the event or matter concerned, to seek instructions timeously over the expenditure of provisional sums, provided he can tell what they are for, from the sometimes enigmatic descriptions attached to them in specifications and bills (see defined and undefined sums immediately below). When the instructions come for variations or provisional sum expenditure, they may cause significantly more work, or lead to work which needs longer to execute owing to its sequencing, setting time, labour-intensive nature etc. They may also lead to alterations to executed work, which the contractor could not have foreseen in this case. Procedures over variations and provisional sums and their evaluation are dealt with in Chapter 6.

Programme effects

Postponement by its very nature holds things up. But like all the other events, it must extend the end date to lead to extension of time, and this often means a very substantial postponement, either in time or quantity of work affected, hence affecting the critical path of the project.

Antiquities are recognised expressly only by the JCT contracts, and incorporated by reference into their subcontracts, but the contractor would need to seek instructions over them under the IFC contract, so that a similar position should arise. The distinction between such objects as a Stone Age toothpick and a full-scale temple is critical to the delay that is likely, especially when the archaeologists become involved, armed with powers to arrest progress for long periods. This is an event or matter that no employer should yearn to see occur, despite the quite short time attributed to it in the case study of Chapter 30, to ease illustration of the main points at issue. It is the one case in which the JCT clause dealing with the procedure also deals with the question of loss and expense, rather than including it in the main loss and expense grouping.

The arrangements for defined and undefined provisional sums given under the present SMM7 for building works and mentioned in Chapter 6 also stand to have a possibly unexpected effect in relation to the present question. Where there are defined sums, those for which specified information is given and not just a sum of money, 'the Contractor will be deemed to have made due allowance' not only when 'pricing Preliminaries' (a proper subject for SMM7 to consider), but also when 'programming [and] planning'. Where there are undefined sums, for which the specified information is not given so that the three areas cannot be properly assessed, 'the Contractor will be deemed not to have made any allowance' for them. Again, the pricing aspect is a proper one for a method of measurement to consider, but the position of programming and planning is more difficult. An inference to be drawn from the wording is that in suitable cases there is room for extension of time, or perhaps even reimbursement of loss and expense, if the work as instructed in expending the sum led to this. As it is the whole effect of the sum which is in question and not some part that had led to a marginal misapprehension, this could be a substantial matter in the case of a large sum relating to work critically placed within the whole.

It must be questioned whether such a provision within a document of less standing contractually than the conditions themselves can legally introduce this effect, even though it is the case that the conditions are simply silent, rather than opposed to the concept, and even though the purpose may well be regarded with sympathy. If it cannot, then an uncertainty of what is to apply is introduced by the ineffective attempt.

Nominated subcontractors and suppliers, named subcontractors

Only in the case of nominated subcontractors and suppliers and named subcontractors, is the architect empowered to give instructions. Domestic subcontractors are entirely under the contractor's control, once they have been approved. Even for the permitted categories, the present reference is to instructions about these persons, rather than to instructions given to them through the contractor, which are included here in those given to the contractor. Instructions about them are broadly about their appointment and cognate affairs, as listed in Table 5.2. They must be distinguished from other related actions by the architect, such as 'directing' the contractor over payments to nominated subcontractors and 'certifying' over completion of their work. These and their consequences do not fall within the present scope as events or matters. (See also 'Delay on the part of nominated subcontractors and suppliers' and 'Suspension of work by subcontractors' in this chapter.)

An effect is most likely to occur when the architect instructs a change over nomination or naming arrangements when these are proceeding. This is to be distinguished from a late initial instruction, which falls into the general category of late instructions taken below. The case of *Rhuddlan* v. *Fairclough* (Chapter 7) indicates some of the distinctions of responsibility here, particularly that the contractor will probably not receive an extension or reimbursement when the architect is acting with reasonable promptitude, but having regard for the proper commercial interests of the employer.

Instructions over named subcontractors are covered in the lists of events and matters by references to IFC clause 3.3, and 'to the extent provided therein', which warrants some repetition here. That clause and those related to it give six possible options:

3.3.1 There are three options if the contractor cannot enter into a subcontract initially:

(1) Changing the particulars included in the contract documents and relating to the named subcontractor, leading to the possibility of either remedy.
(2) Omitting the work, with the further option of engaging a direct contractor, but in either case leading to the possibility of either remedy.
(3) Substituting a provisional sum, leading in its expenditure and only then to the possibility of either remedy.

3.3.4 By referring back to clause 3.3.3, after determination of a named subcontractor's employment, for reasons not the fault of the contractor, this covers three options:

(1) Naming another person, leading to the possibility of extension of time only.
(2) Instructing the contractor to make his own arrangements, leading to the possibility of either remedy.
(3) Omitting the balance of work, with the further option of engaging a direct contractor, but in either case leading to the possibility of either remedy.

These options should be set against the comments in Chapters 7 and 18 over the appointment of named subcontractors and arrangements after determination.

Comments made about delay by the architect under the topic of renomination in Chapter 7 are also applicable here if, as appears likely, the courts were to take a similar view over delay in a renaming situation. The differences in the present patterns are self-explanatory but should be noted closely.

Between the nominated and named subcontracts, the references in the lists of events and matters are the same, but there is a distinction between the two lists in each subcontract. The event is given as 'compliance by the Contractor with the Architect's/Supervising Officer's instructions . . . deemed to include compliance by the Sub-Contractor', while the matter is given as 'the Architect's/Supervising Officer's instructions . . . and the Contractor's directions consequent thereon'. No reason for the distinction can be deduced, it being hoped that the nuances involved will not lead to dispute. As always, actual delay or loss must occur.

As between the main contract and the subcontracts, another distinction exists. Although the main contract gives the various patterns listed above, the subcontracts simply state that compliance with or the issue of 'instructions . . . under clause . . . 3.3.1 [or] 3.3.3' constitute the grounds for a remedy being possible. The distinction introduced by IFC clause 3.3.4 when another person is named is thus not applied, so there is the possibility of loss and expense being available to the subcontractor, but not the contractor, in such a case.

Opening up work or testing work or materials

The architect has the right under JCT clause 8 or the more complex IFC clause 3.12 to instruct that any opening up or testing shall be performed by the contractor, whether or not anything has been specified in the contract documents. The liability over cost of direct work is discussed in Chapter 5, the contractor usually being reimbursed when the items are not found to be faulty, unless the cost of testing and replacement is specially covered in the contract sum. This latter approach is suited only to such routine and regular matters as testing concrete cubes and runs of installed pipework. However cost is actually dealt with, the possibility of extension of time or reimbursement goes with the question of liability in principle, so the contractor is protected if he is in the right, but vulnerable if in the wrong. With the routine tests specified in the contract documents, the contractor is able to make proper allowance in his programme and avoid delay or disturbance. Any which occurs he should bear, even though the precise wording says otherwise. A major item of defective work here can give rise to quite unanticipated results, which he would also have to bear.

Late instructions etc

As may be suspected, late instructions etc is the most frequently invoked event or matter, unless it is beaten over extension of time by adverse weather, used with rather less justification on some occasions and perhaps as an excuse. Here the emphasis is on the simple lateness of instructions as such, rather than on their subject matter, which has been taken under 'Compliance with architect's instructions' immediately above. The provision is given more teeth now that there is a contract requirement for an Information Release Schedule setting out what outstanding information is to come with dates for its delivery to the contractor (Chapter 5), referred to below.

Information concerned

JCT 80 clause 5.4 and IFC 84 clause 1.7 have been changed to take account of the arrival of Information Release Schedules in JCT contracts after April 1998. The content of this category of information will be items without which the contractor cannot proceed to complete the works – as distinct from such matters as variations, which he can well do without to finish his work! Also, it is not contract documents which are to be provided 'immediately after the execution of this Contract' in the JCT 80 instance, and at an unspecified but reasonably early stage under the IFC 84 arrangement. Nor is it information for which the contractor may need to apply, as still envisaged in the new JCT 80 clause 5.4.2 and IFC 84 clause 1.7.2, which allow for the architect to issue from time to time 'such further drawings or details which are reasonably necessary to explain and amplify' what has been unclear. Information in Information Release Schedules is the remaining bedrock data, not provided in the tender/contract documentation, without which a contractor will not know what to do to complete the works. Certainly, the contractor cannot proceed without information like ironmongery, finishings and decoration schedules, which in the main convey data that cannot sensibly be given on drawings and often can be included in bills of quantities only with difficulty.

The orderly and systematic provision of outstanding information has now been given teeth in the JCT contracts by Information Release Schedules and it may be that this requirement gives rise to more disputes if the requirement is not met. Failure of the employer to ensure that the schedule is complete at tender and that information is supplied to time in accordance with dates on Information Release Schedules will mean probable extension and loss/expense provisions will be used.

Under the drawings and specification contracts, JCT or IFC, far less information should be needed during the construction period, as indicated in Chapter 5. If much is still needed, it may mean there are unstated variations or errors being corrected, with the possibility of redress over time or money. Under any of the JCT contracts, it is always necessary to apply for instructions over the expenditure of provisional sums, which are likely to involve drawings etc in their own right.

A list of all items for architect's instructions under the two main contracts is given in Table 5.2, where those which are 'necessary' in the sense described may be identified. Although variation instructions are not themselves 'necessary', incompleteness or uncertainty introduced by them may create a need for other and therefore necessary instructions.

Applying for information

Although there is no clause requiring the contractor actually to apply for any missing information, or even to notice that anything is missing, JCT 80 clause 5.4.2 and IFC 84 clause 1.7.2 hedge the position somewhat by stating:

> Where the contractor is aware and has reasonable grounds for believing that the architect is not so aware of the time when it is necessary for the contractor to receive such further drawings or details or instructions the contractor shall advise the architect in advance to enable the architect to fulfil his obligations under the [clause].

It seems that the principle of the contractor being the expert in construction applies here, presumably to the same extent as about errors in the design and warnings over them considered in Chapter 5.

This may be fine, in so far as the contractor notices some point of uncertainty when he applies his distinctive analysis to the problem of actual construction, using different criteria from those of the architect or the quantity surveyor, working precontractually. But it leaves doubt over those errors of a more subtle character and, in the present context, over those which do not reasonably come to light until almost the moment of execution. Even more onerously here, the contractor is required to apply for information about which neither he nor the architect can be in any doubt is necessary, such as the whole construction of a tank room covered by a provisional sum. It is assumed that such important matters will now be subject to listing under 'Information Release Schedules'.

The view taken in *Stanley Hugh Leach* v. *The London Borough of Merton* (1985), where the contractor had provided two programmes, at and soon after commencement on-site, which he argued and had accepted were requests for necessary information under the JCT contract, will presumably not be held to be necessary now that a contractor, post May 1998, will either have all the information in the tender documents or in a specific schedule in 'information release' form from which he can programme accurately. Experience in use may or may not prove this to be the case.

Challenged instructions

It is obvious that any instruction of the architect, the validity of which the contractor queries under JCT clause 4.2 or IFC clause 3.5.2, cannot be a 'necessary' one for the purposes of the present clause. A delay of any consequence, caused by following the procedure of either of those clauses, is in practice likely only if the alleged instruction is issued closely before it should be implemented. There may then be a case that the contractor has not received instructions which he has effectively asked for, by querying their status.

The contractor is under no express obligation to hold work while the procedure is operated, although it will be prudent to do so and avoid a complaint that he has exacerbated any difficulty. Usually things will be cleared quickly by a straightforward exchange of letters. Should he delay his query until the last minute, he will lose himself any cause for redress. If, however, the architect is slow in replying, but then gives an adequate substantiation, the contractor may find his programme set back. This would appear to fall under the present clause. If the matter is taken immediately to adjudication, as it may be, a delay should not occur due to such a reference but only due to inherent uncertainty. If so, the question of who bears the effect of the delay should go with the decision, so the contractor would secure an extension or reimbursement only if he won in the reference.

Even without adjudication, the outcome for extension or payment will be affected by whether the architect is able to substantiate his instruction, or whether the contractor appears to have acted frivolously in raising his challenge.

Work not part of the contract

It is sometimes the practice for an employer to engage persons to perform work under direct contract with him, at the same time as the works under a building contract, or even to perform work with his own organisation (Chapter 4). Both options are mentioned in the clauses. Provided this work is physically distinct from that of the contractor, there is not likely to be a risk of conflict between the two sets of activities. This is a distinctly less complicated situation than produced in the multiple-contract pattern discussed in Chapter 13 or than what will almost certainly be present in construction management and management contracting.

JCT clause 29 and IFC clause 3.11 give the employer, not the architect, the right to introduce other persons, so long as there was a mention of their work included in the contract documents. An employer's presence is not specifically catered for, except by placing a different interpretation upon 'employed' from that usually given in contracts. In the JCT clause there are requirements about giving rather more detailed information. An employer may also introduce during progress persons not mentioned earlier, unless the contractor has some good reason to object. Without the clauses, the employer cannot act to deny the contractor the uninterrupted, sole possession of the site, which should otherwise be his under the contract.

For present purposes, a contractor may become entitled to an extension or payment, if he is impeded more than he might have deduced from the data in the contract documents. This may occur if the detail given is inadequate or misleading, if no warning is given at all in the documents, or if the direct contractor fails to perform as expected. The last instance gives meaning to the inclusion of the simple 'execution of work' in the clauses, as well as relating to the direct contractor's 'failure to execute'.

In the second case of no warning, the contractor is entitled to make his consent to the entry of another to the site subject to conditions, rather than just to say yes or no. He might therefore require an extension of time or premium at the outset of negotiations, as a condition of allowing the direct contractor on at all. Although he cannot refuse consent unreasonably, he would appear to be on fairly firm ground in stipulating his own ideas over an extension or sum before consenting, within sensible limits, and not having to rely on the architect's 'opinion' or 'ascertainment' (Chapters 16 and 17).

Materials or goods provided by the employer

Provision of materials or goods by the employer parallels the previous section by dealing with items which the contractor is to fix; similar points can therefore be made without need for particular comment.

There is no clause in either contract dealing with the employer's right to supply, as this does not involve any question of disturbance of the contractor's possession of the site. Suitable clauses should be included in the contract documents to cover questions of accounting, wastage and other issues outside present discussion. There should also be a programme for delivery or, better, a procedure for the contractor to schedule deliveries as he needs them. This should cover such points as requesting materials at dates that are reasonable, similar to what has been discussed over details and other information during progress. It is not apparent how straightforward supply by the employer (which is what is actually mentioned), as distinct from lateness etc, can lead to delay. The situation envisaged by the clauses may also arise during progress, by the employer having then 'agreed to supply' particular items.

There are sometimes pressing reasons for the employer to supply materials, as when he can obtain very large special discounts, or he is himself the manufacturer or sole distributor, or long lead times mean ordering before the contractor is appointed. Usually, there is everything to be said for him avoiding this sort of complication, which can so easily lead to a hitch and hence to delay or disturbance. He should not be drawn into promising to obtain materials which the contractor finds are in short supply unless, again, he is in a particularly strong position. This has to be balanced against the possibility of a delay leading to extension of time under later parts of the present clauses. Often the contractor can be pointed in the right direction, without incurring any responsibility for the timetable.

Failure of the employer over ingress and egress

Failure over ingress and egress are rather problematic clauses, in that provision of means of ingress and egress is not dealt with distinctly in other clauses, and there is a potential overlap with the provisions about obligations and restrictions mentioned under the provisions about variations instructions, but again not defined further (Chapter 6). This is due to the mention of access in the other clauses, as a matter subject to possible restriction. Here the main thrust is that the employer is making something extra available, by allowing access through land etc, other than the site but for which he is responsible. In as much as this access, or alternatively this way out, may not be given at all times (so he has 'to give [it] in due time'), there is an element of restriction over it. The reference in both this and the variation clauses is to at least part of what is set out somewhere in the contract documents, subject to the complications of wording noted in each case.

In the present clauses, the main part is clear, provided it is ensured that any notice of an event or matter is related to the correct clause to avoid any confusion. In practice the effect should be the same, but clarity can only help in this difficult procedural area. Reasonably, there is a limitation to such land etc as is 'adjoining or connected with the site'.

Although the statement about such an area in the contract documents may be quite explicit about when it is to be available, there is also the possibility of the contractor needing to work out when within his programme he is going to need it, if it is not continuously available. His use of it may therefore be dependent on the giving of a suitable notice, which then becomes a condition precedent to his right of use and so also to any question of an extension or payment. The need for some form of notice may also arise out of some extension already granted which has thrown the employer's timetable and grasp of events out of gear. In this case the contractor may not be liable for some of the uncertainty element, but he should act carefully to avoid refusal because of inadequate notice.

Less clear is the reference to 'failure of the Employer to give such ingress or egress as otherwise agreed'. It is not easy to construe this as meaning giving ingress or egress in the circumstances of a shift in the programme as just outlined. It appears more natural to place on the words a meaning of something agreed completely de novo post-contractually. If so, the whole thing is by way of variation of the contract provisions, although not precisely covered by the variation clauses themselves. Leaving this lack of precision aside in the present discussion (but see Chapter 6), it follows though that an extension of time is envisaged, because of a failure of the employer to carry out something which itself is part of a variation agreement. Whether the variation itself resulted in an adjustment by way of extension or even reduction, as might be more appropriate if it is possible (see earlier discussion on variation omissions in Chapter 6), affects the logic of allowing anything here. If the contractor has received an additional way in or out quite gratuitously, it is difficult see how his time will be extended by losing it, unless it is because he has rearranged his programme to rely on this gift. A disturbance is more conceivable. Some care is needed in arranging these post-contract changes to give precision of intention.

Quantities not reasonably accurately forecast

JCT 80 and IFC 84 contracts both have clauses to cover the event that quantities have not been reasonably accurately forecast for the work required. The clauses do not say the forecast was too high or too low and so both must fall within this event. They mainly cover situations in which delay is due to performing more work than could have been expected or work of a different character from what has been described. It will become a matter of technical judgment whether a difference in quantity or character within the overall quantity leads to an extension and loss/ expense. To an extent, it may hinge on whether the contractor has the resources to expand his capacity, if needs be, and so to be using his best endeavours to maintain the programme.

The combination of provisions mentioned gives the contractor some protection against gross inadequacy in the quantities and misleading drawings, by comparison with the work eventually carried out. It may also aid him against excessive variations which may not be easy to segregate, although strictly the provision over architect's instructions covers this aspect. The routine valuation provisions, including those over preliminary items, should deal with financial effects, but the contract allows for loss/expense to be recovered.

There is no specific reference to what happens if the drawings and quantities substantially overplay the extent of work required. The contractor has the same time to do less, if he wishes, and the employer has to accept this as part of the risk of using approximate quantities. Strictly, there is no 'omission' as such of work when the design is finalised, so that no reduction of an existing extension can be made.

Construction (Design and Management) Regulations 1994

If compliance or non-compliance with the Construction (Design and Management) Regulations 1994 affects progress of the works, the contract provides for an extension and possibly loss and expense.

Suspension of work by the contractor following non-payment

A long overdue right has been introduced under JCT 80 clause 30 and IFC 84 clause 4 for a contractor to suspend work after notice when he has not received payment due. This may lead to both extension of time and loss and expense, provided suspension by a contractor has not been frivolous. This provision is thus quite strong in its effects, but less final than the more drastic alternative right of determination. This is easier to operate here, as the allocation of responsibility is clearer.

In summary it should be borne in mind that JCT contracts provide mechanisms which require the architect to act and provide information to time, particularly with the advent of Information Release Schedules, and allow the contractor a range of reasons for which he may obtain an extension of time. Any extension may be with or without associated payment under valuation or loss and expense provisions. Proper notification of the circumstances and reasons for extensions of time are required under the JCT contracts with backup information, otherwise a contractor (or subcontractor) may lose his contractual entitlement, but it may be, as in *Stanley Leach* v. *London Borough of Merton*, that there may also be a claims entitlement under common law damage provisions.

CHAPTER 16

DELAY AND EXTENSION OF TIME

- Late completion
- Procedures for extension of time

This chapter is concerned solely with the ways in which the standard contracts and subcontracts listed in Chapter 15 deal with delay and its effects. Some preliminary remarks serve to introduce the subject in this context, most of them being expanded during subsequent discussion:

(a) Delay is treated only so far as it leads to failure to achieve the completion date and not in its effects on the programme before completion.
(b) No direct connection is established between delay and disturbance of regular progress, as one can occur without the other, common though it is for both to be caused by the same set of circumstances.
(c) On the entirety principle (Chapter 2), delay is assumed to be the responsibility of the contractor, so leading to him having to pay liquidated damages, even if the delay is not necessarily his fault, unless the contract specifically provides otherwise.
(d) When the contractor is not to be responsible for delay, the contract provides for him to receive an extension of time, but still no reimbursement for the delay in isolation from any other disturbance effect, and even then only for some causes (Chapter 15).
(e) There are stipulations over the timing of granting extensions and over related procedures, to lessen the possibility of either party suffering through the laxity of the other, or having events manipulated against his interests.
(f) A list of reasons for granting extensions of time is given, outside which the architect cannot step when exercising his power to grant extensions (Chapter 15).
(g) Little consideration is given to possible overlaps between causes of delay, whether or not they are such as to lead to extensions, although they will be taken into account in calculating extensions.

Quite a number of issues lie dormant within these basic points and fall to be examined in this chapter and others. In anticipation it may be said that the tidy contractual positions given by the contracts are often disturbed in practice by such things as the interaction of several causes of delay – to say nothing of the failures of the participants to carry through the procedures laid down. It is therefore useful to remember the basic legal positions set out in Chapter 9, while moving to the detailed contractual situations and on into some of the practical problems, which are what disputes so often concern.

LATE COMPLETION

Main contract works

JCT 80 and IFC 84 deal with liquidated damages under the main contract very similarly. JCT 80 has clauses for this purpose, followed by a clause on extension of time, so that the latter is presented by inference as a means of overcoming the former. IFC 84 reverses the order, so that the contract date and its adjustment are treated first, and then what happens if the resulting date is overrun.

The requirement of JCT clause 24.1 and IFC clause 2.6 is for the architect to issue a certificate stating that the due date, extended as it may be, has been overrun. This is to be done quite baldly, there is no call to express an opinion over where the blame lies. It leaves the possibility of revising the date under the extension of time clauses and, if this leads to a sufficiently later date, removing any liability to damages. The possibility of an earlier date is also present under JCT 80, but not under IFC 84, as pointed out under the clauses concerned. In the case of IFC clause 2.6 only, there is a requirement for the architect to issue a cancellation of a certificate of late completion, if an extension of time renders it no longer correct. An amended certificate is then to be issued, if there is still a lateness in completing.

JCT clause 24.2.1 and IFC clause 2.7 then oblige the contractor to 'pay or allow' damages at the contract rate to the employer, but only after the employer (not the architect) has required this in writing. There is thus no automatic pressure on the contractor to pay without being asked. If the monies outstanding for the contractor and coming through under interim certificates are sufficient, the obvious course for the employer is to deduct amounts from them, periodically as the damages accrue and as the certified amounts allow. The architect may not anticipate this by himself making deductions from or in what he is certifying. The employer thus has a clear right of set-off.

Alternatively, he may ask the contractor to pay the amounts to him independently of any passage of certificates. If amounts due are inadequate, this may be the only clear course. It may be the most hopeful if the contractor appears likely to drop into insolvency, a state usually preceded by delayed progress. Once insolvency happens, though, there may be little to retrieve anyway. The alternative, while there is life, is to recover amounts as a judgment debt – the third option under the clauses, but not one to be preferred while the others are feasible.

The issue of an amended and later date for completion which cancels out the liability to liquidated damages means that the employer has to repay to the contractor any sums already received. There is no requirement for the contractor to ask for these amounts and the employer will be in breach if he does not pay within a reasonable time, although none is specified. The contractor has no redress *within* the contract machinery, even the determination provisions (as the extreme weapon) do not help him, and he must rely on his usual legal rights.

If a JCT contract includes a sectional completion supplement, the level of damages will be lowered when this comes into play, whereas JCT clause 18 over partial possession introduces a similar position without a proper hand-over necessarily being anticipated when the contract is formalised. There are potential complications in each case, as illustrated by *Bramall & Ogden* v. *Sheffield.*

Subcontract work

The several subcontracts deal with the subcontractor's failure to complete in the following clauses:

NSC/C	Clauses 2.8 and 2.9
DOM/1	Clause 12
NAM/SC	Clause 13
IN/SC	Clause 12

They follow their respective main clauses fairly closely. All of them require the contractor to act by giving notice about failure to complete either the whole subcontract works or any distinct section of it. All but one require this notice to be given to the defaulting subcontractor, with whom the contractor alone deals. The exception is the nominated clauses in both versions, which requires the notice to be given to the architect.

The reason for this difference is that the architect has to be involved under JCT clauses 35.14 to 35.16 in the following steps:

(a) Granting an extension of time to the subcontractor for a part or the whole of his works, even when the cause is default by the contractor himself.
(b) Certifying that the subcontractor should have completed by the date currently applying for a part or the whole of his works.
(c) Certifying that practical completion of the whole of the subcontractor's works has occurred.

These procedures are discussed hereafter. When the architect has come to a decision under each head, he acts through the contractor, who passes the result to the subcontractor.

The JCT-related forms contain a provision corresponding to that in the main form for the subcontractor to 'pay or allow' any sum by way of loss or damage to the contractor. There is a proviso under the nominated forms that such sums are due only after the architect has exercised his functions.

However, the IFC-related forms contain no such provision. Provided that a subcontract delay causes disturbance of regular progress for the contractor, the position is covered by clause 14 of the forms. To the extent that a subcontract delay causes simple delay to the contractor, the position is not clear. The reasoning behind the omission is not certain.

PROCEDURES FOR EXTENSION OF TIME

Background considerations

The central purpose of extension of time is to keep a definite completion date in a contract, rather than to allow the original date simply to be replaced by a reasonable but uncertain date when there is some default by the employer which places time at large. This acts to protect the employer, so far as the causes for extension are in the nature of employer's defaults in the broad sense of the term. Some may be less than defaults, whereas others are usually included which represent 'neutral' matters that would otherwise put extra pressure on the contractor to counteract them to complete by the original date, and so a high allowance in his tender. None is intended to be defaults of the contractor.

But in any case, when the contractor is likely to receive an extension of time, it is important that he should know this and its amount as soon as possible, so he is certain over what date he is to aim for and can adjust his effort accordingly. This fact has been recognised in the earlier legal decisions over extensions of time and whether they lead to avoidance of liquidated damages. These were related either to non-standard contract conditions or to earlier standard conditions and need not be considered closely here. In particular, they sometimes revolved round whether an extension had to be granted before the existing completion date, original or amended, was passed to allow the contractor to have the certainty which has just been mentioned. Because of differences in wording, the cases of *Miller* v. *London County Council* (1934) and *Amalgamated Building Contractors Ltd* v. *Waltham Holy Cross UDC* (1952) were decided oppositely in this respect, the former against the right to grant such a late extension and the latter (on a forerunner of the present JCT form) in favour of the right. The result of extension of time provisions in a contract is to protect and maintain the employer's rights to deduct liquidated damages. If there is no provision for extension of time to deal with delays caused by the employer (by his agents, architects and engineers) then if such delays do occur any liquidated damages clause in the contract has been held to be invalid. This happened in *Peak Construction (Liverpool) Limited* v. *McKinney Foundations Limited* (1970) where special terms were inexpertly drafted and resulted in the employer, having caused delay, losing his right to deduct liquidated damages.

Standard forms include procedures for granting an extension of time and invariably go through (1) submission of an initial notice giving general details, (2) submission of supporting details and (3) sometimes further details to provide more up-to-date information. Time limits are *sometimes* stated. The responsibility for deciding on the validity and then granting an extension of time to the contract generally lies with the architect, engineer or supervising officer.

A number of cases indicate that procedure may be important but not always so. In *F. G. Minter* v. *Welsh Health Technical Services Organisation* (1980) it was shown that failure by the contractor to serve the proper notice may result in loss of entitlements. In *Temloc Limited* v. *Errill Properties Limited* (1987) failure by an agent (architect, engineer or supervising officer) to make a decision within a stated timescale may not affect an employer's rights. In *London Borough of Merton* v. *Stanley Hugh Leach* (1985) it was held that failure by the contractor to comply with contractual requirements about notice will not necessarily lose a contractor his right to an extension of time. It is stated in *Keating on Building Contracts* (fourth edition) that 'it may therefore be against an employer's interests for an architect not to consider a cause of delay of which late notice is given or of which the architect has knowledge despite lack of notice'. Where the contract conditions state that service of a notice is a condition precedent to obtaining an extension of time then lack of a notice will usually lose the party his rights; see *Bremer Handelsgessellschaft mbH* v. *Vanden Avenne-Izegem* (1978). To be clear and therefore effective, any clause on condition precedent must state that failure to serve notice will result in loss of rights to an extension of time.

Other aspects of extension of time are important. In *Rosehaugh Stanhope (Broadgate Phase 6)* v. *Redpath Dorman Long* (1990) it was held that where a contractor had a fair and reasonable extension of time under the contract, he was not liable to pay damages until the time for completion had been ascertained. Any cross-claims for loss and expense should also be taken into account before damages were deducted.

In *Balfour Beatty Building* v. *Chestermount Properties* (1993) the important point was established that, under a JCT contract, the architect was entitled to grant an extension of time as a consequence of relevant events whether such events occurred before or after the contract date for completion as then currently extended. The appropriate method of extending time was the 'net' method, i.e. the architect should start with the existing completion date and postpone it to the extent which he considered fair and reasonable, having regard to the delay caused by the requirement to execute the variation instructions.

In *John Barker Construction Ltd.* v. *London Portman Hotel Ltd* (1996) it was held and confirmed (as had been done before) that a court (there was no arbitration clause) has no general power to review and revise decisions, opinions, instructions, certificates or valuations of an architect. However, if a tribunal finds there have been breaches by the employer (by his agent, the architect) it may grant relief in these areas to a contractor. In *Barker* it was found that the architect did not act fairly and lawfully. The Official Referee decided that the architect's decision in relation to how he had given an extension of time was fundamentally flawed. It was invalid because it did not follow a logical analysis in a methodical way of the importance which the relevant matters had or were likely to have on the contractor's planned programme. It was held that the architect had made an impressionistic rather than a calculated assessment of time (besides other matters). Because the contractual machinery had become frustrated, the court decided that it should therefore determine a fair and reasonable extension of time.

Very largely because of these considerations and the ways in which the looser provisions for extension of time have been abused or ignored, the present procedural provisions have been introduced into the JCT contract. They aim to give certainty over procedures, as well as over when an extension may be granted. They have been subjected to an amount of criticism second only to that over the nomination clauses, but much of it is misplaced, as the procedures are largely what should be happening anyway. It is unfortunate when good practice has to become contractual. For the larger contract, they are needed, although to be regretted in the 'without quantities' variants. For the smaller contract, the IFC form embodies a simpler clause, whereas the GC/Works/1 contract manages (as usual) to provide a more succinct and reasonably tight set of words.

Extensions of time must be seen against the provision in each contract that the contractor is to use his 'best endeavours to prevent delay . . . and to do all that may reasonably be required to the satisfaction of the Architect to proceed'. The words mean that, at the very least, he must take action to prevent further delay and to contain the effects of what has already occurred or been set in train. He cannot sit back fatalistically and let affairs drift. Some would argue that 'best endeavours' mean recouping the effects so far suffered, when this is at all possible, and at considerable expense if needs be. This goes beyond 'prevention', but at least cooperation and resource is expected. The contractor stands to gain by some action, because he is incurring extra overheads without reimbursement when delay occurs without loss and expense as well. The point of a trade-off is illustrated in the case study in Chapters 30 and 31.

Standard clauses: JCT 80

It is easier to take the longer JCT clause first and then to compare the IFC clause.

Applying for interim extensions of time

JCT clause 25.2 deals with the actions which the contractor has to take when there is delay. It operates 'forthwith', when 'it becomes reasonably apparent' either that a delay is occurring or is likely to occur. It is therefore better for the contractor to act whenever he has reasonable doubt about progress, rather than to wait until he has reasonable certainty about delay. It follows from the list below that the clause is not intended to operate by retrospective action by contractor or architect.

Failure by the contractor to initiate action can only be to his detriment, even though it does not lose him the possibility of an extension, in view of JCT clause 25.3.3. He may cause vital evidence to be lost, so that he receives inadequate extension – he should not work on the basis that confusion over what has actually happened is bound to help him gain more. But it may also be possible for the architect to take action by way of instructions to reduce the delay, given enough warning that it is threatening. If this opportunity is denied him, then the contractor may face a reduced extension because he has not done what he can to mitigate the delay.

Like all such provisions, it requires the contractor to act in writing, here to the architect. There are several things which he has to say:

(a) *Immediately*, what are 'the material circumstances, including the cause or causes of the delay'. He particularly has to name 'any . . . Relevant Event' (perhaps therefore several), the term used to identify those causes which lead to an extension. It would appear that the purpose of the rather repetitive wording is to ensure that the contractor gives some specific detail, but also that he hooks it rigorously to the list of events. He might otherwise do one without the other and perhaps lead to confusion.

(b) *As soon as possible*, if practicable with the foregoing, particulars of the *expected* effects of 'each and every' relevant event. This means making a segregation which the preceding clause does not require, as it is simply an early warning system. This may be a formidable task, if several events come together and overlap in effects, as is discussed hereunder.

(c) *Again as soon as possible*, but possibly not so soon as (b), 'estimate the extent . . . of the expected delay' in relation once more to each relevant event, but with account of any other delay which is involved, concurrently or otherwise. This estimate is again difficult, but the contractor is not committing himself finally by it. It is for the architect to make the final decision, with this estimate as one factor to aid him.

(d) *Thereafter,* as is 'reasonably necessary' or when the architect 'reasonably requires' (hopefully the two are not in conflict), further information to bring matters up to date.

(e) *Parallel information* to all of the foregoing for nominated subcontractors concerned in the delay.

This list does give a lot of work, but properly followed it also nurses the whole question of delay carefully through in terms of the provision of data by the contractor. What sort of data will best meet these requirements is outlined hereafter.

The clause is related closely to the provisions of JCT clause 25.3, which gives what the architect has to do. Apart from the considerations given in introducing the above list, the contractor should note carefully that the architect has no obligation to act over granting an extension during progress of the works, unless the contractor takes steps to notify him of delay. Without this 'triggering' effect, the contractor may lose himself the benefits of early reassurance that adequate extension of time is coming his way.

Granting interim extensions of time

JCT clause 25.3.1 goes on to what the architect must do as a minimum in response to the actions of the contractor. His action is conditional upon that of the contractor and in particular upon receipt of 'any notice, particulars and estimate'. He is required to exercise his 'opinion' to grant a 'fair and reasonable' extension, which recognises the degree of inaccuracy inherent in the process, and not to agree an extension with the contractor. Although he is to have 'regard to the sufficiency' of what the contractor has supplied, he is not required specifically to check it, approve it or in any other way negotiate upon the basis of it. How he uses it is his concern, although sensibly he will discuss with the contractor any clear point of divergence. The architect is specifically required to notify the contractor if he decides not to grant an extension. This is a reasonable practice under any contract.

As a moderating influence, at the end of the process now being reviewed there lies the possibility of adjudication when, no doubt, any coarseness in operating the procedures would be assessed. It is, however, the architect's opinion that is required, both here and in the final revision of the completion date under JCT clause 25.3.3.

What the architect does have to take into account is twofold. Firstly, he has to decide whether any of the causal events adduced by the contractor are in fact relevant events as listed in JCT clause 25.4. He can act only within this framework and must ignore any other cause of delay, even if responsibility may be laid at the employer's door. Beyond the stated list, the contractor must look to direct interaction with the employer for any redress, either by special agreement or by legal proceedings.

Secondly, the architect has to consider whether the result of any qualifying event is that completion is 'likely to be delayed' beyond the date currently fixed for completion. If it appears that a delay will not extend the date, but simply cause some lateness within the programme, he can do nothing. This may seem harsh, but relates to the fact that the whole extension of time arrangement is to protect against liquidated damages, which relate to final overrun. There are several problems here as reviewed in Chapter 9:

(a) Several small, separate and relatively innocuous delays can build up into a total delay sufficient to overrun the completion date, so that the architect must look beyond the immediate event or events, if needs be, to assess the complete picture. It is up to the contractor to give the broader picture when estimating the extent of delay under JCT clause 25.2.2.2, by leaning on the phrase 'whether or not concurrently' used there.
(b) Delays not falling within the category of relevant events may occur within the same time span, or have already caused such delay that achievement of the actual completion date looks unlikely. There is more than one school of thought on this issue, but much is likely to depend on just how and when things happen, or fail to happen. Some guiding principles are given at the end of this chapter.
(c) The completion date may have been passed already when the delay occurs, and this appears not to be within the scope of the earlier clause (by the words 'is likely to be delayed' or of those which follow). Legal decisions look both ways here: both the JCT and IFC clauses now have provisions to remove the doubt and to allow an extension to be made.

The architect has to fix a new date, so giving an extension, and state which relevant events he has taken into account. This allows the contractor to query why any has been left out completely, so that he can narrow any field of complaint, perhaps as a prelude to arbitration. This date is to be such 'as he then estimates to be fair and reasonable', pointing to the final revision which comes after practical completion.

The extension is only interim in its significance and the architect has to act within a definite time. Usually, this is to be 'not later than 12 weeks' from receipt of sufficient information from the contractor to enable the architect to act, which means that it is in the contractor's interests to chase each stage of supply through promptly. Although the architect is not specifically required to take account of what is supplied to him, he can hold out '[in]sufficiency' of information as a reason for not acting. This cannot be over mere technicalities, but only over major gaps which prevent him from forming his interim opinion and fixing a new date.

The modification to this timescale occurs when there is less than twelve weeks between receipt of sufficient information and the completion date currently obtaining. Here the period for the architect to fix a revised date is shortened to that remaining up to the current date, to avoid that date being overrun without an extension. This would create doubt for the contractor and also trigger the possibility of the employer deducting liquidated damages, as he might even if he knew that an extension was pending.

But within this latter situation there is the possibility of the architect being faced with an unreasonably short time in which to calculate an extension, conceivably even just the day before completion is otherwise due, so that a term of reasonableness must be implied here. An extension of time granted just before the previous date was reached could hardly help the contractor to review his programme, which is the major consideration. Equally, a relevant event cropping up close to completion is not very likely to cause a substantial overrun, although there is a saying about things that have happened at sea. In principle, several situations can crop up, and no clause can be expected to legislate with practical precision for them all. They include the following cases:

(a) The '12 weeks' rule applies, because the contractor acts in time.
(b) The '12 weeks rule' should apply, but the contractor does not act in time, so that the twelve weeks are not available to the architect before the existing date. If the contractor has been dilatory, he is caught by the 'as soon as possible' of JCT clause 25.2.2. If he has done what is reasonable, then the 'less than 12 weeks rule' applies, subject to comments just made.
(c) The 'less than 12 weeks rule' applies, with reasonable time for the architect to act, depending on the complexity of the issue.
(d) This rule applies, but time is inadequate.
(e) The event comes to light before the current completion date, but the data for calculating an extension are not to hand until after the current date has been passed. These should be avoided in most instances by the contractor working to the earliest dates practicable for providing information, bearing in mind that he is giving an estimate and other provisional information.
(f) The current date has been overrun when the event comes to light. This has been alluded to above as the most controversial.

Effect of instructions to omit work

Although all the relevant events may lead to an extension of time, there is also the possibility of a reduction in work content occurring due to omission variations, sufficient to shorten the programme noticeably. JCT clause 25.3.6 states that no completion date is to be fixed earlier than the contract date, and this limits what may be done, but there is the possibility of omissions reducing a potential extension by some margin, even eliminating it. Within the clauses already considered, there lie elements to regulate the matter, although their wording does not lead to self-luminous interpretation.

One controlling factor is that the omissions must be of 'work' under the original clause, so that omissions of obligations and restrictions do not lead to any effect, even though they may have quite a drastic effect. JCT 80 rectifies the situation by including 'obligations', although still not explicitly mentioning 'restrictions'. A measure of special pleading may be needed here – by either party. A second factor is that only omissions instructed since the last fixing of a revised completion date can be taken into account: any omissions preceding that fixing should have been taken into account then. If that was not possible under the rules, they will have been lost for ever, as discussed below.

The third factor is in JCT clause 25.3.2 and allows the architect 'After the first exercise . . . of his duty' to fix an earlier date 'than that previously fixed under clause 25'. It gives the architect power to take sufficient account of omissions when fixing a *second or later* revised date to reverse the extension of time that might otherwise arise, or even to reduce the previous extension in the absence of any relevant event justifying an extension. It excludes any effect when fixing a first revised date, by virtue of its opening qualification and remains subject to the overriding provision that the original date is not to be cut down at any time.

A fourth factor is that the architect is to state when fixing a new date how far he has taken such omissions into account. This is in JCT clause 25.3.1.4 and is qualified as relating to 'instructions . . . issued since the fixing of the previous Completion Date'. There is in fact no explicit authority anywhere in JCT clause 25 for the architect to take account of omissions to reduce, but not reverse, what would otherwise be an extension period, although JCT clause 25.3.2 could perhaps be operated to achieve this. The present clause does not give the architect power to take anything into account when fixing a date, but simply requires him to make a statement about how he has done it. It seems necessary to rely upon some implication of a term to make the clause work in this respect.

Subject to this implication being made, the clause apparently contemplates taking in the effect of omissions instructed before the first fixing of a revised date, with this turning upon the words 'previous Completion Date'. 'Completion Date' is defined in JCT clause 1.3 as 'the Date for Completion as fixed and stated in the Appendix' (i.e. the contract date) 'or any date fixed under either clause 25 or 33.1.3'. The completion date accordingly includes a distinctly defined, if identical, date fixed when entering into the contract and also various versions of itself fixed thereafter. It appears therefore that 'the fixing of the previous Completion Date' includes the entry of the original date in the appendix, unfortunate though the contortions of the several clauses may be.

If this is so, the contractor is obliged to have any omissions, instructed before he applies for his first extension of time, taken into account when that extension is fixed. They cannot be carried forward to be set against later relevant events, however, but must be used up at the time and any excess ignored for ever. A similar position arises over omissions occurring before any later relevant event, as noted above. This opens up a number of ploys over the instruction of omissions and the timing of notices etc to secure a particular effect, which need not be detailed here.

Granting the final extension of time

JCT clause 25.3.3 deals with the final extension of time and requires it to be made within twelve weeks of practical completion. It also covers the period before practical completion when the currently fixed completion date has been overrun, and this is discussed further at the end of this section. The procedure does not depend upon any documentation from the contractor, or at least any further documentation. Indeed, it could be dealt with by the architect when no earlier extension has been given and without the contractor knowing that it is coming. It is to be hoped that nothing quite so remote actually occurs.

There are three options given: a later date than existing, an earlier or the same date confirmed. The first may be the result of a comprehensive review of the history of the project and allows the architect to take account of better knowledge about extensions already given and the relevant events on which they were based. But he may also take note of additional events, whether or not the contractor has drawn his attention to them. He has no authority to review any reductions of extensions based upon omission variations, the amounts of which he has had to state explicitly when granting any extension taking account of them.

Secondly, the architect may take into account any omission instructions which he has issued since the last fixing of a date, whether that deferred or advanced the completion date. This is allowed to give an earlier date, but it would appear reasonable to use both the first and second provisions for reaching an earlier and a later date together and obtain a net adjustment. Strictly, the architect may do *either* one *or* the other, but each has to be 'fair and reasonable', which would be impossible to achieve if he could arbitrarily choose whether to give an earlier or a later date.

His third option is to confirm the previously fixed date. No details are given here, and the architect is free to take into account any factor that he sees as relevant. Reasonably, it should be a combination of the other two options, with the results cancelling out. There is no prohibition by silence on him reviewing reductions already made for omissions variations, as there is in the other options.

Several comments arise. The architect cannot review his previous decisions in a downwards direction, unless he does so under the cloak of the last option. This avoids the contractor working to a given extension, only to find after he has met practical completion apparently safely, that it has been eroded and he has become liable for damages. If there is any genuine doubt, the architect understandably may give somewhat conservative extensions during progress, to avoid this situation and remain fair to the employer. He has to give as accurate extensions as he can, and again the contractor's estimates need stressing as contributing to this result.

Furthermore, the architect has only one chance to firm up on his earlier efforts: he cannot go on revising his figures. He should therefore round up all the detail that he needs before he acts, but do so within twelve weeks of practical completion. This is no longer than he had during progress and finality is required, but he can now act with the benefit of hindsight, so that uncertainty over the future has gone.

As a footnote to the foregoing, should a clause not contain anything (as the present clause does) to clarify whether an extension may be granted when the due completion date has been passed, but practical completion has not been achieved, it is widely held that one may not be granted and further that time will become at large against the employer if variations (perhaps quite critical) are introduced extending the time during this period, or if other 'non-neutral' events occur (see cases under 'Background considerations' above). The architect then needs to keep on top of granting extensions in the closing stages to avoid being caught on a technicality by work running past the existing date. It is also likely that the contractor, being in default, could not obtain an extension for 'neutral' events. The JCT provision is more generous to the contractor than that in the IFC clause.

Standard clauses: IFC 84

IFC clause 2.3 reflects the simple project for which it is intended, by having less procedural detail and what is given here should be viewed in comparison with the JCT 80 fuller provisions.

The introductory wording is closely similar, as is its effect. The contractor has to give notice 'forthwith' and in similar circumstances, stating 'the cause of the delay'. This appears as effective as the longer requirement of the JCT clause in securing warning and some information, although this still means that neither contractor nor architect can afford not to act promptly, so prejudicing their position in coming to a proper solution. There is an obligation on the contractor to 'provide such information . . . as is reasonably necessary' to operate the clause at this and all its later stages. This covers supplementary statements, estimates etc. It is again 'the opinion' of the architect which decides the amount of 'fair and reasonable' extension given, as soon as it can be calculated. There is no fixed period of twelve weeks from adequate data within which the architect must act.

The clause refers only to making an extension of time, not to fixing a new date, these being effectively synonymous terms. Its third paragraph allows the architect to make an extension 'At any time up to 12 weeks after the date of Practical Completion', which therefore gives an effect similar to that under JCT 80, covering the periods both before and after practical completion. The paragraph again covers the review of a previous decision and the case in which the contractor has not given any notice under the main procedure above. It differs by prohibiting *any* reduction of an extension of time already made, so the comments made earlier about the architect being careful not to overstate an initial extension apply with extra force here.

There is silence over whether any reduction can be made to allow for the effect of omission variations. From the preceding paragraph it is clear that they cannot be taken alone or against delaying events (these are not termed 'relevant' in this contract) to lead to a reduction of an extension already granted. Furthermore, there is therefore no authority for setting variations which are omissions against other types of events, such as delayed instructions. It is, however, reasonable to set their effect against addition variations which may be under review at the same time as a possible cause for extension of time. In the circumstances of execution to which they give rise, it may indeed hardly be possible to do otherwise when assessing their effects. In the smaller project envisaged for this contract, the whole question of such omissions is less likely to be significant in the programme.

A provision distinct from the JCT provisions is contained in the second paragraph of the clause. Like that clause, it deals with the doubtful area between when practical completion should have occurred and the possibly later date at which it does occur, by making it explicit that the architect can still make an extension of time due to events happening within that period. There is a limitation to those events which are grouped in Chapter 15 as the responsibility of the employer, rather than 'neutral', a limitation not in the JCT version. It thus does not protect the contractor as he is protected while still within the accepted timescale, unless there is a later extension of time which shifts the boundaries of this grey area sufficiently.

Standard clauses: subcontract forms

These provisions, variously in clauses 11 or 12 of the respective forms, follow closely on the related main contract clauses and, subject to these groupings, are similar to one another as well. They thus run through a pattern of notices, particulars and estimates, leading to interim and final extensions of time, hedged about with similar provisos.

It becomes the subcontractor's responsibility to give notices etc, and the contractor does not have to act at all if this is not done. This is no bar to the contractor seeking extension of time himself under the main contract, when the reason lies in the event which the subcontractor has not raised himself. In fact, the contractor would be ill-advised to delay, as his own right to an extension may be prejudiced. He cannot say, 'The trouble lay in something which delayed my subcontractor first and then later, and only in consequence of that delay, delayed me.' When he sees what is going to happen, he must move to secure his own position.

This highlights the point that what delays the contractor may not delay the subcontractor and vice versa. So far as there is a consequential or parallel effect, they will both need to give notice. Neither can rest on what the other has done or hide behind what he has not done. The respective forms place the responsibility on the party seeking an extension to go after it, however obvious the point might appear.

The subcontracts differ in one respect which does not follow from the immediate differences in the main forms. In the nominated forms alone, the contractor has to bring the architect in to decide whether an extension is due to a subcontractor and, if so, how much it should be. This arises by virtue of JCT clause 35.14 which states, 'The Contractor shall not grant to any Nominated Sub-Contractor any extension . . . except . . . [with] the written consent of the Architect.' The subcontract provisions require the contractor to refer any request for an extension from a subcontractor to the architect to decide both principle and amount. In fact, this is a pretty tall order for the architect, asking of him far more detailed involvement with the project programme than is usually needed to deal with a main contract extension. He is unlikely to be able to deal with it without substantial detail from both contractor and subcontractor, and substantial effort on his own part.

The architect's problem may be that, although he was engaged with the subcontractor's programme at the tendering stage, it could well have moved on since, as it has been integrated with that of the contractor more closely, and as changes have occurred without any question of an extension arising. For example, to grant an extension fairly, he may need to know why the subcontractor's whole programme shifted a month later, before one section of his work was displaced a further month all on its own, with only this latter up for extension.

It is probable that many architects are unaware of just how detailed a burden is placed upon them. As a duty, it is not limited to extensions of time arising out of the relevant events alone, but also those arising out of default by the contractor in his own performance, which carries with it default by domestic subcontractors. There is a whole network of relationships into which the architect may be drawn by the main contract provision and its simple reference to the subcontract. Delay by other nominated subcontractors and by nominated suppliers is covered for the subcontractor by a clause paralleling that in the main contract.

The reasons for JCT clause 35.14 being present are several. It indicates the closer link that exists between the architect and nominated subcontractors, despite the lack of any precise contractual tie with the employer. He has been involved in their introduction and, up to that stage, with their programme. If they have design to perform, he too is involved with that. More directly from a contractual viewpoint, he has to operate JCT clause 25.4.7 over granting any extension of time to the contractor because of delay 'on the part of' nominated firms, both subcontractors or suppliers (Chapter 15).

It is that clause which gives the strongest reason for the architect being involved in granting subcontract extensions of time. The contractor is entitled to an extension due to subcontract delay only if he 'has taken all practicable steps to avoid or reduce' such delay. The architect has therefore to check what has happened and so come to know what is a fair extension, subject to the points just made.

To be entitled to an extension in principle, the contractor must have acted to avoid or reduce the subcontractor's own delay, the delay 'on his part'. This should be distinguished from the delay which is caused to him by the subcontractor's delay. Here he must act to mitigate the effects, as with all relevant events, perhaps by reorganising his domestic subcontractors (who have their own rights to extensions) to counter the impact. Failure to mitigate does not debar him from securing an extension for this particular event, but it may reduce its amount. In reviewing the subcontractor's extensions, the architect also needs to look to this area of delay where there are not extensions of time.

CHAPTER 17

LOSS AND EXPENSE

- Nature of and reasons for loss and expense
- Action by contractor
- Response by architect
- Payment of amount ascertained
- Other rights and remedies
- Provisions in JCT clause only
- Subcontract provisions

This chapter looks at the JCT and IFC main and subcontract forms only. In view of the close drafting of the central provisions and their importance, the full wording of JCT clause 26.1 and IFC clause 4.11 is given in dissected form, even though this renders the quotations out of sequence and context. A table comparing the complete clauses in their proper order is given in Table 17.1. Most of the points drawn out of the immediate wording are returned to in subsequent chapters dealing with practical application. The two clauses are quite comparable and may be taken together in the present discussion.

NATURE OF AND REASONS FOR LOSS AND EXPENSE

26.1 Direct loss and/or expense in the execution of this Contract for which he would not be reimbursed by a payment under any other provision in this Contract due to deferment of giving possession of the site . . . or because the regular progress of the Works or any part thereof has been or is likely to be . . . materially affected by any one or more of the matters referred to in clause 26.2.

4.11 Direct loss and/or expense, for which he would not be reimbursed by a payment under any other provision of this Contract, due to
(a) the deferment of the Employer giving possession of the site under clause 2.2 where that clause is stated in the Appendix to be applicable; or
(b) the regular progress of the Works or part of the Works being materially affected by any one or more of the matters referred to in clause 4.12.

The nature of loss and expense is firstly that it is direct. This aspect is discussed in Chapter 10, but in essence means that what is allowable is much the same as would arise under an award of damages by the courts. It is therefore intended to put the aggrieved party back into the position in which he would have been if the breach had not occurred, so far as money can do this. In that it is direct, it does not allow for consequences which are indirect, that is which follow at a distance because there

Table 17.1 Comparison of the main provisions of JCT clauses 26.1 and IFC clause 4.11 over loss and expense

JCT clause 26.1	IFC clause 4.11
If the Contractor makes written application to the Architect stating . . . that he has incurred or is likely to incur direct loss and/or expense in the execution of this Contract for which he would not be reimbursed by a payment under any other provision in this Contract . . . and if and as soon as the Architect is of the opinion . . .	If, upon written application being made to him by the Contractor within a reasonable time of it becoming apparent, the Architect is of the opinion . . .
. . . that the regular progress of the Works or any part thereof has been or is likely to be so materially affected as set out in the application of the Contractor . . .	. . . that the Contractor has incurred or is likely to incur direct loss and/or expense for which he would not be reimbursed by a payment under any other provision of this Contract, . . .
due to the deferment of giving possession of the site . . . or because the regular progress of the Works or any part thereof has been or is likely to be so materially affected by any one or more of the matters referred to in clause 26.2; . . .	. . . due to (a) the deferment of the Employer giving possession of the site under clause 2.2 where that clause is stated in the Appendix to be applicable; or (b) the regular progress of the Works or part of the Works being materially affected by any one or more of the matters referred to in clause 4.12 . . .
. . . then the Architect from time to time thereafter shall ascertain, or shall instruct the Quantity Surveyor to ascertain, . . .	. . . then the Architect shall ascertain, or shall instruct the Quantity Surveyor to ascertain, . . .
. . . the amount of such loss and/or expense which has been or is being incurred by the Contractor . . .	such loss and expense incurred . . .
. . . provided always that: 1. The Contractor's application shall be made as soon as it has become, or should reasonably have become, apparent to him that . . . [progress is] affected as aforesaid, and	[within a reasonable time of it becoming apparent]
2. the Contractor shall in support of his application submit . . . upon request such information as should reasonably enable . . . an opinion as aforesaid, and 3. the Contractor shall submit . . . upon request such details . . . as are reasonably necessary for such ascertainment as aforesaid.	. . . provided that the Contractor shall in support of his application submit such information required . . . as is reasonably necessary for the purposes of this clause.

is some intervening event. Not all loss and expense occurring 'in the execution of this Contract' is reimbursable, as some may be due to other causes, including those which lie in the contractor's own inadequacies. Equally though, it also means that loss occurring outside the contract, but due to events within it, may be allowable on occasions. This is particularly true of loss of profit on lost business, discussed in Chapter 15.

The twin terms 'loss' and 'expense' mean that both a failure to receive and a necessity to expend may qualify. The 'and/or' conjunction simply establishes that either or both may be in prospect in any instance. It is self-evident at law that the contractor is not entitled to be reimbursed under this provision and again under another for the same reason. The expression relating to 'payment under any other provision' also means that the present loss and expense provision is the last port of call, after as much as possible of what is due has been met under, say, the variations clauses. This point is discussed in appropriate places in relation to the attractiveness of the several options to the respective parties, particularly in Chapters 6 and 19.

The mention of 'the matters referred to' covers the set of reasons described in Chapter 9. These are a smaller group than those which constitute relevant events for extension of time (Table 15.1), there being no necessary connection between them on a given occasion. But it is equally true that only items set out in the present clauses produce an entitlement to recover loss and expense by using its machinery. It is permissible and necessary to resort to proceedings to obtain redress on other appropriate issues not listed.

The other reason in the clauses is that of delay in gaining possession of the site due to the employer's action. In each case it is an optional provision, activated by an appendix entry, which the contractor should therefore ensure is made whenever it may be needed. Lack of an entry means there is a breach if delay occurs, so the contractor has to negotiate or use proceedings to seek redress. The consequences should be fairly clear to identify when there is a straightforward shift of the whole programme, rather than a disruption within it.

If the delay is other than nominal whenever the provision does not apply (the appendix appears to envisage a maximum of six weeks when it does), the contractor may be advised to consider agreeing an amount in advance of proceeding with the contract, as a precondition to the waiving on his part of what is a fundamental breach. In practice this may be a question of expediency, according to which party is most concerned about not entering into the contract or by further delay, tempered by the amount at issue, including such considerations as inflation.

Although no specific consequences are given following on this last reason for loss and expense, actual financial deprivation must of course follow for it to lead to any payment when once a contract with the provision is in being. This is also true of the other matters applying in both contracts, but a further criterion is introduced for all of these: that 'regular progress' is 'materially affected' by the matter concerned.

The latter phrase may be taken first. It is intended to exclude the small aberrations which occur within any programme for activities as complex and vulnerable as building works. It is not in the nature of things precise, but suggests an identifiable disturbance sufficient to step up the contractor's costs by more than the margin to be expected either way within routine fluctuations of output.

There is no absolute bottom figure in such a situation, nor one that can be measured as a percentage of the contract sum or any other amount. Given a small-scale matter affecting a restricted part of the works, as is possible under the wording, the actual loss and expense may be quite small, so long as the disturbance is material within the area concerned. This is especially so if subcontractors are involved, with their more restricted interest financially, although it may be argued (unreasonably, it is suggested) that what justifies their claims against the contractor does not justify that of the contractor in the main contract context.

'Regular progress' indicates the element of disturbance. The contractor is under an obligation by JCT clause 23.1 and IFC clause 2.1 that he shall 'regularly and diligently proceed' with the works, as well as the obligation to complete 'on or before the Completion Date'. It has been indicated that what leads to an extension of time does not automatically lead to a loss and expense payment, and the converse is also true. Nevertheless, disturbance of regular progress can embrace at least two distinct characteristics:

(a) Work being performed in an uneconomic order and possibly piecemeal.
(b) Work being extended over time, perhaps by performance in a series of separated relatively piecemeal stages, perhaps by simple attenuation without breaks.

These characteristics and their effects are considered more fully in Chapter 11 as disruption and prolongation, but may be noted here as the necessary indications for the clauses to come to life. It may be said that the term 'disruption' alone also carries a similar connotation to 'disturbance' for present purposes.

The other feature of the immediate clause, using the JCT wording, is that 'progress . . . has been or is likely to be . . . affected', i.e. there is either a recent effect or an element of futurity about the disturbance, or even both. This wording gives the possibility of two basic situations in which loss and expense will be entertained. The first is when it is already present when discovered, perhaps because it occurs suddenly, perhaps because it creeps up through the operation of several factors, which individually seem harmless but combine to produce the disturbance. In a smaller instance it may even be complete when first noticed. The second is when it can be foreseen, at least as an outline problem likely to occur, even if not quite certain in occurrence and maybe less in detail. By virtue of the proviso about giving early notice considered below, the contractor is required to deal with matters as in the second category whenever possible, if he is not to prejudice his chances of reimbursement.

ACTION BY CONTRACTOR

26.1 If the Contractor makes written application to the Architect stating that he has incurred or is likely to incur direct loss and/or expense . . . [because of] any one or more of the matters . . . provided always that:

1. the Contractor's application shall be made as soon as it has become, or should reasonably have become, apparent to him that . . . the regular progress . . . [is] affected as aforesaid, and
2. the Contractor shall in support of his application submit . . . upon request such information . . . [for] the Architect to form an opinion . . . [and]
3. the Contractor shall submit . . . upon request such details . . . as are reasonably necessary for such ascertainment.

4.11 If, upon written application being made . . . within a reasonable time . . . by the Contractor [over loss and expense] . . . provided that the Contractor shall . . . submit such information . . . as is reasonably necessary.

It is obvious that the JCT clause, as relating to larger or more complicated contracts, is more demanding. But also, in part, the IFC clause has been drafted to be less repetitive.

Both clauses require the contractor to make his 'application' in writing and to give *general* warning (see the cases below) that loss and expense has been incurred or is threatening, in the ways outlined already. This allows the architect to take any action which may be possible to reduce the effects of what is going on, in addition to the contractor taking his own ameliorating actions. It also allows the architect, or the quantity surveyor on his behalf, to keep records of what is happening, so that a proper assessment may be made of the loss and expense, in so far as site records can help. The position under both forms should be distinguished from that under a form like the old JCT 1963 form, referred to under 'Interest and financing charges' in Chapter 12.

If the contractor defaults in giving this notice, he may well prejudice his position by preventing information from being gathered. The school of thought which says that it is best to leave everything over until a late stage when memories have faded, is risking a complete rebuttal, or at best a too modest settlement. As it is usually possible at least to foresee the likelihood of loss and expense, it is far better to give warning. The present contract routine is distinctly more precise than in earlier editions, and is to be welcomed. It also has the effect that it is not necessary for the contractor to give a series of 'updating' notices to cover a continuing matter, as was the case before.

The JCT clause is explicit that the contractor is to state the matter or matters causing his loss and expense, and this means that he must make a separate application for each matter which leads to distinct loss and expense. But if two or more matters occur in close conjunction, with inseparable effects, he should then give them in the same application. This is sometimes referred to as a rolled-up or global claim approach since the case of *Crosby* v. *Portland Urban District Council* (1967), where it was ruled that inseparable matters might lead to inseparable effects which might be ascertained together, provided no duplication between them or other matters occurred. This theme had ebbed to and fro in court rulings and global claims are discussed in more detail in Chapter 14.

The *Crosby* case was followed in *London Borough of Merton* v. *Stanley Hugh Leach Ltd* (1985), where the contractor had written a single letter lacking in detail and covering application for loss and expense covering both clauses 11(6) and 24(1) of the JCT 1963 form. The rolled-up approach was permissible when separation was not practical, and the contractor was not required to put in a moneyed-out claim, but simply to give notice of happenings. Even these might be well within the knowledge of the architect, as he had frequent contact with the site. This approach is not to be used when segregation is practicable, and it certainly is not to be used on the plea that 'it is too late now to do anything else: the records are not available'. JCT 80 and IFC 84 are drafted to rely on progressive data whether separation is possible or not. As discussed in Chapter 14, a rolled-up monetary claim *may* be acceptable in some circumstances. But before that stage is reached, clear demonstration of each cause and its claimed disruption/disturbance effect is necessary to show the effect of *each* claimed loss/expense.

It follows that a contractor may be in a particular predicament when there is an incremental build-up of disturbance due to a series of individually minor matters, and he is uncertain whether to make an application. It may be best to wait until the situation is developing clearly, but no longer.

The alternative temptation for a harassed contractor, frequently resorted to, is to put in an application every time that anything happens, be it even a revised drawing issue of the most minor variety. At the very least, this creates the atmosphere of 'little boy cries wolf', as he is soon not taken seriously. But worse, when his standard acknowledgment always reads as an application about loss and expense, he is giving no effective notice at all, as he is not specifying any precise matter. He is as much in breach of the clause requirements, as if he did not act at all when the real thing emerges.

Thus far, the two clauses are requiring similar action: an 'application' stating simply that loss and expense are present or anticipated and tracing the reason for this back to a matter or matters. There is no obligation on the contractor to put forward anything like a claim document; in fact this may well be impossible at such an early stage. The difference of emphasis is that the IFC clause does not require the contractor to act 'as soon as' there are danger signals, but 'within a reasonable time of' them. This is not quite so onerous, but should not lead to a damaging delay. It probably reflects the realities of life better and tones down the tendency mentioned in the preceding paragraph.

The application is really one for the architect to consider the overall situation, as set out under the next heading. He is the one to decide when to request supporting information to enable him to deal with the principle of the application and then with the evaluation aspect. These two elements are spelt out separately in the JCT clause, whereas the IFC clause is again less specific, but can be read just as demandingly.

It is always wise for the contractor to put forward as much properly detailed information as he can from time to time, and not to wait until he is asked. This issue is taken further in Chapter 25. When the contractor has applied, however, he is not obliged to take the initiative, so he cannot be faulted for failing to supply data which has not been requested or warn of further developments, unless there is in effect some extra matter which comes up to exacerbate the situation, hence requiring a fresh application. The strongest express obligation is that in JCT clause 30.6.1.1 and IFC clause 4.5 for the contractor to provide 'all documents . . . for the purposes of the adjustment of the Contract Sum', but these clauses are quite ambivalent as to whether this is to be done before or after practical completion. There is a weakness in the wording, by comparison with that over extension of time where the contractor is required to continue to give data without being asked, as well as any which may be requested. It is therefore prudent for the architect or the quantity surveyor, when acting on his behalf, to request details with as much precision as is suitable, and so activate the provisions of the clauses. This will be in addition to anything which either of them is obtaining by direct observation on-site and, preferably, notifying to the contractor as being on record.

There are cases in which a contract gets so out of hand that the submission of individual applications and the segregation of the resulting records becomes an almost academic exercise. Until this impossible stage is reached, the formal procedure should be maintained, and even thereafter some tailor-made scheme should be instituted that will prevent the project veering off into little more than thinly disguised prime cost.

Rolled-up, global claims are discussed in Chapter 14. In essence the calculation of loss/expense for *part* of a contractor's total submission for loss/expense *may* be allowed but only if *each* cause and effect is identified in the first place. It is not sufficient to trust that a claim will succeed by saying that 'everything went wrong on the contract and combined to make separation of causes and effects impossible', that so much changed it was no possible to isolate individual causes and show individual effects. Cause and effect must be separated and only then, in some cases, may a rolled-up part of loss/expense be entertained, leaving the remainder to fall for assessment against each cause of disruption/disturbance.

RESPONSE BY ARCHITECT

26.1 If and so soon as the Architect is of the opinion that the regular progress . . . [is, or is likely to be] materially affected as set out in the application of the Contractor . . . then the Architect from time to time shall ascertain, or shall instruct the Quantity Surveyor to ascertain the amount . . . which has been or is being incurred.

4.11 If . . . the Architect is of the opinion that the Contractor has incurred or is likely to incur direct loss and/or expense . . . then the Architect shall ascertain or shall instruct the Quantity Surveyor to ascertain . . . the amount.

This heading and that immediately before are deliberately chosen to indicate that the contractor must take the initiative, whereas the architect acts only in response to the action of the contractor. Even if the architect sees a very clear loss and expense situation boiling up, he need not even suggest to the contractor that he should begin affairs with an application, although desirably he should to avoid acrimony later and the possibility of proceedings. Proceedings are the more difficult alternative to what is effectively a negotiation, which comes about when the present clauses are used.

It is always open to the architect to instruct postponement of all or part of the works under JCT clause 23.2 or IFC clause 3.15 respectively, if he considers this will reduce the loss and expense being incurred. Except in the clearest of circumstances, he is not advised to do this without consulting the contractor fully.

Although the two clauses are worded somewhat differently at this point, their effect is the same. The JCT version requires the architect to form an opinion about progress, whereas the IFC version requires him to form his opinion about whether loss and expense has occurred for the reasons given, loss and expense being in the nature of end-products. It could be suggested that the latter version raises the contractor's hopes rather more! But until actual ascertainment is accomplished, there is in the nature of things no commitment to any particular sum, which may still turn out to be zero.

Both clauses place the whole responsibility upon the architect for forming an opinion on whether a case for disturbance exists at all, and this he cannot shed, however much he may seek advice from the quantity surveyor or others over matters of fact. With the HGCR Act this has been done.

The clauses also place the primary responsibility for ascertaining the amount upon him. This arrangement is one of the most commonly criticised aspects of these clauses, because the architect is saddled with being judge and jury in matters which frequently arise out of his own actions or inactions, as reference to the list of matters will show. It may be that the real cause lies 'further back' with the employer or with forces which even he cannot control, but often it is the architect's own administration which is in question. In the case of the ascertainment aspect only, the clauses do allow the architect to instruct the quantity surveyor to perform this function, and it is to be hoped that he will always require this. There have been calls for the inclusion of an adjudicator within the contract system, to whom disputes could be referred. This has its difficulties, but also considerable merit.

The architect's instruction to the quantity surveyor to ascertain loss and expense must be made in respect of each matter which is notified, or of each group of overlapping matters, unless he chooses to refer all such items to the quantity surveyor by a blanket instruction. He should sensibly refer a complete item, if he does so at all. Apart from the question of the architect's own potential conflict of interests or perception of the situation, the advantage of involving the quantity surveyor is that he has all the other financial affairs under his control and can avoid any double counting or gap. Once he does instruct the quantity surveyor to ascertain an amount, the architect relinquishes control and the quantity surveyor is empowered to reach finality, subject to his continuing need to refer back about any overarching elements within the architect's opinion. Equally, the employer is bound by the results, subject to his final right to proceedings, even though employers are often consulted in practice during negotiations about loss and expense matters.

In *Croudace Ltd* v. *London Borough of Lambeth* (1984), under a JCT 63 contract, the contractor had given regular notices during progress of loss and expense and had made application for its amount within a reasonable time. The council's architect had retired a few days after the application and no successor was appointed, nor was anyone else nominated under the contract for the purpose. The council had annexed ascertainment to their staff and not to the private architect or quantity surveyor acting, but were then dilatory in dealing. It was held that the council could not stay proceedings and seek arbitration (a delaying tactic), as there was no arbitrable dispute, and that the inability of the architect or quantity surveyor to ascertain was due to breach by the employer. Furthermore because of the breach, the absence of a certificate for an amount on account was no bar to the contractor recovering in the action.

A separate application has to be made about each matter as it arises. This means that the architect cannot take in further matters without the contractor first raising them, however obvious they are. This is to be distinguished from the position under the final settling of extension of time, when the architect may choose to take in fresh relevant events at his own discretion. He may well wish to nudge the contractor's elbow to save covering similar ground more than once, if the contractor is slow off the mark. It is also to be distinguished from the position under the JCT 1963 form, where it was necessary for the contractor to make a further application or more about the same matter, if some of its effects were still future when the first application was made. This is emphasised by the *Minter* v. *Welsh Health* case considered in Chapter 15.

Similarly, the architect or the quantity surveyor in evaluating the loss and expense cannot go beyond the matters already raised by the contractor. It is the contractor's responsibility to supply information of two sorts 'upon request': that needed for the architect to form an opinion, and that needed by him or the quantity surveyor to ascertain loss and expense. Strictly, he cannot make a further application via one of these supplies of information, but must do so separately. It is always tempting to throw in quite openly some further item when drawing up details supporting an earlier one, so saving extra administration in what the contractor did not ask for in the first place. If the new matter directly follows on from the old, or overlaps with it substantially, it may be sensible to treat it as an extension of it. If not, it is best to formalise the position by seeking a fresh application.

No particular structure is required for any data which the contractor supplies on request. As these items are qualified by 'reasonably enable' and 'reasonably necessary', the architect (or in the second instance the quantity surveyor) has a right within these limits to request information broken down in sections and generally laid out to facilitate his work, or to refer it back for emendation and clarification. There is no power to require the contractor to forward a complete claim document in elaborate style, of the type which has become so fashionable. Whether he chooses to provide the first statement of the position in such a document or something less formal, is his concern and largely a matter of psychology. The delay may count against the validity of his claim within the contract procedures. A sample format of such a claim is outlined in Chapter 26.

The clauses simply require the architect or the quantity surveyor 'to ascertain' the amount of loss and expense, aided by what the contractor produces on request. This suggests something of a prising out of detail to lead to a complete picture assembled like a jigsaw. Although someone has to make the first move, there is often room for negotiation by stages, rather than a sweeping assertion waiting to be cut to pieces by 'the other side'. The term 'ascertain' has its problems in implementation, as discussed in Chapters 11, 12 and 14. It gives the impression that the loss and expense are some fixed amounts lying somewhere waiting to be discovered in a purely objective manner. Life is not always so easy.

PAYMENT OF AMOUNT ASCERTAINED

Payment of the amount ascertained is one of several provisions subsidiary to JCT clause 26.1 and IFC clause 4.1. It occurs as JCT clause 26.5, whereas IFC clause 4.1 contains it in itself. The JCT clause is the more explicit by referring to 'Any amount from time to time ascertained', whereas the IFC clause says 'the amount thereof' only. In each case there is authority to add the amount to the contract sum. Under JCT clause 3 there is further authority to include the amounts of all adjustments of the contract sum in interim certificates. The IFC form does not contain such a provision, although both JCT clause 30.2.2.2 and IFC clause 4.2.2 allow inclusion of loss and expense amounts in certificates without deduction of retention.

The general philosophy of interim payment is that all amounts allowed may be included on a firm or approximate basis, according to how far the work etc and calculation have proceeded. JCT clause 3 allows inclusion 'as soon as such amount is ascertained in whole or in part'. The latter part of this may be interpreted to include 'some part as ascertained finally and accurately, or some part or the whole as ascertained approximately or provisionally'. It is suggested that all of this enlarged reading is correct and also fair and reasonable. Although the question of delay in making payments is an area of legal uncertainty, whatever its commercial dubiousness, there appears no contractual reason for failure to include amounts when once some fairly secure basis of approximation has emerged. It certainly cuts down on the level of financing charges otherwise entering on an ongoing basis into the calculation of loss and expense under the current contract forms. There is further discussion of the general withholding of amounts due in Chapter 8.

OTHER RIGHTS AND REMEDIES

JCT clause 26.6 and the last paragraph of IFC clause 4.11 recite that the provisions preceding them are 'without prejudice to any other rights or remedies which the Contractor may possess'. They might have added 'or which the Employer may possess', but do not, presumably because the whole of what goes before is for the contractor's benefit.

The effect of the present clauses is that they stand in order of recourse to available remedies as follows:

- *First*: To all other provisions of the contract over payment.
- *Second*: To clauses 26 and 4.11 variously for what is not 'reimbursed . . . under any other provision'.

- *Third*: Following the Housing Grants, Construction and Regeneration Act 1996, JCT contracts have now had to provide that the first recourse of parties in dispute must be to adjudication (unless *both* parties agree not to use it). Then, if the dispute still remains unresolved, they may go to arbitration or to the courts (only in the case of JCT 80 amendment 18 provisions is there a contractual option to litigate, which must be decided at contract stage). JCT 80 provides a contractual option, alongside the other option of arbitration, for what are called 'legal proceedings' but this option is not provided in IFC 84 and could only be activated by the deletion of the arbitration clause. This is discussed in more detail in Chapter 28.

PROVISIONS IN JCT CLAUSE ONLY

There are two items peculiar to the JCT form. The first is an enigmatic reference to extension of time in clause 26.3, requiring the architect to state what extension he has granted for any of the relevant events under clause 25 which correspond to matters under clause 26, 'If and to the extent that it is necessary for ascertainment . . . of loss and/or expense'. This reflects the possibility that one extension of time may embrace several relevant events, so that segregation of the separate subextensions is not automatically given when the extension is granted.

Why this provision is made is not clear. As stressed in several places in this book, no inevitable connection between extension of time and loss and expense is to be assumed, as either can occur without the other. Even when an extension is granted and loss and expense is also due to be reimbursed, it does not follow that any prolongation costs will be measured by the extension granted. There cannot be any question of acceleration in view, as nowhere is the contractor required to make up lost time as mitigation of effects. In any case the present stipulation is in the context of ascertainment, not counteraction.

For the contract to imply some sort of connection here can only create confusion in the minds of its readers. It would appear better, and on wider grounds sensible, for there to be a requirement in clause 25 for the architect to give a full analysis of any composite extension into its constituent parts, if the contractor should request this at any time. This would enable the contractor always to reassure himself or otherwise over what has been granted. It would also deal with the fact that the architect's statement in isolation for any one relevant event may make the extension appear more or less than it should be, when it has originally been compounded with other relevant events during calculation (see the mention of overlapping causes at the end of Chapter 10). The contractor may form a different detailed interpretation within the overall position, given all the facts, and this is his prerogative if he wishes to do it for any purpose, but no immediate concern of the architect or the employer.

The second item peculiar to the JCT clause is the question of nominated subcontractors and loss and expense in clause 26.4, which needs mention because of the architect's involvement, and perhaps the quantity surveyor's, under the subcontract forms in those loss and expense issues which are not solely between contractor and subcontractor. This issue is discussed under 'Subcontract provisions' below, although any amounts which are the responsibility of the employer will be included in the nominated accounts and so come into the main reckoning. Again the puzzling question of extension of time comes into the provision.

SUBCONTRACT PROVISIONS

The preceding sections are as relevant to subcontractors as to the main contractor, but some additional aspects also apply. For these purposes, subcontractors divide into two groups:

(a) Nominated subcontractors.
(b) Domestic and named subcontractors.

The point of the distinction is that nominated subcontractors receive special attention in the main contracts when loss and expense is due to one of the given matters, whereas the other types of subcontractors are not recognised by the main contracts for this purpose. There are, however, other causes of loss and expense which are effective solely between the contractor and all types of subcontractors. These causes are treated in the subcontracts but they are not mentioned in the main contracts.

Nominated subcontractors

The line and direction of communication set out in the JCT forms over subcontract loss and expense are quite clear: the subcontractor in all cases deals primarily with the contractor. Indeed, if the cause of loss and expense is not a 'matter' as defined under his subcontract, he deals solely with the contractor. Even when it is such a matter, he must raise it with the contractor, for the latter to pass on to the architect. It is not the contractor's responsibility when giving a notice of his own to make reference unprompted to any trouble of a nominated subcontractor. He acts under main clause 26.4.1 'upon receipt of a written application made . . . under clause 4.38 of Sub-Contract NSC/C'. That clause requires the subcontractor to take the initiative, just as much as must the contractor under main clause 26.1 over his own concerns.

Two distinctions may be noted here. One is by comparison with domestic subcontractors, over whom the contractor at his own discretion includes any mention in his initial application, or defers it until later, and correspondingly gives his own presentation of figures in his own application and supporting details. This is so, even though these figures may be the same as, or close to those of the subcontractors, on the net direct loss principle, and even though he still has no obligation towards the subcontractor to include anything until approached.

The other distinction is by comparison with extension of time, when the contractor has to take account of delay effects on subcontractors in any notice which he initiates. This is the effect of main clause 25.2.1.1, under which the contractor has also to notify any subcontractor of what he is about. In the case of loss and expense, the contractor has no obligation to start off anything for the subcontractor. The reason for the distinction is that delay can affect numbers of those on-site, because activities interact, and so seeking an extension of time affects them all. Loss and expense is a more 'individual' affair, in that it flows immediately from the causal matter to the sufferer, and not from its financial consequences on another. One person may be losing while another is unaffected by the same matter: site work still proceeds, at least to a point.

It is therefore necessary for a nominated subcontractor to take all the specified steps, as would the contractor over his own loss and expense. The steps in question are the same in all essential features and need no further elaboration. It is then the responsibility of the contractor to pass the subcontractor's application on to the architect. If he omits to do this, so that the subcontractor loses his right to payment by being out of time in applying, the contractor is in breach of subcontract clause 4.38.1. The subcontractor may then proceed against the contractor for this breach, but cannot seek reimbursement directly under the mechanisms provided within the subcontract.

When the procedure is properly operated, it becomes the responsibility of the architect to ascertain the amount or to instruct the quantity surveyor to do so, just as for the contractor's own amounts. Although the point is not express, the principle is for all dealings after the application to be dealt with through the contractor. In practice, as with other financial settlements with nominated persons, it is more expedient for negotiations to proceed direct, although the contractor should be kept informed. This is especially necessary if matters are overlapping with loss and expense amounts which are entirely between contractor and subcontractor, whatever the direction of indebtedness, so creating a three-way situation.

The resulting amount ascertained is included in the total set against the prime cost sum in settling the final account, so the contractor has no real concern over its level, only over whether any extraneous amounts are still to come his way as a result of the three-way pattern.

Subcontract clauses 4.39 and 4.40 deal with the cases of loss and expense caused by contractor or subcontractor to the other, owing to 'any act, omission or default' of the one. They simply require the aggrieved party to give notice 'within a reasonable time' to the other and for 'the agreed amount' to be 'recoverable . . . as a debt' or, in the case of what is due to the contractor only, to be 'deducted from any monies due or to become due'. There is no provision for adjusting the subcontract sum in either direction. Often settlement will be achieved by the most direct and suitable means available.

What is missing, by comparison with subcontract clause 4.38.1, is any further detailed procedure (or indeed any detail), as under the main contract. Obviously, something must happen. In the case of claims by the contractor against the subcontractor, some form of quantification has to be made to operate the set-off procedures of subcontract clauses 4.26 and 4.27, if these are invoked, as is most likely (Chapter 8). Otherwise, the parties must work out their own destinies. The pattern of dealings when the architect is involved gives a sensible guide to follow.

A procedure for the three-way case is not covered in the clauses, because it cannot be included sensibly. In principle the amounts which may occur can be separated into elements, lying variously between the following:

(a) The employer and contractor, alone.
(b) The employer and contractor, but passed up from the subcontractor.
(c) The contractor and subcontractor, alone.
(d) The contractor and subcontractor, but passed down from the employer.
(e) The employer and subcontractor, under Agreement NSC/W, possibly reflecting amounts under (a).

It is up to those dealing to separate out the elements which they do not consider to lie within their own area of action, quite apart from responsibility, and pass them resolutely back, so the relevant procedures apply to each element. The temptation to express an opinion or negotiate over some bordering issue should be avoided studiously, until a potential liability is identified, and even then it may almost create an unwarranted liability. The best way in practice may be for the three sets of negotiators (as they usually become) to meet and seek to allocate items in principle between them. This should be done without offering any trade-off if something is taken elsewhere, as this can have a nasty habit of rebounding later. The other side of this coin is to agree a division 'without prejudice to review', if the various responsibilities are not entirely clear at the stage reached. The considerations and procedures raised in Chapter 13 over multiple-contract situations are relevant here too.

Domestic and named subcontractors

When there are domestic and named subcontractors, the contractor alone deals with the subcontractor concerned in any instance. As noted above, the architect deals immediately with the contractor and receives any representations about loss and expense due to 'matters' as from the contractor. These may well amount to similar sums as they are intended to be settlements equivalent to common law damages, and so to reinstate the sufferer into the position in which he would have been apart from the disturbance. It is well known that claims settled at different contractual levels may differ in practice for an assortment of reasons, but this is to depart from principles or theory.

The two subcontracts therefore deal with all loss and expense amounts of the subcontractor in the same way, that is they do not provide separate clauses for 'matters' and for 'defaults etc', as do the nominated clauses, on the basis that the parties to the subcontract must settle between themselves alone. They appear to differ between themselves over other points. Domestic clause 13.1 requires the subcontractor to apply within a reasonable time and to submit information of effects and details of loss and expense, so paralleling the main contract. Named clause 14.1 has the reasonable time requirement, backed up by an 'information as necessary' requirement. These should not cause any serious practical differences.

A three-way claim situation involving a subcontractor is outlined in Chapter 32, while Chapter 13 looks widely at multiple-contract claim situations.

CHAPTER 18

DAMAGE TO WORKS AND DETERMINATION OF EMPLOYMENT

- Damage to works, and related insurance
- Determination of contractual employment

The twin aspects taken in this chapter lead to several areas of dispute in their own right. They also link with other aspects such as extension of time and disturbance of regular progress taken in other chapters in more detail than the present topics, which are considered in context from a limited standpoint. References mentioned below should be followed up for a rounded view of the material.

DAMAGE TO WORKS, AND RELATED INSURANCE

The wider subject of liability for injury to persons and property, including damage to the works, cannot be pursued here, any more than can issues of insurance. Discussion is limited to the impact of these issues on the main thrust of the present work. JCT Practice Note 22 contains extremely detailed comment and guidance on the standard contract and subcontract clauses referred to below. There are numerous aspects which cannot even be mentioned in the present highly selective treatment.

Liability of the contractor to carry out works

Responsibility for reinstatement after damage

As the JCT and IFC standard contracts are entire contracts (Chapter 2), the contractor is placed under an obligation to complete which is very demanding, except to the extent that the contract terms amend the basis. Fortunately for him, and in many cases in practice for the employer as well, they do amend the basis. Over the obligation to complete the works on time and at the contract price, as well as to the contract quality, there are several provisions relevant to what happens when there is damage. Without these provisions, the contractor would be obliged to replace the works and complete them without extra payment, even in the event of total destruction during progress. Determination of the contract would still be a fundamental breach. It appears that he would also be responsible for delivering the works on time, even though this might well be beyond the limits of the possible, so that he could face liquidated damages during a period of overrun.

In *Charon (Finchley) Ltd* v. *Singer Sewing Machine Co. Ltd* (1968) alteration work was almost complete when it was damaged by vandals. The cost of reinstatement fell on the contractor, despite the manner of damage. The contract was not in the JCT form.

Delay due to damage

The standard contracts allow extension of time when there is damage to the works, due to the same contingencies as are covered by the insurance provisions described below. The contracts also allow extension when there is delay due to *force majeure*, which remotely could lead to damage to the works beyond the normal scope of insurance, although it is a much wider concept, so wide as to be almost limitless. Otherwise, delay due to damage remains the contractor's responsibility in cases not covered by the insurance requirements of the contract. In these instances he still faces liquidated damages should he miss the completion date as currently fixed.

Although the contract insurance provisions help the contractor by covering reinstatement costs, including extra overheads, they do not afford him protection over the costs of disruption and delay as such and he is not entitled to any further payment from the employer over any other costs of reinstating damage. He must therefore look to his wider insurance cover for any protection against these elements, and indeed over liquidated damages.

Within the provisions, there is a clause giving an option (subject to an appendix entry) to the *employer* to have insurance included to cover his loss of liquidated damages. This is done post-contractually and the premium becomes an addition to the contract sum.

Insurance provisions of standard contracts

General scope of the provisions

The overall pattern of the main contract clauses, giving the JCT and IFC numbering, is the same:

(a) *Clauses 20 and 6.1*: Injury to persons and property and indemnity.
(b) *Clauses 21 and 6.2*: Insurance in general against last injury.
(c) *Clauses 22 and 6.3*: Insurance of the works, in particular against loss and damage:
  (i) Works insured by the contractor.
  (ii) Works insured by the employer.
  (iii) Works, existing structures and contents insured by the employer.
(d) *Clauses 22 and 6.3*: Insurance against loss of the employer's right to liquidated damages.

In the options for the works under clauses 22 and 6.3, the requirement is for insurance of 'the full reinstatement value', an expression discussed under 'Loss and expense' below, but adequate to cover the immediate costs of putting matters right.

The arrangement is for the insurance to be in the joint names of the parties, and further for subcontractors to be substantially included by a recognition or a waiver procedure (see below).

Insured risks

Clauses 22 of the JCT form (and the IFC clauses are similar) use three concepts:

(a) *All-risks insurance*: This is defined in clause 22.2 as providing 'cover against any physical loss or damage to works executed and Site Materials', with defined exceptions relating to such elements as wear and tear, defects in design etc, consequences of war, unidentified loss and excepted risks (b). There are further limitations for work in Northern Ireland.
(b) *Excepted risks*: These are defined to cover those risks due to radioactivity, sonic booms etc, which are uninsurable but are covered by statute.
(c) *Specified perils*: These are a limited set of risks covering long-established categories such as fire and storm, but excluding the categories mentioned above of impact, subsidence, theft and vandalism.

The reference in the conditions is to the more limited set of specified perils in respect of several important aspects:

(a) Insurance of existing structures and contents, when there are works of alteration or extension (which themselves are covered by the all-risks level of insurance).
(b) The benefit of the joint names policy which is provided for nominated, named or domestic subcontractors, with a further limitation in the case of domestic subcontractors that they are not entitled to benefit from the policy at all over existing structures.
(c) Extension of time due to the delaying effect of damage.
(d) Determination of the contractor's employment due to a prolonged, continuous suspension of the works.

It is therefore up to the contractor or subcontractor concerned to decide whether he should be covered against the excess of liability, which the contract itself does not require to be covered at these points. It remains the case, as a matter of normal insurance pattern, that there is cover against the negligence of whoever is at fault among them. The insurance not only provides cover for damage to the work of all these persons, but it is so cast that the insurer does not have a right of subrogation against any of them to seek contribution, because of negligence. It is very much a matter for those on whose behalf insurance is taken out, to check that they are being properly and adequately covered. An instance relevant to the employer over design is given under 'Some further aspects of the damage to the floor' in Chapter 31.

Negligence was in issue in *Scottish Special Housing Association* v. *Wimpey Construction (UK) Ltd* (1986), where the contractor caused damage by fire to housing which he was renovating. Under clause 20(C) of the 1963 JCT conditions incorporated into the Scottish Building Contract, the contractor was to be paid for reinstating after his own negligence and the employer sought to avoid this. The Scottish court agreed with the employer, but was overruled by the House of Lords. Insurance was by the employer, so that all concerned were covered, as though they had insured themselves.

Extension of time and determination of employment

For a range of reasons, extension of time and determination due to damage are dealt with under the general clauses of the contracts that cover them. There is the distinction that extension of time has to be granted when justified by a case of damage, and without reference to whether there has been negligence by contractor or subcontractor. Determination can be invoked only in the absence of such negligence, and loss and damage does not apply for the contractor's benefit. The parties bear their individual loss and damage.

In the case of damage under the third option of the employer insuring new works and existing structures, there is separate provision for an optional determination by either party when this 'is just and equitable' and this links to the more extensive all-risks provisions. It is related to damage to the works and materials, rather than to the existing structures and their contents. It would appear, though, that the primary reason for the determination provision must be the existing, rather than the new. Given damage to the new only, particularly if it is a clean-cut extension, it is no more arduous to replace the work than it is for a free-standing structure. The only extraneous consideration is that the employer may be more prejudiced in his plans, because it is the combination of the two sections which he needs, and the delay may be that more biting. This is one of the reasons why he should seek insurance over loss of liquidated damages initially.

But there is the possibility that when the new is damaged, so is the existing. Reinstatement may involve types of work which the present contractor is not equipped to tackle, so that either he does not wish to proceed or the employer does not wish that he should. Again, the extent of reinstatement of the existing may far outweigh that of the new, so that the scope of works amounts to a variation of the contract which the contractor is not obliged to accept, and which the employer may not wish to place with him for the same reason. An extreme case would be refitting by a contractor, specialising in joinery and finishings, of the top floor of an existing framed building which is razed to the ground during progress of the refitting. Unless the existing were replaced, replacement of the new works is impossible: the contract is frustrated by the destruction of its essential context, the building which was to be refurbished.

It appears therefore that a determination by either party would be viewed as just and equitable only when something more was at stake than normal reinstatement. When notice of determination is given, the other party may choose to accept it by doing nothing within the timescale stipulated. His alternative is to seek arbitration which proceeds at once, and needs to, so that the issue may be settled. If the determination is upheld, settlement follows that discussed when the contractor determines under, for instance, JCT clause 28.

Payments for reinstatement after damage

When there is joint insurance under the first option of the provisions, the clauses specify the immediate destination and routing of insurance monies and that they are to be paid to the employer with the agreement of the contractor. They are then paid to the contractor under certificates issued alongside the routine interim certificates for the works proper. These sums represent all the contractor is to receive for reinstatement work, so that he has a prime interest in agreeing the amounts to ensure he is adequately covered. Indeed, he should resist any influence from the employer's side which might reduce his entitlement, although he might welcome any support that tends to increase it.

It is then the duty of the architect and the quantity surveyor to arrange how the amounts are paid out, to keep in step with the progress of reinstatement. The amounts need bear no fixed relationship to the original figures for the work which has been damaged, quite apart from the extent of removal of debris. No retention is to be held, as that already exists for the work originally executed. The employer is naturally out of pocket on-site until the work catches up with its previous position, but he has the insurance monies in hand to cover this. Despite the apparent duplication of payment, further amounts for entirely fresh work are still payable under the regular certificates during this period. The only major problem will arise if the contractor becomes insolvent during the time in which work has not caught up. The insurance funds are not intended to cover this sort of eventuality and all must rest on the insolvency provisions.

In the second and third options, that is when the employer insures, the monies clearly should go to him. Because of the joint names arrangement, it is necessary for the contractor to agree to this happening. The basis of settlement is quite distinct here, as the contractor is due to be paid for the reinstatement as though it were a variation instructed by the architect. The contractor will therefore receive all that is due to him as part of his contract payment. The position is the reverse of the previous, with the employer particularly concerned that the settlement of the insurance claim is adequate, as he will receive no more, whatever he actually pays the contractor. No special certificates are needed, as the amounts of variations are reflected as usual, but under IFC 84 the contractor will face retention on the amounts for this very reason. Under the JCT clauses, he is relieved of having to bear retention.

In view of the uncertainties so often inherent in demolition and repair in circumstances of damage, settlement of amounts is frequently progressive, so reducing the problems outlined and leading to closer reckoning. This may well be an advantage to the insurer as well.

Should a determination occur because of a prolonged suspension after damage or, for that matter, because of any other provision in the contracts, the benefit of the insurance monies will accrue to the employer, subject to any work which the contractor may perform. The overlap of the twin problems of damage and determination may well complicate settlement. The insurer will not expect to pay out a higher amount because of the interruption caused by the determination, and this margin must fall into the determination settlement. If the contractor has determined due to the suspension, the employer will therefore remain out of pocket, as in other instances of the contractor determining. If the contractor has taken out the insurance, the employer will need to take a much greater interest in the insurance settlement, as he (rather than the contractor) will be recovering the money in final settlement.

Loss and expense when damage has occurred

It is a matter of vocabulary that the clauses mentioned use the term 'loss or damage' to mean physical damage to work and materials, whereas the determination provisions use the term to convey broadly the equivalent of loss and expense when there is disturbance of regular progress. There is, however, the question of the interrelation of payments under the insurance arrangements and loss and expense.

Directly, JCT clause 26 and IFC clause 4.12 do not refer to the effects of damage to the works etc in the way that the corresponding clauses about extension of time make it a relevant event. This means that such damage is not in itself a 'matter' under the clauses, and so does not lead to possible reimbursement of loss and expense.

Whichever party insures is entitled to receive payment related to 'the full reinstatement value', even though the basis of assessment varies as has been described. This term covers performing work at current price levels and in the conditions of working associated with reinstatement, but it is not to be taken as covering associated loss by way of disruption or prolongation of other work, not itself damaged. This means that normal loss and expense is not covered by the insurance settlement. For the reasons given in the next two paragraphs, it is for the contractor to ensure that he insures elsewhere or by an extension of the contract policy against this aspect.

All three of the alternative arrangements include the following provision: 'The occurrence of such loss or damage shall be disregarded in computing any amounts payable to the Contractor under or by virtue of this Contract.' This must be read as embodying some such qualification as 'other than the monies payable from the insurances'. It precludes any reimbursement of the contractor by the employer for loss and expense on work beyond what is damaged. When the contractor has insured under the first alternative and this is taken with the expression 'not entitled to any payment . . . other than the monies received', the possibility of reimbursement for loss and expense is closed.

When the employer has insured under either of the other alternatives, restoration etc 'shall be treated as if they were a Variation . . . under clause 13.2' (quoting from the JCT clause). This must mean 'shall be valued in themselves as if . . .'. The wording cannot readily be stretched to allow them to be regarded as variation instructions for the purpose of clause 26, in the light of the provision 'The occurrence . . .' quoted above, so that here again loss and expense cannot be reimbursed.

Subcontractors' loss and damage

The philosophy of the standard contracts is that insurance is taken out under the main contract to cover subcontractors, so avoiding a series of complex and perhaps conflicting insurances, with the danger of cross-claims, subrogation and confusion. The case of *Petrofina (UK) Ltd & Others* v. *Magnaload Ltd & Others* (1983) may be consulted for an instance when subrogation could not be pursued by the insurer, when the employer had insured all property on-site in the names of all contractors and subcontractors.

In the foregoing discussion, some of the key elements regarding subcontractors have been mentioned, and here they are in a summary with some extension:

(a) The term 'the Works' includes those parts performed by subcontractors of all types.
(b) The clauses allow for subcontractors to be included in the policies by one of two methods. There may be a 'recognition' of the subcontractor, individually or as part of a blanket arrangement, as a joint insured. Alternatively, there may be a waiver by the insurer of any rights of subrogation, that is to proceed against the subcontractor to recover amounts paid out to the person actually insured.
(c) The scope of reimbursement is limited to the 'specified perils' (which is narrower than what is available to the contractor). For domestic subcontractors, there is a further restriction when dealing with existing structures. Subcontractors must take out such insurance as they require to cover the rest of their potential liability.
(d) Payments for reinstatement are always valued as the expenditure of provisional sums or as variations, whatever is happening at main contract level, so that the position over loss and *expense* is as described under 'Loss and expense' above.
(e) Amounts for these putative variations are to be recovered as a debt, that is outside the principal settlement of the subcontract. The reference does not allow loss and expense amounts to be similarly relegated, and these fall to be included in the subcontract settlement proper within the main final account.

As with the main contract provisions, there are numerous matters of significance with subcontract insurance which do not fall within the boundaries of present discussion. Again, the JCT practice note may profitably be consulted over the provisions.

DETERMINATION OF CONTRACTUAL EMPLOYMENT

The area of determination overlaps with others taken in varying detail in this book, such as ownership of materials at an insolvency, subcontracting and renomination or its equivalent. It is treated in this chapter only over aspects which relate strongly to the main themes of the book. The clauses mentioned may be followed up in more detail elsewhere, and should be if other aspects are at stake. Detailed commentary on them is given in *Building Contracts: A Practical Guide*.

For simplicity of reference, it is only the JCT clauses that are quoted expressly. Those in the IFC form have different numbering but they are structured identically, so direct comparison is possible.

Main contract determination

Three variants of this possibility are given, depending upon the reasons leading to it:

(a) Determination by the employer.
(b) Determination by the contractor.
(c) Determination by either the employer or the contractor.

The subsidiary elements in each case are taken side by side by way of comparison.

Reasons for determination

These break down into determination on account of specified reasons in (a) or (b) which are the default of the other party or in (c) which are neutral in character (most of them are a smaller selection of what is in the list of relevant events for extension of time) or which are on account of insolvency of the other party. The reasons which are also 'events' are given in Table 15.1 and are considered in Chapter 15.

The following reasons for determination *available to the employer only* are not included in other clauses in the case of default by the contractor:

(a) Wholly or substantially suspending the works without reasonable cause.
(b) Failing to proceed with the works reasonably and diligently.
(c) Failing to remove defective work etc as instructed.
(d) Unauthorised subletting.
(e) Corruption at main contract level only, but otherwise worded very widely indeed.

The following reasons for determination *available to the employer only* are not included in other clauses in the case of default by the contractor:

(a) Not paying the amount properly due on a certificate.
(b) Interfering with or obstructing the issue of a certificate.
(c) Failing to operate his side of the subletting procedures.

The first two are discussed under 'Certificates and payments' in Chapter 8 and there are also procedures and notices, which should be carefully observed to avoid dropping into unwanted problems.

Procedures and terms for determination

Each of the clauses gives procedures for settlement of affairs when determination is a fact. The standard position is that the employment of the contractor is determined, not that the contract is determined. This means the contract provisions over determination remain in place and still apply, so the parties are not left to sort affairs out from a position of complete uncertainty. Provided the procedures are followed, the key features of the clauses rest in the terms for settlement, which may be summarised.

When the employer determines, there are several avenues down which events may be traced, according to the cause invoked as shown in Figure 18.1. In all of them the contractor has to vacate the site without further payment until final settlement and the employer may choose whether or not to complete the works, and if he does perhaps utilising the contractor's plant etc temporarily, and engaging other contractors to do the work. Settlement embraces the balancing of the various amounts which have been lost or actually expended against those which should have been expended, had the determination not occurred. This involves quite an amount of calculation of actual and notional accounts.

A minimal framework of payments is shown in Figure 18.2, and some of the elements may also be mentioned with brief comments here:

(a) Original contract interim payments made and any payments in tidying up accounts, such as nominated subcontract retention as obligatory and perhaps payments to particular subcontractors for special reasons as optional.
(b) Temporary measures on-site during the transition period.
(c) Amounts under the completion contract or contracts, and by way of extra fees.
(d) Loss and damage for delay, discussed under the next heading.
(e) Hypothetical account for what should have been incurred, this being deducted from the other elements to give the net indebtedness of one party to the other (usually in the one direction, and probably worth little to the employer if the contractor is insolvent).
(f) An account of what the contractor actually did and provided up to determination is not strictly needed here, but may be useful for collateral purposes.

When the contractor determines, again he has to vacate the site. Thereafter, the employer may or may not choose, or be able, to carry on to completion. What is provided under clause 27.6 is that accounts are to be settled, as between the employer and the contractor, without reference to what the employer may do next (if he does not have the works completed, this is dealt with under clause 27.7, noted separately below). This involves settling the existing contractual account, while taking in the effects of the determination itself:

(a) All work completed or in progress and materials delivered or otherwise the employer's responsibility.
(b) Loss and expense during the works as performed.
(c) Costs of removal and leaving the site tidy.
(d) Direct loss and damage, discussed under the next heading.

Clause 28A prohibits the contractor from determining when one of the 'neutral' reasons applies, and he or his subcontractors etc have been negligent. This is covered only for damage to the works, although this is the only likely case. It is

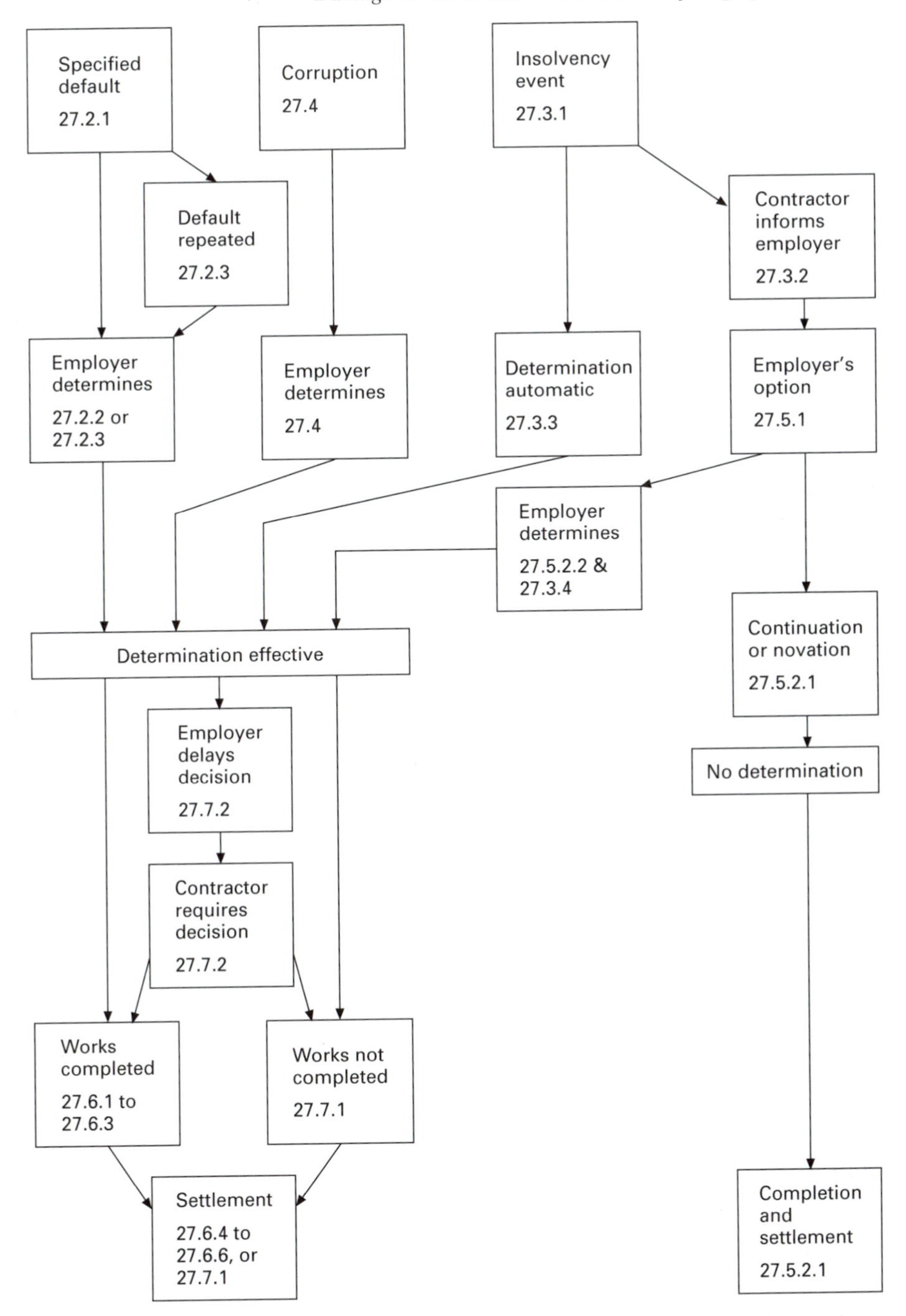

Figure 18.1 Patterns of contractor's default or insolvency and subsequent activities under clause 27 of JCT form (Adapted from *Building Contracts: A Practical Guide*)

Position as it should have been		
Contract sum		£1,000,000
Value of variations etc (net additions)		20,000
Hypothetical final account		£1,020,000
Position at determination, and payments eventually made		
Value of work executed		£410,000
Interim certificates (net of direct payments below)	£380,000	
Less 3% retention	11,400	
	368,600	
Nominated subcontractors paid direct before determination (including retention)	10,000	
*Nominated subcontractors paid after determination (including retention) and already in the £380,000 above	16,000	
Payment of retention on nominated subcontracts actually paid by contractor and already in the £11,400 above: Three per cent on £120,000	3,600	
		£398,200
**Margin on work up to determination (or £27,800 if contractor insolvent)		£11,800
Position at final settlement		
Paid at or about determination		£398,200
Paid for completion contract		690,000
Employer's loss and expense		50,000
		£1,138,200
Less hypothetical final account		1,020,000
Net indebtedness of contractor		£118,200
(Reduced by £16,000 to £102,200 if insolvent*)		
If the contractor is still solvent this debt should be met in full. If he is involvent and only able to pay his creditors, say 20%, then the position could be as follows:		
Payment due to employer (20% of £102,200)		£20,440
Payment due to contractor for plant sales (not in figure above)*** Gross realised	£14,000	
Less expenses	2,000	
		12,000
Net payment made by contractor		£8,440

*These payments may not be made contractually if the contractor is insolvent.

**The margin of £11,800 (or £27,800) is 'absorbed' in the final settlement and in effect is kept by the employer.

***The full benefit of plant sales goes to the contractor; the employer cannot retain the £12,000 nor even deduct it from the £102,200 and thus benefit by 80% of £12,000.

Figure 18.2 Financial results of a contract determined under clause 27 of JCT form (Adapted from *Building Contracts: A Practical Guide*)

always necessary for a party to be sure of his grounds before determining: the case of *Burden* v. *Swansea* mentioned in Chapter 8 illustrates the danger of choosing the wrong reason. Furthermore, a trigger-happy attitude is counter-productive, as determination is always a last resort. It is not be used 'unreasonably or vexatiously', as the clauses put it, and an arbitrator or judge is likely to take an adverse view of determination over a mere technicality of little effect.

Direct loss and damage

Here the distinctive feature is 'direct loss and/or damage', as clauses 27.6.5.1, 28.4.3.4 and 28A.5.5 express it. It is often held that this expression means the same in practical terms as 'direct loss and/or expense' in the corresponding clauses about disturbance of regular progress. This may be so, but all the clauses make some distinction by adding 'direct loss and/or damage' on to the lists about either the cost to the employer of finishing the works or the amounts due to the contractor for what he has done on the works and in leaving them. 'Loss and damage' is thus what is still due to the party concerned, once the equivalent of 'expense' has been calculated. By virtue of clause 28A.6, reimbursement for loss and damage due to the specified perils is made to the contractor when, and only when, it has been caused by the negligence of the employer. Otherwise such loss and damage lie where they fall for the parties.

When the employer determines, the liquidated damages provisions cease to be effective, although any damages already paid or deducted need not be repaid; they help to reduce the ultimate indebtedness. Instead, the employer becomes entitled for the period of delay to actual amounts which he can prove as flowing from the determination on a basis similar to that for damages at common law (Chapter 9). According to the nature of the employer and his activities, the following areas are most likely:

(a) Higher costs of alternative accommodation retained or obtained while the new accommodation is not available.
(b) Associated staffing, equipment and overhead costs due to less efficient operating conditions.
(c) Loss of return on capital locked up in the uncompleted project.
(d) Loss of profit or other revenue from markets or sectors not being satisfied.
(e) Financing charges due to prolongation.

These items are listed as suggestive only, as they do vary with employers (according to whether, say, they are operating on a commercial, profitable basis or not), and there is obvious overlap between them. The first two go together as elements of higher operating costs borne elsewhere in the enterprise due to inefficiencies, arguably having started from when it was first proposed, continuing while the project is uncompleted throughout its life, but becoming a source of relevant loss only for the excess period due to the determination. The loss needs some care and analysis for its calculation, as the new premises may well be providing more accommodation, as well as better accommodation. Only the relevant part of the improvement should therefore be allowed. Even so, it is also necessary to avoid double counting against the next pair of items.

The next two are alternative ways of expressing the one phenomenon: loss of what the enterprise exists to obtain for its shareholders or other owners. Which is the more suitable expression, however similar the resulting finances, depends heavily on the type of employer, according to whether he exists to obtain a return, perhaps notional, or make an actual profit. It is subject to the propositions in the law cases in Chapter 9 about the level of loss of profit by the other party, which the defaulting party to the contract might reasonably expect with his actual knowledge to flow from breach, taking any cause leading to determination (including insolvency) as equivalent for the present argument to a breach.

The last expresses a drain on the enterprise's resources, avoiding which should result in greater return, and it may well correspond to or be absorbed in the third item. Usually, it is better for an employer to settle for one of the other heads of claim, rather than financing charges which are less cogent in these circumstances. Such charges are also subject to a greater burden of calculation, as discussed in the context of loss and expense in Chapter 12.

Whatever the items drawn into the calculation, they should all be allowed for the same period. This is to be calculated on the one hand as the original contract completion date, with any extensions which occurred during progress of the original contract and any granted during the completion contract period. These, it must be presumed, would equally have been granted to the original contractor had he continued. On the other hand, the actual overall period taken to completion should be allowed, including periods of cessation between contracts, so that the full difference may be used.

When the contractor determines, the calculation is quite distinct. Any loss and expense incurred before the determination is included in what is really the final account for the truncated works as performed, so that prolongation costs, as well as disruption, are covered if they have occurred. There should be nothing else of substance left except loss on work not performed, which becomes an admissible head of recovery here, following *Wraight Ltd* v. *P. H. & T. Holdings Ltd* (1968). This case was brought on a contract under the JCT 1963 form, and it was held that the contractor was entitled to recover for both loss of profit and overheads on work not performed because of the determination.

This is quite a different position from that obtaining when the contractor claims for loss of profit and overheads when there is disturbance of regular progress in a continuing contract. There the position is that set out in Chapter 12, in that he does not secure profit on the extra costs he incurs. Instead, the contractor stands to recover reasonably demonstrable profit foregone on work elsewhere which he could equally reasonably be held not to have obtained because of the disturbance, and also overheads specifically increased or reasonably attenuated by the disturbance.

Here he stands to recover the profit on what he has not been able to carry out, irrespective of whether he has a full order book despite the determination itself, as there has been breach of contract by him being denied the opportunity to perform. However, as with loss and expense under the other arrangements, the *level* of profit is that which he was likely to achieve, given his actual efficiency and actual market conditions, rather than what was hoped for in the tender assumptions. This leaves a number of questions to be asked, over elements like sublet work and materials purchases, beyond the relatively simple questions about site productivity. The overheads element represents the actual commitment to the project which cannot be neutralised when it fails to proceed in the expected way, because of the determination. This needs a careful assessment of the implications, following the lines set out in Chapters 12 and 31, although it is suggested that a somewhat more generous allowance may be appropriate in all the circumstances. Whereas profit is recoverable because it has been denied, overheads are recoverable because they have been incurred.

Non-completion of the works

Clause 27.7 deals with the option of the employer not to complete the works after determining against the contractor. There is no corresponding provision in clause 28A when he may determine for 'neutral' reasons. The clause gives less complex provisions, as none of the completion costs are incurred. The contractor is to be reimbursed for the work performed and other contractual amounts, which is more in line with the case of him determining against the employer, but without payment for materials and removal from site. He also has to meet the employer's direct loss and damage.

There is no requirement for the employer to justify such an action to the contractor, but he does have to notify him of his decision within six months of the determination, failing which the contractor may require him to state in effect whether the preceding clauses for settlement still apply. These aspects are reasonable and the only sensible course in most situations, although nothing is said about what happens if the employer states that the preceding clauses still apply. The contractor, who in principle is in the wrong, then has to wait until the employer acts.

What may be problematic is the provision for the employer to be reimbursed for his direct loss and damage. Although there is always an element of discretion available to the employer in deciding commercially how to complete the works, there may be much wider financial implications in deciding whether or not to complete at all. This is one matter when some part of the works is not started at all and can be omitted with minimal effect, or is sufficiently close to completion as to be usable as it stands. Even here a grey area exists in which most may be left uncompleted, while a little is done. This might be argued not to be non-completion at all, but completion with substantial variations. However, it is another matter altogether to decide to sell off the partly constructed works or even to leave them standing and unusable. Here it seems inevitable that the courts would imply some term of reasonableness into any loss and damage they were asked to consider, while arguments over market value and so forth might be raised. It is likely that a view would be taken over what was reasonably within the contemplation of the parties when entering into the contract.

There may even be situations in which the employer will be relieved at not having to pay for completing works which, since entering into the contract, he can see are an overall liability for his purposes. This should be viewed as in the nature of an otherwise ill wind blowing some good for the employer, rather than as a relief to the contractor from liability.

Subcontract determination

At this contractual level, the basic systems for determining and then settling are quite similar to those at main contract level. The main difference lies in the systems for providing another subcontractor when this is appropriate, and these have been tackled in Chapter 7, so far as nominated and named subcontractors are concerned. In the case of domestic subcontractors, whether and how the contractor sublets is entirely in his own hands, subject to obtaining the architect's approval of any subcontractors. Many of the various facets taken here may therefore be given together for all types of subcontractors, with differences emphasised.

Four ways of determination

There are four ways in which determination under a subcontract may occur, giving the JCT references, but not the IFC references (which compare in content), with any main references:

(a) Determination of the subcontractor's employment by the contractor, *JCT clauses 7.1 to 7.5*: the subcontractor is at fault or insolvent.
(b) Determination of the subcontractor's employment by the subcontractor, *JCT clauses 7.6 and 7.7*: the contractor is at fault or insolvent, or problems are being experienced at main contract level, which are affecting the subcontract.
(c) Determination of the contractor's employment by the employer, *JCT clauses 7.6 to 7.9 and main clause 27*: the contractor again is at fault or insolvent, but the employer has acted first.
(d) Determination of the contractor's employment by the contractor, *JCT clauses 7.6 to 7.9 and main clause 28*: the employer is at fault or (if a private employer) insolvent, and the contractor has acted.

The last two result in automatic determination under the subcontract.

In the first situation under a nominated subcontract, the architect has to be involved with the contractor in allowing a determination, because there is going to be a change in the amount payable for the subcontract by the employer through the contractor, and also because of the potential effect of renomination on extension of time and loss and expense matters. This involvement under clause 35.24.4 does not extend necessarily to giving direct assent to the determination or to giving an instruction to effect it, but it does mean the architect must ensure that the contractor gives proper notice of default to the subcontractor, with an option to give backing instruction before the contractor may determine. This leaves the contractor still with an option of whether to determine at all if, say, the subcontractor improves performance. The clause modifies somewhat the position under *James Longley & Co. Ltd* v. *Reigate and Banstead Borough Council* (1982).

In that case a nominated subcontractor under the previous Green Form subcontract (with amendments) had delayed considerably. The supervising officer (SO) in that instance, was prepared to give an extension of time, but not to give the contractor permission to determine, even when the subcontractor left the site.

He said that his consent was unnecessary, but that he would renominate when determination took place. The subcontractor went into liquidation and an early renomination was made. The contractor argued that the SO was obliged to instruct a determination when this was in the interests of the work, just as he was obliged to renominate, but it was held that this was not so. The contractor could and should determine without the SO's consent, who had no power to interfere. The employer was not liable for the costs of disturbance or for the delay.

In the first situation of determination, when there is a named subcontract, the contractor acts on his own, but he is required to keep the architect informed of events. The amount payable is not affected in this situation, but the other contingencies affecting the employer may apply. The reason for the difference is therefore not entirely clear.

In the second situation of determination, under a nominated subcontract or a named subcontract, the subcontractor is required to give the architect notice of any default by the contractor possibly leading to determination, as the contractor is not likely to do so in the circumstances.

When there is a domestic subcontract in the first and second situations and whatever the type of subcontract in the other two situations, the architect is not involved in the determination process.

The reasons for a determination under each of the above patterns flow from the underlying logic of events:

(a) When the contractor determines, the defaults are those which would enable the employer to determine against the contractor, as listed above and also in Chapter 7, in discussing reappointment of subcontractors.
(b) When the subcontractor determines, the defaults are those which would allow the employer to determine against the contractor, these being suspension of the works or failure to proceed regularly.
(c) When there is a main contract determination, the reasons that apply are those already given.

The place of the architect in procedures has already been mentioned. Otherwise affairs follow predictably from the main contract pattern, over the subcontractor vacating the site and over obtaining any replacement subcontractor. The terms for settlement also parallel the equivalent set in the main contract, although the parallel is somewhat distorted in places, as is noted below.

Procedures and terms for determination of nominated subcontracts

The terms for nominated subcontracts are the most complex. When the contractor determines, they are effectively as when the employer determines under the main contract, with one major modification. The architect remains responsible, with the quantity surveyor, for settling with the subcontractor, and is required to issue a certificate which includes an amount for any balance of work by the subcontractor. This is work not included in certificates up to the determination, because from then onwards nothing has been paid. The employer may deduct from the certificate, when he honours it to the contractor, any expenses in completion of the subcontract and loss and damage which he has incurred through the determination. In turn, the contractor may deduct these amounts and any loss and damage which he has incurred himself. By comparison, under a main contract determination against the contractor, he receives no further certified amount calculated in original contract terms, but the balance of indebtedness is still calculated from the twin summations of all amounts due and actually paid. Obviously, the amount of deduction by the employer under the two systems fluctuates against the other amounts to keep the final answer the same in principle.

Strictly, the present clause is misworded as it allows the employer to deduct the whole cost of completion, rather than the excess over what would have been payable had there not been a determination, and is also suspect over the provisions about deductions for loss and damage. If entrenched positions develop, these details should be followed up in other works (such as *Building Contracts: A Practical Guide*), otherwise sensible interpretation should prevail.

When the subcontractor determines, he is paid in just the same manner as is the contractor when he determines against the employer, that is for all he has done, for removing from site and for loss and damage, which includes the loss of profit on work not performed. To the extent that these amounts are in excess of what the subcontract works would have cost, the contractor has to reimburse the employer to meet the difference. How the amounts are set against one another, and when, is not made clear in either the subcontract clause or in main contract clause 35.24.6. Clearly, a larger payment to the subcontractor by the employer will result in a greater element of loss.

When the contractor's employment is determined by either party to the main contract, this may be due to the 'neutral' reasons in main contract clause 28A. In these circumstances, to keep the main and subcontracts in step, the subcontractor does not receive any payment covering loss and damage.

A major consideration in each of these cases will be whether the contractor who is in the middle is still solvent, as this affects the adequacy of settlement to be expected by the end party who is initially out of pocket. In the employer's case, there is also what value any employer/nominated subcontractor agreement in force has for him. This gives him a direct line of redress over loss and expense and extension of time which he may have to grant the contractor because of subcontractor default.

Procedures and terms for determination of named and domestic subcontracts

The absence of the architect from financial settlement of either type of subcontract has been noted above, as correspondingly Chapter 7 noted his part in selecting a replacement named subcontractor only.

The procedures for removal and completion under the two types of subcontract are both similar to those for nominated subcontracts. Most of the elements in settlement are the same between the subcontracts and follow what has been said above. They allow for a difference-based settlement when the contractor has determined, so that actual total costs are set against what should have happened. These therefore take account of loss and damage in the way described above. When the subcontractor determines, except when the 'neutral' reasons apply, the elements allow payment for what has been done and for loss and damage, without regard to how the contractor then proceeds, if at all. They are all adduced without reference to the architect over the employer's position.

A point of significant difference is in named subcontract clause 27.3.3. This requires the subcontractor to pay to the contractor an amount which the contractor has undertaken to recover under IFC clause 3.3.6 (b), when a determination by the contractor against the subcontractor takes place. This amount is then to be passed on to the employer. It represents the following:

(a) Loss and expense payable by the employer to the contractor because of the determination and subsequent events.
(b) Liquidated damages which the employer would have recovered from the contractor for delay on the same counts, had the contractor not been entitled to an extension of time.

This arrangement therefore provides an equivalent recompense for the employer to that which he has under the employer/nominated subcontractor agreements. If the contractor fails to take reasonable action to recover, he becomes liable for paying the amounts concerned to the employer.

PART 6

DELAY AND DISTURBANCE: ALTERNATIVE CONTRACT PROVISIONS

CHAPTER 19

COMPARISON OF ALTERNATIVE CONTRACT APPROACHES

- Outline comparison of contracts
- Outline comparison of key issues

The following chapters in this part of this book each deal with a form of contract or a contractual method in relative outline compared with the treatment afforded the two JCT forms underpinning much of the discussion in preceding chapters. The aim is show how these contracts and methods deal with some issues, especially delay and disturbance, covered in detail in earlier chapters, doing this partly by comparison with the JCT contract (referring for simplicity to the main form with quantities alone and leaving aside the intermediate form entirely), but relating to the special structure of the forms and their provisions, which vary from the JCT contract. This serves to guide when working with the particular contracts taken and also to afford a view of alternatives, so giving a perspective between approaches. It is always the case that a principle applying in one contract may be modified or even negated in another, for reasons that are compelling or sometimes baffling.

OUTLINE COMPARISON OF CONTRACTS

The JCT contract is used as the main basis in earlier chapters because of its wide use and so relative familiarity. It is also the form which has attracted the most scrutiny by the courts, an accolade of somewhat mixed value. The detailed stance of the various contracts taken in this part is given in the chapters following, but a brief introduction to the contracts is made here.

GC/Works/1 contract

This contract is intended for both building and civil engineering and there are no alternatives within it relating to one or the other out of these types of work. This is in distinction from the JCT and ICE forms, and the present contract operates none the worse for its generality of wording. There are several versions of the contract; the firm quantities version is considered in the next chapter, and there is limited reference to the design and build version in Chapter 24.

The contract has been unilaterally prepared, is tightly drafted and has some bias towards the client body for whom it has been prepared, in this case usually within or associated with central government. It is clearly confrontational in character, but aims to be clear in intention (and mostly is). It is likely to be construed contra proferentem by the courts where ambiguous, although it seldom reaches the courts. Earlier settlement appears to be the common practice, for reasons official or otherwise which must be conjectured.

ICE contract

This contract is intended for civil engineering work only and would be difficult to use for other work, although subsidiary building work can be accommodated by it. It has been prepared between consultants (who will act for the client in some respects) and contractors, who may employ professionals of the same discipline as the consultants. It is considered in the chapter hereafter in its major version, which is based solely on the use of approximate quantities and so on complete remeasurement. There is limited reference to the design and construct version in Chapter 24 on design and build.

Although it is confrontational in principle, it does not have special bias. It tends to be drafted fairly loosely in places and has a rather verbose tone and some variation in its use of terms which render it less precise over some issues. It has been aired in the courts less than the JCT contract, but more than the GC/Works/1 form. Usually this has been over associated practice rather than close interpretation of wording.

NEC contract

This contract is intended for all forms of construction. The contract is a 'system' of interlocking contracts or a family of contracts. It is original in its style by being written entirely in the present tense, in everyday language and it may be courageous in its intentions. The stated objectives for the system are that it should be flexible, it should be a stimulus to good management and it should be expressed in more simple and clear language than other forms of contract. It has been prepared to be used in its various options with design by a client, by management contracting or prime cost. It is discussed in more detail in Chapter 22.

It is non-confrontational in philosophy and aims to reduce conflict by steering a distinctive route through procedural matters. It seeks to resolve disputes before they develop into serious divisions. Its drafting style is held by some to make it imprecise in significant areas. Its use has not been anywhere as widespread as other standard forms, perhaps because it is innovative, and there is no knowledge of any case law on it.

JCT minor works contract

This is the smallest member of the extensive JCT family and is considered here because it has very extensive use in practice, despite its size, or perhaps for that very reason. It is intended for projects of the scale of a quite small house or less, but is used for many much larger schemes, whether its fairly skeleton provisions are adequate or not. The issues it raises are treated in summary in their own chapter and need no introduction here.

Design and build contracts and subsidiary contract provisions

Several approaches are grouped here, coming from various drafting bodies, including those responsible for the JCT form and the other forms above. They have usually reached the courts over issues other than their design provisions and effects, which are all that are touched on very selectively in this book, and so are in a rather distinct category from the other approaches taken here.

OUTLINE COMPARISON OF KEY ISSUES

This comparison is made between the JCT contract followed extensively in earlier parts of this book and the contracts outlined above, with less reference to the design and build set and none to the minor works form. The sections of comparison are in the sequence of the main key issues taken in each chapter hereafter, although less so for design and build.

The treatment of issues in places is uneven, because so are the contracts. Thus, the GC/Works/1 contract is the strictest (some might feel the most adversarial), whereas the ICE contract has something of a looser, almost 'old boy' air about it and the NEC contract seeks to encourage (can there be a stronger way of expressing it?) cooperation between the parties as against the commonly experienced confrontation. Design and build implicitly expects more cooperation and possibly achieves it, whereas the minor works form exists principally in a realm where many of the regular boundaries are not so well recognised by most of the participants.

Contract basis and administration

These aspects are broadly similar between the contracts, except for the NEC contract which pursues a quite distinct path due to its aims of cooperative working. The NEC has rigidly separated the design function from the contract administration function to the extent that it does not mention 'designers' in its construction contract. Its points are dealt with further within its own chapter.

Site and ground conditions

The JCT contract ignores these aspects as physical entities, even referring almost always to 'the Works' rather than 'the Site'. It leaves all questions of inspection of the site, the supply of information about it and what its ground conditions might be and how they affect such things as progress and payment to be given in or by reference from the contract bills. This usually means there are soil surveys and rules about payment for working in rock, running sand, water etc, but nothing is said about time implications, which need to be pegged on variation instructions.

This follows through into the JCT with contractor's design form, with silence over the design and resultant payment implications, unless matters are defined further in the employer's requirements or contractor's proposals.

Both the GC/Works/1 and ICE forms recognise the special significance of ground conditions for progress and adjusted payment, while also spelling out the otherwise implied responsibility of the contractor to check on-site and surrounding conditions more widely.

The NEC form provides for the 'experienced contractor' test to be applied to the contractor when there is any application for a matter in difference that arises from site or ground conditions.

Programme and time

The JCT contract takes little account (and that uncertain in significance) of the standing of the contract programme and refers only to any such programme prepared by the contractor. It does, however, deal extensively with responsibility for completion on time, the repercussions of this not happening and the question of damages. It has several provisions about related matters, such as obligations to request and supply information necessary to construct the works. The main emphasis is on the contractor's right to organise the works how he will to achieve the end result. This is subject to limited powers afforded the architect to instruct changes of sequence etc by way of variation, and to the possibility of some intermediate date or other restriction which is critical and is given in the contract documents. Nevertheless, the parties on projects of any size will have a pattern for progress review and other such matters embodied in regular meetings, whether these are mentioned in the contract or not, these usually being considered an advantage by all concerned.

The GC/Works/1 and ICE forms give the client and his representatives defined powers to check what is happening and to modify aspects of the programme, over the completion date or dates and over sequence, although it is provided that actual acceleration to give early completion may happen only by agreement between the parties, as can always happen under the JCT system despite its silence on the subject. In the case of the GC/Works/1 form, the emphasis on the programme represents the concern lying behind this contract to keep that firm grip on the programme of events, as is so frequently not done within current contract arrangements.

To some degree the powers under the GC/Works/1 and ICE forms touch on methods of working. For instance, they may extend into the design of temporary works under the ICE form, where there may well be time and cost implications. These are areas where design and build approaches tend to hold off, as such interaction can also have significant effects on ultimate design liabilities.

The NEC form also gives the client and his representative, the project manager, powers to control what is happening and to modify aspects of programme, for instance to ask for acceleration to be implemented by agreement. The philosophy of the NEC is that the contractor is in the best position to take risk for changes in requirements if a change is proposed. The programme is an essential constituent of the NEC contract and the contractor must produce a new programme whenever he thinks that the completion date or the contract price will not be met.

Instructions and payment

The JCT form specifically restricts the power of the architect to issue instructions – so leading to adjustments of what is payable – to those matters expressly given in the contract, and it very tightly proscribes the powers of the employer to act other than through the architect.

The GC/Works/1 contract allows the project manager to issue instructions without question on any matter he considers appropriate, reflecting the presence on occasions of questions of national imperatives. The ICE form gives the engineer wide powers to instruct, but is silent about limitation to what has been given, which probably has the legal effect of limiting him to a position somewhere between the other two contracts. Design and build arrangements usually allow the employer to require a change in the design, while leaving the detailed solution to the contractor. When they are separate main contracts, they also provide for instructions on other matters paralleling what is given in the forms from which they are adapted.

Under the JCT, GC/Works/1 and ICE contracts there is a broad similarity in the pattern of payment, in that there is a starting sum for the works, which is then adjustable for variations and for other defined causes arising, such as disturbance, and for other optional matters, such as fluctuations. The main qualification to this statement is that in the case of the ICE contract the 'starting sum' is related to a bill of approximate quantities and is theoretically to be ascertained as affairs proceed by complete remeasurement of the scope of works included in the original contract, although the adjusting provisions to cope with what is not within the original scope (such as variations) are worded as though the starting sum was known from the beginning.

The provisions for valuing variations are similar in effect between the three contracts. The major difference in what is payable for disturbance and other somewhat cognate matters lies between the JCT form, which allows reimbursement for *both loss and expense*, and the other two, which allow reimbursement for *cost or expense only*. Even here, these two differ, in that the ICE form allows recovery of financing charges, whereas the GC/Works/1 form denies them (its condition about financing charges relates to recompense for late payment of amounts already due).

There are also differences in how the forms provide for the calculation of disturbance, with the JCT forms mostly and the ICE forms in all cases segregating it from other adjustments in all cases, but with the GC/Works/1 form requiring those elements of it which are associated with variations to be calculated as part of the variations valuation, as a first option by a quotation and in advance of the work being performed, and in any case split between each variation instruction. Other disturbance elements are to be given separately, as under the other contracts.

The NEC form has reference to instructions and there are provisions for the project manager to instruct on any matter. 'Change' and 'variation' are not terms used in the NEC, whereas 'compensation event' is the equivalent of a variation which is issued by the project manager when he agrees that there has been a significant change to the contract due to his requirements.

Settlement of disputes

Methods of settlement of disputes in standard contracts are detailed in Chapter 28. Although each standard contract has provisions (a) for agreeing *within contract provisions* for matters such as loss and expense (except for the NEC form) to be paid and (b) for contract provisions for dispute resolution to follow if necessary, *it is quite possible*, by recourse to the courts or to arbitration, that general damages may be the basis for sums claimed to be due, if the circumstances of a project permit.

CHAPTER 20

GC/WORKS/1 WITH QUANTITIES CONTRACT

- Contract basis and administration
- Site and ground conditions
- Programme and time
- Instructions and payment

The form considered here is used widely for central government projects and is commonly referred to by its short title, Form GC/Works/1. It is also offered for use in the private sector, when a number of amendments are desirable. It is taken in its 1998 edition and in its with (firm) quantities version, which has supplementary conditions and appended examples in a separate Model Forms and Commentary document covering all versions of the contract.

The differences occurring in other versions of GC/Works/1 are not of consequence when comparing the form with the JCT form, as is done here, over present issues only and so on a very selective basis. A note on the distinctions in the design and build version is given in Chapter 24, as there are important differences there, not affecting present matters.

The inclusion of supplementary conditions is common practice. These often relate to the specific site or government department's requirements and, as such, should be read assiduously on the assumption that they have been included because they at least are actually intended!

CONTRACT BASIS AND ADMINISTRATION

A prominent feature of the contract is that the present edition has been prepared primarily for the Property Advisers to the Civil Estate. Although there has been discussion with various sections of the industry, professional bodies and other interested groups, in the end the decisions on drafting are those of that body, so the documentation is formally unilateral in origin. This sometimes makes it a bit clearer than the JCT form, because it does not have to adopt similar convolutions as a means of compromise.

The result is a contract which in places is weighted fairly heavily against the contractor by comparison with other forms. But it also follows that it may well be construed contra proferentem, so the contractor could have his redress at law. This is of limited value, as going to law may win a pyrrhic victory. In practice there does appear often to be settlement off-stage with the contractor, which is more favourable than might be expected from the wording of the form. This may represent fairness, or it may relate to a saying about dirty washing.

Following recommendations of Latham, the contract now includes a condition 1A requiring that the parties 'shall deal fairly and in good faith and in mutual co-operation with one another'. Implication of such terms is no longer necessary.

Persons

Several persons are identified by the conditions (as they are named rather than as 'clauses') who are on the client side of the contract and who have important functions:

(a) 'The Employer', who is more active in the present contract than under JCT arrangements and can make decisions which affect both parties. Very importantly when it comes to disputes, some decisions are identified as 'final and conclusive', so they are not subject to arbitration or proceedings. Some are peculiar to government contracts, but not all, and they are listed under 'Arbitration' below.
(b) 'The Project Manager' (abbreviated to PM) is the equivalent of the JCT architect, although he may not be a designer. However, he is located very clearly within the authority's organisation functionally, not as an independent professional person operating between the parties. Again he may give decisions that are final and conclusive in defined areas.
(c) 'The Quantity Surveyor' (abbreviated to QS) is distinguished from the PM and appears to have an independent standing, whether or not he is within the authority's organisation. But again, he may give decisions which in some cases are final and conclusive.
(d) '[A] Clerk of Works or Resident Engineer', who will exercise the PM's powers over quality control etc under condition 31. These include powers to reject work and materials and to introduce independent experts, whose powers go further than those of the JCT clerk of works. They also cover authority to be satisfied that the contractor is working diligently and keeping to programme. Nevertheless, the PM is still able to reassume any of his powers.

Sometimes when the contract is used for central government projects, at least (b) and (c) are treated to considerable numbers of departmental circulars which constrain their activities significantly. All of this is particularly relevant to such matters as site control, affecting the contractor's ability to sort his operations freely and such elements as works by direct contractors.

Contract documents and instructions

The contract uses a broadly similar set of documents to those in the JCT form with the addition of the specification as a contract document; this effectively takes the place of the unmentioned bill preambles in the JCT system.

Permitted instructions by the PM are listed in condition 40(2) and there are additions to those by the architect under the JCT form. There are powers to instruct specifically over 'hours of working and the extent of overtime and nightwork', presumably by increasing or even decreasing them. Furthermore, there is a catch-all provision for instructing over 'any matter which the PM considers to be necessary or expedient'. Potentially they open all manner of avenues, including disturbance effects on the programme and progress over which other conditions seek to exercise particularly firm control. It may be suggested that 'any matter' does not extend to items which would produce a variation to the nature of the contract.

Conditions 41 to 43 deal with the valuation of instructions in some detail and give a number of differences from the JCT pattern. As these conditions contain elements akin to other extra expense conditions, they are all looked at together under the major heading of 'Instructions and payment'.

Limitation period

As the contract is always executed by submission of a tender entered upon the standard form, accepted by letter and recorded in a contract agreement executed by the parties, the option of a speciality contract does not arise without special action. As a result, the limitation period for bringing actions over defects etc is six years instead of twelve years.

SITE AND GROUND CONDITIONS

Site

The contractor is 'deemed' under condition 7(1) to have taken into account when tendering a wide range of matters about the site, its approaches and its surroundings. Some of these matters are somewhat fussy; they are not spelt out by the JCT form as they are self-evidently the contractor's responsibility unless he is relieved of them, such as means of communication and access. As with instructions, the list is open-ended; it concludes with unspecified 'any other matters or information', which is oddly in contrast with most of the list.

Ground conditions

Of more consequence is the listing in condition 7(1)(d) of 'the nature of the soil and material (whether natural or otherwise) to be excavated', so the contractor has to allow for what he knows, subject to any requirement for the bills of quantities to give distinct items for any category, as discussed in Chapter 4 in relation to SMM7. This is usually limited to the very difficult to remove, the very difficult to support and the very wet.

Following on from this, conditions 7(3) to 7(5) deal with 'Unforeseeable Ground Conditions'. The contractor is required, as a condition precedent to receiving suitable allowance for the ground conditions, to inform the PM of the conditions and state what action he proposes to take. This is important for several reasons:

(a) Extension of time due to their simple presence may arise under condition 36.
(b) The value of work 'properly' carried out or omitted as a result of such conditions is to be valued as a variation under condition 42, although little seems to be left by condition 7 for this beyond what is routinely measurable under most standard methods of measurement.
(c) Disruption or prolongation may arise in drastic cases of these conditions under condition 42, such as with the discovery of an unknown large structure or a watercourse. In the major case study in Chapter 30 this became a serious question affecting more than groundworks, owing to the late stage at which the water problem came to light and its effect on permanent internal construction nearing completion.

Unforeseeable ground conditions are those which the contractor 'did not know of, and which he could not reasonably have foreseen' from information supplied or within his reasonable ability to ascertain. This definition is thus related to his actual ability alone and not strictly to that of an experienced contractor as the ICE conditions have it (Chapter 21), or 'an experienced and competent contractor' as the present condition 31(2) has it over conforming to quality requirements. In practice it may be contended that he has put himself forward as having these qualities by tendering, so the distinction may not be significant. Usually there will be a borehole report or other information provided by the employer, although this may be qualified by statements that the contractor must draw his own conclusions as to its meaning and how typical it is of the site as a whole. This suggests that the risk of unexpected material rests with the contractor, except where the method of measurement backing up the bills of quantities states that it is to be measured separately, and so paid for at a distinct rate whether or not foreseeable.

The definition is qualified by the inclusion of artificial obstructions, which would be measured, and by the exclusion of conditions caused by weather. The latter could mean, for instance, water on the surface or surface water entering the excavations, or extremely dry or frozen ground. Particular strata may change their characteristics drastically in such circumstances. The contractor will be held to have been aware of the strata and to have borne the risk of how such weather conditions affect them. By contrast, most methods of measurement and bills of quantities take account of varying conditions due to groundwater (even though it will originally have come out of the sky as a constituent of the weather), such as softness of the soil and removal of the water. This leaves the problem of deciding the origin of the water: the expedient of colour-coding it is not known to have been refined to the point where it has become practicable!

The contractor is required to give notice of these conditions and of how he proposes to deal with them prior to the PM certifying them as unforeseeable ground conditions. The PM is given no express power to affect how the contractor does deal with them and it would appear his authority would be limited to banning methods that would contravene specific contract requirements or prohibitions. This gives a position similar to the silence of the JCT contract, but differing from that under the ICE contract, where the engineer may well become involved in designing work and giving instructions. Under the NEC contract, the contractor is under an obligation to perform according to the 'experienced contractor' standard. This will mean that he will be judged by what it is held that a contractor carrying out works of the nature involved could be expected to have judged to be necessary from the tender documents and from his experience.

Other aspects than actual ground conditions may come into the reckoning. In *Holland Dredging (UK) Ltd* v. *The Dredging & Construction Co. Ltd and Imperial Chemical Industries plc (third party)* (1987) delay occurred over determining the proportion of backfill that required to be imported due to a shortfall from excavated material that had been indicated would be available to the contractor from information supplied in tender specifications and bills of quantities. This caused a delay for which the contractor was reimbursed.

Other aspects of groundworks and their valuation are explored in Chapter 4, particularly in relation to methods of measurement.

PROGRAMME AND TIME

These issues are particularly related to parts of conditions 33 to 39. They are affected in various places by the closer control established over the programme for the employer and the PM. The generalised discussion of the site and programme matters in Chapter 4 may be referred to as a background to the present specific provisions.

Programme

There is a requirement for a programme to be included in the contract under condition 33 and for the contractor to warrant that it properly represents his intentions. It is to show details not only of the sequence of the works, but also of resources proposed to be used and of critical events for 'satisfactory completion'. In view of the possible evidential use of the programme over extension of time or disturbance of sequence, it is important for both persons to give the programme close attention before agreeing to it as part of the contract. The PM may require elements here, so he can monitor and to an extent control progress and correct deviations. The contractor may also put on record elements which may prove important for him, even if some of them are by way of forecast at this stage. When there is a post-mortem, a programme produced as a document intended to be followed and accepted by the PM as a working guide, starts with more credibility than a document produced entirely after the event.

With this said, it is the contractor's overall responsibility to complete on time under condition 34, as with other contracts. There is a provision for the employer to notify the contractor of when he may take possession of the site or parts of it, subject to constraints set out in the abstract of particulars, which is equivalent to the JCT appendix. This gives more flexibility than under the JCT arrangements over the beginning of the contract period and also its end, something which cuts both ways, especially in a fixed price contract. No extension of time results from the proper exercise of this provision, but an overrun (as a default by the employer) will lead to extension of time and perhaps payment of expense for prolongation and disruption.

Progress meetings

Condition 35 requires 'regular progress meetings', usually to be convened monthly by the PM and attended by the contractor. The pattern of these meetings is not directly stipulated, but may be discerned from the documentation to be produced before and after, respectively by the contractor as a report on how things are going and by the PM as a response, embodying his opinion and proposed actions. The same set of five points occurs (with some variation of order) in each case and may be summarised as follows:

(a) Progress of the works, in relation to the programme and instructions, with the resultant effects, as both see it.
(b) Outstanding requests for drawings etc with the PM's response, which may indicate a satisfactory situation or otherwise, according to when information is due.
(c) Circumstances in the contractor's view threatening since the last meeting to cause delay and extra cost and those which the PM sees as having this effect, apparently running from the commencement of the contract.
(d) A similar set of statements over extensions of time requested and granted.
(e) Proposals from the contractor to amend the programme 'to ensure' completion by the current due date for completion, with 'the steps which the PM has agreed with the Contractor to reduce or eliminate the effects' of delays.

The first three represent exchanges of information and that general running over issues new and old which can simply pass the time. They come to a focus in the last two over moving the completion date on to accommodate what cannot be avoided and of making changes in the programme to keep to the course currently agreed. The effect of all this is to formalise and to an extent dramatise what happens under other contracts with or without meetings. It is useful as it controls the making of decisions about the programme, often so easily put off.

The reference to agreement between the PM and the contractor over steps to combat delay and extra cost needs care in interpretation, as it is not clear as to whose steps are meant. On some issues it may be either. The question of supply of drawings etc is covered by (b) and they will either come or will not, so that responsibility for decisions and actions lies with the PM, so long as the contractor has given notice of all needs. The question of what the contractor is to do is more complex, as there is the risk for the PM that he will be drawn by implication into matters of the contractor's site organisation and production methods where he should normally keep to one side. Both should establish demarcations so far as possible at an early stage to seek to avoid this sort of overlap.

The direct question of whether any such steps can be introduced to achieve the current completion date remains and it may not be practicable to do this.

Furthermore, there is the question of whether the contractor is to be paid for measures he takes. He should ensure that he receives an instruction under condition 40 over such matters as hours of working or under condition 38 over acceleration. The latter is something dealt with by the employer and so beyond the power of the PM to initiate alone. Instructions and acceleration are discussed below.

This contract may be compared with the JCT at this point. The JCT form lacks a framework within which things may happen, but is explicit over such procedures as giving notice of missing information within reasonable time limits and in setting out a detailed system for granting extension of time. The present form concentrates on the framework, within which the PM exercises defined control, but less is said about the rules on how problems are to be resolved. This tends in practice to give the employer under Form GC/Works/1 more power, as has been indicated. Sometimes some of the rules, as the employer sees them, may be spelt out more explicitly post-contractually and this may or may not tend to help the contractor. There is much to be said for the architect under the JCT form setting up clear arrangements about keeping progress under review, so that blockages are removed. The contractor is often ahead of him here, and this is better than the 'let it drift and make capital out of it' strategy.

Extension of time

This topic is covered separately in condition 36, even though progress meetings may provide a forum for building up some of the detail. The condition is far shorter than its JCT equivalent, although it achieves a number of similar effects, a point which the JCT itself might note.

Extension may arise by a notice from the contractor requesting it and giving grounds as distinct from effects and estimated delay, or by the PM himself considering it necessary to look into a delay that has happened or is threatening. The former is less onerous than the JCT provision, whereas the latter is something not envisaged by the JCT form and perhaps something not even valid under it, except after completion. The condition allows for 'extension of time for . . . a relevant Section', whereas the unmodifed JCT form only considers the whole works. The latter then allows for early possession of a part at the employer's discretion and by agreement with the contractor during progress, although the supplement provides for sectional completion to become part of the initial contract.

The PM is to notify his decision 'as soon as possible' and within forty-two days when the contractor has made a request. He is protected from undue pressure by the option of either an interim or final decision, although he has finally to decide all extensions within forty-two days after completion of the works. The contractor in turn cannot request any extension after completion of the works, although before this he is given no time limit. This compares with the stipulation about the contractor acting 'forthwith' when actual or threatened delay 'becomes reasonably apparent' in the JCT case, which does not allow for him giving any notice after completion but allows the architect to consider any extension (notified or not) at that stage.

The PM may not withdraw or reduce 'any interim extension' in his final decision, as with the JCT arrangement, except that he may vary one 'to take account of any authorised omission' of work. This is difficult to interpret; presumably it is limited to the case of a subsequent omission which falls within the scope of work affected by an interim extension granted before such an omission was authorised.

The contractor is given the right to appeal within fourteen days against the PM's decision over an extension of time, whoever initiated the proposal. It appears possible for him to keep up a series of appeals, provided he can muster enough arguments against successive decisions within the time limit.

All these procedural elements allow a fair amount of flexibility, as does the JCT set, and the attitude of the courts (Chapters 9 and 10) tends to introduce yet more flexibility. This is generally to the good, if people are seeking to reach sensible settlements, instead of taking one another up in strangleholds.

The list of causes possibly leading to extension of time is considerably shorter than in the JCT form:

(a) Modified or additional work.
(b) Act, neglect or default of the authority of the PM.
(c) Strike or industrial action: these are defined widely enough to include strikes etc not directly related to the works, but are limited to what is outside the control of the contractor or subcontractors.
(d) Accepted risks or unforeseen ground conditions: accepted risks are defined to cover the excepted risks and broadly also the hostilities risks under the JCT contract; unforeseen ground conditions are discussed earlier in this chapter.
(e) Any other circumstance outside the control of the contractor or subcontractors, limited to what could not have been reasonably contemplated 'under the Contract', and so by the contractor rather than by subcontractors of at least the nominated variety, it appears.

Weather conditions as a cause are excluded under (e), so the complete risk of delays due to all weather, not just the exceptional, falls upon the contractor. If some other allowable cause pushed work into a period of less favourable weather, the resulting extra delay might well qualify as an effect; see discussion of *Walter Lawrence & Son Ltd* v. *Commercial Union Properties (UK) Ltd* (1984), page 258. Among the cases in Chapters 9 and 10 is *Balfour Beatty Building Limited* v. *Chestermount Properties Limited* (1993), which should be considered with other cases when establishing extension of time periods.

The much shorter list of causes under this contract is nevertheless equivalent in scope to much of what is in the JCT form, by virtue of the width of the provisions in (b) and (e) above. They are likely to cover, for instance, many categories of PM's instructions under condition 40 (itself containing an open-ended list as does the present condition), impossibility of obtaining labour or materials (perhaps even plant) and government action. The main area of comparative deficiency is in regard to nominated subcontractors and suppliers, where delays on their parts or due to renomination are not covered, even though renomination itself is covered in condition 63.

The condition ends with a provision about the contractor endeavouring 'to prevent delays and to minimise unavoidable delays', much as the JCT form gives.

Early possession

Condition 37 may be noted for completeness. It deals with early possession covering the ground of sectional completion and partial possession as used in the JCT contract. It therefore allows for a section defined for this purpose (among others) in the contract and a part of the works of which possession is given by agreement between the parties. But it goes further by including a part of the works over which the PM gives the contractor an instruction to give possession.

When early possession occurs, the expected results flow over the part: transfer of responsibility for damage, beginning of the maintenance period, a reduction of liquidated damages and of the reserve, as the retention fund is termed here. The provisions avoid the proportioning anomaly of the JCT form.

Acceleration

This is an arrangement quite beyond anything in the JCT form. That form is based on the principle of completion by the defined date, subject to any extension for defined reasons, that is things either stay the same or get worse. Any agreement about acceleration constitutes a modification to a basic contract term, is supplemental to the original contract and neither party is obliged to enter into one. If one is proposed, the approach set out in the present condition 38 contains the key elements necessary for its reasonable introduction.

The condition provides a procedure which allows the introduction of an acceleration within the contract framework. Although there is a procedure, there is still no contractual obligation upon either party to allow or accept an acceleration. For this to occur, there must be an agreement as the condition provides.

The procedure has two prongs. In condition 38(1) the authority (not the PM) may wish to advance the date for completion of the whole or a section of the works. Then the authority is to direct the contractor to submit priced proposals and programme alterations or, alternatively, explain why he cannot achieve the date required. If the latter happens, the matter rests unless the authority comes up with another date or otherwise keeps the matter alive. If the former happens, the contractor may still choose to offer proposals which price him out of acceptance.

In condition 38(3) the contractor may propose such an advance at any time and the authority undertakes to consider it, presumably on the basis of similar supporting information or by declining it after a more superficial consideration of its virtues. The reasons why either party should wish an advance are of course their own concern, but may generally be inferred.

Whichever prong applies, the authority may accept the contractor's proposals, which may be after any negotation, as is left silent. Under condition 38(2) the authority is then to specify the following:

(a) The accelerated date for completion.
(b) The programme amendments with supporting details.
(c) The increase in the contract sum.
(d) The revised milestone or stage payment chart or charts (the system for payment used under condition 48 and based on an S-curve approach).
(e) Any other agreed contract amendments.

INSTRUCTIONS AND PAYMENT

This is the precise heading of the section of the contract which deals with the whole range of these matters, including those corresponding to what the JCT form terms valuation of variations and loss and expense. The appropriate elements are picked out here for comparison, whereas others are left aside. There is not a direct correspondence in the elements between the two contracts and some quite distinct differences in principle.

The present approach may be summarised as follows:

(a) There are two types of PM's instruction, general and variation instructions (VIs); VIs are issued when the finished works are affected mainly in quality or quantity (condition 40).
(b) The two types of instruction are valued within the same broad framework, but with distinctions in detail (condition 41).
(c) A quotation or valuation for an instruction of any type is to include the direct cost of performing the work and also amounts broadly equivalent to JCT expense over prolongation and disruption, but not loss (condition 41).
(d) The PM may require the contractor to give a lump sum quotation for a VI before carrying it out and the PM may accept this after checking by the QS (conditions 41 and 42).
(e) Alternatively to a quotation for a VI or after not accepting it, the PM may require valuation by the QS within a tight timetable, although not necessarily before performance of the work (condition 42).
(f) General instructions are to be valued on the basis of expense incurred or saving made (condition 43).
(g) Payment for prolongation and disruption, again at an expense level, may be made in restricted circumstances not arising out of instructions, including some lapses on the part of the authority or the PM (condition 46).
(h) Finance charges may be payable due to late payment of other sums, these charges therefore not being equivalent in nature to what may arise under the JCT form (condition 47).

Conditions 44 and 45 deal with labour tax and VAT and are not relevant to present discussion.

PM's instructions

Unlike the JCT form, this contract lists all instructions in one place, condition 40. The items given are mainly equivalent to the JCT items. Condition 40(1) refers to 'Instructions, including Variation Instructions (VI)' (sic) but nowhere in the list is a precise segregation given between the two types of instruction, general and VI. The importance of this segregation lies in the method of valuation, as discussed below. Under condition 1(1) 'Definitions etc', a VI 'means any Instruction which makes any alteration or addition to, or omission from the Works, or any change in the design, quality or quantity of the Works', whereas the definition of an instruction at large includes a VI, except where expressly stated to the contrary. These definitions are indicative, but not in themselves precise.

From the list in condition 40(1), the following items are clearly VIs in many or most cases (the letters of the condition are used, with brief wording):

(a) Variation etc of documents or design, quality or quantity of the works.
(b) Resolution of discrepancies.
(c) Removal from the site of 'Things' (the endearing contract term for goods and materials for incorporation).
(d) Removal and/or re-execution of work.
(i) Opening up of work for inspection (when it turns out to be in order).
(j) Use or disposal of material obtained from excavations.
(k) Cost savings resulting from changes agreed between the parties to effect efficiencies in the design (how these are to be evaluated is by no means always apparent, this being outside present considerations, but see *Building Contracts: A Practical Guide*).

However, other items are likely to lead to expenditure by the contractor without usually falling within the definition of a VI, but being reasonably likely to qualify for some form of reimbursement:

(e) The order of execution of the works.
(f) Hours of working and extent of overtime.
(g) Suspension of execution of the works.
(l) Execution of emergency work.
(n) Actions following discovery of antiquities etc.
(o) Measures to avoid nuisance or pollution (at least in some instances).

These items leave only the following which are unlikely to attract reimbursement, unless their circumstances lead to dispute:

(h) Replacement of any person employed in connection with the contract.
(j) Making good defects.
(p) Quality control accreditation.

There remains just one more item, which may fall either way according to its content:

(q) Any other matter which the PM considers necessary or expedient.

The second group may be seen largely to parallel items in the JCT form to some extent and, most importantly, often those relating to obligations and restrictions in clause 13.1.2 over 'limitations of working hours' and 'work in any specific order'. However, the present items go beyond 'limitations' by allowing the *instruction* of working hours and the extent of overtime. This is different again from the ICE position, where *approval* is needed for night and Sunday working. Furthermore, the last item of 'any other matter' gives the PM wide scope in what he may instruct so that, for instance, the aspects of JCT obligations and restrictions not paralleled explicitly are within its orbit. It may be compared with the relative vagueness of the ICE contract, which may still give the engineer some scope to act in this area.

It may be suggested that any items not mentioned in this list, but governed expressly by other provisions of the conditions, are not to be read into 'any other matter'. Thus, working hours and overtime are mentioned here, but acceleration to achieve earlier completion is dealt with and restricted by condition 38. On this basis the contractor could not be obliged to finish early simply because, for example, he was being paid for overtime, even if this were enough to effect a significant overall programme shortening. On the other hand, an acceleration agreement arising in its own right might well take account of a shift in this same overtime pattern. The whole matter of payment for overtime is riddled with questions about the non-productive element and balancing of hours worked against total direct expenditure and overhead savings. It is touched upon in a limited context in Chapter 11.

The PM should ensure that he distinguishes the two varieties of instruction (variations and the rest) in the documents on which he issues them. If then the contractor receives one classified in what he considers the incorrect way, he should raise the point quickly to avoid later dispute over the method and therefore the amount of payment. The question of quotations for VIs is looked at in the next section, where the similarities and differences between the present provisions and those in the JCT clauses may be noted (Chapter 6).

Principles of valuation of instructions

Condition 41 acts as a preliminary to the following conditions covering valuation of the separate types of instructions. Three of its four provisions are routine: the following conditions are to be followed for valuation, the result is to be added to or deducted from the contract sum (unless due to the contractor's fault) and the contractor is to supply the QS with information so that he can perform the valuation.

The other provision bears comparison with the new provisons introduced into JCT contracts as set out in Chapter 6, at the beginning of which the similarities and distinctions between the several provisions are outlined. The present provision is that 'The value of any instruction shall include the cost (if any) of any disruption to or prolongation of both varied and unvaried work'. This requirement assumes that costs arise in isolation and can be segregated completely between individual instructions, so there are no overlapping results flowing from distinct causes. Equally, it assumes the possibility of complete segregation from similar costs referred to in condition 46, which deals with prolongation and disruption due to what are essentially malfunctions or lapses from the authority's side. These in turn may lead to related instructions covered by the present condition.

This pattern is the absolute antithesis of the rolled-up or global claim approach described in Chapter 14 and it will be plain from the discussions and examples throughout this book how difficult such extreme separation may be, especially of secondary amounts as defined in Chapter 12. Some fairly arbitrary apportionment of more global sums may be expected on occasions. Only clearly identifiable amounts should properly be included on the way along and there is no catch-all provision anywhere to allow for the reimbursement of any unallocated balance not taken up in the series of valuations which are likely to occur.

Even an apportionment will present an additional layer of problems when the contractor is required to give a quotation for a VI in accordance with condition 40(5) and for this to be submitted to the QS for evaluation under condition 42(2) ahead of performance. This aspect branches even further still and so is taken under the next heading. It may simply be observed here that the requirements are likely to work out in application quite distinctly from the JCT loss and expense concept.

Valuation of variation instructions

Condition 42(1) provides two distinct approaches to the valuation of variation instructions; the first approach is the submission and acceptance (as a condition precedent to the VI becoming effective) of a lump sum quotation from the contractor. Condition 40(5) allows for a quotation required at the same time as the VI is issued and provided within twenty-one days of receiving the VI. Acceptance or otherwise is to be by the PM within a further twenty-one days, but first the QS is to 'evaluate' the quotation under condition 42(2). This process specifically includes the provision of other information required by the QS and possible negotiation before acceptance.

Under condition 42(2) the contractor has to show separately in any quotation two elements: 'the direct cost of complying with that Instruction' and 'the cost (if any) of any disruption to or prolongation of varied and unvaried work'. There is no reference to the price level of the first element of directly performing the work, such as whether it is at the same level of competitiveness as the bill prices. In fact, by the mention of 'cost', there could be a hint of something other, even though cost proper is not available until after the event. It is unfortunate that 'cost' is employed here at all, as the quotation is of a price or prices, inclusive of profit, whereas in conditions 42(13), 43(1) and 46 'cost' is apparently to be interpreted as what is paid out net to achieve the result, and 'expense' to be likewise interpreted as having the same meaning.

The second element is even more problematic, as disruption and sometimes prolongation are factors which lead to somewhat unconsidered results and which it may be necessary to evaluate by what has been incurred or expended, as conditions 46(3)(b) and 46(6) recognise. There is the old argument for pricing variations other than at bill prices to allow for the change factor, which may be there even when no disturbance element occurs. In any case there is the problem under the JCT formula of how ascertainment is to be performed without recourse to a measure of assumption in the calculations (as discussed at various points throughout this book, such as in the early part of Chapter 11). As these factors fluctuate, 'the cost' of disturbance becomes the more uncertain to attempt to determine in advance.

Presumably the contractor may either seek to avoid quoting at all on this basis, or he may give a high enough quotation to ward off its acceptance. There is no express provision in the conditions for what happens over a breakdown in negotiations; presumably there is a reversion to the second approach to valuation.

The relationship between the timescale given in conditions 40(6) and 42(3) is established in condition 42(7) as depending on when the QS requests information which will be about the execution of the varied work or, perhaps more critically, unvaried work affected by it. The unvaried work in particular may not be easy to assess and provide within such a timescale.

If under condition 42(4) there is failure by the contractor to provide a quotation or by the parties to agree upon it, the process reverts to the second approach, which also applies if a quotation is not required. This is the traditional one, as in the JCT form, given in condition 42(5) of using measurement priced at bill rates or derivatives. It proceeds primarily through the usual options of bill rates, pro rata rates and star rates to daywork, with the application by condition 42(10) to rates of any percentages included in the bills. It relates properly only to additions as against omissions, as may be seen from such expressions as 'prices for . . . similar work [in the bills]', 'prices deduced or extrapolated' and 'fair rates and prices', when these are given their accepted meanings in practice. In the absence of any more applicable provision, the bill prices will usually be applied to omissions, as they will be in effect when there is a *net* addition to a bill item.

There are several riders following over pricing and the timetable for agreement. Over pricing, condition 42(6) provides for adjustment of rates for 'work not within the direct scope of the VI' to cover any prolongation or disruption. Presumably the term quoted is intended to mean the same as 'unvaried work' used before.

Whether work is or is not within the scope of the VI, it is not always practicable to deal with disturbance effects by rate adjustment, as a wide-ranging cost-related assessment may be needed, some of it applied to single events (such as familiarisation time), rather than evenly spread effects. It also may not be possible to do it within the timescale laid down later in the condition, as the work may not have been performed, a problem mentioned earlier. The wording permits the QS to use his discretion by the term 'considers appropriate', perhaps adding allowances for full overheads and proift, not barely ascertaining net amounts, as under JCT clause 26. Internal government procedures and memoranda may find this difficult to assimilate, but they could be negated by the contra proferentem rule. There is a reverse-effect provision in condition 42(13) noted below.

Condition 42(11) allows 'fair rates and prices' instead of bill rates, to cover situations in which varied work is performed out of sequence or in odd-sized parcels. This does not allow for disruption, simply for steady working in special conditions.

The timetable for agreeing valuation covered in conditions 42(7) to 42(9) is a subject not usually tackled by contracts, which assume that all comes out in the wash, at the latest by the time of the final certificate or related proceedings. Although condition 41(5) gave no period for the QS to ask for information, condition 42(7) now picks up the trail by requiring the contractor to provide it within fourteen days, the QS to notify his valuation within a further twenty-eight days and the contractor to notify any disagreement with a still further fourteen days, giving reasons and his alternative valuation.

After all this, the trail runs out, as there is no statement about agreeing after there is a difference, reasonably enough. The contractor is held to have accepted the amounts and is bound by it, if he does not disagree within the fourteen days, but otherwise there is no sanction imposed on him for failing at any point in the timetable. On the other hand, the QS is under pressure if he delays or changes his mind, by virtue of the possibility of finance charges arising under condition 47(1). The times are all very tight by normal standards of settlement, especially when the VI is far-reaching or complex, or when disturbance is involved. The process relies on the QS not starting it off too soon, and ignores the normal process of negotiation which frays the edges of tidy parcels conceived here as in separate time divisions.

The whole of condition 42 thus far is worded on the basis that all VIs will result in an addition to the contract sum, or at least in a net addition, as noted above over using rates for measured work. But condition 42(13) in closing allows for a decrease in the contract sum by the amount of any 'saving in the cost of executing the Works . . . as determined by the QS' and resulting from a VI. This is straightforward if it refers to normal measured omissions, as it counterbalances 42(5), which appears to be referring only to additions valuation, and if the QS's determination then uses bill rates and also takes account of any unvaried work executed under changed conditions because of the VI, both of which are covered explicitly in the JCT form.

However, the term 'the cost of executing' does not convey this meaning precisely, or perhaps at all. Rather it supports the idea of assessing the contractor's actual costs of performing work and setting them against what it would have cost without the VI having been issued, then deducting the difference from the contract sum. As discussed in Part 4, the 'what it would have cost' question is the problematic one here, as it has not happened in the nature of things. By whatever means it might be arrived at, the sum in question presumably should be calculated net of profit, but inclusive of overheads. Indeed it may sometimes consist of little more.

Over and above the presumed meaning of the provision is the question of its commercial fairness to the contractor. He may wish to resist it being operated in the way that has been described. If his costs are within his presumed profitable estimating margins, he may even stand to have less deducted than he was due to be paid for the unvaried work, but the reverse situation may arise. A reduction in this instance would go against the usual principle of not going behind the contractor's pricing when dealing with variations, as is implied by JCT clause 14 of the limits to adjustment of the contract sum. As there is no necessary connection here with any disturbance, but perhaps solely with variations, this is not a matter of setting a saving of cost against an increase when considering the overall effect of disturbance, as happens under condition 43.

Valuation of other instructions

Although VIs are likely to constitute the bulk of instructions in most projects, there remain those not leading to physical changes of the works but likely to lead to financial changes, as listed earlier under 'PM's instructions'. Several of them correspond to changes in obligations or restrictions in JCT terminology, so that a similar method of evaluation is predictable. In these instances there is no provision for a quotation to be provided in advance. Usually, but not always, that approach would be unsuitable.

Two subsidiary elements in adjustment by the QS are given in condition 43(1), broadly corresponding as a basis of valuation with parts of those given in condition 42:

(a) Expense 'properly and directly' incurred beyond that provided for or reasonably contemplated by the contract.
(b) Saving in the cost of executing the works.

The term 'expense' is used here for the first time in the present set of conditions along with 'cost', but both recur in condition 46 over prolongation and disruption.The practical or even intended difference between the two is not clear from their general definitions or from the several contexts in the conditions, so the distinction is not pursued here. Bill rates will seldom be relevant, but there is the possibility that daywork may be appropriate (at rates desirably given in the bills).

Condition 43(4) defines 'expense' as 'money expended by the Contractor', but then excludes 'any sum expended, or loss incurred, by him by way of interest or financing charges however described'. This may be compared with the JCT position where decided cases show that interest and financing charges may be reimbursed in particular circumstances. So too, under those provisions, there may be 'direct loss' as money not received and including loss of profit on work forgone on other contracts not obtained elsewhere, due to the burden of complying with the instruction, as distinct from profit on extra work done. The present restriction to direct expense as money expended is therefore more wide-ranging than just to exclude what is mentioned expressly. The corollary is that the definition here (and it recurs in condition 46) allows a far more limited reimbursement than the JCT definition; the comments on the JCT definition should be closely consulted, even though the claim for actual reimbursement there does have to be well supported in principle and in demonstration of its quantum.

It is suggested that 'saving in the cost' in the present context must be interpreted as the reverse of 'expense [incurred]', that is not to expend money that otherwise would have been expended. But it appears equitable that the term should then also be interpreted as excluding any surrender to the authority of the amount of interest or financing charges for which the contractor might otherwise have been liable.

Again there is a timetable, not quite so tight, for agreeing or disagreeing the amount for any one instruction. The contractor is to provide information within twenty-eight days of complying (perhaps finishing complying) and the QS is to come back with his amount within a further twenty-eight days, followed by fourteen days within which the contractor has to notify disagreement and the amount of his alternative, or be held to the QS's amount, as happens with a VI. Much will depend on the nature of the instruction. It may be a compact operation or something which has a diffuse effect across a substantial section or even the whole of the works. This may effectively divide the whole process into a series of subprocesses, each with its own timescale. As before, there is no provision about what happens when there is not agreement. However, failure by the contractor to notify disagreement within the time given will again lead to loss of his right to further claim.

A great deal of the action here will depend on principles of the types covered in the chapters about JCT loss and expense in Parts 4 and 8 of this book, and these should be consulted and compared.

Prolongation and disruption

Condition 46 is the closest equivalent to JCT clause 26, although some of the issues taken there are covered in conditions 42 and 43 of the present contract. It opens with the same expression as in condition 43(1) 'properly and directly incurs any expense' and ends with the same definition of 'expense', and the comments about those terms in that condition are fully relevant here.

The causes leading to a qualifying expense and paralleling the matters in the JCT clause are set out in condition 46(1) and continued in some cases in condition 46(2). The first is the execution of works by other contractors on the site and contemporary with the works, but only if they lead to disturbance as stated. If the contractor has been given warning in the contract documents that particular works are to proceed in this way, he will not have cause to claim unless some divergence from what he expected leads to the unexpected effects stipulated.

The second cause is delay in being given possession of the site. Here the contract gives a more flexible margin over starting than in the JCT case, but once it is transgressed the same principle applies. The third cause is *delay only* over any of the bunch of matters given in condition 46(2):

(a) Provision of drawings or other design information by the PM.
(b) Execution of work or the supply of 'Things' ordered other than from the contractor and where such ordering is not due to a prior default of the contractor in not ordering or doing it himself.
(c) Action by the authority to allow or procure persons to be on-site: this covers a range of actions from nominations to issue of passes, but not in circumstances of renomination as in condition 63(8).

Although there are no individual stipulations about requests for any of these items by the contractor or about their timing, condition 46(3) provides that reimbursement flows only when the authority has failed in relation to a date agreed with the contractor or a reasonable period specified by the contractor in a notice. This parallels the JCT position and would particularly come alive through the progress meetings of condition 35. The requirement is for notice to be given when 'regular progress . . . has been or is likely to be disrupted or prolonged', which recognises the two situations mentioned in the JCT clause.

The fourth and last cause is 'any advice from the Planning Supervisor'. The requirement for an employer to use a planning supervisor was introduced by the CDM Regulations which came into force in 1995. If the information given to a contractor, who has to act as the 'principal contractor' under the regulations and implement the health and safety plan drawn up by the employer's planning supervisor, is not correct (or is incapable of implementation), it is possible that the contractor will be delayed or disrupted.

Again, under condition 46(4), there is a time limit of 'within 56 days' for the provision of details of the expense incurred from one of these causes, with a further twenty-eight days under condition 46(5) for the QS to notify the contractor of his decision. Once more and reasonably, there is no provision for what happens if there is disagreement.

A summary

The terms 'prolongation and disruption', and their verb equivalents, are used in this contract as between them carrying a similar meaning to 'disturbance' and 'disturbed' in the JCT contract. For instance, condition 46(3) refers to 'regular progress' as being 'disrupted or prolonged' (the order of the words being casually reversed from time to time). Earlier chapters of this book have also used the two terms in this way. The ICE contract is less precise over this and other matters, as the next chapter indicates in various places.

However, discussion to this point shows that the ways in which reimbursement for their effects is treated is on the whole more fragmented under the present contract than under the JCT form. This may be noted against the background of the procedural list at the beginning of this section on instructions and payment, which outlines the conditions under that heading in the contract. The several fragments may be drawn together and highlighted from preceding discussion in the following list, using the single term 'disturbance':

(a) Valuation of instructions, principles: condition 41 as the general introduction states that the value of any instruction is to include for any disturbance cost.
(b) Valuation of variation instructions: condition 42 provides two options. The first is valuation on the basis of a quotation from the contractor, which must show separately (i) the direct cost of compliance and (ii) any cost of disturbance affecting the variation or related unvaried work. There is no statement about the price level of this quotation, although it is subject to examination before implied agreement and acceptance.

(c) Alternatively to (b), condition 42 provides for valuation in the traditional way, based upon the data in the bills of quantities and so at the price level of the bills, but including price adjustment for disruption of unvaried work affected.
(d) Valuation of other instructions, not producing variations: condition 43 does not refer to disturbance elements, although these may be present and condition 41 allows for them, but to when the contractor 'incurs any expense' or 'makes any saving'. Expense is so defined as to exclude interest or financing charges.
(e) Increase in the contract sum due to particular causes essentially stemming from the authority's side: condition 46 allows direct expense due to these causes to be determined and reimbursed.

In items (a) to (d) instructions are required to be valued individually with inclusions for disturbance or something rather similar (sometimes specifically given separately), when there may be numbers of instructions possibly, even probably, with overlapping effects. The problems of allocation or juggling this is likely to produce have been alluded to in discussing them. Only under condition 46 does it appear that a more global approach (Chapter 14) may be adopted. In all cases any reimbursement is limited to either the cost of disturbance or expense incurred, these elements being *exclusive of loss* in the sense in which it figures under the JCT provisions over loss and expense.

Not only are interest and financing charges as allowable under the JCT arrangements excluded in these cases generally (and under condition 43 specifically), but condition 47 (not discussed here) entitled 'Finance charges' defines and uses the term to relate to payment to cover the extra cost due to delayed payments to the contractor in given circumstances elsewhere within contract operation, hence it is outside the scope of expense due to disturbance.

CHAPTER 21

ICE MAJOR WORKS CONTRACT

- Contract basis and administration
- Site and ground conditions
- Programme and time
- Instructions and payment

The form taken in this chapter is fully titled Conditions of Contract and Form of Tender, Agreement and Bond for Use in Connection with Works of Civil Engineering Construction, but it is very often known simply as the ICE contract, by inference 'major'. It is the most common form used for major civil engineering works in the United Kingdom and is considered in its sixth edition of 1991. In addition there are variants for minor works and for design and construct projects, and a form based on the major form is used for projects internationally. Only the major UK version is considered here. Although it is intended for civil engineering projects and would be fairly difficult to apply to other projects, it is included in a book about *building* claims and disputes because of the background comparisons which it introduces in places by its variant methods.

CONTRACT BASIS AND ADMINISTRATION

The contract is prepared jointly by the Institution of Civil Engineers, the Association of Consulting Engineers and the Federation of Civil Engineering Contractors. It is thus a consensus document – similar to the JCT form and different from the GC/Works/1 form – so it is not likely to be construed contra proferentem as the second form might be. The ICE form has been modified relatively little over its life since 1945 and it is worded much more verbosely and in places loosely than the other forms. It has featured in a moderate number of legal cases and the decisions have not usually been based on the minutiae of the wording, as has happened to its more judicially battered JCT cousin.

Short of legal proceedings, the contract is often interpreted with a degree of tolerance, which probably springs from its origin between bodies of comparatively like-minded members and its more chatty wording. Nevertheless, there are areas where quite strict interpretation is needed and these, as might be expected, are particularly those to which this book is directed.

Persons

The key figure for design and supervision of the works for the client (again known as 'the Employer'), for acting between the parties over the standard of the works and performance issues and for exercising the functions allocated to the quantity surveyor by other contracts, is 'the Engineer'. He also has more powers affecting the technical conduct of the works, particularly over temporary works and groundworks, reflecting their significance in civil engineering works.

However, he may and usually does delegate his powers widely to 'the Engineer's Representative', usually the senior person on-site. There are a few exceptions and the engineer may also at his discretion take his powers back. The precise state of affairs is to be given to the contractor in writing and he should check them carefully. The engineer and his representative may have 'any number of persons' as 'assistants' but their powers are limited, so they are effectively clerks of works who may issue instructions related to their own roles only. Here the distinctions are ignored.

Contract documents and instructions

Documents are governed by clauses 5 to 7, but are defined in the Form of Agreement. They are like those for the GC/Works/1 contract, in that they include the tender and acceptance and the specification, as the JCT contract does not. The contract is always based upon approximate quantities subject to complete remeasurement, because of the nature of civil engineering works. Clause 5 states that 'the several documents . . . are to be taken as mutually explanatory of one another', and the engineer is to instruct over any 'ambiguities or discrepancies'. This means there is no defined order of precedence for the documents, which is probably as well since their technical detail may be quite incomplete as originally issued.

There is not the same precision about instructions as a class in this contract, although the engineer or his representative, if there is delegation, may issue instructions. For instance, under clause 7(1) he may 'supply . . . instructions as shall in the Engineer's opinion be necessary' for performing the works. The term 'instruction' is used comparatively rarely in this contract, often some other term is used, such as 'require', 'order' or 'direct'.

These instructions, however named and like other matters in the conditions, may lead to delay and extra cost. An instance is in clause 7(4), following on from clause 7(1) when instructions and other information are late. There are several references in the conditions to delay and cost, and these are considered together under 'Instructions and payment', along with the financial effects of instructions and related matters. The question of differences in definition of the term is considered under the heading of 'Extra cost' at the end of this chapter.

SITE AND GROUND CONDITIONS

Site

Clause 11(1) states that, in advance of tendering, the employer is 'deemed to have made available to the Contractor . . . all information on the nature of the ground and sub-soil including hydrological conditions obtained by or on behalf of the Employer', so there is to be a high level of disclosure. Should information be withheld or misrepresented so as to influence the level of the tender, the contractor will have grounds for a claim. However, it is up to the contractor to interpret the information provided on the basis that it is correct and adequate.

Counterbalancing this, under clause 11(2) the contractor is 'deemed' to have performed his own inspection of the site and its surroundings before tendering, 'and to have satisfied himself so far as is practicable and reasonable' about relevant matters. These include 'the ground and subsoil', so that he is to take account of what the employer has provided, but not necessarily limiting himself to this information, as is summarised in clause 11(3)(a). There remains a conceivable area of uncertainty between what the employer might have provided but did not, and so be responsible for, and what the contractor could in any case have discovered and so have taken into account without being advised.

Ground conditions

Aspects of clause 11 form part of the basis for clause 12 'Adverse physical conditions and artificial obstructions'. The substantive body of the clause does not employ the word 'adverse', but clearly that is what is relevant here. It is not expressly limited to ground conditions, but it is difficult to envisage other physical conditions without straying into the artificial. The clause thus has an overall scope similar to condition 7(3) of GC/Works/1 and the comments made on that condition may be followed and differentiated here, and those made in Chapter 4 in a JCT context and in relation to SMM7 may be also noted so far as appropriate in civil engineering measurement.

The present clause is also in step with the effect of the GC/Works/1 condition by excluding from its scope for redress 'weather conditions or conditions due to weather conditions'. In all of this it is what 'an experienced contractor' should have foreseen that forms the criterion for what is deemed to have been included in the tender and so in the contract rates. This may be seen as explicitly slightly stronger than the GC/Works/1 term. The NEC contract follows the same philosophy of referring to the 'experienced contractor' standard.

The clause here allows for net reimbursement of extra cost arising, although much of the reimbursement is likely to be through measured or daywork items based on the bill of quantities, which will therefore automatically include for profit. By comparison, the GC/Works/1 pattern is to use its condition 42 valuation of variations method to arrive at the reimbursement, either by a quotation which includes both 'direct cost' and 'disruption . . . or prolongation' or by 'measurement and valuation' with prices adjusted if appropriate. ICE and GC/Works/1 approaches should each be used to give comparable results. A case that showed the need for consistent and appropriate information on ground matters was *Holland Dredging (UK) Ltd* v. *The Dredging & Construction Co. Ltd and Imperial Chemical Industries plc (third party)* (1987), where delay occurred over determining the proportion of backfill that required to be imported due to a shortfall from excavated material that had been indicated would be available to the contractor from information supplied in tender specifications and bills of quantities. This caused a delay for which the contractor was reimbursed.

PROGRAMME AND TIME

Programme

A programme is to be provided for the engineer's acceptance within twenty-one days of the award on the contract under clause 14(1), supported by a general statement of construction methods. The engineer has authority under clause 14(2) to reject the programme only and require a revision within a further twenty-one days or to seek further information. This process has several loops built in and may take quite some time to complete in view of the complexity of much civil engineering work.

During progress, the programme is significant in that the engineer may require the contractor under clause 14(4) to revise it to bring progress of the works or any section back on time when they are falling behind. A high degree of realism in the original programme preparation is therefore necessary, and the contractor needs to be alert to any unreasonable modifications which the engineer may seek at that stage. It is also reasonable that revisions during progress should be sufficiently flexible to allow for changes found to be necessary in view of the way in which the project has unfolded. A review of these several aspects may even indicate that some apparently necessary revisions are actually superfluous. The possibility of a constructive acceleration may occur if adequate use of the extension of time arrangements is not made. Acceleration of the works or a section can occur contractually only if the contractor agrees to it within an agreement made under clause 46(3). Again, there may be need for flexibility in trading off one subsidiary adjustment that shortens the programme against another that extends some part. These provisions are not matched at all in the JCT scheme and are worded so as to give the engineer more obviously directive powers than under the GC/Works/1 contract, although in practice this is softened by the existence of the procedures for settlement of disputes. There is no pattern of progress meetings as under the GC/Works/1 contract to deal with these matters; they would be dealt with as usual by arrangement between the parties, or as set out in the other contract documents.

The requirement for the contractor to give the general statement of construction methods is added to in clauses 14(5) to 14(7) by provisions about amplifying this statement in various ways throughout construction, including over design matters which fall within the contractor's orbit. These are outside present consideration, but may in themselves lead to programme effects and also to cost effects. Clause 14(8) therefore provides a 'delay and cost' remedy (discussed under 'Extra cost' at the end of this chapter) but it does not apply to the programme provisions as such.

Several clauses deal with the inevitable elements of the programme. The commencement date is covered in clause 41 by three options: the specified date in the contract appendix or, if none is given, within twenty-eight days of the award of the contract or such other date as is agreed. The contractor would be advised to look for the last if the first does not apply, as the second might creep up on him after a period of delay not of his making and when the balance of his commitments had shifted. There is provision for phased possession of the site, but this is to be in accordance with the contractor's programme as accepted, which adds to the significance of the programme. If the works are spread across a wide area, this can be a very necessary arrangement. Failure to give possession leads to a right for the contractor to be reimbursed.

Completion under clauses 43 and 48 is of the whole of the works or of any sections specified within it and the contract differs from others by using the concept of substantial completion. This is less onerous than practical completion by allowing suitable elements of work, such as landscaping dependent on seasonal and special weather conditions, to be still going on during what is here called the defects correction period. It is more onerous by requiring that any work included in a certificate of substantial completion shall have passed any prescribed final test.

Extension of time

Like the GC/Works/1 contract, the ICE contract has a much shorter set of provisions than the JCT contract, to deal with extension of time, but it may be none the worse for that. Under clause 44(1) it requires the contractor to apply to the engineer for an extension 'within 28 days . . . or as soon therafter as is reasonable' of a cause arising, but in any case giving 'full and detailed particulars in justification' of what is claimed. This may be noted as an example of the looser provisions of the contract, although it does recognise thereby the problem of early assessment. The 'particulars' refer to calculation of the extension claimed and in that context will also state the cause.

The various causes given in the clause are as follows:

(a) Variations (which are very widely defined under clause 51(1)).
(b) Increased quantities (over the approximate ones in the contract, rather than over what the design may actually have shown all the time).
(c) Any cause of delay referred to in the conditions; these are all of the causes leading to reimbursement of extra cost, hence they are causes which originate from the employer's side; they are listed under the heading 'Extra cost' at the end of this chapter.
(d) Exceptional adverse weather conditions, so paralleling the JCT contract, unless any distinction be drawn over the arguable reference of 'exceptional' to weather here, rather than 'exceptionally' to adverse there.
(e) Other special circumstances of any kind whatsoever.

The last item may be noted as another loose provision, in the sense that it is wide enough to cover all manner of contingencies, to the point where its legal definition is difficult to pin down and it may be effectively useless. Thus it may be expected to cover the excepted risks under the contract (the risks for which the contractor is not liable to the extent that they cause loss or damage, as in the JCT contract) and misdeeds or failure to act by the employer (such as not making part of the site available on time), but probably not shortages of labour and materials except when due to circumstances on the broad scale of a national emergency. There is no reason to expect it to cover subcontractors of any type, including nominated subcontractors.

Granting of an extension by the engineer is dealt with in clauses 44(2) to 44(4) and there is recognition of sectional completion. There are three stages to his actions:

(a) Granting an interim extension as soon as he can.
(b) Reviewing the position when the due or extended completion date arrives, and advising the employer and the contractor as to whether he is proposing a further interim extension. The question of issuing a variation order or similar action after the due or extended date for completion is covered in clause 47(6) in relation to liquidated damages.
(c) Granting the final extension upon issuing the certificate of substantial completion.

Once the engineer has granted an interim extension, he may not reduce it in the final extension. This goes part way, because it works in one direction, to resolving the doubt which may arise when an extension must be retrospective. It needs care in its operation, but it is probably not reasonable to try to be closer to an ideal scheme.

Rate of progress and acceleration

Clause 46(1) shows a difference in principle from the JCT contract by allowing the engineer to notify the contractor that his rate of progress on the works or some section is too slow so that he is going to complete late, after taking any extensions into account. Presumably the engineer will relate his opinion to the programme which he has accepted and which was produced by the contractor under clause 14. The contractor is then 'to take such steps as are necessary and to which the Engineer may consent to expedite the progress'. If the contractor incurs extra expense as a result, he has to bear this and so needs to weigh up carefully the correctness of the engineer's opinion.

There is no procedure for the contractor to appeal against the engineer's action, but equally nothing which the engineer can do contractually if the contractor does not comply. There has not yet been an actual breach by lateness and this is covered, as usual, by the liquidated damages provisions. The engineer has formed an 'opinion' and dispute settlement under clause 66(1) could perhaps be invoked, but his immediate action is simply to 'notify' the contractor of the opinion and not precisely to instruct him to act on it. However, if the contractor is eventually late in completing, then events at the present stage could yield substantive evidence against him over damages.

With this topic may be noted clause 45 which in general prohibits night and Sunday work, except for technical reasons or cases when it is customary. It is affected by clause 46(2), under which the engineer may concede such working (still at the contractor's expense) to make up time lost by the contractor's own fault, effectively negating it in this case.

Clause 46(3) deals briefly with the question of an acceleration to give completion ahead of the date currently fixed. It simply requires any special terms to be agreed before action is taken, which is the implicit position, since acceleration gives a modification of a basic contract term. The GC/Works/1 contract and the JCT management contract have acceleration provisions with rather more detail about procedures, but they end with a need for agreement and so amount to much the same as the present short clause.

Liquidated damages

Clause 47 goes about its business mainly in a similar way to its JCT counterpart, and it secures a similar result, in that it relates to an original completion date and extensions due to the specified causes to maintain the right to damages. It allows for deduction of damages from payments or for otherwise recovering them, and it covers reimbursement of any excess damages levied if a review leads to a later date being substituted when, by way of difference, interest is payable. But it also has an elaboration for relating the amount of damages to progressive handing over of the works, supported by details in the contract appendix. Alternative clauses are provided for this:

(a) For the whole of the works when there are no designated sections for completion, under clause 47(1) there is an overall sum for delayed completion, which may be reduced in proportion to the values concerned if any part of the works is certified complete early.
(b) For the works in designated sections for completion, under clause 47(2) there are individual sums for each section, which again may be reduced proportionately.

Clause 47(3) attempts to provide that all the sums constitute liquidated damages and not a penalty. As set out in Chapter 10, this has no effect, the question being one of the reasonableness in any instance of the sum inserted as a pre-estimate of likely loss for the employer.

Clause 47(4) is entitled 'Limitation of liquidated damages' and in a way is as odd as its title promises. It refers to the sums to be entered in the form of a tender appendix to support the main provisions of whichever out of clauses 47(1) or 47(2) apply, and it may contain several possible pairs of sums:

(a) One pair for 'the whole of the Works'.
(b) Further pairs for individual sections of the works.
(c) A final pair for 'the Remainder of the Works'.

Of these the first is for clause 47(1) and is an alternative to the rest as a group, which are for clause 47(2). The appendix gives the pairs of sums as 'per day/week' and as 'limit of liability'; with the possibility of the second being deleted or left blank (with the same effect) in any or all cases. The clause indicates that the limit is that of the total amount of damages for the whole or section concerned, effectively capping the number of days or weeks to which damages will apply. It seems rather pointless for the employer to contemplate this arrangement unless, remotely, he is concerned that damages will constitute a penalty beyond some time. Alternatively, it may be conjectured that he will consider the reverse to apply and wish to seek unliquidated damages for stepped damages beyond the time, in which case some contract entry would be more likely to succeed by giving the contractor positive warning.

It may be that the group covered in (b) and (c) will add up to the sum which would otherwise have applied for the whole of the works, which is the assumption made by the partial possession provisions of the JCT form. But it may well be that some section of the works taken alone has an effect for the employer of greater or less consequence than as part of the whole, as the JCT form does not expressly provide, although it can be accommodated with a small measure of plastic surgery. With this goes the further possibility of two or more sections overlapping in their effects, so that some adjustment along the lines facilitated by the limit of liability options may be desirable to avoid an objection that the combined liquidated sums are jointly a penalty. These immediate comments relate not so much to dealing with disputes, as to seeking to ease them in the first place, always a desirable concept.

Clause 47(5) allows the employer to deduct damages from payments which he makes, but also covers the payment back to the contractor of damages which turn out on review to be in excess of what is due. There is then an addition for interest to the net amount.

Clause 47(6) deals with the position when variations are issued before completion but after liquidated damages have started to be payable, that is during culpable delay. It provides for a suspension of the payment of damages for a period sufficient to compensate for the additional delay to completion caused by the extra work, but without this suspension affecting the contractor's liability for damages in the preceding or succeeding periods. This gives the equivalent of an extension of time, as may happen with variations in general.

INSTRUCTIONS AND PAYMENT

As already noted, the term 'instructions' is rarely used in these conditions, although several other terms are used with, it may be inferred, intended similar effect. Furthermore, there is no single definitive list of items which fall within the scope of these terms, as in the GC/Works/1 form, and no network of matters specifically labelled 'instructions' or 'instructed', as occurs in various clauses of the JCT form.

Alterations additions and omissions

Clauses 51 and 52 cover in turn, but with some overlap, several main topics:

(a) Ordering variations and the related question of change without variations from original to final quantities in this all-remeasurement contract.
(b) Valuing of variations as ordered and of their effect on work outside their immediate scope.

(c) The procedure when the engineer and the contractor differ over prices relating to variations.
(d) Notice of claims and keeping records, that is essentially administrative matters, with claims meaning two distinct things: the contractor seeking a higher price for a variation-related item, but also extra cost items covered in various places in the conditions and treated under the heading of 'Extra cost' (in most contracts this second group is not associated directly with the wording in the heading to this section, even here its substance is still elsewhere).

The process of measurement or remeasurement is covered by clauses 55 to 57, considered hereafter.

Clauses 51 and 52 use the term 'order' or its variant grammatical forms to convey what is done to bring variations into effect. An order here may be made by the engineer for two sets of reasons within his opinion: when 'necessary for the completion of the Works' (afterthoughts, perhaps) and when 'desirable for the completion and/or improved functioning of the Works'. These reasons give him plenty of scope, but conceivably may preclude other options, so limiting him. Other contracts allow and define variations, but wisely do not attempt to justify them.

Ordering variations is dealt with in clause 51 and they are defined in character quite widely to include the terms in the heading to this section and more relating to changes in the physical works produced, and also 'changes in any specified sequence method or timing of construction'. This definition thus broadly parallels in effect equivalent wording in the JCT and GC/Works/1 contracts over matters of working order and so forth, although it only permits changes to what has been 'specified', that is in the contract documents as such, or in preceding orders. It goes further by the inclusion of 'method' which may conceivably cover changes in type of plant or other constructional techniques, especially when they relate to temporary works where the engineer has noticeably more powers, particularly under clause 14(6), but again there is a limitation to what has been specified.

A distinct addition to what is permissible under the JCT and GC/Works/1 contracts is that variations may be ordered 'during the Defects Correction Period', but this does not mean the extra work can necessarily be completed within that period. This is due to the greater tendency in civil engineering works for additional work to be found necessary when they start to function. Under the other contracts, the contractor is not obliged to take on further work once completion has occurred, even at enhanced prices.

Increase or decrease in quantity of work does not require a variation order when due to a simple discrepancy between what is needed on-site and what is in the bill of quantities. This needs to be seen against the background of the use of approximate quantities as the sole contract basis in this form, where accuracy is almost always a relative term and where it would be pointless to trace every up or down through the variation process. Strictly, it frequently leaves a grey area between what is clearly in the bill (even if inaccurate) and what is clearly not, so constituting a variation. In most instances this will be taken up by a variation omitting the whole of the original portion of work shown on the drawings and described in the specification, and substituting what is required.

Valuation of variations is the first subject of clause 52 and, by implication, remeasurement of work included in the original bill (conceptually distinct from variations as just discussed) is to be priced at the prices in the original bill. Rules for the valuation of variations in clauses 52(1) and 52(2) embrace the usual options as follows:

(a) Bill rates are to be used for 'work . . . of a similar character executed under similar conditions' to bill work, thus rendering a distinction from unvaried, but remeasured work effectively superfluous. The term 'similar' carries overtones, so that pro rata pricing may be applicable in some but not all cases.
(b) Where work is not similar in both character and conditions of execution, pro rata pricing will be the basis if possible. Specifically, work ordered during the defects correction period is to be treated under this rule.
(c) If neither of the foregoing can be used for pricing other measured work, a fair valuation is to be made.
(d) If any variation by its 'nature or amount' relative to the contract work, in whole or part, renders any original price 'unreasonable or inapplicable', then that price for the work is to be varied. This relates to situations such as changes of scale, timing in the programme and intricacy effects (see also clause 56(2) mentioned below). In any such case either the engineer or the contractor is to notify the other of the desire to seek a price change.
(e) Daywork is available upon the engineer's orders to cover work where he considers it 'necessary or desirable', whatever the distinction implies. The working details for calculating daywork are in clause 56(4).

These rules broadly run down the same cascade as in the traditional JCT set but without so much elaboration, rather than that in the GC/Works/1 set, where disturbance is included. Although they refer throughout to prices, there may well be instances (especially under (d) above) where lump sum or other adjustments may be the reasonable way to deal with cost implications which do not relate directly to the quantities of measured work as such. These points are brought out in Chapter 6, which discusses the various detailed issues in the JCT clauses.

It is the procedure that the engineer is due to notify the contractor of the prices which he is 'determining' or 'fixing', if they are unable to agree on them. However, by an unusual use of the term, the contractor may still notify a 'claim' within twenty-eight days under clause 52(4)(a) that he is seeking a higher price. Nothing else is said immediately about further discussion and the provision is possibly reserving the right (which would be there anyway) to seek resolution under the disputes procedure, although sensibly a further review between the engineer and the contractor is likely to happen.

The rest of clause 52(4) deals with claims for 'any additional payment' in more nearly the regular sense by setting down the procedures for such calculating and for making such payment 'pursuant to any Clause' other than clauses 52(1) and 52(2). The procedures are more precise than those in the JCT form by requiring notices and records, the records either at the contractor's discretion ahead of notice or as required by the engineer thereafter. Progressive accounting is required, although it is recognised that not all records will be of direct costs, as by implication there will be other records to be used for the equivalent of JCT ascertainment at the end of the period of extra cost. As discussed in Chapters 11 and 12, 'ascertainment' is not always a precise description of what can happen in practice.

Measurement

Clauses 55 to 57 complement those just discussed and, especially as they say nothing unexpected, could well have been combined with them to avoid the reader referring between them. Clause 55 clears the ground by rehearsing that the original quantities in the bill are estimated only and that any error in description or omission from the bill (other than a pricing error) is not to vitiate the contract, but is to be corrected. The value of work is then to be dealt with under clause 52, presumably after clause 56 has been followed to give measurements.

Under clause 56 the engineer has the duty to 'determine by admeasurement the value' of the work performed. Here 'admeasurement' appears to include pricing action, as already set out in parts of clause 52. There is provision in clause 56(2) for changing bill prices when this is made suitable by increase or decrease in final quantities compared with the original bill. This shadows the provision of clause 52(2) that there is to be similar change when the effect of variations on unvaried work justifies it.

Otherwise, the clauses deal with procedural matters, the pricing basis for daywork and the applicable standard method of measurement.

Extra cost

Clause 1(5) defines 'cost' for the purposes of the contract and a number of clauses rely significantly on this definition, which is 'all expenditure properly incurred or to be incurred whether on or off the Site including overhead finance and other charges properly allocatable thereto but does not include any allowance for profit'.

This definition lines up closely in scope with the several categories set out as primary and secondary loss and expense at the beginning of Chapter 14, dealing with JCT provisions, and that chapter and the next may be followed for detail. The reference to profit is to the addition of a profit margin to defined costs otherwise net, rather than to loss of profit on other work foregone, so lining again with the JCT situation. However, like the GC/Works/1 provisions, the definition does not stretch as far as covering 'loss' (whether of profit or other items), because it uses the term 'expenditure' as central in its definition of 'cost'. That other form employs the term 'cost' in condition 42 and 'expense' in conditions 43 and 46, whereas it employs 'finance charges' in condition 47 to refer only to a quite distinct category of expense from what is due to the causes now being considered.

Despite the absence of 'profit' in the definition of 'cost', most of the causes attract at least a partial profit allowance under the individual clause provisions. This is profit on 'any additional permanent or temporary work', which will usually be dealt with on the basis of measured items or daywork and so automatically include a profit element. Beyond this, 'cost' attributable to these causes, such as overhead elements, will not attract a profit allowance. The exceptions to this, giving more or less areas in principle for profit, are noted under causes (a), (b) and (e) below.

All of these causes leading to extra payment also permit extension of time, by virtue of the inclusion under clause 44(1)(c) 'any cause of delay referred to in these Conditions'. There are other causes leading to extension under clause 44(1), but they do not lead to extra payment. This may be noted as showing that there is no automatic link between delay and disturbance in these respects, as is emphasised throughout this book.

The various causes are as follows:

(a) *Delay in the issue of drawings, specifications and instructions (clause 7(4)(a)):* No addition for profit arises in this case, as the permitted reimbursement is limited to 'the amount of such cost as may be reasonable'.
(b) *Adverse physical conditions and artificial obstructions (clause 12(1)):* The follow-on in clause 12(6) provides for reimbursement of costs 'with a reasonable percentage . . . in respect of profit'. This must be read as applying directly only to elements reimbursed on the basis of costs incurred, as some elements, perhaps most, will be reimbursed through measured items (see under 'Ground conditions' above) which will include a profit margin.
(c) *Ambiguities and discrepancies in the contract documents (clause 5) and legal barriers or physical impossibility in constructing the works (clause 13(1)):* These are treated together under clause 13(3) and clearly the second set also must relate to the requirements of the contract documents.
(d) *Furnishing consent to the programme (clause 14(7)):* Extra cost may be reimbursed under clause 14(8) if it arises out of aspects of clause 14(7). These are delay in the engineer giving consent to the contractor's methods of construction and design limitations he may impose which the contractor 'could not reasonably have foreseen' when tendering.
(e) *Variation entailing public street works (clause 27(6)):* The introduction by variation of work in a street or similar situation will mainly be valued like any other variation. However, any extra cost is covered here although, as above under clause 4(7)(a), no profit is added.
(f) *Facilities for other contractors (clause 31(2)):* If possible, the works of others whom the contractor is to allow on the site should be identified in the contract documents, so that the contractor can provide for them in terms of what he will have to cover in cost or in programme organisation. If this is not done, the contractor still has to allow them to operate if they are introduced during progress, with the cost covered here. This gives a total arrangement similar to that under the JCT form.
(g) *Suspension of work (clause 40(1)):* The engineer may order suspension of the works or some part in time and manner which is at his discretion. The contractor is entitled to reimbursement of extra cost, except when the

contract provides otherwise or when suspension is on account of weather, the contractor's default, safety of the works (unless the risk is due to the employer, the engineer or the excepted risks). Suspension for more than three months gives the contractor a right to treat a part suspended as omitted as a variation or the whole suspended as an abandonment of the contract.

(h) *Failure to give possession (clause 41(3)):* This again relates to the whole or a part. Although this cause is somewhat similar to suspension (but occurs before work even starts), it does not give a right to treat work as omitted or abandoned.

CHAPTER 22

THE NEW ENGINEERING CONTRACT

- Contract basis and administration
- Site and ground conditions
- Programme and time
- Instructions and payment (compensation events)
- Settlement of disputes

The NEC is a new form of construction contract that is radically different from other standard forms. It is relatively new to the UK construction industry. It can be used for building and civil engineering projects. The New Engineering Contract (NEC) is a suite or family of contract documents and the acronym NEC is properly used to refer to all documents published in that family. This chapter deals only with the contract for construction works, entitled the Engineering and Construction Contract (ECC), and even though this contract is still generally known as the NEC contract, it is referred to in this chapter by the abbreviation ECC. The abbreviation NEC is used to refer to the total family of documents or to the philosophy or concept behind the NEC documentation.

NEC documentation uses capital letters for defined terms (referred to later) but this is not followed in this chapter; after the first use of the term with capital letters subsequent references are in lower case. The aim of the chapter is to compare principal provisions in the ECC main contract with those in the JCT main contracts.

CONTRACT BASIS AND ADMINISTRATION

The NEC documentation is published for the ICE. It cannot claim to be a consensual document, that is one which has arisen from consultation with a number of recognised industry bodies leading to an attempt at a balance of a range of contractual and consultancy interests, although consultation took place before its authors, the NEC Panel, produced it.

The family of documentation seems comparatively large, when compared with other forms of standard contract and when viewed at first sight in its 'boxed set', containing fourteen separate NEC documents, in total around eight hundred pages.

In 1998 no record was found of a published law case on the NEC documentation. The future may produce some, particularly if clients other than those who have embraced 'partnering' and other 'long-term relationships' procurement styles, begin to use the form. Ultimately this will be necessary in order to make possible comparative comments based on experience of a wide cross-section of consultants, contractors and clients, unless the form is to remain one used by a few clients.

The following comparison has been made by considering aspects of only the JCT main forms and only the ECC documents for option B (using bills of quantities that are remeasured on completion of the project) with occasional reference to the ECS (the NEC subcontract form) and the NEC adjudicator's contract.

Persons

The NEC guidance notes explain, in their words, 'The NEC System – key players and contractual links'.

Apart from the usual contracting parties, 'the Employer' and the 'Contractor', there are also roles and duties for four other organisations or people in the NEC documentation:

(a) The 'Project Manager' (PM) who manages the project on behalf of the employer with the intention of achieving the Employer's objectives for the completed project.
The role of the PM is crucial to the success of the ECC. Ultimately his action or inaction can be referred to the adjudicator. In *Balfour Beatty* v. *Docklands Light Railway* (1996) it was emphasised that the engineer may be strongly identified with the employer and has a duty to act fairly. The employer may replace the PM after notice to the contractor.
(b) The 'Supervisor' who acts for the employer in maintaining quality control. Projects will not justify his full- or even part-time presence on a site. Disputed actions of the supervisor, like those of the PM, can be referred to the adjudicator (clause 90.1, settlement of disputes).
(c) The 'Adjudicator' who can be referred to by the contractor if he feels that a decision by the PM or the supervisor is not in accordance with the contract, or by the contractor and/or the PM if either has a difference on any other matter arising from the contract.
(d) The 'Sub-contractor' of whom there may be none on small projects and several or many on large projects.
Designers are not given a status or direct role in the ECC.

Contract documents

A criticism that has been made of the ECC is that it is not possible to be certain what the contract documents are. Although there is no definition of 'contract' in the NEC it can be inferred that the conditions of contract are the core, main and secondary clauses possibly together with schedules and supporting contract data. But it appears unclear how far the supporting detail will be interpreted as part of 'the contract documents'. The ECC, in the section Contract Data, clause 1 General, states: 'The conditions of contract are the core clauses and the clauses for Options . . . of the second edition (November 1995) of the NEC Engineering and Construction Contract.'

Taking the NEC at its word, it would therefore appear that the contract documents comprise core, main and secondary option clauses with supporting schedules and contract data. All other documentation referred to above, whether produced by the NEC or by the employer, is not necessarily part of the contract documents. Each set of documentation needs to be looked at in its context and on its merits. This is unfortunate, and during the course of some dispute, perhaps clarification will be given.

The ECC has several elements:

(a) Core clauses that are used in all of the six options of the ECC contracts, grouped into nine sections.
(b) Six procurement options:
   (A) *Priced contract with activity schedule*: A lump sum contract ideally suited to employer's design but could be used for employer's design or divided design.
   (B) *Priced contract with bill of quantities*: This requires that employer has design carried out before seeking construction tenders.
   (C) *Target contract with activity schedule*: Similar to option B.
   (D) *Target contract with bill of quantities*: Some design required before work starts and final figure is cost-based.
   (E) *Cost reimbursable contract*: Allows development of design as the works proceed.
   (F) *Management contract*: Allows some split of design between consultant and subcontractor secondary option clauses.
(c) Fifteen secondary options, some of which contain matters that other contracts treat as essential, and should arguably therefore be within the main option clauses, such as retention from interim payments on account and delay damages (liquidated) for late completion.
(d) A numeric numbering system governed by the nine sections of the ECC core clauses listed above. For instance, section 1 'General', commences at 10.1 and each section within the core clauses has its own number, e.g. 11.1, 11.2. If there are subclauses within a clause, for instance subclauses of clause 11.2 are numbered 11.2 (1), 11.2 (2) etc.
(e) The standard contract usually contains definitions and the ECC contains, in core clause 11.1, something approaching this, namely 'identified' and 'defined' terms:
   - *Identified terms* are particular to the contract, such as names, details of the site etc, which are identified in the 'contract data'.
   - *Defined terms* are terms which are general to the NEC contracts and given a particular meaning in clause 11 of the ECC.

   There are not as many terms and words as seem to be necessary or helpful in order to achieve better understanding of the documentation. For instance, 'instruction' and 'compensation event' are not included, either as an identified or as a defined term. Compensation events are crucial to the ECC and are insufficiently defined in the contract. Such terms if not identified or defined can be deduced, in part, by reference to text in other ECC clauses or by reference to the guidance notes, which are of course not part of the contract.

(f) Several other schedules, data sheets etc.
(g) Bill of quantities (for use with option B) are produced by the employer from design that exists at the stage when tenders are sought. Option B is a priced contract with a bill of quantities in what is known as a remeasurement contract.

Only this option is considered below in relation to the JCT main contract forms, under a similar framework to that used for comparing the ICE, GC/Works/1 and design and build forms of contract in Chapters 20, 21 and 23.

SITE AND GROUND CONDITIONS

Site

Under core clause 11.2 (6) the employer must supply 'Site Information', within the 'contract data', and this will probably include borehole reports, test results, physical conditions within the site and/or its surroundings. This is normal information given by an employer to a contractor from information that the employer has had to find out in order to progress the project thus far. If it turns out to be sufficient or not contradictory, there may be no problems. If not, there are provisions in the contract to deal with this.

Section 6 deals with 'Compensation Events' and within these are included site (and/or ground) conditions that turn out differently from that given at contract stage and how these will be taken into account within the provisions of clause 60. The effect of 'compensation events' is discussed below.

Clause 60.1 (7) requires an instruction for dealing with objects of value or historical or other interest found within the site. If the contractor encounters physical conditions within the site, which are not weather conditions and which an experienced contractor would have judged had a small chance of occurring such that it was unreasonable to have allowed for them, then the PM, under clause 60.1 (12), must give instructions on how to deal with them.

Clause 60.2 states that the PM will, in judging whether or not he should issue a compensation event, will assume that the contractor will have taken into account the site information, publicly available information, information available from a visual inspection of the site and any other information that an experienced contractor could reasonably be expected to have or to have obtained. These are wide-ranging assumptions for the contractor to have made, particularly when judged against the 'experienced contractor' standard of assumptions, and it is possible they could lead to disputes between the PM and the contractor, disputes which ultimately may require the services of the adjudicator or a tribunal, in accordance with the contract provisions. This is similar to the provisions/stance in ICE and GC/Works/1 contracts.

Clause 60.3 states that if there is inconsistency within the site information, it will be assumed the contractor has taken into account the physical conditions more favourable to his doing the work.

Section 7, clause 73.1 states that the contractor has no title to an object of value or of historical or other interest found within the site.

Ground conditions

The largest area of risk for the parties, but under the ECC principally the contractor, is generally in the ground conditions. Provisions for the site, outlined under 'site' above, apply also generally to ground conditions. Piped and other services below ground, publicly available information on the ground and its surroundings will probably be referred to in the contract data section of the tender information. Information on adjacent building structures, or other buildings on the site that are not part of the works, and so on, may be given if these constrain or require the contractor in his methods of working. Interpretation and judgement of the background and reasons for considering the need for a compensation event, when circumstances are found not to be the same as stated or implied in the contract, will be made against the caveat (as stated under 'Site' above) of 'the experienced contractor' rule. Ground conditions often turn out to cause more difficulties than a contractor has assumed he will encounter. The 'experienced contractor' test is a concept that is contained in the ICE and GC/Works/1 contracts and, like the word 'reasonable', may become open to some interpretation. Building works usually encounter less uncertainty in the ground but there may be some 'experienced contractor' tests waiting for the unwary in the superstructure of many a building contracted under the NEC system.

PROGRAMME AND TIME

Programme

The ECC section 3 is headed 'Time' and deals with several important matters in addition to the programme and extensions to the programme. However, not all provisions on time and programme are contained in section 3.

Section 6, dealing with compensation events, also contains provisions on programme and time. The ECC contains a principle that the contractor is managing the works and is therefore the party best able to know when the programme is not proceeding to plan. If the contractor considers that some event has changed or will change the completion date, he must notify the PM and revise his last 'accepted programme' in detail to justify obtaining, in the JCT terminology, an extension of time.

Clause 31.1 deals with the programme and details what the programme should contain, when it should be revised and submitted by the contractor for acceptance by the PM. Programme information will probably be a compilation of documents including bar charts, networks, method statements and resource histograms, almost certainly all generated on computers.

Section 1 General, clause 11.2 (3) defines the 'contract date' as the date when 'this contract came into existence'. (It seems inconsistent in an ECC document which is generally written in the present tense not to say 'comes into existence'.). This is not as precise as having a date agreed by the parties or one given in the tender documents. Naturally, it is important to know the 'contract date' because, as with the JCT and all other standard contracts that have time provisions, it is the datum from which other matters are measured and judged.

It is emphasised that the contract date should be established quite clearly. In a contract claimed to be drafted in clear, simple language, the provisions for time and the effects of changes on programme and time are complex, and this area of potential uncertainty should be avoided. In the NEC documentation there are thirty-nine flow charts that explain, among other things, programme and time.

Clause 30.1 sets out the contractor's obligation for completion of the contract as 'not to start work until the first possession date and does the work so that Completion is on or before the Completion Date'. (This extract from the ECC text is an example of the present tense that is used in drafting the NEC documents.) Thus there is a clear contract obligation for the contractor to carry out the works required and complete on or before the completion date.

Clause 31.1 requires the contractor to submit and update the accepted programme. Clause 33.3 requires the PM to respond to the submission of a programme within two weeks. If the PM does not respond on time, his late response is a compensation event. The PM can reject the programme with reasons. Clause 31.2 requires the contractor to identify in the programme dates when he will need matters to be provided by the employer.

Clause 30.2 provides for the PM to decide the date of completion and he is required to certify this within one week of completion. There is no provision for deemed completion.

Clause 33.1 requires the employer to give possession of the site on or before the relevant possession date.

Clause 10.1 requires the employer, the contractor, the PM and the supervisor to act in a spirit of mutual trust and cooperation. This is the philosophy of the NEC and for once has been written into the contract as an obligation on both parties and the employer's agents to act to a common purpose and trust.

The accepted programme under the ECC is an important document and is the basis for the PM making judgements on revisions to the completion date and compensation events. From the information supplied by the employer in the contract data (part one), the contractor submits a 'first programme'. Quite convoluted provisions (and wording) are made for operation of the programme. Clause 11.2 (14) says the accepted programme is 'the programme identified in the Contract Data or is the latest programme accepted by the PM.' It goes on to explain that 'the latest programme accepted by the PM supersedes previous Accepted Programmes.'

If a programme is not identified by the contract data, it appears there is no accepted programme until the contractor first submits a programme and it is accepted by the PM. Clause 50.3 provides a draconian sanction if the contractor fails to submit and have his programme accepted. One-quarter of the Price for Work Done to Date (PWDD) is retained until the contractor does submit such a programme. Other provisions exist if the content of information in the programme is not as required. Nothing exists in the JCT that requires the detail suggested from these clauses on the construction programme.

Clause 31.1 provides that if a programme is not identified in the contract data, the contractor should submit a programme to the PM for acceptance within the period stated in the contract data. It may be tautology to say there is no accepted programme until there is an accepted programme, but that is what these convoluted ECC clauses state.

Clause 31.2 gives reasons why the PM may decide not to accept a programme or revisions to a programme. The PM can reject a programme if he considers it:

(a) is not practicable
(b) does not represent the contractor's plans
(c) does not show provisions for float and risk
(d) does not show provisions for procedures required by the contract
(e) does not provide for health and safety issues
(f) does not provide for other procedures contained in the contract
(g) does not include the information that the contract requires

Clause 31.3 requires that the accepted programme should include

(a) dates that are stated in the contract data
(b) dates decided by the contractor
(c) method statements
(d) the order and timing of events
(e) any other relevant information

From this it can be seen that any programme and any revision to that programme must be very detailed. Clause 32.1 requires that a revision must record the actual progress made to date for each operation plus the reprogramming intended for future work. The effect of compensation events issued to date should be shown. Each revised programme has to be accepted by the PM.

Clause 32.2 states the points in the contract when the contractor must submit a revised programme. The need for continuous submission of revisions to programmes by the contractor puts the PM in a position of having to respond rapidly to a large amount of information, possibly on large contracts delivered weekly or even more often. The contract requirement to do this for 'compensation events' would seem to have the effect of making a contractor committed to programming and also to ensuring that he shows compensation events whenever they may appear likely. With all of this going on, opportunities for 'claims' are brought constantly to the forefront of a contract. Compensation events are meant to be quoted for in advance, and this puts programming into a very high priority, both for the contractor and for the PM.

Unlike the JCT contract, under the ECC the PM has an ability to interfere with the contractor's progress. Clause 34.1 gives the PM authority to stop and start the work, perhaps for perceived reasons of risk to persons and/or property. If the programme is affected in this way then this instruction may constitute a compensation event, but if the reason for the stoppage arises from the fault of the contractor, he must bear the cost of the compensation event.

The ECC provides for the PM to accelerate the work. If he is concerned about delay that has already occurred to the work, he can instruct the contractor to produce a revised programme, under clause 32.2, to show how, if at all, the lost time can be recovered. Clauses 36.1 and 36.2 provide for the PM to obtain a quotation from the contractor for accelerating the work. The contractor cannot be forced to agree to an instruction to accelerate and there is no remedy if a quotation for acceleration is not produced or if the contractor's quotation is found to be unacceptable to the PM. This last point should go without saying, in a contract that is promoted for its non-confrontational aims. As an aside to this, if there is any possibility that the programme is so important that, if delays do occur then it will be important to try to catch up lost time, options C, D or E of the ECC should be used; this is because they envisage that time is a priority, more than in the other options.

Sectional completion can be included in the contract by detailing each part, called a section, of the contract and the dates for its completion, ahead of overall completion.

The overall provisions for programme and time are that the contractor manages the programme, because he is the best person with the knowledge to take the risk for completion.

Because the ECC has the objective to encourage cooperative working, clause 16 provides for 'early warning' by either party if they foresee possible problems on the programme. The provisions of clause 16 may be significant, where an onus is placed on both the PM and the contractor to notify the other as soon as either becomes aware of any matter which could delay completion, increase prices or impair performance of the works in use. The duty should be exercised by either or both parties when it is apparent, or should have been reasonably apparent, that an instruction to proceed with compensation events may cause a delay or other significant effects on the contract. If, in retrospect, a party wishes to maintain that the effect of an event should reasonably have been foreseen, such that early warning procedures should have been followed, the test for this will be the 'competent contractor test' that is applied, for instance in the contractor assessing allowances for adverse weather. The early warning test may also be applicable to the PM.

Compensation events, 'variations' under the JCT, mainly emanate from the PM and the onus for early warning must start largely with him. The ECC provides that either the PM or the contractor may instruct the other to attend an 'early warning' meeting, along with other persons that can be likewise 'instructed' (it is not known how this provision can this be enforced, without the power of the courts). At an 'early warning meeting' it is intended that by cooperation the parties will consider proposals for how the effect of the event can be minimised by taking agreed actions. The test for calling 'early warning' is whether or not the total of the cost and/or the date for completion and/or the performance of the works in use may be impaired. Here are some examples:

(a) The employer fails to supply information or material to programme.
(b) Failure to communicate within the specified time in the contract.
(c) Worse than average weather conditions (judged by the ten-year test).
(d) Failure of a subcontractor to provide design information.

If the contractor does not comply with the early warning provisions, he can have all or part of the effect of a compensation event withheld from him.

Damages for delay

Provisions for damages for delay, called Delay Damages, are included in secondary option clause R. The amount of damages per day is inserted in the contract data and needs to cover circumstances where the contractor fails to complete the works by the completion date. In the ECC, damages due if delay to completion occurs are not called liquidated and ascertained damages because there are also other provisions for damages, possibly due under circumstances other than delay. These are for Low Performance (option S) and for interest for Delayed Payments (clause 51.2).

INSTRUCTIONS AND PAYMENT (COMPENSATION EVENTS)

The term 'instruction' appears in the ECC but is not a defined or an identified term. The word 'variation' does not appear in the NEC. Instead the equivalent of a variation is called a 'compensation event' and such events can have an effect on both time and money. Compensation events are not listed as an identified or defined term, under core clause 11.1. As sometimes, the ECC guidance notes clear this up by giving, it would seem, a definition, as copied below. Under clause 60.1 there are 18 'events' (given without definition), here listed in abbreviated form:

- (a) The PM instructs a change in the works information.
- (b) The employer does not give possession of part of the site.
- (c) The employer fails to provide something which he must provide by the date required in the accepted programme.
- (d) The PM instructs stopping or not starting work.
- (e) Other organisations do not work to the time shown on the accepted programme and/or within the conditions stated in the works information.
- (f) The PM or supervisor does not reply to a communication from the contractor within the stated contractual period.
- (g) The PM instructs on dealing with valuable, historical or other objects of interest found on the site.
- (h) The PM changes a decision previously given to the contractor.
- (i) The PM withholds an acceptance for reasons not stated in the contract.
- (j) The supervisor instructs the contractor to search and the result is that no defect is found.
- (k) A test or inspection done by the supervisor causes unnecessary delay.
- (l) The contractor encounters physical conditions within the site other than weather conditions which at the contract data an experienced contractor would have judged were unreasonable for him to have allowed for.
- (m) Weather is recorded before the completion date for the whole of the works, weather which has occurred on average less frequently than once in ten years.

(n) An employer's risk event occurs.
(o) The employer uses part of the works before completion and the completion date.
(p) The employer does not provide materials, facilities and samples for tests stated in the works information.
(q) The PM notifies a correction to an assumption about the nature of a compensation event.
(r) Breach of contract by the employer which is not one of the other compensation events.

There are other compensation events included in the Optional Statements section of the contract data.

Unlike the other major standard contract forms, both time and money are dealt with under the provisions for compensation events. So extensions to the contract period will also allow for any additional costs, if appropriate. The PM's authority is wide in changing the works information, considerably wider than is given to an architect or engineer under the other standard contract forms. No authority is given to the employer or to the supervisor to issue changes. Strangely, there is no requirement for a change to be issued in writing.

Clause 80.1 defines Employer's Risks events. These cover actions or inactions by the employer that give reasons for a compensation event.

Valuation of compensation events

For such an important matter, it is surprising there is no inclusion of 'compensation event' as an 'identified' or 'defined term'. The ECC guidance notes state that 'compensation events' are 'events which, if they occur, and do not arise from the contractor's fault, entitle the contractor to be compensated for any effect the event has on the Prices and the Completion Date'. The notes go on to say that the assessment of an event is always of its effect on both prices and on time.

The guidance notes state that a principle of the contract is the PM should be given options by the contractor for the PM to decide how to deal with any problem that arises. It is said that the contractor should be indifferent to which option is chosen; although this seems in theory possible, in practice it may not follow. Perhaps this view arises from experience of the construction industry when a contractor, under the more common standard forms, may make the problem work to his advantage. Under the ECC it is the philosophy that if the contractor is compensated for a particular event he will not then make a problem worse in order to increase his remuneration for reacting to the problem.

Clauses 60 to 65 deal with valuation of compensation for Compensation Events. Once valuation has occurred, the total of the 'compensation' become the mechanism for payment to the contractor for the change that has occurred or been introduced into the contract. Valuation of each compensation event is based on a forecast of the cost and also of the time consequences of solving that event. In the case of option B, which is the most comparable with the JCT main forms, the 'prices' which are the basis of the contract are not necessarily the basis for evaluation of an 'event'. At this point, the contract provides that 'cost', although it will have to be foreseen, is the method of evaluation of compensation events. In many, perhaps most, instances the event is required to be priced and agreed *before* the work is carried out, hence the whole risk of that event is passed to the contractor. Perhaps the uncertainty of taking on all of the risk will mean that a contractor will have to price the foreseen consequences in his favour, if he has any doubt, otherwise he will make a loss, with no opportunity for redress. Although it is argued that the contractor is able to price risk and is in a position analogous to that when he is tendering it may not be so simple. It may be much more difficult to forecast the total consequences of a compensation event, not least the effect on time. The combined effects of several events may not become apparent until some time later in the contract.

Clause 60.4 provides that a change in quantity is not in itself a compensation event. This is as the ICE contract but not as the JCT contracts. A compensation event is only triggered by the changed quantity satisfying the two tests stated in the clause. This clause only applies to changes in quantities which do not result from changes to the works information. A change in the works information is always a compensation event, regardless of the effect on quantities.

Clause 60.5 provides that a difference between original and final quantities is not in itself a compensation event. The final amount due to a contractor includes the PWDD, which is based on the actual quantities of work done. However, any difference in quantities which causes completion of the works to be delayed is a compensation event.

Clause 60.6 provides that if there are mistakes in the bill of quantities because the bill does not comply with the method of measurement or because of ambiguities or inconsistencies it may be a compensation event.

Clause 61.1 provides for the PM to notify the contractor of the compensation event at the time of the event.

Clause 61.2 enables the PM to ask for a quotation from the contractor before a proposed instruction is issued. Naturally the contractor does not put a proposed instruction into effect until it has been agreed by the PM.

Clause 61.3 enables a contractor to notify the PM of an event which has happened or which he expects to happen if he believes the event is a compensation event or the PM has not notified an event to the contractor or it is less than two weeks since the contractor has become aware of the event.

Clause 61.4 provides that the PM will not issue a compensation event to change the prices (the basis of the contract) and the completion date if he thinks the event was the fault of the contractor, or the event has not happened, or it is not expected to happen or it is not one of the compensation events given in the contract. If the PM wants, he can ask the contractor to submit quotations within one week from the time the contractor has notified him (or within a longer period that he can agree with the contractor).

Clause 61.5 is most important as it deals with the 'early warning' concept, mentioned earlier. If the PM considers that the contractor did not give the employer early warning of an event, in the way an experienced contractor could or would, then the PM notifies the contractor of this omission when he asks the contractor for quotations. This will enable the contractor to make his case when he submits his quotation to the PM.

Clause 61.6 enables the PM, in effect, to make a provisional assessment of a claimed compensation event and then review it when he has more information.

Clause 61.7 provides that no compensation event can be issued after the defects date. The concept of asking for quotations before deciding on the issue of an instruction or the value of a compensation event is defended by proponents of ECC. They say that the contractor is in the best position to assess risk and to therefore to take the risk. Sometimes the contractor is asked to submit alternative quotations and then the PM chooses which course to follow. If the programme will be affected by a compensation event then the contractor submits a revised programme with his quotation.

Clause 62.3 requires the contractor to submit his quotations within three weeks of being asked and for the PM to reply within two weeks. The PM either accepts the quotation or asks for a revised one or states that he will not issue a compensation event (i.e. the PM will not go ahead with the proposed change) or he notifies that he will make his own assessment.

These timescales are short and clause 62.5 provides for the timescale of submissions and approvals to be extended if the PM and the contractor agree.

Clause 63 deals with the assessment of the value of compensation events. Clause 63.1 requires that the prices are assessed for the effect of the compensation event upon the actual cost of the work already done, also its effect on the forecast actual cost of work not yet done.

Clause 63.2 provides that if the effect of a compensation event is to reduce the total actual cost, then the prices are reduced only if the compensation event is a change to the works information or a correction of an assumption stated by the PM in assessing a previous compensation event.

Clause 63.3 states that an extension to the accepted programme is to be the amount caused by the compensation event.

Clause 63.4 allows the PM to assess a compensation event where the contractor did not give an early warning as if the contractor had given early warning. Other clauses provide that the assessment of a compensation event should include for time and money, that the risk of items the contractor could see would have a significant chance of occurring should be at the contractor's risk, and that the contractor reacts competently and promptly to a compensation event.

Clauses 64.1 and 64.2 allow the PM to assess the effect of a compensation event in four circumstances. If the contractor has failed to submit a quotation and details to the PM in the allowed time, if the PM decides that the contractor has made incorrect assessments of time and/or money, if no programme changes have been submitted or if the PM cannot accept the effect shown on the accepted programme then, under any of these circumstances, the PM will make his own assessments. This seems to be an area of potential difference between the parties.

Clause 65.1 provides for the PM to implement each compensation event on acceptance of the contractor's quotation or on the issue by the PM of his own assessment.

Clause 65.2 provides that if a forecast turns out to have been wrong, the assessment of the compensation event will not be revised, even though it was based on wrong information.

The provisions for 'early warning' are important in the assessment of compensation events. As 'early warning' is not an identified or defined term, under clause 11, some uncertainty is present. It may be that the obligation on either party may turn out to be onerous on the employer, through his agent the PM.

In summary, the ECC has a quite different approach to evaluation of changes to the works from that adopted in the JCT contracts. The JCT has rules for valuation of variations within a structured framework of work carried out under the same, similar, different or fair valuation circumstances. If the JCT provisions for valuation are inappropriate then provisions for recovery of loss and expense are contained in the contract. This is to follow the mechanisms of English law that provide for compensation in a contract if a breach has occurred. The remedy for breach is damages and the intention is to place the innocent party in the position, as much as money can do that, as if the contract had been performed. *Hadley Baxendale* rules arose from this concept. The second mechanism available is for a quantum meruit to apply where the contract may have become so radically different that its payment/compensation terms no longer should apply – *Laserbore Limited* v. *Morrison Biggs Wall Limited* (1993). Valuation according to rules is similar to valuation on a quantum meruit basis and compensation by loss and expense is the equivalent of compensation by damages.

The ECC has adopted a hybrid of these approaches. It has provided that a compensation event will be paid for from calculations of actual cost plus percentages for overhead as included in the contract. This is arguably reasonable for evaluation of past events. But the ECC requires quotations to be given for compensation events before they are carried out. When this happens, estimates of resources are required by the PM and the contractor has to estimate them. He is not able to obtain more if his estimate proves to be inadequate. It would seem that the contractor, under these circumstances, will have no option but to submit a figure that he thinks is certain to be sufficient. It seems likely that the PM may find that the quotation is not agreeable and then he may instruct, under clause 64, the work to commence without agreement, based on the PM's evaluation of the event. Adjudication may have to follow if negotiation is not successful.

Measurement

There is no provision for measurement according to a recognised standard method of measurement for building or civil engineering works. This is understandable for a contract that is meant to be suitable for all projects. If a standard method of measurement is not used, it is important that the method adopted for the contract is clear such that disputes arising from differences of interpretation are avoided. The doctrine of contra proferentem may be applicable in the case of doubt, and this may go against the employer.

Payment

The ECC core clauses for an option B (bills of quantities) contract contain the following payment provisions. (Option Y(UK)2 The Housing Grants, Construction and Regeneration Act 1996 has been produced solely for contracts that will be subject to the Act; this is expanded on later in this section.)

Clause 11.2 (21) defines prices in the bills as (a) the lump sum and (b) the amount obtained from multiplication of rates by quantities of work actually done (a remeasurement contract).

The final amount paid to the contractor is therefore calculated from the rates in the contract bills multiplied by the final quantities as remeasured, any lump sums and the total of any compensation events, which generally have been assessed in advance. In option B the only use of actual cost is as the basis for quotations for compensation events when actual cost is calculated from the schedule of cost components for work done by the contractors and its subcontractors. In effect, the work is based, in advance on a schedule of plant, materials and labour together with agreed levels of profit and overhead.

Amounts paid on account of work carried out are governed by clause 11.2 (25), which has the effect that the price of work done to date (PWDD) is calculated by multiplying rates by quantities of work done, plus lump sums and compensation events. This is usual practice in other standard contract forms.

Clause 50.1 provides that the assessment procedure for interim payments is not as prescribed in the JCT contracts and is flexible to allow the PM to make periodic payments. Now that the HGCR Act is in force, the NEC has an additional Option Y(UK)2 to make the provisions of interim payment comply with the Act. Notice is required if the employer intends to withhold payment of a certified amount and the contractor is entitled to suspend performance, all this being similar to provisions that have been drafted in all standard forms following the HGCR Act.

SETTLEMENT OF DISPUTES

Following the first edition of the NEC documentation, extensive comments were made on the provisions for settlement of disputes. Provisions for dispute resolution were not in keeping with the overall philosophy of the documentation and provided very little opportunity for disputes to be resolved until the construction works were completed and serious disputes were left to be resolved by litigation. The current edition of the ECC has been amended to take account of the Housing Grants, Construction and Regeneration Act 1996 and is committed to a two-stage process of resolution: adjudication is tried first; if the adjudicator's decision is contested, the second stage is 'review by a tribunal'. This is explored in more detail in Chapters 27 and 28.

The nature of the tribunal has to be identified in the contract data and agreed by the parties, presumably before they enter into contract. The tribunal can be arbitration, expert determination, litigation or a dispute resolution panel.

ECC critics argue that provisions remain unclear and that reference to a tribunal is not specific. At the root of the discussion is the NEC proposal that adjudication forms a major part of resolution and it is easy to obtain the impression that disputes are not expected ever to go beyond adjudication. The ECC explanatory notes state, 'It is the intention that all disputes should be resolved by the adjudicator who . . . is able to act independently.' 'Resolved' suggests that the adjudicator may need to exercise his function in such a way that he acts as an arbitrator not as an adjudicator, and critics have stated this may lead to decisions of an adjudicator being held to be invalid. This is discussed in more detail in Chapter 27.

CHAPTER 23

JCT AGREEMENT FOR MINOR WORKS

- Contract basis and administration
- Site and ground conditions
- Programme and time
- Instructions and payment

This small document is intended for small works, where it is a very useful document to have around. It is frequently used for schemes of a size well beyond that of the moderate house or extension envisaged and its sometimes briefly worded provisions may then be under some strain over the problems which this book is discussing, by virtue of the scale of subject-matter and the unexpected intricacy which it may assume. The issues discussed here are taken in the same order as in most of the preceding chapters.

CONTRACT BASIS AND ADMINISTRATION

The contract is produced to suit four contract characteristics:

(a) Minor works on a lump sum basis and with an architect or similar supervisor.
(b) A lump sum based upon drawings and specification, without quantities, although a priced document (a schedule of lump sums or a schedule of rates) may be used.
(c) At 1987 prices, projects up to £70,000 value.
(d) No fluctuations provisions, as the contract period is intended to be too short to warrant them.

SITE AND GROUND CONDITIONS

There is no provision for the contractor to have full and exclusive possession of the site. When the works are by way of alterations, as they often may be, this is understandable. It need hardly be recorded that there is no reference to ground conditions, by virtue of scale and the usual JCT approach.

PROGRAMME AND TIME

The works 'may be commenced on' and 'shall be completed by' the dates entered and practical completion is to be certified. There are no stipulations about any form of programme. However, the possibility of instructing changes in the phasing of the programme follows from the instructions provisions mentioned below.

Extension of time may be granted after the contractor has notified the architect, although not necessarily in writing. He is to do this when 'it becomes apparent that the Works will not be completed' by the due date, which means that he should make his notification before the completion date has been overrun and, in most cases at least, before actual delay has occurred. There could be a reduction in his entitlement if the architect is denied the opportunity to take avoiding action.

The extension granted is to be 'reasonable' and in writing, with no time limits given over when it is to be given. Cases bearing on this given earlier in this book are relevant and uncertainty for the contractor could only be held to give him the benefit of any doubt. No other procedures are given over estimates, interim extensions or the other matters set out in other JCT forms. Hopefully there will not be time for them to be needed within the intendedly short programme.

Causes leading to an extension of time come within the blanket term 'reasons beyond the control of the Contractor'. This catch-all expression is given as including compliance with architect's instructions not due to the contractor's default, these being given properly and possibly having effects on the time required to completion. This removes doubt over whether any such instruction would otherwise put time at large by leading to an overrun of the due date.

Beyond the express illustration given of instructions, the term 'reasons beyond the control' may be construed quite widely, extending to cover many of the relevant events of other JCT forms (see 'Instructions and payment' below). It affords considerable potential, apparently more than the provisions of JCT contracts generally do, in view of *Scott Lithgow* v. *Secretary of State for Defence*. In this case a contractor was entitled to relief from actions of a supplier, who supplied cables of incorrect quality, because it was held that failure of that supplier fell within the meaning of 'any other cause beyond the contractor's control'. The provision may be compared with the similar wording in the GC/Works/1 form (Chapter 20).

INSTRUCTIONS AND PAYMENT

There is no restriction in the conditions over the scope of instructions, although the most widely occurring are likely to be those over variations, provisional sums and inconsistencies. Furthermore, there is no provision for the contractor to query whether an instruction is valid, so he must comply with anything reasonable that did not subvert the nature of the contract.

Variations are defined to cover changes in the physical works and a more limited version corresponding to the obligations and restrictions provision in other forms. This relates to 'change in . . . the order or period' of carrying out the works. 'Order' clearly relates to changes in phasing or of subsidiary operations within the programme. 'Period' is more difficult if it relates to the overall programme, as it appears to, rather than to some minor facet, as it has suggestions (but not more) of postponement or acceleration, neither of which has any mention elsewhere in the contract. On general principles, it would appear unlikely that a fundamental term like the contract period could be changed unilaterally, particularly from the side of the party not constructing the works. A restricted instruction to move activities about or to speed or slow them within the same contract period is more appropriate as an interpretation. Even here, the contractor could well be securing an extension of time, while payment may flow over such instructions. Valuation of variations is to be made on 'a fair and reasonable basis', which may be interpreted as fitting the usual pattern, so far as there is any data upon which to act accordingly.

Loss and expense occur solely in the context of the valuation of variations, as to be taken into account with them. This leaves any other categories of loss and expense undefined as to their nature and unprovided for over procedures, such as notice and ascertainment, and indeed excluded from the provisions. It is therefore the case that any matter which is of the nature of breach of contract (as some of the matters leading to loss and expense in the other JCT contracts are) strictly have to be resolved by legal proceedings, should they arise. In practice it is obviously better for the parties to sort things out less formally, but the gap exists nevertheless.

CHAPTER 24

DESIGN AND BUILD AND RELATED CONTRACTS AND PROVISIONS

- Some design and build contract arrangements
- Design
- Ground conditions
- Delay to and disturbance of programme

This chapter differs from others in this part of the book by looking at certain principles of the design and build system as commonly employed, but not by going into the precision of how each of these contracts operates, apart from taking one of them as an example of some points of fairly general applicability. This is intended to give a broad appreciation by which the parties may form an opinion over what they should investigate more thoroughly for a particular project in relation to the form of contract actually proposed. Common features are thus ignored so that, for instance, the general question of payment, which is essentially the same for a JCT or other contract for design and build as it is for other comparable contracts, is not discussed here. Each of the themes taken here may be followed up in detail in the context of a rounded discussion of all major aspects of design and build work in *Design and Build Contract Practice.*

However, the approach to design and build has a distinct methodology, and that methodology determines the elements needing treatment here. (See *Design and Build Contract Practice* and *Building Procurement* for detailed examination of the concepts involved in this procurement method and standard contract provisions.)

SOME DESIGN AND BUILD CONTRACT ARRANGEMENTS

There are at least two extremes to which those espousing design and build may go. One, possible from the contractor's stance, is to suggest that the client needs no professional advisers at all, perhaps on the basis that the contractor already has them. For the basic small shed (which is some people's total concept of design and build), this may be suitable; but for anything more, at least one adviser and usually several is an advantage, to the extent that most design and build contractors will recommend that there be an appointment. Even so, it is the case that any consultant, for say design or cost, who is at the client's elbow as adviser is just that, adviser. He does not have the position that is his under other contracts of being able to issue instructions or deal immediately with the contractor, except as the client's adviser. He does not have any role between the parties to decide issues; even to act as inspector is simply a matter between him and the client, again to advise not to direct the contractor in any way. The absence of the JCT 'architect', for example, gives the client a more active part in the drama.

The other extreme relevant here is any practice of the client seeking to place design responsibility on the contractor without using a properly drafted set of provisions or by amending a standard set in such a way as to increase the contractor's risks (as distinct from assuming a level of responsibility) over what he is actually designing. This in itself is in danger of leading to a number of risks which may act against the better interests of the client and should be studiously avoided. This is also the case even when, or perhaps particularly when, any provisions given require the contractor to have his proposals checked technically by the appropriate consultant of the client, this being likely to pass the onus from the contractor to the client, with or without redress for him against the consultant.

Here are some reliable, if not impeccable, standard forms providing in whole or part for design by the contractor:

(a) JCT with contractor's design contract: a thoroughgoing design and build form, as are the next two.
(b) GC/Works/1 single stage design and build contract.
(c) ICE design and construct contract.
(d) JCT designed portion supplement: an additional document for use with the JCT with quantities contract to allow part only of the works to be performed as a design and build subproject, whether the part be a separate structure or some element within a structure which is otherwise being dealt with through design by the client's consultant and on a regular quantities basis.
(e) JCT provisions within the standard forms with quantities for performance specified work, these being intended, it must be inferred, for smaller cases of what would otherwise need to be arranged under (d).

Occasional reference is made to several of these forms in this chapter, from which some major principles may be generalised to the rest. This reference is very selective to highlight only potentially risk-laden issues which fall within the scope of this book.

DESIGN

The fundamental distinction between design and build contract patterns and all the other patterns taken in this book, lies in the locus of design responsibility. It is assumed here that the basic principles are known. A short statement is given in Chapter 2 under 'Design and build', and the system is explored fully in *Design and Build Contract Practice*. The tendency of design and build arrangements is to place as much design responsibility as possible upon the contractor, both at the tender stage and during progress, while maintaining an unaltered price commitment between client and contractor. Provided the client's requirements are clear and remain unaltered, this is what the contractor is usually prepared to carry over most of the design in return for the advantages of regular working etc which the system brings him. In essence the responsibility for performing the design lies with the contractor, along with the responsibility for the consequences of design inadequacies and faults.

Procedure for design

These responsibilities lead to the question of what the design is, as comprised within the contract and the procedure for producing it. Using the terminology of the JCT contract, the division of responsibility there may be traced. The start is with the employer's requirements, which need to set out all that is essential for him about the project's design, but not necessarily anything more. This means they should give a brief with any detail which cannot be left to the contractor to provide. Thus any special internal environmental requirements, such as over humidity control, must be given if they could not be inferred or be provided as a matter of course. Broadly, the requirements may be anything from a simple schedule of accommodation to a closely defined layout and supporting technical specifications, at least in performance terms. A guiding principle is that the more is left to the contractor's discretion, the more should he be able to exercise his ability to provide the optimum full design solution at the most advantageous price.

The contractor's proposals, as part of his tender, should indicate in sufficient detail how he intends to meet the employer's requirements. Again, how precise these need to be will depend in part on how unusual the project is. Here is the hinge point on which future disputes may turn. In all cases, but particularly if he is tendering in competition, the contractor has an interest in not doing possibly abortive work at this stage, while wishing to present his scheme in a winning way. This is recognised in the contract by the term 'appear to meet the Employer's Requirements' which is used of the contractor's proposals as embodied in the contract. This is a somewhat odd expression at first reading; surely the contractor's proposals either meet what the employer wants or they do not? However, it has to be read in the light of the contract policy that the proposals are to be just that, proposals of what the contractor intends, but has yet to develop into a fully-fledged scheme. The contract then proceeds on the basis that the contractor will keep within the proposals and do the developing as the works go along. If there is a dispute over what should be provided within the contract sum, the contractor's defence should be that what he is providing is in line with his proposals as meeting the employer's requirements. In effect, the accepted proposals overrule the requirements if there is a discrepancy.

Risks in design procedure

Here is the risk element, that the resulting design development may lead to dispute over whether it actually does meet the employer's requirements. This is not really helped by the lack of any contract provision for the contractor to put forward full drawings and specifications before going on with any part of the works. Two precautions are needed here. The first is for the employer to see that he obtains enough information before entering into the contract, so that the proposals are sufficiently explicit, at least at the performance specification and aesthetic levels. The second is to include in his requirements a procedure for provision by the contractor of final drawings and specifications for inspection by the employer before any work is put in hand. At least, this should avoid any dispute not surfacing until materials have reached site or work is executed, when more will be at stake. Any such provision about inspection should not be so worded as to give the contractor approval of the work intended, but simply provide that it still 'appears to meet' at a more advanced stage of affairs. This is in line with not formally approving site work under any regular contract, to maintain the employer's rights over defects.

This pattern of what prevails under the JCT contract is the opposite of that in the GC/Works/1 and ICE contracts, where the requirements overrule the proposals. This acts more positively for the employer's benefit, provided that his requirements are sufficient and unambiguous. It remains the case that any requirements or proposals under any of the contracts will be construed contra proferentem if they are uncertain. It may be worked out that it is possible for an unscrupulous or even initially unobservant party to come back at a dispute stage and play the contents of their prevailing requirements or proposals to their own advantage when they diverge from the other's documents. For instance, an employer with his requirements prevailing might note that the proposals did not meet his requirements by inadequate provisions with a corresponding tender price; but having noted it, he might make no comment. Later he might insist on his requirements prevailing to gain higher quality or quantity, but within the same price, now embodied within the contract sum. This could in turn lead to action alleging mistake or misrepresentation. (Chapter 3 looks at the importance of avoiding mistakes in documentation.)

Although all this is true of the GC/Works/1 contract, it differs from the other forms by a requirement that design is to be completed after receipt of tenders and before entry into the contract. This leaves the question of which documents prevail unaffected, but should mean that the scope for dispute is reduced by the clearing of uncertainties during the intermediate period.

Another stage at which dispute may emerge is when introducing changes (as the JCT form has it), otherwise variations. There is a whole area of uncertainty if they come about when the works are not fully designed, but still represented by the proposals, whatever prevails. This raises the question of what was initially there to be changed by any instruction given. Even without this uncertainty, the instruction should be cast in the form that the employer wishes the contractor to vary the design to meet amended requirements, rather than as a specific technical change. There is a risk otherwise that something technically specific will have effects that subvert the contractor's design in some part, so that overall responsibility for its integrity is split.

Often there is a question in valuing a change about quantity, or even quantity of what, and another related one about price allocation from the total contract sum or a known subdivision of it for any particular part of the works. These aspects do not raise special forms of dispute, but need somewhat unusual treatment and care in their resolution, as may be pursued in detail in *Design and Build Contract Practice.* The illustration under 'Ground conditions' below of an inadequate allowance for substructures is also illustrative of the present matter, indeed there is an overlap.

A particular consideration in design development is over statutory matters, such as building regulations and planning consent. Building regulations are so closely tied up with the design solutions selected by the contractor that all responsibility for compliance is regularly borne by him. Planning is more complex in that the client has a more direct interest and may already have been in negotiation before the contractor is introduced to the process. It is therefore commonly the procedure for the contractor to be responsible for obtaining approvals and for their consequences, even if he takes on from where the client has reached and whether he obtains approvals before or after entering into the contract. For instance, this is the express position of the JCT contract and may be inferred (so far as not express) to be that of the GC/Works/1 contract. Each of these then provide that changes in statutory requirements of all kinds made after tendering qualify to introduce variations.

Liability for design

Standard contracts for design and build usually restrict the contractor's design responsibility to that carried by an independent designer under the normal conditions of engagement for his professional body. This represents a move away from that existing at law for a contractor under an unqualified design and build situation, where the contractor has the much more onerous duty of strict care, rather than a duty of general care. Under a consultant's regular duty, he is required to exercise the level of care expected of the average member of his profession, rather than the standard of an exceptional member. Clients who are unaware of the effect of the change introduced may see the wording given in the standard forms as affording them extra protection, whereas it is actually eroding what they would otherwise enjoy. Contractors, not unnaturally see things differently.

It is not uncommon, to say the least, for knowledgeable clients to seek to delete or amend contract terms to reintroduce strict liability and for contractors to be aware of this and to resist it. Should he engage consultants, they in turn are affected by any enhanced liability which he will wish to pass on to them. Collateral warranties, to allow clients and others to whom they may assign or sell to act against the contractor and his subcontractors etc, are another avenue of potentially increased liability. Again, these areas are covered in more detail in *Design and Build Contract Practice.*

GROUND CONDITIONS

So far as the superstructure is concerned, all normal design considerations are unaffected by the presence of anything physical at the site other than (fairly) fresh air and weather, and unusual features like overhead lines. Beyond this clearly apparent area, there lies the more uncertain area of design and ground conditions.

Risk under contract arrangements other than design and build

The cost of substructure and external works is affected by such considerations as the nature of the strata and water conditions, as has been noted under 'Site and ground conditions' in preceding chapters. In the case of quantities-based contracts with the design produced by an architect or other consultant and solely constructed by the contractor, these risks are fairly well contained for him. The risks can be divided into four main areas:

(a) The actual cost of working within the ground conditions encountered against what is to be paid for such working, which is usually to be reimbursed under a defined method of measurement using fixed measured rates, so the contractor's risk lies in the adequacy of the rates tendered or which he can negotiate for work not covered exactly or at all by the contract rates.
(b) Which conditions encountered qualify for reimbursement as priced quantities or daywork at these rates and which are held to be included within the contractor's general risk. The applicable standard method of measurement read with any qualifications in the bills of quantities will set these out, the most common being types of hard materials (such as rock and existing structures), soft, unstable materials needing special support and handling (such as running sand) and water within the excavations as opened up. Here again the contractor's risk lies in his rates, whereas the client's risk lies in the extent of the conditions encountered.
(c) Whether the structures to be located within the ground need to be redesigned to be adequate for the bearing capacity, hydrological and other characteristics of the strata. This usually means that there will be more substantial structures, although occasionally they may be narrower or shallower, and either way these will be valued as variations, so that the risk lies with the client.
(d) Extension of time or reimbursement for disturbance may arise in severe cases because of the instruction of variations, these allowances depending to some extent on the particular contract provisions, so that the client stands to bear the risk in the main.

The overall risk for the contractor therefore lies in his original pricing and what may be derived from it. Beyond this, the cost of dealing with ground conditions is mainly at the risk of the client and the effects on the cost of the permanent works is also his risk.

Risk under design and build

However, when unexpected ground conditions are encountered under design and build, the unqualified position is that the contractor is responsible for taking all necessary steps to overcome the problems and for absorbing the cost within the sum already agreed as payable to him.

Thus, for instance, the JCT contract is silent on the whole issue, so the contractor is totally at risk. This may be satisfactory on the most innocent of greenfield sites, but otherwise it may be far from it and lead to inflated tendering. As a result, it is prudent and common for the boundary of responsibility to be expressly shifted. This will happen to the extent that a site investigation report is provided by the client as data upon which the contractor may rely and which he will be held to have taken into account in preparing his design and tender. A departure on-site from what the contractor has anticipated in preparing his design and from the design included in his proposals expressly or by inference will then be clear. Even without such a report, there is the practice of the contractor stating the assumptions on which he has worked, with these too being embodied in the contract. If either of these or something similar is done, the basis exists from which to calculate a variation adjustment. To facilitate this, it is desirable to have a detailed analysis of the substructure in the contract sum analysis (a set of lump sums equivalent to the contract bills under a with quantities contract) or to have a schedule of rates of substructure items, or even both suitably interrelated. A more drastic change, unsupported by enough rates, may arise if conditions render it necessary to change the nature of, say, the foundations in some way, such as substituting bored piles for load-bearing strip foundations, but here the problem is in essence only what it is under other contract patterns.

The GC/Works/1 contract uses provisions very similar to those in its firm quantities version to deal with the question of ground conditions, that is by the category of 'Unforeseeable Ground Conditions' considered in discussing that other version. These are beyond what the contractor could reasonably have foreseen and taken into account. This gives a position like that under the JCT form when modified by such statement as has been mentioned. There is one special rider as in the other version: this contract excludes from any adjustment 'ground conditions . . . caused by weather'.

The ICE design and construct contract differs from its parent form, the main ICE form discussed separately, by apparently being a lump sum contract like the other design and build forms (a point not made expressly, but to be inferred from the way the contract wording proceeds throughout) so that the contractor has to include for firm design of the works in the ground (where most of them may be). It proceeds broadly along the lines of the GC/Works/1 contract under the preceding paragraph, so that a defined basis exists. However, it includes two extra points. Firstly, the client is to have made available to the contractor 'all information . . . obtained . . . relevant to the Works', so giving the client a duty of disclosure, but not one of obtaining information to pass to the contractor. The contractor may therefore be exonerated from liability over inadequate or erroneous design if he can demonstrate that he lacked available information which he would have taken into account. Secondly, he is also deemed to have taken into account all 'information available in connection' which he could have obtained before tendering. This in itself is nothing extra to what a contractor might expect as inferred under any contract, but it includes reference to 'the form and nature [of] the ground and subsoil', although this is qualified by 'so far as practicable'. This could leave a difficult area over whether the contractor should have undertaken any ground exploration work before tendering, as often he cannot. He is advised to state with his tender what he has been able to do and its limits.

A potential problem exists whichever of the arrangements sketched above applies, or if there is another like it. This is that the substructures included in the design may have been inadequate for the works, even if no unforeseen and difficult conditions had been discovered. This may take one of two forms: there may have been inadequate design indicated on whatever the contractor submitted in his tender (what the JCT form terms 'the Contractor's Proposals') or the work may not have been designed, while a sum for its value may or may not have been shown separately within the tender. Either way, the client and any advisers are not obliged to have checked or accepted the adequacy of design or price for such work: the liability is on the contractor to have included for work sufficient for the task and to produce it in due course without extra payment, subject to the points already made.

To illustrate the problem in one of its possible forms (and as already stated, to expand on the more general points made under 'Risks in design procedure'), let it be assumed that the contractor originally included work carrying a low price of £20,000, whether he had simply designed it inadequately (if at all) within the technical solution proposed or had simply stated too low a price or whether he had done a bit of each. If it were the case that adequate work for the original scheme properly priced would have cost £30,000, there is no contractual problem if nothing untoward occurs, as with any other error in design or pricing within the build-up of a lump sum contract: the contractor performs £30,000 worth of work, including work to make up any deficiencies, but only receives £20,000. If the necessary work when actual ground conditions are discovered is to cost £40,000, how much extra should the contractor be paid? If more work of the same character is required (say deeper or wider foundations), it may be argued that the initial allowance should have been £30,000 and the increase to £40,000 would then have been one-third. The increase should now be one-third of £20,000 and so about £6,700. This treats the error as essentially a money error and the solution follows the principles applied to unit price errors in Chapter 6. On the other hand, if the redesign calls for work of a different character (say piles in place of strip foundations), the allowance for the original work will be omitted and the revised work should be priced at its normal level, as would happen with a similar case under a quantities contract, where the contractor is effectively relieved of the results of his initial error by the change in design principle which overrules the money error.

What will be paid in practice may well depend on whether the original work has been designed at the time of finding the problem on-site and then on whether the original inadequacy is noticed at all. If it is, it is then a question of how accurately the work can be valued which would have been required had ground conditions not been troublesome. This problem is not unique to substructures, but is present in any element of design and build work so far as the question of valuing variations, otherwise known as changes, is concerned. It is just obscured and compounded here by the presence of an aspect that is not present elsewhere, the unforeseen nature of the ground.

DELAY TO AND DISTURBANCE OF PROGRAMME

Delay to and disturbance of programme are aspects which arise for the usual reasons with slight modifications in a few cases. The question here is whether the contractor's design activity may become involved at some point or other.

The contractor is usually (as under the JCT contract) to 'proceed regularly and diligently' with the works. This is relevant to design, but harder to check and so to take action over from the client's side, as design does not have to be, and perhaps cannot be continuous in the same way that site activity should to be. Different criteria should be applied before any action is taken over an apparent failure to proceed regularly and diligently. There is also the question of whether the contractor in using his 'best endeavours' (again JCT terminology for simplicity) to mitigate a delay, which may not be his direct responsibility should be carry out redesign which might recoup any part of the delay. This is a doubtful point and in any case should not be done without the agreement of the client, who is entitled to receive the works for which he has contracted. Certainly the client cannot insist that the contractor should do this, unless he is prepared to instruct him accordingly and meet the excess cost which is likely to follow. The only cause of delay where the contractor might be under more obligation to effect a change without payment is over the question of shortage of labour or materials, if his choice in the design is open to criticism for not foreseeing clear problems of availability.

The general question of mitigation also comes up when there is disturbance of progress. In an extreme case, the contractor might be under criticism if some mitigating change were feasible that would reduce the disturbance effects markedly. Usually, the time required and other practical issues will tell against any fire brigade action of this type being of value.

PART 7

PRESENTATION AND SETTLEMENT OF CLAIMS AND DISPUTES

CHAPTER 25

EVIDENCE AND NEGOTIATION

- Evidence
- Negotiation

It is the pattern of most of this volume to be concerned with the more precise elements, legal and otherwise, of the subject of contract disputes, as these are what must be observed whenever it is considered that a rigid settlement must be obtained. Even so, there are numerous places where it is necessary to hint at, or point out explicitly, areas in which precision is not obtainable or where practitioners may have differing emphases or views about what should apply. In practice there are also areas within which factual precision is not always easy to achieve or where there exists uncertainty over the interpretation or embodiment of facts. These areas may often be summarised as those of evidence and negotiation.

How these twin facets are to be looked at frequently depends on which person in the construction process is involved, as each has different perceptions of what has happened and different accesses to information about how it occurred and what have been the results. It must also be admitted that each has a different interest in the outcome, this being one reason why recourse to proceedings by way of arbitration or the courts is available as the last resort, which it always should be. It lies beyond the scope of this book, which is concerned with settlement within the contract machinery. It often pays to settle rather less favourably than had been hoped, to avoid the risks, costs and delays involved in proceedings.

Throughout this chapter, the pattern is more often followed of expressing matters rather ambiguously, so they may apply from the viewpoint of employer, contractor or subcontractor, to save undue repetition. The desired flavour should be imported by the reader.

EVIDENCE

Evidence is information tending to establish the truth of facts, the facts themselves or the opinions based on those facts. In court there are rules of evidence and also as to what may be produced. 'Relevance' of evidence is the criterion before evidence is 'admissible'. 'Admissibility' used to be more difficult and care was necessary over hearsay evidence but now there is little that is not admissible, provided it is relevant and rules on submission of hearsay are observed. (Hearsay is something which a witness has heard others say or which is contained in documents produced by someone other than the witness.) In practice there is now little evidence that cannot be admitted.

Evidence is required for all types of dispute over construction work. As is clear from preceding chapters, it is often technical, detailed and extensive. Most disputes lie between either the employer and the contractor, or the contractor and one or more subcontractors, although some span the complete set of persons. Evidence of fact should not remain an issue in most disputes but interpretation of 'the facts' and other 'expert opinion' is usually where the dispute lies. (Chapter 27 defines some terms used in dispute resolution.) Often the effective difference of opinion is with the architect or quantity surveyor. The evidence needed is not affected radically by these variant possibilities, so that reference can usually be made to main contract situations without omitting what is equally applicable in the other situations.

Gathering evidence

In some situations, especially extension of time and disturbance of progress, the contracts are explicit in requiring the architect to deal on the basis of information provided to him or the quantity surveyor at his request by the contractor. These procedures are detailed in Chapters 16 and 17. Even so, it is regularly necessary for these recipients to gather other information of their own to supplement or confirm what is given to them. In other situations it is up to the several participants to make their own arrangements and collect what best fits their operations.

Although some standard contracts make explicit provisions in certain instances, those persons concerned with contracts should sensibly be accumulating reasonable evidence during progress of the contract, against the contingency that it may be required. As is indicated by what follows, this collection can hardly be avoided so far as some items are concerned, although proper classification and storage does ease actual use, should this become necessary. When affairs begin to run into major problems, usually the situation becomes quite obvious fairly soon, and any warning lights should be heeded. It is often the small individual problem which may miss the net and cause trouble later.

It is almost inevitable that some piece of information will be missed or cannot be isolated from the mass of what is known. In these circumstances, those coming to an agreement must sort out affairs as best they can: a lack of data does not allow either party to plead that the item being claimed should be dropped.

Pricing variations constitute an area where evidence is not always required. This is because the pricing is to be related to the pricing in the contract bills in the first instance and, when that is not relevant, to 'fair and reasonable pricing' (Chapter 6) and whether actual costs are higher or lower. In this situation the actual costs incurred are frequently more hindrance than help, unless they can be suitably broken down to show the effects of working in changed conditions or whatever is involved. Only when 'valuation' is by daywork does the position change radically.

There is an attitude emanating from some contractors that it is better not to have precise evidence, as its absence makes it easier to manoeuvre and gain a higher settlement. This may be so on occasions, but it can become a highly risky path to follow. This is particularly so under the current editions of contracts, where early warning of a claim to entitlement supported by evidence is required. In these cases, and indeed all others, in which a party suppresses evidence even by default, he cannot expect to gain the benefit of the doubt, but probably to suffer a reduced settlement.

There is equally an attitude emanating from some consultants that no contentious item should be considered until after completion of construction, so that facts are to be avoided for as long as possible. This may be simple concern with getting the job finished, or it may be a hope that the problem will go quietly away if the remainder of the outcome of the project is satisfactory to the contractor. Whatever the reason, it passes the benefit of the doubt in settlement clearly across the divide to the contractor. It may also lead to additional disturbing items such as financing charges, as shown in Chapter 12.

Sources of evidence

Site investigation

Site investigation is perhaps so obvious as to need mentioning simply for completeness, but it is primary and one which may easily be neglected. Although reports and assorted forms of feedback are valuable and may record firm data, a comparatively casual, but planned, routine stroll round the site may throw up questions or provide impressions for someone who is not concerned with the direct task of production. This sort of activity can be the springboard for mounting a claim, or it may supply the lead for confounding one. Many matters may come to light, for example:

(a) Labour and plant doing little or appearing to do what is fruitless, such as moving materials an extra time, leading to the question as to whether this is inefficiency on the inside or interruption from outside.
(b) Work being removed, when apparently not defective.
(c) Concentrations of effort, showing signs of inefficient working by undue use of labour when plant might be expected, or just everyone in everyone else's way.
(d) Work out of normal sequence at the stage the project in general has reached.
(e) Undue activity by supervisory staff in some quarter.

There may be quite sufficient reasons for these and many other clues which are discovered, but it is always as well to enquire. It is a matter of who is doing the enquiry which will decide how it is to be conducted. Usually it is easier for the contractor's surveyor to ask than his opposite number, and usually he has a more pressing reason to ask. But the architect or quantity surveyor should not avoid finding out; this may prevent problems building up.

It is only too easy to become chair-bound and miss the flavour of the project as well as the details of what is happening, and so be unable to discuss matters with any authority or conviction during later negotiations. Just where oral enquiries are suitable may be left until the role of documentation has been reviewed.

Contract documents

The great advantage of what is on paper, or perhaps in the computer, is that it is on record for all to see, however accurate it may or may not be, and however open to contrary interpretations. Ask for it in writing, confirm it and confirm it back; this remains a permanent wisdom, and the same is true of graphic and photographic information.

The starting-point must always be the contract documents, so that a careful perusal should always precede any advance or rebuttal. This is obvious over allegations of divergence or discrepancy, but should also be applied when the apparent problem lies out in the practical operations in progress. It is especially true over reading those documents which are clearly unusual, but can follow for those which are routine, but specifically drafted. It is no excuse to say, 'Well, they don't usually say that in their specifications.'

When there is a possible doubt, it is wise to go back to what happened during the period of tendering or assessment, whether there was a substantial element of negotiation or not. Although what is expressly incorporated into the contract is of prime importance, there are situations in which representations made during the preceding activities will be held to be effective – known as parol evidence. The basic rule is that the written documents have precedence over matters which may be in writing, but not in the contract, or matters oral and therefore not in a written contract, except by specific incorporation. In *Mottram Consultants Ltd* v. *Bernard Sunley & Sons Ltd* (1974) it was ruled that words deleted may be looked at but that 'surrounding circumstances' is an imprecise phrase which can be illustrated but not defined. Parol or extrinsic evidence is generally not admissible unless misrepresentation or the existence of a collateral contract is claimed. This is true of the attitude of the courts in a sufficient case (Chapter 3), but it may also become a forceful argument between those negotiating later, especially if they were also involved at the earlier stage. The position here is no different from the position in court, but often a reference back to early events will give the edge in negotiation or allow the other person to concede a difficult point of bargaining without loss of face.

Post-contract documents

Here are the main types of document likely to be of use:

(a) Drawings, schedules and other technical data issued to amplify the contract drawings or to support instructions.
(b) Instructions issued by the architect or directions issued by the contractor.
(c) Confirmations of oral instructions or directions.
(d) Notices and other formal documents issued to comply with contract requirements.
(e) Architect's certificates, especially on matters other than payment.
(f) Interim valuations in support of architect's certificates for payment.
(g) Minutes of meetings.
(h) Wage sheets, time sheets, plant records, materials invoices and other details of costs incurred.

(i) Statements prepared for the calculation of fluctuations on the traditional basis.
(j) Programmes and method statements, whether prepared in response to a requirement in the contract documents or purely for construction purposes.
(k) Correspondence between those now involved in resolving the dispute or in creating the conditions for it.
(l) Reports on special aspects which have arisen, perhaps not in a dispute context.
(m) Diaries and other individual and private compositions, such as notes of site visits.
(n) Photographs of the site at large or of special pieces of work.

This list follows a broad pattern of setting out documents in a descending order of formal status, although this is only broad and sometimes the imputed formality depends on the source of the document or perhaps on whether it was ratified for some purpose (which may not be the present) by both sides to the dispute. A number of comments may be added.

When a document has been issued specifically for the contractor to act upon it, such as a drawing or instruction, its evidential value is clear, although this is modified by when it was issued in relation to other information and progress of the works, and how far it changed the preceding position or introduced a conflict. The contractor was entitled to act on it and to be reimbursed or otherwise compensated for the results. When a document is required by the contract, such as a certificate or notice, it has a defined effect and may be read accordingly. An arbitrator and now a court, following the *Beaufort House Developments (N1)* case, has the power to reopen a certificate, although persons in negotiation may well refer a certificate back to the architect for his comment and possible agreement to amend in a particular situation.

Interim valuations and minutes of joint meetings between the contenders fall into an intermediate category. Both enjoy a measure of existing agreement or at least acceptance between the parties to give them a status. It is, however, a status achieved in quite distinct circumstances. Interim valuations are produced as a means of calculating payments which are declared to be approximate and open to revision (Chapter 8). They may therefore be quite misleading about, say, progress of an element of work at some past stage or may embody 'swings and roundabouts' to avoid too much elaboration. When the valuations have been prepared to serve the additional purpose of calculating fluctuations by the formula method, they will usually be more accurate, as they are final calculations in this respect. Like all figures, valuation details need care in interpretation and, subject to this, may be very useful data.

Minutes are prepared to record for posterity what was discussed and agreed. The main caveat is that usually they are produced by one party only, even if they were then 'accepted' by the others at a following meeting. At best, they may be vague over the points later at issue; it is surprising how often a clear and accurate record does not state what is eventually needed. At worst, they may have been worded in such a way as to be bland over the contentious or misleading over the factual. The standardised recording of matters of progress on-site, information flows, errors and warnings of problems helps considerably, but a close examination of unagreed minutes, perhaps additionally by someone not directly involved, is vital. They should have been prepared and circulated hard upon the date of meeting, to make this realistic. Like valuations, their present service is to give additional reminders of and tracks into formal sources of data; by themselves, they cannot be taken as embodying architect's instructions or other documents which are issued separately.

Wage sheets etc and fluctuations statements have also been prepared for other purposes, but differ from the sources just mentioned in that many of them have been prepared for reasons needing financial accuracy. They also should have been prepared without a view to misrepresenting facts in the interests of later disputes. This is always possible and in the extreme circumstance should be watched. It is usually unlikely for at least two reasons: the sheer complexity of mounting a large-scale, convincing deception, free from obvious discrepancies; and the normal incidence of human error, which tends to unbalance any tampering with reality. Such things as cost allocation sheets, prepared by a mass of rather disinterested employees, are often difficult to interpret and reconcile in any detail at the best of times, without a conspiracy being mounted to distort them in a particular direction.

This last point indicates the dangers which may exist in taking even the most innocent of information and giving it credence beyond what its origin warrants. The charge-hand who is keen to be off to the hostelry or to obscure odd happenings on-site, may well present the claims sleuth with an utter absurdity if scientific analysis is attempted. This is one reason why 'ascertainment' requires judgment and not just a Geiger counter. The bald calculations set out in the case study chapters could conceal any amount of reassessment of crude data.

Programmes and method statements are usually open to discretion in their use for the reasons given already, particularly that they were prepared for working purposes which may diverge from the exercise of investigation. They may have been made overoptimistic for production reasons, they need not have been followed in detail, even apart from any element of delay or disruption. Again, they may omit the very facts which are crucial or subtly misrepresent them. All of this is quite apart from any intentional drafting to create an exaggerated effect when a disturbance does occur. Programmes and their status (or lack of) receive treatment in Chapter 4. They are often not contract documents in the normal course of events, but treated properly, they are still very useful evidence.

Correspondence, reports and diaries are again usually the product of individuals, probably of those now on one side or other of the dispute. They must be assessed against the background of their production and the motives of those producing them. An external report may be utterly unbiased, a letter may have been written to set up grounds for the very contention now being advanced. Diaries are open to the same problems. Often they are intended to be fairly private documents and this may enhance their value, especially if they provide evidence in a direction contrary to what was to be expected. In this case they may never be produced! There are limits to what may be demanded, and what is produced voluntarily should be viewed in that light.

Photographs are valuable and probably underused, although this is changing. The camera is not exempt from lying, but it does give an added dimension.

Regular sequences of shots build up a time effect, as a single picture cannot. All photographs need to have substantiated dating to be useful and acceptable.

Interviews

When visits have been made and documents perused, it is frequently necessary to amplify or clarify what has been found by recourse to those involved. At this stage, the question of whose side of the actual or threatening dispute they are upon becomes important, as affecting their evidential value. In general, the contractor's enquiries are limited to questioning their own colleagues, whereas the architect or quantity surveyor has a wider opportunity by virtue of his position under the contract. This does not necessarily produce greater objectivity, if he has already been deeply committed to a point of view by earlier events. This is a weakness of the contract assumptions and is noted in earlier chapters.

More broadly, when conducting enquiries, it is necessary to consider the standpoint of the interviewee. This person may be constrained to defend an action that was intrinsically weak, even without the disturbing effects which are being assessed. He may do this without regard to, or despite, the position in negotiations of the interviewer. As a result, there may be a playing up or down of particular elements of the case to provide an excuse or a smokescreen. It is always helpful to interview several people over the same set of 'facts', just as much as in road accidents! Whatever emerges, the results should be noted and shared so far as practicable with the opposite negotiator, who may even be afforded the opportunity of conducting his own interview. This can be very salutary for all concerned, including the second interviewer!

The immediate source of information should be the person suffering the apparent loss of time or money, usually the contractor or a subcontractor. There is usually no problem in obtaining a response here – indeed, it is commonly anticipated! He is under an obligation contractually to provide some information and may offer far more than is needed.

For whoever is responding to these persons, there is a choice of others to interview. The architect may wish to interview the quantity surveyor; or the quantity surveyor, if dealing, will need to see the architect. As just indicated, there can be a problem here, more often with interviewing the architect, who has been deeply involved in the management of the project procurement overall and who is subject to the possibility of prejudging issues accordingly. Nevertheless, it must be done with skill, care and thoroughness. The option of proceedings is always round the corner and can always be mentioned to any reluctant interviewee.

Consultants are an extension of the architect for present purposes, although they cannot undertake any enquiries in support of a settlement, except informally on behalf of the architect. They are given no contractual authority to decide or act, as the contracts do not name them in any way. Nothing added in their specifications changes this position. Their statements are still very useful, especially as they may throw light on matters affecting subcontractors, in ways which the architect or contractor may not be able or willing to do.

The employer himself may be able to contribute information, although he again seldom has an active part to play contractually. He too may have sufficient interest in the outcome to be less than impartial, but if needs be, he should be assured of the facts of proceedings. A project manager appointed by the employer is another person not recognised by the contracts, but who takes an important part in activities at points where the disputes aspect can be prominent. He is less likely to be affected by his prior involvement in providing details, but should be interviewed just as searchingly for all that.

Last but not least, there is the clerk of works. Although he lacks authority under the contracts and exists, if at all, solely as an inspector, he is very well placed to provide comments. He is often full-time, and so constantly on-site; he sees the project through in detail and has dealings with all the other members of the team. His diaries, if conscientiously kept, may be the fullest available. Even so, he should be questioned as carefully as anyone else.

Outside those directly and regularly concerned with the project, there is the contractor's head office. It is most likely that evidence obtained will fall into the oral category, once the formal analyses and financial statements have been obtained. This is because these statements etc are usually of such a blanket nature that they need explanation for an outsider to glean much from them.

Any evidence of this oral type can never take the place of more formal and objective information, but it may provide just the crack in the cliff face which is needed.

Various sources

A nose for the assorted categories of information useful in special circumstances is needed, particularly when the larger, perhaps ill-defined claim is under consideration. A few of these categories may be listed, to illustrate the variety of what may be drawn upon:

(a) Ordering times required for materials, plant or even labour, as the result of variations or delays imposed at short notice.
(b) Shortages of inputs caused by regional or national circumstances, not by the disturbance or other factors emanating from the employer's side of the project.
(c) Inefficiencies in execution by consultants, the contractor or subcontractors, which have rebounded upon others in such ways as to exacerbate effects. These then may or may not result in extra payments from another person.
(d) Weather conditions, which may explain what has happened and negate the claim. On the other hand, they may explain it through indicating how work was affected by being pushed into adverse conditions.
(e) Other 'neutral' matters – weather is in principle a neutral matter – which may come into play in such a way as to be causal or consequential. If causal, they precede the direct cause of disturbance or whatever is in question and combine with it. If consequential, they follow it and amplify the effect, without necessarily provoking it; see the discussion of consequential/indirect damage in Chapter 9.

Like some other points suggested above, this group is rather more a matter of instinctive development than simple listing, so the few examples given should be regarded only as examples.

Use of evidence

Whatever is collected, the person reviewing it is faced with the task of using it. A great deal will be discarded, as in any research. The temptation to use everything should be resisted, especially in preparing a claim. It can arise out of regard for the effort expended, for those who contributed it and averred its usefulness and for the principle that the more one throws in, the stronger the case. Often it is just the reverse.

The logic of the situation should always be kept to the fore. This covers the basic legal position and what is specific in the contract, as well as any arguments advanced by anyone, including the investigator. It may thus be necessary to ask whether figures in the contract documents or supplied collaterally, such as prices or programme durations, are valid in the calculations being performed, either absolutely or as a means of regulating the use of other data. This position is argued out in several earlier chapters, where what is relevant in principle to particular cases is discussed, so that items from those set out in the present chapter may be deduced.

Broad areas of loss

In the nature of the case, however, there are frequently times when the precise detail required is not available, a situation also considered in Chapters 11 and 12. This is quite likely to occur when the matter under scrutiny is a wide-ranging or umbrella type of dispute, such as a financial claim over loss of productivity. This may span a substantial period of the contract and much or all of the work performed during that period, or it may embrace the whole contract period and scope, although this should be very unusual. It may become impracticable in the former instance to segregate with assurance the relevant costs from all the costs for that period or even other periods, and very close attention should always be given to the reasoning for figures produced to assert a drop here. It may be impossible in either instance to draw up comparative costs for what the work would have cost without the disturbance element, because no similar work has been done that way. The use when possible of this approach is argued in Chapter 11. But now it is necessary to establish figures from some broader base.

This should not be a simple comparison of tendered figures and actual costs (discounted in Chapter 11), although what is done may sometimes come close to this, by starting with an analysis of both sets of figures to establish general feasibility. Assuming a contention for some drop in productivity of labour and plant as a whole for some period, which is seen as probable in all the circumstances, this might mean the following:

(a) Taking the tendered figures and assessing them to see whether they represent a fair level for the work. It may be necessary to consider whether there has been 'front loading' or any other distortion of pricing which would give this, and how the preliminaries amounts have been allocated. Adjustment should be made, including removing all preliminaries amounts.
(b) Without or even with internal adjustments, which cannot change the overall total of what has been tendered, making a further adjustment to allow for the apparent deviation of the tender from what might have been expected as economic at the time.
(c) Within these figures, with any adjustment, to attempt to isolate the net provision for the elements which it is contended are subject to loss of productivity. Refinements may be made to allow for fluctuations adjustments and any allowances elsewhere in the final account, such as different pricing for the same elements as varied.
(d) Taking the costs incurred and allocating them to correspond to what has been calculated under (a) to (c).

It is obvious that this process so far is subject to making a whole range of assumptions and to considerable theoretical criticism over its inaccuracy and basis. With the data obtained, there comes the problem of interpretation. The amount claimed may well have been derived from quite distinct reasoning, applied to an equally distinct set of starting figures. Almost certainly it will be different from the crude difference between the figures derived above, even if essentially the same line of reasoning has been applied to precisely the same set of figures.

The best that can be done is to 'form an opinion'. This, it may be noted, is what the architect has to do over practical completion and extension of time, even if in the case of extension he then has to 'estimate' or otherwise produce figures based on his judgment. An opinion is not to be excluded from 'ascertainment', when this is what is being done. The opinion may be formed by looking at either or, more likely, both ends of the data. It may be considered on the basis of the figures, judged by experience, that the possible drop in productivity cannot amount to more or less than particular percentages of the initial or actual figures. The resulting ranges may overlap, suggesting a working limit to what should be allowed. But perhaps they do not, giving more problems in reaching a probability area. Here is an example:

Tendered figure inclusions	£200,000
Percentage excess 10–15%	£20,000–£30,000
Costs as incurred	£280,000
Percentage allocated 12–17%	£33,600–£47,600
Difference between tender and cost	£80,000

Here the higher percentage allocations from the higher costs cause the problem of the ranges not overlapping and the reasoning may be reviewed as suspect at some point. The actual amounts obtained by adding or subtracting the top-of-range figures to or from their base figures are close, being £230,000 and £233,000. These are not significant in achieving a reconciliation, but serve to indicate a gap not explained in what has been obtained in the data figures of £200,000 and £280,000. To give just some possibilities, there may be some excess loss due to inefficiency on-site, there may be some undiscovered disturbance or there may be an error in the original or adjusted bills of quantities.

These illustrative figures are deliberately wide apart to emphasise the points made, but are not utterly unknown. They indicate the uncertainties of using such theoretically unacceptable derived data, but also underline the need to look critically at other more acceptable derived data for the calculations described in earlier chapters, where marginal differences may lead to significant errors in ascertainment. It is only too easy to make false assumptions in using information obtained from disparate sources, such as cost returns, site observations and 'standard' figures from outside the immediate contract.

Seldom, if ever, will all sections of the workforce or their plant suffer the same percentage drop in productivity, so it should be possible to refine the example calculation and obtain a better result. But the problem element remains.

Single areas of cost subject to overall loss

Another way in which a broad effect may occur is when a single element of cost is subjected to loss of efficiency throughout, say the element of hoisting. This is really a special case of what has already been illustrated, there being a restriction to a single category of input affected. In the instance given, it is to be expected that labour would be affected as well, but this is ignored here.

There is little value in seeking the inclusion for hoisting in the tender, as the amount may be spread and lost inside the estimating calculations, and is in any case subject to heavy oscillations in practice, due to changes in method and the actual progress of work. Hoisting is something of a helper to much of what goes on. There is a value in knowing the total cost of hoisting as actually incurred. Given this is £57,000 (and noting the comments about plant costs in Chapter 11), it becomes difficult to sustain a loss of £36,000, which suggests the undisturbed cost would have been only £21,000. An increase of over 170% means that something catastrophic has happened, which can hardly have been missed by the most casual observer. Something more modest is probably to be sought.

This approach does not establish an amount, but it does help to contain the situation. From this viewpoint, a knowledge of broad costs can be useful, even though it is insufficient to provide answers.

Variations

Variations have already come into the reckoning as fitting into the category of instructions to the contractor from the architect, but they have particular significance in evidence for at least two reasons. One is that they are usually numerous and directly affect the works being produced, hence the contractor's and subcontractors' production activities, which are central to present concerns. The second is that they are paid for in their own right by adjustments of the contract sum, upwards or downwards. Chapter 6 explains that the contract provisions for reimbursing variations allow for nearly all attendant circumstances to be taken into the valuation, as well as the direct contents of the work as varied from the original contract inclusion. The exception is disturbance of regular progress, as this is a result of variations, rather than part of their content.

Variations are common, but also vary greatly in their scale and effect on progress. Sometimes a quite small variation can have an effect out of all proportion to its physical scale and the value of additions and omissions which it produces. This is where the timing of a variation becomes important, so it is crucial to check this against the stage of progress when it was issued, making due allowance for lead times for ordering materials or building up labour and plant levels. It is extremely difficult for anyone outside the direct production process to identify all the effects from a heap of cold documents after the event, so the evidence of those directly involved becomes necessary, subject to scrutiny and such corroboration as is available. It is the interaction of a string of small variations which can be damaging, and this does not show in a list of instructions issued.

The contractor's surveyor himself, should he be preparing a claim or other statement, needs to obtain precise information from those on-site before he indulges in too much work, even though the underlying variations can usually be dealt with clearly enough. In the case of *Minter* v. *Welsh Health* (1980), discussed in Chapter 12, mistakes and delay over fixing brackets for windows led to consequences out of all proportion to the cost of the brackets.

A special case is variation to obligations and restrictions which affect work on-site in ways that do not show in the finished construction. This is important as variation of these aspects tends to overlap with elements of disturbance. It is possible to miss something here unless care is taken, or to allow something twice. The cardinal rule remains to allow as much as possible in the valuation of the variation itself, as this is what the contracts intend and also allows for price adjustment inclusive of profit, rather than ascertainment on a loss and expense basis with profit uncertain (Chapters 6 and 11).

Even work priced as variations by daywork needs care. Although the direct cost should be fully covered by this method, an unduly large element of daywork in an account may signal disruption which is not being recompensed.

Timeliness

The intention of the current contracts is to expedite settlement of matters of extension of time and loss and expense, so avoiding uncertainty and the accrual of financing charges. The search for evidence should therefore go on steadily and, when found, it should be assessed and used as warnings or by incorporation as soon as practicable. There is still a 'wait and see' attitude which does not help. All else apart, memories fade and accurate assessment becomes more difficult. This may not be the advantage it is sometimes thought to be. It can be beneficial if those who deal for each party set up a mutual system for recording events, and it may even lead to later evaluation. This is not only efficient, but it also breeds confidence that the other is dealing in a straightforward way and helps subsequent negotiation. If necessary, agreements reached can be contingent on later developments.

NEGOTIATION

Volumes have been written on the practice of successful negotiation, be it an art, a science or just a result of experience. It is not presumed in the present short compass to do more than set out a few points by way of review. Treatment is limited to the context of post-contract negotiation, when the parties are committed and can neither withdraw nor continue without considerable cost.

In principle it might be deduced from most of what has gone before that settlement of contract disputes is really a case of following the rules and totting it all up. It is hoped that this is not the impression created, as life is not like that. The case studies in particular should dispel any such illusions. The following headings indicate a few guidelines during the actual process of agreement.

Flexibility

It may sound like a counsel of compromise to suggest flexibility in negotiating, and so it may be on occasions. But there are times when what is at stake is by no means so clear as cold principles would suggest. Examples may be found by a close reading of the case study chapters and the use of a little imagination. Another is in the discussion of umbrella disturbance costs earlier in this chapter. But with this said, flexibility has a wider applicability.

It relates here to the attitude of being prepared to listen to and understand the viewpoint of others. Construction is based upon a number of long-established routines and segregated specialisms, and the managerial and other professional aspects are no exception, to say the least. In general participants have experience initially of just a single discipline, and it is therefore very easy for them to take a somewhat unbalanced view of the whole enterprise. Successful negotiation is not about beating the other fellow into the ground, but about achieving a result which is fair to all concerned. But this is not the same as achieving what everyone might have wished.

An important aspect of flexibility is not being blinkered, whether as a designer who sees the construction as flowing automatically from naturally unambiguous information provided whenever is convenient to the design office, or as a site manager who sees the designer as someone to be goaded for his lapses, or as a quantity surveyor who sees affairs as represented by his analytical and rather theoretical concepts set out in the frigidity of a cost plan or bill of quantities. Flexibility therefore comes about when the limitations of upbringing and standpoint in working are recognised and there is respect for the integrity of another's work, so that entrenchment is avoided.

Even more painfully, it may mean giving up self-preservation when threatened by the direction which investigations are taking. Negotiation is not just about the final settlement, but also about defining one's own position in a complex pattern of events. It may also mean admitting ignorance of what the other fellow is about and learning on the way through. Seldom is the right conclusion reached without letting down the guard of invincibility and omniscience.

Environment

Although negotiation is often seen as the confrontation of two or more individuals who have a specific arena and tasks, it may also be seen as part of a wider structure of relationships and actions. As a minimum, those directly negotiating are usually operating on behalf of others, those whose money or other resources are at stake. It is therefore useful for a participant to keep in mind what it is that his opposite number is concerned about in coming to an agreement. If the other is constrained by having to go on until a certain level of result is achieved, and if this is beyond what is within the first person's remit, then there is insufficient consensus about what is being done, and consensus is required for two persons to interact usefully. It becomes necessary for one or both to go back to the people they represent to see whether any change of brief is possible, so that continuance has any point.

This situation may come about for several reasons. It may lie in the formal constitution of a party to the contract, probably the employer who is governed by standing orders or similar matters, unchangeable so far as he is concerned. The singular is adhered to, as throughout this book, despite the corporate nature of such an employer. Here the person who 'is' the employer for the negotiator may have no way of breaking a deadlock other than going to some committee. Where they go may not be his concern.

Not quite so formalised, but equally compelling is the deadlock that may be caused by a ceiling on the settlement funds. The budget allocated may not permit movement further or, worse, the party represented may face insolvency if pushed over the line. In any such case the negotiator can do no more than explore where the other side will settle, and even to do this may be to reveal more of one's hand than is desirable at that stage. A provisional offer to settle at what the negotiator considers a satisfactory level, subject to authority, can sometimes ease matters if the authority is forthcoming, indeed it may be a useful counter to play. But if it turns out that authority will not or cannot be given, it can lead to an impossible position for the one putting it forward. He must stop forthwith.

Another way in which the environment can play a part is if it affects the negotiator in what he is immediately doing. He may find that he is effectively negotiating with more than one person at the same time. This is a problem of fielding a team of people, say architect, quantity surveyor and a consultant, who have not resolved internally just where they are going. It is better to break off for a while and resolve the tension, hoping that on resumption the other side does not make too much of what they have inferred from the break.

A variation on this theme is the situation in which the negotiator is acting because of compulsion to conciliate or impress another present, i.e. an act is being presented which draws off force from the actual work of negotiating. To some extent, all groups consist of individuals that are conveying a whole variety of signals, so all groups risk conflicts of interest or even loyalty, risks that must be guarded against. A related form of problem exists if the negotiation is also being conducted with someone who is not present. A common version of this pattern is the contractor dealing with either the employer's representatives or a subcontractor, while knowing that he has to close with the other in due course. Either he must play for time, and this may be non-existent or the attempt counter-productive, or he must decide to settle on the current front in a way he can stand, whatever happens when he faces about.

Even more complex is the situation of negotiating with an absent principal. It can be that the person in the field, as quasi-agent, is at odds with his principal over some aspect. He may consider that he is being pushed too hard, he may occasionally believe he could do better than he is being asked to achieve and resents the overruling of his personal standards or pride. The temptation in either extreme is to seek a settlement which meets the letter of what has been asked for, but also gives something extra which it is hoped will bridge the point of difference without causing a rift. This may be gratifying if it comes off; but if not, it could court disaster.

Authority

The previous section raises the question of authority in negotiations. It is critical that a person acting must have defined authority, which may range from 'settle your own way and at any cost' through a series of gradations to 'do not settle anything, but report back'. No one should act without his brief being clear and the other side should be equally clear what that brief and authority is. So far as the architect and quantity surveyor are concerned, when they act in the capacities given them under the contracts, the position is defined for them and the employer. Strictly, they are not acting as agents of the employer to negotiate for him, but to implement various terms of the particular contract. In this position they are to act impartially between the parties. They do not have to and should not seek authority for their acts within this framework. If they do, there is the danger that they may lay themselves open to charges of bias.

That these persons appear to be acting in the polar position from the contractor on some occasions is because of some of the functions given them. The architect has, for instance, to inspect work and any adverse judgment is bound to be against the contractor. The quantity surveyor has to settle measurements and prices, and this may mean assuming an adversarial stance while sifting the contractor's arguments over something unusual, until assured of where the final position should be. These factors are the result of authority given by a contract, to which they are not parties, and accepted by the parties.

Although each negotiator should have clear authority, and especially know his limits, it is also highly desirable that those engaged opposite each other in a single phase or segment of negotiation should have matching authority. It can be frustrating when one negotiator has authority to settle but the other has to report back over comparative trivialities, even though it may give the more autonomous negotiator an edge. And the less autonomous negotiator may find his position psychologically inferior, almost humiliating, unless it has been set up that way to allow agreement to be reached gradually, uncertainly so far as the other side is concerned, and apparently grudgingly. It can be unnerving to the opposition to have to deal through one person with someone else who is always one step removed and inaccessible. For all straightforward activity, straightforward dealing is to be preferred, and if necessary sought by rearranging the pattern of negotiations.

The more individual aspects of authority are a mix of what has been outlined along with personality and conduct. The negotiator has to define his own authority out of the circumstances of discussion or even dispute. He must show that he is firm but reasonable, and that he is open but has his own undisclosed thoughts. He should make it clear when he is holding and when he is conceding, to avoid the 'cloak and dagger' atmosphere. Above all, he should be fair to the subject-matter and to the integrity of the other person. If several people are involved, he should avoid playing them off against one another so as to belittle anyone. Tension has its place, but an appropriate sense of humour is invaluable.

Methodology

There is no single way to negotiate; much depends on the matter under discussion and the personal approaches of the negotiators. It may also depend on how much preliminary work has been done to establish the factual basis. In the case of much construction-related negotiation, there is no absolute demarcation between the stages; the same persons often deal with both. This means that the exercise of finding and sifting facts can have a negotiating air about it. Earlier in this book, the principles affecting settlement have been prominent, so the processes of analysis and synthesis leading up to settlement have not been stressed. They usually contain elements of marking out the areas of discussion which will occupy the end phase. The investigators need to be watching for clues all through.

Although this is true, the discussions should proceed as informally as possible among the activities of aggregating data and the following segregation and classification. It is desirable to agree as much as possible during these stages, so that decisions are not unduly deferred. This allows the general drift to be identified and warnings to be given to those for whom the investigators are acting, even before there may be formal negotiation. The question of agreement can be a thorny one; the feeling may arise that one person can be lured unawares into a net, only to find that later findings give the earlier agreements unexpected significance. An instance is the agreement of prices for measured work of apparently minor importance, when later it turns out that considerably more work of the general type exists. To use the prices agreed in the one situation as applying to the other would give quite the wrong answer, but it seems that all has been committed.

If the work is quite distinct in extent, and perhaps in situation as well, then the first set of prices need not be held to for later work, just as a set of prices in the contract bills may be adopted when suitable following the principles in Chapter 6. And any agreement made during the progress of settlement should be subject to the possibility of review. This does not mean reneging frivolously on all manner of details all the way through, but accepting that a broader view of the situation may justify some revision. All settlement agreements should therefore be subject to this sort of proviso, even though it is not explicit. Initial claim documents should be read in this light, a point which recurs in Chapter 26. Sometimes it requires a statement to explain that agreement is contingent on specific, foreseen but unquantified eventualities.

There is a distinction between this type of agreement and the infusion plant agreement in Chapter 30, which involved a supplementary contract. Even had it not been entered into quite so formally, the effect should have been the same: it was concluded as an inducement to proceed with work. But an agreement over work already past is in a different category, when it forms part of an ongoing negotiation of an account. When there is extension of time, a similar arrangement is made express in the contract provisions, where the architect has a power of review. There is also an implication that the contractor may be bringing forward fresh evidence.

The aim of negotiation should be to clarify issues not to confuse them. Sooner or later there has to come a time of weighing all the matters together; it can only be eased by understanding those matters along the way. Then the avoidance of predetermined attitudes, already implied in this chapter, is important in easing progress. False assumptions of superiority clog progress, even when the assumptions are embodied in platitudes. But this is where the authors themselves are in danger of falling into the trap – and so this chapter ends.

CHAPTER 26

PRESENTATION OF COMPREHENSIVE CLAIM DOCUMENT

- Principles of presentation
- Sample claim presentation

Some comments are given here on a single all-embracing claim document of the traditional type presented as one entity at or after the end of construction. The arguments against doing this as of necessity are reviewed in other chapters, particularly the likelihood of losing reimbursement and the current requirements of most contracts aimed at earlier, progressive interim payment and final settlement. Nevertheless, the 'big bang' still occurs when a total view is needed before finalisation is possible. There are also those who, quite wrongly, will not entertain a claim 'until we see how things are going', so that the contractor is forced into the overall presentation. Some contractors prefer it to confuse the issue by mixing details and perhaps fiction with fact, but this is another, risky and dubious matter. This chapter sticks to the general line taken elsewhere that the parties are still fairly friendly, despite all that has happened.

A sample claim document and comments are given at the end of the chapter in relation to the minor factory project situation under the JCT intermediate contract, as described in the case study in Chapter 32 and based on the various arguments and settlements in principle reached there. However, it is assumed here that everything was left until the end to be settled, rather than that there was a progressive presentation and settlement as in that version of events.

The claim is presented under a contract in the JCT form, but shows a typical form which might be adopted in many cases. The claim statement gives only such information as should be needed by those, such as the employer, the architect and the quantity surveyor who have a close acquaintance with the project. It also means that undue work is not performed that will not be needed unless and until matters go on to legal proceedings. The additional work required for those not previously familiar with the project and its history can then be performed and will also rank as costs for recovery if the action is successful (Chapter 27). It is most likely that work carried out by a contractor in presenting what is necessary for 'a claim', but which would have been necessary in any event to agree loss and expense in a final account, will not be reimbursed under a head of claim. However, if additional work is required to be done by a contractor that can be shown to have been necessary during or in contemplation of tribunal proceedings (arbitration or litigation) then this may well be 'costs' payable by the other party, provided the case has been lost by him.

PRINCIPLES OF PRESENTATION

General arrangement

The golden rule is to say enough, but not too much. Sheer verbiage is not worth so much a page! The 'too much' bit means what is said when, as well as what is said at all. The crux of the case and the amount of claim should be given early on in the document, with the supporting detail following as just that – support. The general pattern of the explanation in Chapter 32 may be taken as a framework: if it serves to communicate to one class of reader, it should do to another. But what has been written there also contains introductory background material about the contract for the reader of this book. This should be familiar to those receiving the claim and should be excluded, as apart from a brief statement to give a rounded 'feel': the document is not intended for an arbitrator or similar who, as approaching matters from outside, needs to be led into the facts. The wrong material only irritates.

The crux section should be contained within at most four or five sides of A4 paper (using single-spacing, the sample given is shorter) for a project of comparative simplicity, as was the factory extension and alterations. It can be limited to three sections:

(a) A narrative of little more in scale than that given under 'Changes during progress', in fact worded rather similarly. The aim is to remind the reader of what happened in sequence, what were the main dates, whether instructions were issued and what were the outline consequences. A neutral tone is desirable, precise and firm, but not recriminatory. This gives an account of what happened in terms which should not be open to radical contradiction.

(b) A summary of the heads of claim fairly similar to items (a) to (h) given under 'Contractor's overall analysis', with total amounts only against each item or minimal subdivision. This immediately sets out the financial effects, without pages of meandering and suspense before all is revealed. This section may be combined with (a) preceding, according to which arrangement gives the more immediate clarity.

(c) Supporting technical data, of at most a sketch site layout (no more detailed than Figure 32.1) and a key programme (as in Figure 32.2) showing 'before and after'. This data should ease description in the narrative section, by allowing references to be given, such as 'main cable P'. It also makes the facts more vivid. If these are given as two sheets, the rest should be only three. If they are not included, the rest may be stretched to four, but only if description cannot do justice to events in less.

Level of detail

The rest of the document should contain the necessary supporting detail. The temptation here is to throw in anything and everything which appears to add fuel to the conflagration being produced, such as letters complaining about delayed information and changes, extracts from minutes in a similar vein and copies of incorrect drawings. These may have contributed to the problems, but seldom allow of being pinpointed as causing a specific expense. They should therefore be used sparingly as cogent examples, or summarised in annotated schedules to show their frequency. Sometimes they may be open to the suspicion that they were originally prepared to support a potential claim which has now crystallised.

What should be given is concise information under each of the heads of claim on how the situation developed and how the excess costs etc were then incurred. There should enough detail to show the general basis of ascertainment and reasonableness, but not necessarily every minute calculation. After the architect has read the introductory section, he needs to be able to turn to these more detailed sections to discover what happened (if he does not already know in enough detail) and why it has cost so much – or little! Although he is responsible for adjudicating on the principle of entitlement, he may wish to pass the detail over to the quantity surveyor. Segregation of principle and detail aids the task and perhaps enhances the favour with which the claim is viewed initially.

One reason for still limiting the detail at this stage is that the contractor is required by the contract to provide 'such information . . . as is reasonably necessary' for the architect or quantity surveyor to perform the ascertainment. They will ask for anything extra they require, so long as the contractor has pointed them clearly in the right direction. Equally, they will indicate if they think the direction is incorrect. Another reason for keeping the document within limits is that the contractor does not stand to recover the costs of its preparation as part of his claim settlement (Chapter 9).

It may be questioned whether it is wise from the contractor's point of view to give a breakdown of the claim into the heads shown, because this is restricting his freedom of movement. There are several major reasons for doing this:

(a) The broader question of global claims is important (Chapter 18).
(b) The claim does not look credible without adequate structuring. It looks like a try-on, which is what freedom of movement may mean.
(c) Those looking at the claim will very soon ask for detail and so fix key areas.
(d) Until the claim is settled, the contractor retains the right to change his approach if it turns out to be flawed, just as much as he may change his submission in an arbitration. He will not wish to appear unsound at every turn and so lose credibility, but some measure of reallocation or reassessment can often be seen as sensible by both sides.
(e) If the claim has been approached and is being considered in moderation, those receiving it should be prepared to guide the contractor over areas which are unsoundly presented, but which have a substratum of entitlement. A claim should not be any more adversarially considered than any other part of the final account of which it is part.

Contingency

A final point suggested by the foregoing list and studiously ignored throughout the rest of this book is, What allowance should the contractor include for what is going to happen in settlement? Is he going to be knocked down, or will he be asked to split the difference? In all equity, if he puts forward an absolutely straight claim, he should receive 100% of it. In many instances, he sees this as a pipe-dream, especially if the employer knows there is a claim and expects to see his interests 'guarded'. Often the contractor is expected to have inflated his claim. (These considerations are an incidental set of arguments for progressive claims; they reduce the tension and the employer's concern and they should lead to less inflation, if the first figures arrive realistic and later ones stay that way. They are also related to the discussion of negotiation in Chapter 25.)

It is therefore not at all unusual to find that a claim is inflated, but by how much? Taking expectations a little further, there are those who expect a midway settlement and inflate or aim to reduce accordingly, or rather further. It becomes, alas, a question of knowing the opposition, if that is how they are seen. Very often, an unduly inflated amount for a reasonably clearly defined happening on a construction site looks just that, unduly inflated, and may provoke the very reaction which the contractor would rather avoid. If it is felt that some concession must be expected, it is probably better to aim to have one or two throwaway points in principle, which will come to light early in the negotiations, rather than to have everything subject to heavy and justifiable criticism. Much may depend upon the contractor's prior knowledge of who will deal with the claim: it may be better to have a little in most places, so there is a fairly well spread trimming, but to have a few places where no impact can be made, to sustain the reputation of the contractor untarnished, but unpredictable.

SAMPLE CLAIM PRESENTATION

This claim sample is given as indicative of how a claim for a moderately sized project might be presented. It is based upon the factory case study in Chapter 32, which should be read before this document is studied and where the full account of what happened and other, more detailed comments are given. The case study relates to a JCT form although this does not seriously affect the claim presentation. Relative detail, such as 'loss and expense' rather than 'expense', if in a GC/Works/1 setting, is incidental.

There are many ways of styling a claim from the cryptic to the effusive and from the antagonistic to the chummy but, style aside, the basic structure here is put forward as giving a logical flow and allowing the reader to start at the beginning, read on without skipping and stop where his or her own concerns run out, leaving the rest to those who must pursue it.

The claim has been worded as though the employer will see it before it is agreed. This often happens in view of the frequently sensitive nature of the subject-matter and its causation, although it is not a contract requirement, the settlement being simply another part of the final account. As the architect (in this case) is the person charged under the contract to decide on the issues, it is also desirable to word it tactfully, if relationships have not been soured during progress! In this case, matters seem to have remained fairly sweet.

The general presentation has therefore been given in a form intended to offer a reasonable and moderate way forward. It presupposes a measure of consensus that grounds for most of the heads of claim exist. It could have a more openly adversarial tone in other circumstances, but nothing is to be gained by aggressiveness, sarcasm or other emotional colouring. A neutral tone to the claim is better, however potent the facts enshrined within it. Here, as distinct from when appearing before a tribunal of some sort, the aim is still to seek agreement between those acting. Any suggestion by the contractor that the employer has acted flagrantly or that the architect has failed to administer his aspects of the contract fairly or efficiently should not be brought to the fore, unless there is already a complete breach of relations. Much the same may be said of the probity of the activities of the quantity surveyor in settling what is already in the rest of the final account. Disagreement over fact is another question and should be put forward as strongly as the situation requires.

Looking at this claim as a document received, it appears to be worthy of serious consideration in principle by the way it has been drawn together and also in the light of the history lying behind it and set out in Chapter 32. Whether the figures are overplayed or even understated is another question yet to be determined. It remains the duty of the architect in the first instance to 'ascertain' and this document represents evidence in that quest. He may even have requested it for that purpose. Hopefully he and the quantity surveyor at his behest have been accumulating other data as things have gone along, on some of the lines suggested in Chapter 25.

The contractor has not expended too much time and effort in producing the sections of the claim which are set out here. How detailed the section of 'Details of calculation' is must be conjectured. There needs to be enough to give cogency, but on the other hand the architect can always ask for more. There is also the underlying question of whether there is going be agreement rather than an appeal to arbitration or other tribunal. This is nearly always best avoided in view of the likely costs and the difficulty of adequate recovery, so that more detail now may secure avoidance. Against this, there is the fact that costs in proceedings are recoverable, as those for the present claim preparation are not (see earlier note on this).

Although a book exercise never lends itself to ascertaining quantum, a number of broad questions may suggest themselves to those reading such a claim as comments and potential lines of enquiry about it. The history in Chapter 32 brings out numbers of suggestions which may be compared with those here, precise correspondence between the two sets not having been sought. Comments here which are not part of the claim presentation are scattered throughout its flow and are given in italics.

Construction of Extensions and Alterations to Factory

- at -

Jim's Alley, Yonderville

- for -

Messrs Goodsmakers Ltd

Statement of Direct Loss and Expense Incurred

Outline of contract

The works consisted of the following main elements, as set out on site drawing 333H in Appendix 1:

(a) Construction of the steel-framed research and development bay with a free-standing structure inside, containing the 'maxi-machine' (provided under a separate contract), a control facility, a substation internally and a system of floor ducts
(b) Alterations to area X of the existing offices over the stores building, removal of a corner of the building and construction of a first-floor bridge to the R&D control facility.
(c) Demolition of the machine parts shed and construction of a range of maintenance garages and a fuel pit.

This follows the pattern in Chapter 30, but is less full, as the employer and his advisers are familiar with what was done. It acts as a quick reference and also to inform those who do not know – or may have forgotten! This is important if there has been a time lapse since the works were performed.

The contract period was 30 weeks overall, with intermediate target dates written in. There were also particular restrictions placed on access, working space and order of working to allow for interaction beyond factory and construction operations. These various factors are shown on or with the programme given in Appendix 2.

The contract sum for the works as adjusted by the final account has been agreed, covering all variations to the design etc as instructed by the architect, but is subject to the further agreement of the amount for loss and expense now put forward, so that the total financial statement is as follows:

Contract sum	£610,000
Adjustments for variations etc	£58,000
Loss and expense	£72,000
Total account	£740,000

All agreements of rates and other amounts are actually interim until any final account is totally agreed, as progressive agreement is not called for. This tends to be forgotten. This contractor is sensibly making the point clear to those who may not be expected to know, especially the employer who must be regarded as lay for present purposes.

Narrative of events leading to loss and expense

The construction period finally extended to 35 weeks for the various causes over which extensions of the contract period have been granted by the architect. In addition the sequence of the elements of work was modified. Bar charts for the programme achieved are in Appendix 2, along with the original programme data. The several items of extra work, changed sequence and delay occurred in succession and during progress, so that our attempts to rearrange our programme were more difficult to coordinate than would have been the case had we known before work started on-site.

(a) A major change in sequence was introduced when the installation of the maxi-machine was brought forward by 10 weeks. This was done by agreement with ourselves and with the understanding (see correspondence in Appendix 3) that reimbursement would be made when the effects were known. As a result, the steelwork and cladding at the roadward end of the new building had to be erected earlier than planned and ahead of the rest, rather than the reverse. This threw out the order of and access to foundation and other work away from the road and also affected our steelwork subcontractor (named under the contract). It was also necessary to provide protection for the sensitive maxi-machine when installed. While screens etc have been included in the main final account, work in the main area of the R&D bay was impeded, causing loss of production and extra work.

This is the key to all else and so logically comes first.

(b) Revisions to the floor duct layout and design followed from the technical nature of the maxi-machine chosen and also from late revisions to the intended arrangement of related machines within the R&D bay. The effect of these items was to cause further expense at the stage that the earlier disturbance was beginning to work itself out. It was heightened by having to work inside the newly erected building envelope, requiring uneconomical plant and careful working over the maintenance of unusually clean conditions for the maxi-machine.

This disturbance followed from the previous set, but it is pointed out that it can be separated out in its effects.

(c) During the work in the floor, an existing live power cable (marked P on the site drawing) to serve another part of the factory and of which we were not aware was discovered crossing the R&D bay diagonally through the ducted area, leading to the need for piecemeal working around it.

Work to the cable, such as protection and ducting has been measured, but the disturbance effect belongs here and flows from the instruction about the cable itself.

(d) As a result of the early introduction and commissioning of the maxi-machine, it was necessary to provide the bridge between the offices and the machine control level early as well. This restricted the use of access 1 into the site, both while work proceeded and after, when the height of plant able to use the access was reduced.

This could perhaps be related more closely to item (f) which makes some reference back to it.

(e) Work in the offices was affected by the provision of the bridge early, while additional offices to the contract requirement were altered, involving weekend working (covered by daywork in the main account) and uneconomical working in occupied offices.

The extra costs covered already are stated clearly, thus suggesting that the contractor is dealing in a straightforward way.

(f) The employer found it necessary to retain the old machine parts shed for over 12 weeks beyond the programmed date, to store items related to the maxi-machine being installed early. This meant that access 2 to the site was not available, so we could not circulate our vehicles as intended; this was especially difficult when the bridge was constructed.

This and the earlier item could have been highlighted as a failure to give ingress or egress under the contract or as a variation of a contract restriction (perhaps better, here, of a facility).

(g) Consequent upon the last matter, we had to defer construction of the garages, as they overlapped the shed site. We had intended to use them for storage of vulnerable materials for part of the contract period.

This intention had not been declared at any stage to the employer, but a contractor has a general right to store his materials etc where he will on-site, unless specifically restricted.

(h) The fuel pit was delayed by the foregoing events and then modified in major respects in its design, leading to extra costs particularly over supervision.

Most of these costs could well fall into the measured account and perhaps still may. Supervision is the odd item here, which might go either way.

(i) As a result of all these disturbance effects, our programme was extended by 5 weeks, leading to additional supervisory and overhead costs.

This looks like a partial global claim. It may need to be proved accordingly, or reallocated to some of the heads of claim set out above, but see an alternative view under the summary below

Notice was given as far as possible about the occurrence of these various events. We were advised that some elements would be covered in the main final account and so should delay putting forward a loss and expense claim until the account was finalised.

This is a fair statement of what probably has happened, but what need not have happened. Equally, it quietly leaves aside the question of who ascertains and whether any payment could have been made on account. This contractor is tactful, always helpful in these situations. The comments under 'Timing of changes' in Chapter 25 are relevant.

Statement of heads and amounts of claim

The following summarises the financial effects of what has been narrated above. Details of calculations indicating how the individual amounts have been ascertained follow below.

This introduces the items with condensed wording, usually with a single figure for each. The sequence follows that of the preceding narrative and there may be short claim documents where the two could be combined to avoid undue repetition. Given that there are more than two or three heads of claim, there is some advantage in separating the bare facts of happenings and consequences to maintain a flow. Detail like '20% drop in productivity for a gang of eight men affected this area between weeks 5 and 14, but during weeks 7 to 11 there was a further lump sum effect due to introduction of new plant (a smaller crane and two dumpers) with costs of transport to and from site' should always be reserved for the backup following. This present part is most likely to interest the employer and possibly the architect, who will usually consider the principles and then want the quantity surveyor to burrow into the detail.

(a) Change in sequence due to early installation of maxi-machine.

Reorganisation of workforce and drop in productivity in foundations and floor ductwork	£6,300
Amending design and bracing, reorganisation and drop in productivity of steelwork subcontractor	£2,900

How the steelwork subcontractor was affected was a distinguishable issue. In Chapter 32 it is assumed that the subcontractor was dealt with as the work proceeded over the question of design which occurred. If this were still so, it would raise the possibility of overlap with the present claim. It has not been mentioned clearly here.

(b) Revisions to floor duct layout to suit amended machine layout, including those due to the maxi-machine.

Reorganisation of workforce and drop in productivity	£4,600
Uneconomical working in space available and measures to avoid damage to maxi-machine by dirty activities	£4,900

This was the major disturbance and the heads of claim given are obvious enough in themselves.

(c) Delay obtaining instructions and working around power cable £1,600

This is also straightforward, with the delay element a question of fact and a clear matter under the contract. The working around element could possibly have fallen within the measured account. Items within the first part of the major case study in Chapter 30 bear comparison.

(d) Earlier construction of the bridge.

Rescheduling the construction, including less economical ordering and additional gang costs	£5,600
Working through access 2 alone during bridge construction and to a greater extent than planned thereafter, affecting circulation, unloading and plant movement	£2,700

Although rescheduling construction on-site of the bridge was a fairly clear matter, the effect on ordering was not so immediately certain. In any case it might overlap with the claimed overhead costs under item (i), even though that was stated as due to prolongation.

(e) Work to the offices extended by earlier construction of the bridge and by increased scope.

Additional supervision initially and during work in the partly occupied offices £1,700

Reduced productivity working in occupied areas £ 5,600

It may be wondered how the earlier start on work in the offices led to more supervision, as distinct from that arising later in the occupied section.

(f) Retention of machine parts shed by employer, preventing early use of access 2 and leading to additional costs of transport entering and leaving site £5,200

The non-availability of access 2 was definite, but the question might be how did it relate to the earlier problems over its restriction under item (d), the bar chart being interesting here. There appears to be almost an understatement of the position between the two items. An enquiring mind might wonder whether this suggests there has somehow been an understatement or an underclaim.

(g) Non-availability of garages for storage of materials.

Purchasing in smaller lots, off-site storage and stock-keeping costs £2,200

Double-handling from store to site £3,400

The activities here could well have been covered by daywork, apart from any charges for storage facilities themselves. A routine check could be made back into the final account.

(h) Delay time over fuel pit, engaging fresh gangs and special supervision £3,600

This item is likely to have occurred when personnel involved in concrete work are likely not to have been around, even though delayed work went on the R&D bay. Here and elsewhere, time sheets should provide names.

(i) Extension of overall programme time, affecting site supervision and head office costs £9,800

This is a moderately sized job and the amount here is small in relation to likely annual turnover, but then this is a moderately sized contractor. The site element is the easier to ascertain and to justify. The head office should be justified by specific costs, rather than by use of a formula, in view of the size of the contractor – and probably can be just that way. This avoids the global approach, although something of this may be present in an attenuation claim like this.

Details of calculation of additional costs ascertained

Here would follow calculations for each of the elements itemised above, following the principles set out in Chapters 11 and 12 on loss and expense and Chapters 30 and 31 containing the major case study. These calculations have often been set out there in part only and (to ease the emphasis on principles) not in the normal layout of an account, which should show here how the detail has been derived with orthodox calculations.

Appendices

These would contain the items referred to earlier: site drawing, programme and correspondence and, as their titles suggest, they would end the documentation.

CHAPTER 27

METHODS OF DISPUTE RESOLUTION

- Definition of terms
- Perceptions of dispute resolution
- Negotiation
- Litigation
- Arbitration
- Adjudication
- Expert determination
- Mediation and conciliation
- Mini-trial

Previous chapters have discussed the background to procurement and tendering for construction works. Following that, contract provisions for the entitlements contained in a range of contracts for negotiation and settlement, of what is generally but erroneously known as 'claims', have been discussed. More correctly, these are agreement of items under heads of claim arising from entitlements to loss and expense types of provision contained in construction contracts.

From a client or consultant's viewpoint, 'claims' may be those parts of settlement of a final account with a contractor that often have not led to early or to easy resolution. If it were otherwise there would be many fewer 'claims'. From an honest and knowledgeable contractor's (or subcontractor's) viewpoint, and ideally from a similarly honest and knowledgeable consultant's/client viewpoint, 'claims' should not be viewed as such but as entitlements under provisions of a contract. Strictly, the true meaning of claims at common law arises as one for damages for a breach of contract, and inherently will not have been provided for within a contract. 'Entitlements' have been provided for within most of the contracts discussed (not for a breach of contract but in recognition of the occurrence of one or more of certain 'events' listed in standard construction contracts.). Recognition of these may be one thing but (mis)interpretation, computation and disagreement over those entitlements in the context of a particular contract often translates into 'claims'. This is what keeps the construction industry in need of an industry within an industry – the dispute resolution industry.

'Claims' may appear not to need to encompass those areas of settlement contained in all the contracts in this book, except for the NEC form, which are governed by any valuation rules that may apply, particularly in JCT, ICE and GC/Works/1 contracts. But even here, on occasions, there may be considerable areas of disagreement, that have led to 'claims' over the application of just such 'valuation' rules. This was demonstrated in *St Modwen Developments Ltd* v. *Bowmer & Kirkland Ltd* (1996), where valuation of preliminaries and valuation of expenditure arising from provisional sums in contract bills of quantities was the subject of dispute over a considerable sum (Chapter 6).

It is perhaps more likely that disputes will arise over the claimed effects of the issue of many variations, over extensions of time claimed but not given, over disruption and often prolongation of the works, over loss and/or expense claimed to have arisen as an effect of the prolongation and/or disruption, over liquidated or unliquidated damages etc. In the midst of 'expense based' claims, disputes may arise from different interpretation by the parties over extensions of time, deduction or not of liquidated damages, valuation of variations, even interpretation of instructions as variations or not, and so on. In summary it is not easy to list briefly the range of categories over which parties may find, and have found, grounds for disputes.

In passing, in the NEC form – a contract that sets out to put cooperation between the parties at the heart of its procedures – there does not seem to be an inherent place for 'claims', in the sense referred to above, that may arise from a contractor seeking recovery of loss and/or expense; there is no apparent procedure in the NEC for recovery of 'cost based expenditure', unless an NEC form of cost or target cost contract has been used. In usual procedures within the context of the NEC form, an estimate for a 'compensation event' will be priced for its effects on time, cost of work and quality of work, submitted to the project manager and accepted by him, so in theory there should be no need for 'claims' under the NEC form. It would appear that adjudication and, maybe, reference to 'the tribunal' provided under the contract, may turn out to be more common than the theory suggests.

Previous chapters have also counselled the wisdom of endeavouring to settle each dispute as quickly as is possible, within the context of whatever machinery and procedures the particular contract provides. Advice in earlier chapters has been to avoid the presentation of 'a claim document' at an interim point in a contract and, even more so, to try to avoid presentation of a 'claim' at the end of a contract, without a party having endeavoured to exhaust all means of presentation and negotiation that are available to the parties and their representatives. Settlement of each claim on its own, before it is allowed to be caught up with other differences is the ideal.

Chapter 26 has set out a framework for presentation of a comprehensive claim document, in the context of negotiation during or at the end of a contract. The context of that chapter is that the parties are still 'negotiating', in the time-honoured ways that, in most cases, still lead to agreement of a final figure to be paid to a contractor by his client (or to a subcontractor by his client). Indeed, if there is much substance in any 'claim', it is inconceivable that the reason for it and much of the detail and calculations in support of it will not have been discussed, perhaps several times, between the parties during the course of the contract. The circumstances of the 'claim' will, or certainly should, have been apparent to both parties during the contract and both should or will have had opportunities for resolution by use of the usual procedures. But if all these opportunities have not been taken, other measures will probably then be necessary.

This chapter sets out a view of alternative ways in which parties may resolve disputes by the engagement of another body, beyond solely the actions of the parties and their consultants as named and appointed in a construction contract. Formal dispute resolution may need to be used if usual methods of negotiation, leading to agreement of a final account or settlement of an issue during construction have been exhausted. By that stage in a dispute it is accepted that it may be necessary to proceed to litigation or other methods during the course of construction, provided the contract allows for this or does not exclude it.

The following may be useful either for general information or for consideration of possible ways forward when faced with a dispute in practice. However, it is stressed that this chapter provides only an overview of dispute resolution methods and procedures available. Recourse to one particular form of dispute resolution by a party in dispute should not be taken lightly nor solely on the basis of the following text. Once a dispute has arisen, or appears likely to arise, informed advice is almost always necessary, probably at the earliest stage possible. Such advice would need to take into account the nature of the dispute and circumstances peculiar to it, the form of contract under which it has arisen and the options available under the contract for dispute resolution. Although there is a range of organisations (ranging from general consultants to companies specialising in construction claims through to lawyers) that can provide advice on dispute resolution, much of this section of the construction industry has become a specialist area of expertise and advice. It is suggested that advice be sought only from those that can show they have such expertise. Generally the context, nature and size of the dispute will influence the choice of specialist chosen to act for the claimant or respondent and, very probably, the choice of the method of dispute resolution. Also, it is probably necessary that once a dispute is recognised or even suspected then objective advice should be obtained from an organisation that is not currently involved in any relationship on the project. However, at this point it is most likely that, once another, outside organisation becomes involved then the costs of resolving the dispute will steadily begin to mount. It is at this stage there is often the best opportunity for resolution of the dispute, while the parties have the chance.

It has become fashionable for the newer forms of dispute resolution, other than litigation, to be referred to under a banner of 'alternative dispute resolution', not least because this indicates there are 'alternative' methods, that is all other methods are seen as alternatives to litigation (and maybe to arbitration). Rather than dealing only with the 'alternatives' in this chapter, the most common ways of settling a dispute are also discussed below. These embrace litigation, arbitration, expert determination, adjudication, mediation and mini-trials.

Traditionally, until now, there have been three ways of sorting out disputes. The following is an extract from a column in *Building*, contributed by Tony Bingham, their legal correspondent, that came to hand in the very last stages of completing this chapter. It is a quotation from John Parris, a lawyer and previously editor for Longman.

> First, there is self-help: this consists of beating the opponent over the head with a club. It tends to be counter-productive and lacking in finality – especially if your opponent has relatives and friends. However, it is still highly popular in international affairs.
>
> The second method is not greatly superior to the first. It consists of submitting the argument to be settled by somebody who has a bigger club than either of the disputants and who is powerful enough to belt both over the head. That is the principle of the English legal system. It may be effective, but all too often both disputants end up feeling as if they have been belted over the head.
>
> The third and most civilised method of settling a dispute is for those concerned to agree to submit it to a third person in whom both have confidence, and to undertake to abide by his or her decision. That, in essence, is arbitration.

Now there are more than three methods that may be used to determine a dispute.

DEFINITION OF TERMS

Before considering methods of dispute resolution, it is helpful to understand a number of the titles and terms that are regularly used; some of these terms are explored in more detail throughout this chapter.

Ad hoc reference: A reference (see below) that is arranged *after* a dispute has arisen.

Adjudication: A system for interim determination of a construction dispute, ideally not for a series of issues. In construction, adjudication is now governed by the statutory right to have adjudication if either party wants that, which cannot be excluded by any contract exclusion, arising from the Housing Grants Construction and Regeneration Act 1996. Section 108 of the Act provides eight features that must be complied with in any contract adjudication scheme, otherwise the Government Scheme for Construction Contracts (the Scheme) will apply by default. Virtually all construction contracts made after 1 May 1998 are subject to the Act (see the major section on adjudication later in this chapter).

Appointment: The act by which either the parties or a third-party appointing body establishes the identity of an expert, adjudicator or arbitrator.

Alternative dispute resolution (ADR): Strictly, ADR covers any methods of settling disputes other than by recourse to litigation in the courts. But in the vernacular that has grown up in the construction industry, ADR has come to mean dispute resolution by other than litigation or arbitration, perhaps because both these systems are adversarial. Adjudication and arbitration appear to fall under ADR, although they are not strictly consensual, which some feel must be present in ADR.

Arbitration: A form of private dispute resolution between two parties which is subject to the law of arbitration, currently the Arbitration Act 1996, but is consensual, in that the parties have to choose that form of resolution.

Arbitration rules: The expression given to any rules that govern procedure for arbitration. The current model arbitration rules that apply to JCT contracts let after 1 May 1998 are those produced by the Society of Construction Arbitrators, first edition published in February 1998, known as the Construction Industry Model Arbitration Rules (the CIMAR Rules). Previously the JCT rules applied.

Award: The title given to a decision of an arbitrator on issues put to him by the parties. It is usually final and binding, subject to statutory reasons allowed for appeal or review.

Burden of proof: The general rule is that a party alleging a fact must prove the truth of that allegation. When a party has put sufficient evidence to raise a presumption that what he asserts is true then the burden of proof shifts to the other party to rebut that presumption.

Claimant: The title given to the party in arbitration proceedings asserting a right in respect of any claim.

Claims consultants: A title given to consultants who are not lawyers. They may prepare claims, defend claims, prosecute or defend claims on behalf of their clients. In smaller arbitrations and some litigation they act for parties and claim to give a service at better value for money than a client would obtain by engaging lawyers and, possibly, support specialists.

Counterclaim: A claim asserted by a defendant or respondent against a plaintiff or claimant. In principle a counter-claim stands in its own right, as opposed to a defence (see below), and if a claim had not been made earlier, may or would itself have been the claim.

Decision: The title given to what an expert gives in writing as his ruling on the issues as referred to him by the parties. It is sometimes called his 'determination'. It is also applied to an adjudicator's decision.

Defendant: The title given to the party in litigation proceedings denying a right claimed.

Defence: A statement denying a claim. In principle, anything in a claim not specifically denied is deemed to have been admitted. Detailed rules exist for admitting parts, denying parts, doubting parts, neither admitting or denying parts etc in litigation and arbitration proceedings.

Determination: See Decision.

Difference: For the purposes of this chapter, see Dispute.

Disclosure: The term used to describe the process of revealing to the other party, once discovery (see Discovery) has finished, the documents that the other side wants to see to help it in furthering its case or defence. There are exceptions to the need in principle to 'disclose' all documents, if the other side wants to see them. Discovery and disclosure can be a very expensive process, sometimes involving the review, listing and copying of thousands of letters, drawings, agreements etc.

Discovery: A process whereby the parties to a dispute list to each other the documents in their power, custody or control relating to the dispute. It is sometimes discretionary in arbitration but is compulsory in litigation. Not all documents 'discovered' may need to be 'disclosed' to the other party.

Dispute: For the avoidance of doubt 'dispute(s)' is held to include any difference between parties that, in the context of dispute resolution, requires the use of a third party for that resolution. The exact definition of what is a dispute and what can therefore be referred for resolution to a third party will need to be studied carefully in the context of the particular contract and the circumstances giving rise to the dispute or difference. The standard forms that have been reviewed previously contain different provisions over what matters are referable to what method of resolution and at what stage and sometimes in what sequence reference may be made. Disputes and differences have existed over when 'a dispute' is different from 'a difference'.

Disputes clause: A clause in a contract which controls how disputes are to be processed, for instance by reference to adjudication followed by arbitration or litigation.

Expert determination: The process by which an expert decides questions referred to him.

Expert witness: See Opinion evidence.

Formulated dispute: An issue on which the parties have taken up defined positions, which are likely to require arbitration rather than expert determination. Adjudication as now provided in the construction industry is on 'formulated disputes'.

Nomination: The act of one of the parties in a dispute proposing an arbitrator, expert or adjudicator for settlement of the parties' dispute.

Opinion evidence: Generally, only witnesses of fact can testify in litigation or arbitration proceedings but witnesses giving their expert opinion, called expert witnesses, may be admitted to assist a tribunal to understand technical matters. Courts and arbitrators are now encouraged to control the use of expert evidence because of the cost involved, which can sometimes be considerable. Limitation of the number of experts is encouraged in pursuit of modern dispute management both in the courts and in arbitration.

Plaintiff: The title given to the party in litigation proceedings asserting a right in respect of a claim.

Pleadings: This is the term given to the submissions exchanged between the parties that describe the general cause of action and defence but in such a way that the submissions leave the widest possible area for manoeuvre at the trial, while scattering as many hurdles into the manoeuvring area of the opposition. Drafting of pleadings is an important function of counsel and, unlike a statement of case, they are designed to tell little more than the very basic issues in dispute.

Reference: The title given to the procedure by which an arbitrator, expert or adjudicator decides a dispute or issue. Arguably, the word can apply to reference to any tribunal.

Reply: A reply is necessary by a claimant or plaintiff to set out its challenge to a defence. No new head of claim can be produced at this stage, strictly speaking.

Respondent: The title given to the party in arbitration proceedings denying a right claimed.

Rules: See Arbitration rules.

Schedule of terms: An expression sometimes used to detail the contract arrangement between the parties and a tribunal.

Set-off: A claim made by a defendant as a defence to a claim.

Standard of proof: In civil litigation and arbitration the standard of proof required for most claims is that of proving the evidence is true 'on the balance of probabilities'.

Statement of case: An expression used particularly in arbitration rules to describe a party's pleading that comprises arguments plus the documentary evidence upon which those arguments rely and state the remedy sought. An example of this was the JCT Arbitration Rules, which have now been replaced by the JCT Construction Industry Model Arbitration Rules (the CIMAR Rules with a JCT amendment). Statement of claim procedure now applies in the CIMAR Rules for a 'full procedure' and statement of case procedure where there is a 'documents only' arbitration. Statement of case procedure does not imply that anything not denied is admitted and often matters that come to light during the proceedings may be raised.

Statement of claim: An expression used in litigation to cover the basic requirements of a claim setting out a short description of the claim and a succinct statement of the facts relied upon. The claimant should state that he believes the statement to be true and state the remedy sought in order to identify which court should have jurisdiction.

Terms of reference: An expression used to describe the terms agreed between the parties and an expert or adjudicator to detail the issues to be decided by him and any procedural methods that they require him to follow. The term is sometimes used in a similar way in arbitration.

PERCEPTIONS OF DISPUTE RESOLUTION

Before looking at the range of methods used in the construction industry to settle disputes and differences, reference is made to the perceptions that are held of the effectiveness of methods of dispute resolution and the possible risks that each method may have for its users. Until recently there has been little information available on the amount of use and the effectiveness of each method of resolution.

Table 27.1 Actual experience of using dispute resolution techniques

	Number of times respondents had participated in the techniques during the preceding 12 months	Number of times respondents had participated in the techniques during their careers
Mediation/conciliation	251	1,024
Mediation/arbitration	28	71
Executive tribunal	32	151
Expert determination	178	943
Adjudication	193	1,096
Arbitration	711	6,324
Litigation	890	8,950

Source: This table is reproduced by permission of the authors, Nicholas Gould and Michael Cohen; it was originally published in April 1988 in Volume 17 of *Civil Justice Quarterly*.

The authors acknowledge and refer below to research, published in the *Civil Justice Quarterly*, Volume 17, April 1998, carried out by Nicholas Gould and Michael Cohen, who were senior lecturer and research assistant respectively at the University of Westminster when they carried out their survey. Their work was conducted with industry support and funds from the Department of the Environment, with additional funding from the Society of Construction Law and Masons and Nabarro Nathanson, two firms of solicitors. Questionnaires and interviews were used to obtain results which revealed, among other things, the extent to which the industry currently uses each form of dispute resolution and the perception that users have of each method.

Table 27.1 shows the use that was made of each method of dispute resolution in the preceding 12-month period and also over the career of the respondent to the survey. The sample on which the replies are based was around 400. It is clear from this research that litigation and arbitration were the methods most commonly used by the respondents to the survey. Table 27.2 shows how the survey respondents perceived the effectiveness of the various methods of dispute resolution. The survey sought to establish how respondents 'perceived' the types of dispute resolution technique rather than how they had 'experienced' them. The respondents were asked to rate the effectiveness of a range of dispute resolution techniques with a number between 1 and 5, where 1 indicates very ineffective and 5 indicates very effective. In Table 27.2 the first figure indicates the average response, and the figure below indicates the percentage of respondents who completed the question. The article summarises its findings as follows:

- Most respondents perceived negotiation as the most effective method in all respects.
- Most respondents appeared most able to answer questions relating to negotiation.
- Most respondents were slighly less confident about arbitration and litigation.
- Most respondents were particularly unsure about the other processes of dispute resolution.

Table 27.2 Perceptions of dispute resolution techniques[a]

Attribute	Negotiation	Mediation[b]	Expert determination	Adjudication	Arbitration	Litigation
Reduces time necessary to resolve dispute	3.9 (96.3%)	2.4 (73.6%)	2.3 (71.9%)	2.2 (73.3%)	1.6 (90.6%)	1.3 (90.1%)
Reduces the cost of resolving disputes	4.2 (95.7%)	2.5 (73%)	2.3 (71.6%)	2.3 (73.3%)	1.4 (90.3%)	1.2 (89.2%)
Provides a satisfactory outcome of resolving disputes	3.7 (95.7%)	2.3 (72.2%)	2.1 (71%)	2.0 (72.2%)	2.4 (90.1%)	2.3 (88.9%)
Minimises further disputes	3.2 (94%)	2.1 (70.2%)	1.9 (70.2%)	1.8 (71.9%)	2.2 (88.6%)	2.1 (86.9%)
Opens channels of communication	3.9 (94.9%)	2.5 (70.7%)	1.8 (69%)	1.8 (71.0%)	1.6 (88.4%)	1.4 (86.4%)
Preserves or enhances job relations	3.8 (94%)	2.3 (71%)	1.9 (69.3%)	1.8 (71.3%)	1.4 (87.2%)	1.1 (86.4%)

[a] Each entry is an average of the respondents' perceptions, where respondents used a five-point scale to rate the ability of each technique to deliver each attribute: 1 = very ineffective, 5 = very effective. The figure in parentheses is the percentage of respondents who completed that question.
[b] Including mediation/arbitration and executive tribunal.
Source: This table is reproduced by permission of the authors, Nicholas Gould and Michael Cohen; it was originally published in April 1998 in Volume 17 of *Civil Justice Quarterly*.

- Most respondents who were consultants considered expert determination as effective whereas contractors considered it to be ineffective.
- Most respondents who were consultants were indifferent as to the effectiveness of arbitration whereas clients and contractors perceived arbitration to range from ineffective to very ineffective.
- Clients considered litigation to be ineffective and contractors were indifferent or considered it to be very ineffective.

But remember these are 'perceptions' rather than 'experiences', although it may be that respondents based their perceptions on only one or several isolated incidents which result in a strong belief that the process is always ineffective or, more likely, effective. This would seem predictable – if there was only limited experience of a method it may lead to generalisation of perception. Arbitration remains the formal default dispute process but slowly 'mediative' methods are beginning to be used, though not to any high degree in comparison with other methods. Table 27.3 shows the authors' view of the range of risk that may be involved in the methods of dispute resolution.

Table 27.3 Assessment of risks between methods of dispute resolution[a]

Method	Length	Speed	Management resources	Control	Cost
Negotiation	1–5	4	5	5	1
Litigation	5	1	1	4	5
Arbitration	4	2	1	4	4
Adjudication	1	4	1	2	2
Expert determination	1–2	4	1	2	2
Mediation or conciliation	1	5	3	1	1
Mini-trial	1	5	3	1	1

[a] The table assumes that, once started, the process goes to completion rather than breaks off or is abandoned: 1 = low or little, 5 = high or large.

NEGOTIATION

Before considering the employment of any 'outsiders' to settle a dispute, it is worth repeating that a further attempt at negotiation may still be appropriate, at some time during the period when other methods of dispute resolution are being conducted. Negotiation should not be forgotten as it may be appropriate and could be less expensive than continuing with or contemplating other methods. If negotiations between the parties during the course of the contract have taken place, it will generally have been between the contractor and the employer's agent – his architect, engineer, quantity surveyor, contract administrator, client representative, project manager or whatever other term has been used.

The philosophy of all the contracts described previously is that the person or organisation named in the contract will have the best interests of his client in mind in the administration of the contract. Standard contracts assume that an employer's agent, in making decisions and carrying out negotiations over contentious matters, will act with expertise and integrity and that in most cases agreement will be reached. Anyone who has taken part in the usual processes of contract administration, let alone litigation, arbitration, expert determination or adjudication may sometimes doubt this. The theory of an agent administering a contract, in the objective manner that the drafters of contracts require, a contract in which he may have been instrumental in causing or contributing to delay or disruption, does not always happen in practice. Also, the expertise of some professionals is simply not sufficient to obtain a settlement by administration of the contract and negotiation within the usual contract procedures. Perhaps this is one of the reasons that has led to separation of the design function from the administration function in the GC/Works/1 and NEC contracts.

But it should always be kept in mind that negotiation should be the first step, and that it is available to the parties even when other methods of resolution have started. If there appears to be no progress in negotiations, it is often necessary for the employer to become aware in sufficient detail of the dispute, perhaps by having an outside adviser look at the dispute fairly early in its life. Also, it should be the concern of any person who is appointed as the employer's agent (architect, engineer, surveyor, manager) in a construction contract to know when *he* may need to seek advice or to advise his employer to seek advice (or possibly both). *West Faulkener Associates* v. *London Borough of Merton* (1994) held that an architect who had failed to understand the obligation of a contractor to proceed regularly and diligently with the works, and did not so recognise, and therefore failed to serve a notice of default that would have entitled his employer to terminate the contract, had failed in his duty of skill and care to his employer. Many a professional may put himself and his employer at risk by failure to act, for instance by seeking or advising his employer to seek further advice. Also *Chesham Properties Ltd* v. *Bucknall Austin Project Management Services Ltd and (Others)* (1996) has established that a professional may have a duty in contract and tort to advise his employer of actual or potential deficiencies in the performance of other organisations (perhaps other consultants or contractors) if his conditions of engagement require that he does this.

Lastly, negotiation should always be in the mind of the parties in relation to 'offers' made to the other party. 'Offers to settle' by one or both of the parties, at any time during the course of a dispute, may be made and may also be significant in relieving a party of significant costs (i.e. legal and associated costs of the party that made the offer) in a tribunal process (litigation, arbitration or presumably adjudication). This is because, in simple terms, if the amount that was in dispute between the parties is eventually awarded at a figure *lower* than the amount a party can demonstrate it has made in an offer to settle during the course of tribunal proceedings (which offer has not been accepted by the other party), then the party that 'refused' the offer will have to bear the other party's costs that were incurred in carrying on with the tribunal *after* the date the offer was made. 'Offers' are a specialist area of negotiation, beyond the scope of this book, but offers, arising from continuous review of a party's negotiating position, should always be kept in mind in the need to reach commercial settlements.

LITIGATION

Litigation as a method of dispute resolution, sometimes called legal proceedings in dispute resolution clauses in standard contracts, is reviewed first because it is the most traditional method. It is not necessarily the most appropriate or successful way of resolving disputes but it is the one, perhaps together with arbitration, to which other methods are then seen as 'alternative'. It is the 'fall-back' process in all construction contracts that will obtain if all other methods of resolution do not apply (even though some of them may have been tried). Arbitration will not be the 'default' process that applied in standard construction contracts before adjudication revisions were published to the JCT family of contracts in April 1998. Litigation has become one of the alternatives in JCT 80 that parties may choose instead of arbitration. This is not so in IFC 84 or the Minor Works form.

The process and the decision

Parties in litigation have not, as such, agreed to enter into litigation (except as now provided as an option in JCT 80). It is implicit that anyone entering into a contract may ultimately end in court in the United Kingdom. Unless the contract has precluded reference to litigation by making arbitration the only option or unless the process has excluded recourse to *any* further tribunal (see later under 'Expert determination' and 'Adjudication'), then a dispute that started in mediation, adjudication or arbitration may end in court, if an appeals process provides for it. Appeals from arbitration will end in court. When persons or corporations within the jurisdiction of the United Kingdom enter into contract, they ultimately may fall within the jurisdiction of the civil courts and will then be bound by the civil justice system, if all other methods of dispute resolution are not appropriate. This is so unless they have made provisions in their contract, or subsequently agree to 'stay', that is not proceed with, dispute resolution by other (alternative) methods, such as adjudication, arbitration or mediation as described below.

Written pleadings or statements of case are submitted and replies, further and better particulars are exchanged, discovery carried out, an 'agreed bundle' of documents to be used in court is produced (probably at least ten or fifteen copies for interested participants) and a hearing date arrives. A court is used and a judge presides and decides from the evidence presented to him which party is successful.

Litigation, because it is adversarial, will require the parties to engage each other in a contest before the court by revealing enough evidence to try to establish that one party's case is stronger than the other's. It will involve the use of lawyers who are trained and paid to discover and present information to help to win their client's case. A case is alleged and facts are presented in an attempt to prove that case. The rules do not require a search for the truth, at least not all the truth, but only sufficient factual (and, maybe, opinion evidence) to meet the objective of 'winning the case'. A decision that a party has proved its case or defence to the standard of proof known as meeting the criteria of 'on the balance of probabilities' is all that is required.

A decision of a court is final, leave of appeal being necessary before appeal to a higher court will be permitted. Leave to appeal is probable only when a significant point of law is at issue that may have general application to other circumstances, or a matter of public policy may be involved. Appeal is not at all likely, if at all, on interpretation of the facts. Appeals from a court's decision are possible, according to rules that are not necessary to expand on here; and if the appeal process becomes extended, the 'certainty' of having the dispute resolved by legal process is removed, until the last court of appeal has given its decision. Appeals can and have reached the House of Lords from relatively minor matters that started in the lower courts. *Ruxley Electronics Ltd* v. *Forsyth* (1996), the swimming-pool case referred to earlier, concerned a dispute over a sum of around £25,000, where costs of each of the parties would have been significantly higher than this.

The parties and costs of litigation

Generally the parties will be represented by solicitors and sometimes also by barristers (counsel), if the case is complex and matters of law have to be argued. In preparation for and defence of a case, construction consultants may also need to be engaged to assemble and prepare documentary evidence to support allegations made and defended. Expert witnesses may be used. The process can be expensive, often very expensive, principally because of the amount of time involved in assembling documentation and preparing submissions, taking witness statements, going through the process of 'discovery' and then 'disclosure' of information to the other side that will probably be needed. Once a case proceeds to trial in court, the daily cost of lawyers, their assistants and witnesses of fact and expert witnesses can quickly amount to considerable sums for each day the trial lasts.

After the court has given its decision on the substantive issues (the issues over which the dispute arose and which were argued before the court), the parties will reflect (if not before) on the costs they have incurred. It may be that the parties can agree between them how they will apportion the total cost bill of the two sides. If they are not able to agree, they may ask the court to rule on costs. Because the court will have listened to and then decided in favour of one of the parties on the substantive issues, it will know the reasons for its decisions and can make a judgment on the payment of costs incurred by the parties.

There is a general rule on recovery of costs, namely that 'costs follow the event', meaning that a successful party under the English adversarial system is entitled to recover 'reasonable costs reasonably incurred' in the procedure, if his case has been successful. This includes, of course, if a court decides there was a successful defence. 'Costs' is a subject all on its own but it is sufficient here to advise that it is unlikely that a successful litigant will recover all the costs that it has incurred. The process of 'taxing' costs may be used, whereby a taxing master, an agent of the court, will review the fees and expenses of lawyers and experts as submitted and may make downward adjustments to them. In simple terms, a litigant may have contracted to pay his professionals' fees and expenses, some of which it may be decided the unsuccessful party should not, in fairness, have to reimburse to the successful party. Some experts in litigation have advised that often only around two-thirds of a successful litigant's costs may be recovered from the unsuccessful party.

The cost of the court has not until now been borne by the parties but there are now proposals for a 'court fee' to be charged for the use of the court in civil litigation. This will increase the cost of litigation. It is probable this will mean that a judge will also decide on responsibility of the parties for the proportion of court costs they should pay.

Advantages and disadvantages claimed for litigation

Advocates of litigation, invariably lawyers, argue that through the process known as 'pleadings' each side can test its own and the other side's case. Pleadings are often made on a broad front before committing a client to perhaps a much narrower avenue, probably at increasing levels of cost. It is argued that if fundamental issues of law are involved litigation is the correct method, the most likely one for resolution of a dispute in order to obtain for a client his rights. Advocates of litigation maintain that resolution of the interpretation of the law is best done in the first place in court, rather than for a party to have to go through the appeal process allowed in arbitration statute. The Arbitration Act 1996, which follows the previous statutes, provides that there is an appeal on a point of law held in an arbitrator's award, whereas findings on matters of fact by the arbitrator are not appealable. With the arrival of statutory adjudication in construction contracts, under the Housing Grants, Construction and Regeneration Act 1996 (see 'Arbitration' and 'Adjudication'), appeal is possible from the adjudicator's decision either to litigation or to arbitration and this may cause more litigation, not less, if appeals become numerous and parties have not deliberately chosen, within their contract options, arbitration as their tribunal route. *Beaufort House Developments (NI)* v. *Gilbert Ash NI and Others* (1998), where the *Derek Crouch* decision has been overturned, will also probably mean that litigation may run arbitration closer, now that amounts in interim certificates can be looked at by a court.

Critics of litigation argue that the parties put themselves into a contest that at best will secure for one party only part of its disputed money (in simple terms, that the principal, most common remedy sought and obtained in litigation is monetary damages) at a considerable cost to both sides. This is because even the successful party may be faced with considerable costs, both legal costs and those costs that will be incurred within his own organisation, that he cannot recover. The financial costs and emotional disruption incurred within the litigant's own organisation can be considerable in having to respond and brief the professional team before and during the court hearing. In the context of the majority of construction disputes, litigation would seem an expensive way to seek resolution, unless significant issues of law are at issue.

ARBITRATION

Arbitration, or something very similar to it, is a very old method of dispute resolution. Evidence for arbitration can be found, for instance, in Egyptian and Roman times, where reference is found of the use of praetors in ancient Rome. It is recorded that in England, in the mid 1200s, merchants and traders, travelling around fairs and commodities markets, had adopted a dispute resolution service. If they found themselves in dispute, they then appointed a respected third party, whom they accepted had good and wide experience of their trade and whose decision they would accept as a final and binding end to the dispute. There was an Arbitration Act 1697 which attempted to re-reinforce the power of arbitrators to make awards. Further arbitration Acts followed in 1889, 1934, 1950, 1975 and 1979. Arbitration, where the seat of the arbitration is England, Wales and Northern Ireland, has been brought up to date by the Arbitration Act 1996 (referred to hereafter under this part as 'the Act'), which came into operation on 31 January

1997 (at present, Sections 85 to 87 are in the Act but have not been brought into force by Parliament). Scotland has its own provisions that are outside this chapter.

Arbitration as a form of alternative dispute resolution is a process whereby parties that have a dispute agree to appoint a third party and to be bound by that party's decision when acting as their arbitrator. Arbitration was generally provided within standard forms of construction contracts as the chosen first-line method of dispute resolution, at least until relatively recently. Rather than allowing recourse to litigation, parties have contracted out of that option by a specific contract term that bound them to settle, by arbitration, any dispute (as defined by each contract term) that arose between them in the future. Even without having a specific term within their contract that provides for arbitration, once a dispute has arisen the parties may agree to chose arbitration to settle that dispute, subject to any statutory requirements or rights (see below) that require or allow parties to have their dispute settled in the first case by adjudication.

The ability to go to arbitration or litigation as the first step in resolution of construction disputes has now been modified in most circumstances by the statutory right of a party to seek adjudication as the first step in resolution. Adjudication was introduced during the 1980s, as a first step in dispute resolution, and appeared in some of the standard construction contract forms. Now, with the introduction of the adjudication provisions of the Housing Grants, Construction and Regeneration Act 1996 (the HGCR Act) under section 108, applicable from 1 May 1998, there will be mandatory provisions for adjudication in most construction contracts, before recourse to litigation or to arbitration can take place. Notwithstanding the HGCR Act, the parties may still agree to dispense with adjudication and enter into arbitration, if that is what they consider is more appropriate for them to resolve their dispute. Adjudication is discussed in more detail below.

The process and the decision

Once the arbitration process is commenced, unless the parties reach a settlement themselves, it will continue until a final award is made by the arbitrator. (A series of awards may be issued by an arbitrator, if issues can be separated from other issues and disposed of as the process continues.) If there are terms in a contract or an agreement between the parties is entered into during the course of the contract, then unless both the parties agree, a submission of one party to have the issues tried in court will usually be 'stayed' from resolution in court. *Halki Shipping Corportion* v. *Sopex Oils Ltd* (1998) has reinforced this.

Arbitration is governed today in England, Wales and Northern Ireland by 'the Act'. In a similar way to litigation, the method involves presentation of evidence to an arbitrator. Under the Act there are provisions and rules applicable for three Arbitrators to be appointed, the third one acting as an umpire if the other two cannot agree, but arbitration in the construction industry is always before a sole arbitrator. Section 82 of the Act defines 'arbitrator' as including an umpire. Unusually for Acts of Parliament, it sets out the purpose of the legislation that the following sections of the Act enact; namely, under section 1, it states that the object of arbitration is to provide:

(a) fair resolution of disputes by an impartial tribunal without unnecessary delay or expense
(b) that the parties shall be free to agree how their disputes are resolved, subject only to necessary public interest safeguards
(c) that the court shall generally have no power to intervene except as specifically provided under the Act

Arbitration can be conducted by documents only (fairly common in simple construction disputes), by documents and then a hearing or even by a hearing only (rare in construction, unless the issue is simple). Submissions are made by each party, in principle in the same way as described for litigation, but in the majority of construction arbitrations there will be more informality, less need to be bound by strict rules of evidence, perhaps less need for the employment of experts than may be necessary in litigation, if the arbitrator is knowledgeable in technical matters, accepting that he cannot usually 'give himself' evidence.

Once this process is over, either when the hearing is ended or the arbitrator has clarified all matters that he needs in 'documents only' arbitrations, he then considers the evidence and produces a written award (it is virtually unknown for an award without reasons and the Act *requires* reasons, unless *both* parties agree to dispense with them). Writing an award is an exact process and an arbitrator has to show how he has arrived at his decision and what evidence has led him to that conclusion. Each issue will require to be listed and to be a part of the award. A series of separate awards may be issued if the arbitrator deems this appropriate. For instance, if he considers that an amount of money will, in any event, be due to one party, he can award this in advance of determining the final amount due.

An arbitrator's 'award' is final on all matters that it has correctly dealt with and usually is not subject to appeal to a court unless conditions set out in the Act are met (sections 67, 68, 69 and 70). Correction of an award may be possible under section 57 for clerical errors, accidental slips or to remove any ambiguity or supply any clarification. An additional award is possible in respect of any claim that was presented to an arbitrator but that was not dealt with in his award. Appeals may be possible during an arbitration on a preliminary point of law (as such, not an appeal from an award, section 45), on a question of law arising out of an award made in the proceedings (section 69). There are other circumstances, known as 'serious irregularity', contained in section 68 of the Act, where a court may remit the whole or only part of an award to an arbitrator for reconsideration, or set an award aside or declare the whole or part of an award to be of no effect.

The agreement to arbitrate

An essential part of arbitration is that the parties in dispute must agree that they wish to be bound by the decision of their arbitrator. Unlike litigation, where one of the parties responds to the action of the other and cannot avoid defending in the legal process (unless he wishes to agree the claimant's case), arbitration initially must arise from a consensual act, albeit it is often by the inclusion of a provision in a standard construction contract. The decision that disputes will be settled by arbitration is usually made by the parties as part of their construction contract for the works or services they intend to have carried out. Occasionally, after a dispute has arisen, the parties may then agree to enter into an arbitration agreement to settle their dispute. Following the HGCR Act 1996 they may find it appropriate to draft their arbitration agreement to recognise that they have agreed to ignore their rights under that Act.

All arbitration agreements must be in writing in order to fall within the control provided by the Act, namely that an award can be enforceable through the courts. An oral agreement to arbitrate would be valid but it would be unenforceable through the courts. The term 'in writing' is wider than it might be thought; 'written' may mean in common usage and has been defined in the Act (section 5) and thereby produces a wider catchment area of 'written' agreements.

Provided a term has been included in a construction contract that the parties have agreed to settle disputes by arbitration this will establish the jurisdiction of an arbitrator to act as the arbitrator in any disputes between the parties. Failing this, arbitration as the chosen method of dispute resolution intended to be used by the parties must be sufficiently clearly referred to in related documents that will, if necessary, need to be determined to be part of the contract, in order to establish the arbitrator's jurisdiction. In any event it is essential for an arbitrator to have jurisdiction in order to act. Without jurisdiction his actions, including any decisions and any awards he purports to makes, will be invalid and therefore unenforceable. This will mean that all costs incurred by the parties, including fees and expenses of the arbitrator, will have been wasted, or 'thrown away' by the parties, who will not be pleased at the waste in time and money. The arbitrator will almost certainly not receive his fees and costs incurred because he has proceeded without jurisdiction. In all arbitrations that arose after 31 January 1997, when the Act became applicable, an arbitrator is empowered, unlike the position before the Act, to rule on his own jurisdiction and it is advisable that he does so as soon as any challenge or uncertainty arises from the parties to it. His ruling on jurisdiction may still be challenged, for instance if it is held that the Unfair Terms in the Consumer Contracts Regulations 1994 apply to the building contract, but he now has power to make a ruling whereas before he could not rule on jurisdiction. If both parties agree or the arbitrator gives permission (under section 32), an application to a court to determine jurisdiction can be made on the application of a party.

Procedural rules governing an arbitration

It may also be that there are some arbitration 'rules' which apply to govern the procedure and processes that are to be adopted throughout the arbitration process to determine the issues in dispute. Rules have been produced over the years by certain bodies, such as the Chartered Institute of Arbitrators, or by contract drafting bodies such as the JCT. Examples of rules referred to in arbitration clauses in JCT contracts used to be the JCT Arbitration Rules. These have now been replaced in JCT contracts entered into after 1 May 1998 by the JCT CIMAR

Rules, referred to above. Other procedural rules may be referred to in arbitration provisions in other construction contracts, for instance the ICE has retained the use of its own arbitration rules. Standard arbitration rules usually include timetables for each party to submit its case, the detail required to be included in documents submitted, options for short hearings, full hearing or settlement of the dispute solely from the submission of documents etc. Even if any rules are not mentioned in a contract, the parties may subsequently agree to be bound by provisions of such rules. Because standard rules contain codified provisions, embodying good practice that has been drawn up after consultation with the industry's leading practitioners, they may be applicable to use, provided they are not inconsistent with the nature of the dispute, for instance if the dispute were solely about quality of work carried out.

Contract between the parties and their arbitrator

Whereas agreement of the parties, either arising from the contract or subsequently made, to arbitrate is crucial before an arbitrator has jurisdiction to act, it is also essential that a contract containing terms for appointment of the arbitrator by the parties is agreed by them with the arbitrator. The parties can appoint an arbitrator by agreement between themselves or, failing this, they can agree that the arbitrator is appointed by a body that maintains a panel of arbitrators. For instance, standard construction contracts have provisions for appointment of an arbitrator by a president or vice-president of The Royal Institution of Chartered Surveyors, The Chartered Institute of Arbitrators, the Royal Institute of British Architects and, in the case of ICE contracts, by an officer of the ICE. Those bodies maintain a panel of trained arbitrators who are experienced in construction contracts.

Ideally, the contract between the parties and their arbitrator should be concluded before the process of arbitration is started and certainly before it proceeds in any detail. In essence the terms between the parties and the arbitrator (the agreement for an arbitrator to act is at least a three-way agreement) are that the parties accept the arbitrator to determine their dispute and agree to pay his fees and expenses, details of which will be in the agreement. Details may also be set out in the arbitrator's terms that they agree to give him powers to direct them to do specific things, if they are not powers already stated in the Act. The parties were always bound jointly and severally for the fees and expenses of the arbitrator and this is now specifically covered in the Act (section 28(1)).

The arbitrator

The function of an arbitrator is similar to that of a judge. He has to act in accordance with procedures in the Act and his decision, called 'the award', is enforceable in the same way as a judgment of the court. Although procedure in arbitration is regulated to some degree by the Act, there is considerable discretion for the parties to agree, it is suggested always in writing, on the procedures that they wish to adopt, including any relaxation of the rules of evidence. Failing such clear agreement between the parties, the arbitrator must then give his directions to the parties on a procedure that he decides is appropriate and which must be followed by the parties. Failure by a party to act as directed can be enforced by certain sanctions given as preremptory orders by an arbitrator. The process is not a consensual one unless both parties agree, in effect, not to comply with an arbitrator's directions, when they have agreed that their own requirements will be followed.

Costs of the parties and the arbitrator

The costs of arbitration are paid by the parties. If the parties do not or cannot agree between themselves on the apportionment of their costs, they can ask that the arbitrator makes an award on them. The Act (section 61(1)) states that the arbitrator may make an award allocating costs as between the parties, subject to any agreement of the parties. Section 66(2) states that an award on costs shall be on the general principle that costs shall follow the event, unless the parties agree otherwise. The arbitrator has discretion only where he considers that in the circumstances it is not appropriate for the general principle to apply in relation to the whole or part of the costs. The arbitrator decides who pays the total costs of the arbitration, which will include the costs of any representatives of the parties, their experts, the costs of any fees due to an appointing body and also the arbitrator's fees and expenses. In simple terms, as outlined earlier for litigation, the losing party pays the costs of the successful party, except where there is reason for mitigation in the circumstances of a particular reference to arbitration. And any offers made by parties during the arbitration may have to be taken into account in apportionment of the parties' costs.

Advantages and disadvantages claimed for arbitration

Advocates of arbitration argue that the main advantage of arbitration is that an arbitrator can be appointed quickly, in comparison with a court, which may have to wait before it can begin a case. The parties may be able to choose and agree the particular arbitrator they want, but failing this, appointment by an appointing body normally follows quite quickly. Because the process is claimed to be quicker than litigation and may often not involve lawyers, it is held to be cheaper than going to court. The process and any decisions made remain private, no public records of any proceedings are made; and the time, venue and procedure for the arbitration can broadly be fitted to the nature of the dispute and the wishes of the parties. Within limits (see section 34 of the Act), strict rules of evidence can be relaxed, provided both parties are given a reasonable opportunity to explain and to have explained to them the case against a party and any defence to that case. Because many disputes arise out of the 'trade and custom' of an industry, a technical arbitrator will be able to bring his experience and understanding to the dispute, provided he remains an arbitrator and does not act as an expert (note, *not* an expert witness), or an adjudicator. As it has been put, 'It is not necessary to explain to an arbitrator, as it may be to a non-technical tribunal or judge, what is the use of a brick or what is the function of the frog in a brick.'

The timetable and venue for meetings, and any hearings, can generally be set for the convenience of and within the control of the parties, reasonably quickly. The parties can obtain their arbitrator through agreement between them or, failing this, through appointment by a nominating body. The appointment process need not take much more than say three to six weeks from a notice of intention to arbitrate being given by one party to the other; the arbitration then begins. This would include any attempt to agree on the appointment or, failing this, the appointment by an appointing body. Furthermore, it is claimed that, particularly if an arbitrator is skilled and uses the powers given to him under the Act, costs of the parties should be less than if litigation is pursued, principally because the arbitration process from beginning to end should be quicker.

Critics of arbitration, who often seem to have come from the ranks of construction lawyers, question the ability of arbitration to resolve issues where questions of law are at the root of matters. This significant disadvantage is claimed when issues of fundamental law are involved, because some arbitrators may not be equipped to decide on law if the issues are complex. It is claimed this sometimes leads to more appeals from arbitration awards on points of law – which therefore add to the overall cost of resolving the dispute – than there might have been if the parties had gone straight to litigation. However, although the parties may agree otherwise, arbitration has always allowed for an arbitrator to take legal advice, to appoint a legal adviser or assessor. This is expressly provided for in section 37 of the Act. A further disadvantage claimed is that multi-party actions are generally not possible in arbitration (unless *all* the parties agree, which is not often). They also question whether or not savings in costs or in time necessarily arise in arbitration. These criticisms may still exist but were often made before the opportunities that now exist for control of the arbitration process, and therefore of the parties' costs, within the powers given to an arbitrator under the Act.

Although parties can agree on which arbitrator they wish to have, in many instances they are already in dispute over issues arising under a contract. By the stage that arbitration is proposed by one of the parties, their disputatious relationship may also have extended to their being in dispute over whether or not there is any need even to go to arbitration. Against this background, parties often will not agree on a mutually acceptable arbitrator (unless they have agreed the name before entering into the contract) and a third-party appointment must then be made. Occasionally, some criticism has arisen when parties have suggested that they have suffered in the arbitration process from the quality of a third-party appointment if that appointment has not led, in their eyes, to a satisfactory resolution of the dispute. Research has shown, for instance from consultation with both parties carried out by the Royal Institution of Chartered Surveyors after arbitration awards have been issued (as part of RICS quality control procedures), that parties are satisfied with the quality of arbitral appointments in the large majority of cases. Occasionally, some criticism is present in the total context of arbitration and third-party appointments. Perhaps to some degree this is unavoidable, because at least one of the parties, and possibly both, may fail to obtain an award in its favour and may therefore be disappointed by an arbitrator's award. Also, the costs involved may emphasise the losing party's dissatisfaction. Perhaps what has also led to criticism of arbitration is the increasing involvement of lawyers, claims experts and expert witnesses in many arbitrations, which many construction participants consider should be a less structured and more effective tribunal than provided by litigation. This involvement has been seen as often extending the arbitration process, compared with a view of arbitration when 'it used to be for people in construction conducted by people in the industry to settle their own disputes, quickly'.

Having said this, arbitration is very often a satisfactory method of resolving disputes, particularly for those requiring the establishment of factual liability and quantum, which are the majority of the construction industry's disputes. Under the Act, it is the firm intention that dispute resolution by arbitration should be 'without unnecessary delay and expense' in order to make it 'fair resolution by an impartial tribunal without unnecessary delay or expense'.

ADJUDICATION

There is no definition of adjudication in the HGCR Act, referred to above, and few can agree at present exactly what adjudication is or what it may become. Adjudication has been called by a member of the judiciary, before the HGCR Act was on the statute book and not relating to that Act, 'a quick, dirty fix'.

It has been said that 'adjudication is what it is'. Perhaps because of the perceived failings in litigation and arbitration, aided by recommendations in Sir Michael Latham's fundamental review of the construction industry, as published in the report *Constructing the Team* in 1994, adjudication has now been given priority in the construction industry to settle disputes, at least as the method that must be used as the first line of resolution. The HGCR Act has made adjudication a statutory right of parties in construction contracts (with a few exceptions, including where the contract is for the residential use of one of the parties, certain highway work, certain work affected by planning matters, agreements to adopt work under the Water Industry Act and certain private finance initiative work). It therefore follows that all consultancy agreements for design and cost services, main contracts, subcontractor and supplier contracts, unless they do not involve any installation work fall within the HGCR Act. It is the intention that any dispute will be resolved in the first case by an interim determination, which will be final and binding until it is reviewed by arbitration or litigation, which cannot be started until after practical completion of the works. The interim decision of an adjudicator will become final unless it is then challenged. Adjudication will be a prerequisite of any recourse to a tribunal of a court or of arbitration.

The process and the decision

Adjudication is the last in the category of dispute resolution that involves a consensual agreement to go to a third party for dispute resolution. It is consensual but presumably if both parties agree, they may decide not to use it. In the construction industry it had its origins in the attempts to resolve disputes between contractors and subcontractors over set-off. Following agreement, by accepting it as a contractual provision, now backed by a statutory requirement for all construction contracts, adjudication from now onwards has become an 'imposed' solution on the parties.

Once the adjudicator is appointed, either by agreement between the parties or by appointment from an appointing body, as detailed for arbitration or expert determination, the adjudicator will very probably have to act to a given timetable. The procedures and process are meant to reach quick, interim decisions and are not designed to take the time that is required for final determination, such as litigation and arbitration. An informal procedure is followed with written submissions within short time limits and there is no provision for a hearing.

The status of the adjudicator's decision has been considered in *A. Cameron Ltd* v. *John Mowlem & Co. plc* (1990), where the plaintiff tried to have the decision enforced as an arbitrator's award. This was rejected by the court, which refused to enforce it as it held that to do so would be to acknowledge that it was an award. The decision of an adjudicator was described as having an ephemeral and subordinate character. The essential point is that the adjudication must be to determine matters that fall within the legal rights of the parties. In the *Cameron* case it was held that 'the contractor shall not be obliged to pay a sum greater than the amount due from the contractor', as allowed within the contract. It was held that the decision was binding only until determination by an arbitrator. But in *Drake & Scull Engineering Limited* v. *McLaughlin & Harvey plc* (1992) a decision was enforced by a mandatory injunction of the court. It appears there is no clear authority for enforcement of an adjudicator's decision, principally because it is not final. It is interim, whereas an arbitrator's award is final, unless it has been appealed, and an award can be enforced as though it were a court judgment.

The HGCR Act establishes the rights of parties to a construction contract to refer disputes to adjudication. Any contract provision, in order to comply with the Act and therefore to be able to avoid having to use the Government Scheme for Construction Contracts (the Scheme) must meet the following requirements:

(a) Enable a party to give notice at any time of his intention to refer a dispute to adjudication.
(b) Provide a timetable that has the object of appointment and referral to an adjudicator within seven days.
(c) Require the adjudicator to reach a decision within twenty-eight days (or longer, if agreed) from when the dispute was referred.
(d) Allow the adjudicator to extend that period by fourteen days with the consent of the referring party.
(e) Impose a duty on an adjudicator to act impartially.
(f) Enable the adjudicator to take the initiative in ascertaining the facts and the law.

Section 108(3) of the HCGR Act requires construction contracts to provide that the decision of an adjudicator is binding (not final) until the dispute is finally determined by legal proceedings, by arbitration or agreement of the parties. The parties may accept the decision of the adjudicator as finally determining the dispute.

Section 108(4) provides that the adjudicator is not liable for anything done or omitted in the discharge or purported discharge of his functions as adjudicator unless the act or omission is in bad faith; and any employee or agent of the adjudicator is similarly protected from liability.

This gives an immunity to adjudicators that is necessary if the introduction of adjudication is to be successful. If a construction contract does not comply with these provisions, the Scheme will be imposed as the framework for adjudication.

There are rules that give a framework to the adjudication process. Procedural Rules for Adjudication, 1996 Edition is published by the Official Referees Solicitors Association (ORSA) and CIC Model Adjudication Procedure has been published by the Construction Industry Council, together with a model adjudicator agreement.

The agreement between the adjudicator and the parties

Adjudicators may be named in the contract, and possibly appointed, at the time that a construction contract is drawn up. When adjudication was discussed during the lengthy consultation period which led to its inclusion within the HGCR Act, it was often proposed that each construction contract should give the name of the adjudicator, perhaps in the appendix to the conditions. It was suggested that because adjudication would have to be a quick and broad method of resolution, no time should be lost in finding an adjudicator. If one were to be named in each contract, it was also suggested that this might somehow act as a deterrent to the parties calling on the adjudicator's services. This must be arguable.

After consultation and much consideration the industry has appeared to come to the view that some contracts will name the adjudicator but that most may not. JCT 80 and IFC 84 have an adjudication agreement and the parties and the adjudicator agree to use it. There is one for when the adjudicator is named in the contract and one for when he is appointed later. A client may name a person perhaps for major works, then the tenderers will be aware of the name and may comment or accept as appropriate. On large projects, perhaps a retainer may be necessary for such adjudicators to remain available at short notice to take up and resolve issues speedily. After consideration the industry has concluded that it will in many cases be inappropriate to name the adjudicator until the nature of the dispute is known; for instance, is it one of disputed quality of work, of contract interpretation or of valuation of work in progress? The choice of adjudicator will probably be influenced by these factors as it may well be that no one person may be the best to span all subjects. On reflection, it is also apparent that a person named in year 1 may, when a dispute occurs in year 3, not be available to suit the urgent timescale needed by the parties. For this reason it has been suggested it is possible that an adjudicator may be an organisation, unlike an arbitrator that is always a personal appointment.

Similar to expert determination, it is essential for the agreement between the parties and the adjudicator to set out the exact terms of reference and terms of delivery and of payment for the adjudicator. Immunity for adjudicator's acts is in the HGCR Act but the possible effects of an adjudicator's decisions should be considered by them and by their clients.

The adjudicator

The adjudicator receives his jurisdiction either from the contract or from the HGCR Act that establishes the right for adjudication and it should be specific on the areas permitted for adjudication. In the HGCR Act, in the statutory scheme it is intended that an adjudicator is not acting as an arbitrator and that in effect he will be acting as an expert, but his decision is not final, unlike that of an expert.

The requirements for the way an adjudicator should perform his duties are virtually as those for an expert. He is not required to act judiciously but only to act impartially and he need not hear evidence but he almost certainly will need to in order to reach a decision, as the requirements stated above for the HGCR Act outline.

An adjudicator will either be agreed between the parties or will be provided by 'an adjudicator nominating body', as defined in the Act. Adjudicators have been trained and a number of the professional bodies maintain a register or panel of adjudicators. It is expected that most contracts will not insert the name of the adjudicator in the contract. If an appointing body is used, the Scheme provides that it must identify an adjudicator within five days.

Costs of the parties and the adjudicator

There is no statutory restraint on how the parties should pay for the costs of the adjudicator and he has no inherent power to apportion them nor his own fees and expenses. These matters must be agreed by the parties when they make their contract with the adjudicator, if the originating contract or any statute has not provided for this (which it often has).

Advantages and disadvantages claimed for adjudication

Adjudication is claimed to have the advantage of being immediate, inexpensive in comparison with other methods, a means of taking a dispute over one issue and nipping it in the bud, before it festers and perhaps joins with other issues and forms part of a much bigger dispute. Because it is now a statutory provision in construction, it must be used during the course of the works for all disputes that have gone beyond resolution by the parties, or at least by their agents. Because of the very tight timescales involved, it is claimed that the parties will have resolution for little cost to themselves and in fees to the adjudicator.

Critics suggest that adjudication will become just another stage that parties will have to go through before choosing a tribunal that they want, either litigation or arbitration. It is claimed that because of the short timescales required by the process, adjudication may work in favour of the claimant (there is no accepted term for the referring party) because he will be able to prepare his case, over some time, in order to choose his time to ambush the other party. The respondent will have very little time to defend. It has been suggested that only simple issues are capable of being adjudicated, because the interlocking of issues will not be possible. Because there is uncertainty over the enforceability of an adjudicator's decision, critics say that it will lead to litigation when the amounts at issue are large and the 'defaulting' party wishes to avoid paying on a decision that is against him. Others have suggested that adjudicators have no power to decide on matters of law and that means there will be appeals on this very issue.

It appears that the industry has a new challenge, the advantages and disadvantages of which may take a little while to become clearer.

EXPERT DETERMINATION

Expert determination (sometimes called determination by an 'independent expert') is a consensual process under which the parties to a dispute, either in their contract or subsequently when a dispute has arisen, agree to be bound by the decision of a third party who has expert knowledge of the subject-matter in dispute. It is the process by which an *expert*, as opposed to an *expert witness*, decides questions referred to him. Though having some similarities to arbitration and to adjudication, it is not a method that has been used very often in the construction industry, although it has been used on some of the largest projects, such as the Channel Tunnel. In essence it is about deciding an *issue* rather than about deciding *disputes*. The third party is commonly known as 'the expert' and the process is held to be quick and cheap; it is also private.

Agreement to use an expert

Expert determination will arise either from an ad hoc agreement or from a contract clause that requires the resolution of disputed issues by an expert. It is the law that the courts will stay litigation (i.e. not allow legal proceedings to commence) if there is an expert clause in a contract. This was decided in *Channel Tunnel Group Ltd and Others* v. *Balfour Beatty Construction Ltd and Others* (1993). There may be multi-party disputes, unlike arbitration (where all parties must agree to allow joint actions to proceed). Because many construction disputes are multi-party disputes, expert determination may be appropriate.

The process and the decision

There are no procedural rules for expert determination issued by a construction organisation, although there are rules for adjudication (see 'Adjudication' below). It is important that procedure is agreed between the parties and the expert and this should enable timetable and rules of procedure to provide the necessary framework that is required for the expert and the parties to make submissions, produce documents, witnesses etc, if that is appropriate to the work of the expert.

Once the expert is appointed, either by agreement between the parties or by appointment from an appointing body (see 'Arbitration' above), it is essential that the expert obtains in writing the exact questions the parties wish him to determine. Failure to do this will almost certainly result in dissatisfaction of one or both parties and may render the expert open to a claim for negligence if loss results from his wrong determination. Courts take the view that if the parties agree to use an expert, they must be bound by his decision, unpalatable though it may be to one or both of them. Usually, there is no appeal process from an expert's determination, unless there has been some patent miscarriage of the process on the face of the result or the expert has acted in bad faith.

The contract between the parties and the independent expert

It is essential that the exact terms of reference for appointment of the independent expert are concluded between him and the parties, not least because of the liability of the expert. The agreement should be in writing and should specify the exact issues that the parties want determined; it should cover the terms and responsibility between the parties for payment of the expert's fees and any expenses as required by him. The agreement may set out any procedure and timetable that is to be followed by the parties and the expert.

The independent expert

The independent expert's role is one of investigation. Like an arbitrator, he *may* receive and may take into account evidence and arguments from the parties in the dispute, but unlike an arbitrator, he is not bound to do so. Unlike an arbitrator, the expert should not and cannot be bound by any submissions and must come to a decision based on any combination of enquiries made by him and on his existing knowledge and expertise. An expert would be unwise to follow litigation or arbitration style procedures because he may subsequently be held to have acted as an arbitrator. He is not required to act judiciously but only to act fairly.

His decision is known as his 'determination' and it is final and binding, with no right of appeal against it, provided the determination has answered the questions put by the parties. There is usually no appeal against an expert's decision, even if the expert has made a mistake, but only if he has failed to answer the question put to him or, indeed, has answered a question not put to him. If he gives an incorrect answer to the right question, he cannot be liable but the parties will not thank him for having saddled them with his determination. If he gives a correct answer to the wrong question, he can be liable in negligence and suffer damages for any loss sustained by a party through that negligence. This is unlike the position of an arbitrator, who has immunity (statutory, see section 29(1) of the Act) for his award, although arguably not for negligence resulting in poor or incorrect administration of an arbitration that results in loss to one or both of the parties. An arbitrator may have his award appealed, usually only on a point of law, not his findings of fact.

Costs of the parties and the independent expert

At the time of making his agreement on his terms of reference and payment, the expert should establish his responsibility for determining any apportionment of the parties costs, how his fee will be paid and by whom. Each party usually bears its own costs, there is no statutory authority for apportionment of the expert's fees and the parties often agree to pay them in equal parts.

Advantages and disadvantages claimed for expert determination

Advantages claimed for expert determination are speed, cost and possibly finality. Because an expert will generally be asked to decide an issue, rather than a number of matters, as in arbitration or litigation, he will have a short timescale and therefore the costs will be lower than for litigation or arbitration. Procedure should be tailored to the issue and to the time allowed for the expert to decide. Formal hearings are best avoided in expert determinations; there are no powers to subpoena witnesses and no obligations for disclosure of documents, all of which makes for a quicker, less costly process and may be viewed as arbitration, particularly if one party were wishing to find opportunity to appeal the expert's decision, perhaps by saying that the process had amounted to arbitration. An expert may take legal advice but it is his decision. He need not hear evidence or arguments but he should be able to demonstrate that he acted fairly.

If the parties wish an expert's decision to be final and binding, even if wrong, they will be bound by it unless it can be shown to have been made in bad faith or fraudulently. An expert can be sued for breach of duty if, for instance, he fails to answer the question put to him. If he answers the correct question, but in the opinion of one or both of the parties he gives the wrong answer, they cannot have the decision reviewed. The expert does not have to consult the parties nor need he be shown to have acted judiciously, only to act fairly towards the parties, because he is not carrying out a judicial function, as are an arbitrator and a judge.

There is generally no appeal against an expert's decision and it is unclear how easy it is to enforce it, as opposed to an award of an arbitrator which it is clearly enforceable as a judgement of the court. A recent illustration of this was the case of *Dixons Group Plc* v. *Jan Andrew Murray-Oboynski* (1997), where it was upheld that if it was clear from the contract or other provisions that the expert's decision shall be final, the court will not easily interfere. If fraud or collusion were to be found, that would be a different matter. Otherwise it is necessary to show that the expert has departed from his instructions. In the *Dixons* case a citation from *Nikko Hotels* v. *MEPP Plc* (1991) was made: 'If he (that is the expert) has answered the right question in the wrong way his decision will nevertheless be binding. If he has answered the wrong question, his decision will be a nullity.' In *Dixons* the difficulty was that the plaintiffs could not identify the instruction said to have been breached or the breach itself. The interpretation of the agreement was left to the expert and thus whether or not there was any requirement as to the basis of the method to be used was a matter for the expert. The expert was not required to give reasons and so could not easily be challenged to find out if he had erred or not.

MEDIATION AND CONCILIATION

The remaining methods of 'alternative' dispute resolution are non-binding on the parties. They involve a process entered into freely by both parties on a 'without prejudice' basis, that is if they fail in the process it cannot be used against them in subsequent proceedings; perhaps easier said than done. The parties must wish to make a resolution of their dispute and are brought together with an independent third party, often called a 'neutral' or 'mediator'. There are no strict rules of evidence, no rules or set procedures and a decision cannot be imposed on the parties by the neutral. He is an 'enabler', there to hold the ring and 'caucus' between the parties.

Mediation and conciliation essentially mean the same thing and are taken here as interchangeable names for a process where the parties voluntarily agree to explore their dispute using the help of a mediator.

The process and the decision

Mediation is the word given to a private and confidential method which attempts to resolve disputes. It is claimed to have been widely used in the United States, where to avoid the use of litigation and the very high costs thereby involved, it is sometimes made a mandatory first step in dispute resolution in many industries.

The parties to a dispute invite a neutral third party to facilitate negotiations between them with a view to resolving the dispute. It is informal and non-binding. There are no universally accepted procedures for mediation but a process such as this may often occur:

(a) The parties recognise that they have a dispute and, ideally, agree that they wish to try mediation as a means of settlement. At a high level in each organisation, they must want to settle their differences and to restore the relationship between them.
(b) They find a mediator, either from their own knowledge or by contact with one of the mediation bodies, such as the Centre for Dispute Resolution (CEDR).
(c) The mediator will contact the parties and arrange a joint meeting at which he will allow each party say one or two hours to present its case.
(d) Then each party will go into private session. The mediator will move from group to group. There may be more than two parties or there may be 'multi-party disputes' and the mediator will hear more about each party's point of view. If anything comes out in these sessions, often called caucuses, the mediator may not reveal it to another party unless expressly given authority to do so.
(e) It is stressed that the mediator does not, at the stage at least, impose his views. He may express opinions and test the information he is hearing in an attempt to cause each party to evaluate its case – to assess its strengths and weaknesses..
(f) The mediator will travel from party to party to convey any offers or to pass on information that he has been given, all the time attempting to find a mutually acceptable position that will lead to agreement between the parties.
(g) If and when an agreement has been reached the parties are called together by the mediator. He will draw up the terms of *their* agreement, because it is one that *they* have reached and the parties should sign it. It may be sufficient to be 'the agreement' or a more formal one may be drawn up, which must of course contain exactly the same conditions.

The mediator

Once the parties have agreed to try mediation, they must find a person who is acceptable to both and who has the abilities to help them. A mediator should be an agent who helps the parties. He should be good at solving problems, able to listen and to communicate with the parties. His function is not to take sides in the dispute, not to try to impose his views or his solution on the parties. Instead his role is to assist the parties in finding their own solution to their dispute.

Mediators need to have been trained in understanding the qualities they will need and in the process of mediation as they travel through its different phases. It is often a lonely, individual role and at the end the parties may not seem to have needed the mediator, but they should realise they may not have reached agreement without him.

The parties

Because the process is non-binding, the representatives of the parties must have clear authority to settle the dispute and not need to refer back to others not present at the mediation. Participants at a high level in each organisation are therefore essential.

Costs of the parties and the mediator

There are no fixed rules on costs. The process is voluntary and, unless the parties agree otherwise, the fees and expenses of the parties are usually borne by them, and the fees and expenses of the mediator are usually borne by the parties in equal shares, paid before the process starts.

Advantages and disadvantages of mediation

Advocates of mediation claim that it helps to foster and reinforce long-term relationships of the parties and enables them to seek resolution without confrontation, which will often be present in the adversarial methods of litigation, arbitration and possibly in some instances of adjudication. It is relatively inexpensive in comparison with other forms of third-party assistance or control, but the tribunal costs of reaching a third-party decision or obtaining third-party assistance are usually the smallest fraction of a party's costs.

Critics say that each party leaves an unsuccessful mediation having revealed important aspects of its case to the other side. Precisely because the process is non-binding, the amount of commitment to achieving success must be evident for each party to see in the other and trust must be there to reveal significant parts of the 'case' to the other side. If mediation fails, resolution by other means may be that much harder.

It is difficult to evaluate and compare the relativity of costs versus benefits by pursuing a conciliatory and cooperative path; this is because perceived 'rights' are surrendered for 'gains and losses' in a theatre that will be changing throughout the conciliation, which usually lasts a day or perhaps a little longer. Because there will be a pressure of time to make decisions, to make concessions, the mediator will not allow the parties to think about things, sleep on them and come back tomorrow. It will appear that the immediate costs of staff time combined with those of the conciliator are much less than those involved in adjudication or arbitration. This should be so because the cost of a day with a mediator and the costs of the high-level personnel from the companies involved may be relatively small. But it is not a true comparison of the overall benefits gained or the possible overall losses averted, because the amount surrendered from the *substantive* issues that are being claimed by a party that perceived it had a 'winning' case may be considerable.

On balance, it may be that a party that perceives it has a weak case will be advised to try to reach a settlement by a conciliatory path. On the other hand, a party that perceives, or has been led by its advisers to believe, it has a strong case will want to try to obtain as much of its 'just rewards' by using the higher-risk path of third-party imposed settlement. If neither party feels strongly that it has a good case, mediation or conciliation should be worth the risk. If it fails, there are always traditional routes of settlement to try.

MINI-TRIAL

A mini-trial is similar to mediation and is a non-binding process. Generally there will be a panel of people comprising a senior executive from each party plus a neutral adviser. The executives will not have had any previous involvement in the dispute. The neutral is there to advise on points of law and to give objective views of the issues; he should also have some knowledge of the subject-matter. Perhaps a lawyer can best provide this 'neutrality'.

A mini-trial should take place after discovery of documents so that there is a set of common information available. It has three stages:

(a) Each side presents its case for one to two hours. Time for any experts would be included in this and a reply from the other side is limited to say thirty minutes. This forces concentration on to only the essential issues.
(b) The executives on each side retire with their advisers and consider their position. They can call on the neutral adviser to assist and answer any questions.
(c) The parties continue meeting together and hopefully an agreement will emerge after an iterative process.

There are several schemes run by trade associations where conciliation has to occur before arbitration can then be used.

In methods of conciliation and mini-trial resolution, the aim above all else is to find an acceptable agreement. There must always be a joint objective to want to find a solution – in many construction disputes one party often has no inherent interest in reaching a solution. Cooperative, conciliatory methods work best where there is a context of continuing business relationships, such as envisaged in those that may be appearing in the industry, loosely under the umbrella of 'partnering'. The advantages of confidentiality and control by the parties of their dispute, as opposed to surrender to a third party who imposes his decision, are seen as key factors in using conciliation as an alternative method of resolution. Unless the parties want to reach an agreement, ADR will not work. If documents are not disclosed, in case they may not be regarded as privileged in any subsequent litigation or arbitration, it may be that the issues are not made sufficiently 'proven'. Likewise, immunity for the neutral witnesses may not be given if ADR is not successful. In the ICE contract, the conciliator is specifically given immunity from being called as a witness by any person seeking to rely on his evidence.

CHAPTER 28

PROVISIONS FOR DISPUTE RESOLUTION IN STANDARD CONTRACTS

- JCT Standard Form of Building Contract 1980
- JCT Intermediate Form of Building Contract 1984
- JCT Agreement for Minor Building Works 1980
- GC/Works/1 Contract with quantities (1998)
- ICE Major Works Contract 1991
- New Engineering Contract 1994

All standard forms of construction contracts now have provisions for dispute resolution that have taken account of section 108(1) of the Housing Grants, Construction and Regeneration Act 1996, referred to hereafter as 'the Act'.

In some cases the forms have been amended and state that the amendments also take into account matters arising out of proposals in *Constructing the Team*, the 1994 Latham Report. Provisions for the statutory right under the Act to adjudication in construction contracts came into force for all contracts entered into after 1 May 1998. Among other things, the Act gave a statutory right for a party to a construction contract to have adjudication used in the first instance in order to try to settle an issue between parties (there are a few exceptions to this, particularly where the work is carried out for domestic purposes. Only after an adjudication procedure has been used and the parties have failed to resolve their dispute may they then go to the next stage of litigation or arbitration. Although adjudication is now a statutory right of the parties, exercisable by both or more likely only one, both parties may decide not to use adjudication and then use whatever provisions their contract has for dispute resolution. This chapter reviews the features that the forms discussed in previous chapters have for dispute resolution by adjudication, arbitration and litigation.

Before dealing with each contract that has been previously discussed within this volume, section 108(1) to (5) of the Act is repeated below, as it is the statutory requirement that any construction contract comply with that section or, if not, the adjudication provisions of the Scheme for Construction Contracts related to the Act will apply.

Right to refer disputes to adjudication

108-(1) A party to a construction contract has the right to refer a dispute arising under the contract for adjudication under a procedure complying with this section.

For this purpose 'dispute' includes any difference.

(2) The contract shall
(a) enable a party to give notice at any time of his intention to refer a dispute to adjudication;
(b) provide a timetable with the object of securing the appointment of the adjudicator and referral of the dispute to him within 7 days of such notice;
(c) require the adjudicator to reach a decision within 28 days of referral or such longer period as is agreed by the parties after the dispute has been referred;
(d) allow the adjudicator to extend the period of 28 days by up to 14 days, with the consent of the party by whom the dispute was referred;
(e) impose a duty on the adjudicator to act impartially; and
(f) enable the adjudicator to take the initiative in ascertaining the facts and the law.

(3) The contract shall provide that the decision of the adjudicator is binding until the dispute is finally determined by legal proceedings, by arbitration (if the contract provides for arbitration or the parties otherwise agree to arbitration) or by agreement.

The parties may agree to accept the decision of the adjudicator as finally determining the dispute.

(4) The contract shall also provide that the adjudicator is not liable for anything done or omitted in the discharge or purported discharge of his functions as adjudicator unless the act or omission is in bad faith, and that any employee or agent of the adjudicator is similarly protected from liability.

(5) If the contract does not comply with the requirements of subsections (1) to (4), the adjudication provisions of the Scheme for Construction Contracts apply.

All the standard contracts referred to below have included within their dispute settlement provisions for adjudication procedures that comply with sections (1) to (4), in essence by matching the above requirements and therefore avoiding section 108(5) becoming operative by default. Provisions in JCT 81, the ICE Minor Works, the ICE Design and Construct and the NEC subcontract have not been discussed in detail as they are similar in principle to the provisions in the main forms, although they should be checked in detail before assuming any exact provision.

There are no standard procedures for adjudication, unlike the CIMAR Rules for arbitration. The Construction Industry Council has published a Model Adjudication Procedure (MAP) and a model adjudicator agreement, and another is published by the Official Referees Solicitors Association (ORSA). Table 28.1 shows a comparison of dispute settlement provisions in the main standard contract forms.

Table 28.1 Dispute settlement provisions and options in the main standard contract forms (correct at August 1998)

Contract	By agent of employer	Conciliation or mediation	Adjudication	Arbitration	Litigation
JCT 80	*	–	*	*	*
JCT/IFC 84	*	–	*	*	–
JCT Minor Works	*	–	*	*	*
GC/Works/1	*	–	*	*	*
ICE Major Works	*	*	*	*	–
NEC	*	–	*	a	a
DOM/1	*	–	*	*	*

[a] The NEC contract provides for reference to 'a tribumal' to be agreed by the parties at contract stage. A tribunal would include arbitration, litigation or expert determination.
* = provision or option is present
– = provision or option is not present

JCT STANDARD FORM OF BUILDING CONTRACT 1980

Amendment 18 to the JCT 80 form for use with and without quantities was issued in April 1998 with amendments arising out of proposals in *Constructing the Team* and the Act plus a number of other amendments to JCT 80, in total twenty-two items.

The provisions for dispute resolution begin with article 5 of the Articles of Agreement which provides that under the contract either party may refer a dispute or difference to adjudication in accordance with clause 41A. Two new articles have been introduced into JCT 80 to allow (a) by article 7A for any dispute or difference to be referred to arbitration in accordance with a new clause 41B and for the arbitration to be conducted according to the CIMAR Rules, and (b) by article 7B for any dispute or difference to be determined by legal proceedings in accordance with clause 41C.

Part 4 of the JCT 80 contract is headed 'Settlement of disputes: Adjudication, Arbitration, Legal Proceedings' and contains clauses 41A, 41B and 41C. The JCT guidance notes issued with amendment 18 give advice on the choice between arbitration and litigation. The JCT guidance notes state that the provisions have been drafted after advice from leading counsel. It is assumed that this provision was made to place litigation on more of an equal footing with arbitration concerning looking at certificates issued and decisions made and that the effects of *North Regional Health Authority* v. *Derek Crouch* (1984) will be limited or circumvented. The JCT amendment was issued before the *Beaufort House Development* case reversed *Crouch*.

Clause 41A: adjudication

41A.1 Applies where either party refers a dispute to adjudication ('dispute' and 'difference' are referred to hereafter as 'dispute').

41A.2 Provides that an adjudicator shall be an individual agreed by the parties, or by a person named in the contract appendix as 'the nominator'. Clause 41A.2.1 requires that the parties and the adjudicator must agree that the adjudicator is appointed under the 'JCT Adjudication Agreement' which is provided separately from the contract.

41A.2.2 Provides that any agreement of the parties on appointment of an adjudicator must be reached with the object of securing an appointment within seven days of the date of the notice of intention to refer or that any 'nominator' shall appoint within seven days.

41A.2.3 Provides that the adjudicator shall execute with the parties the JCT Adjudication Agreement.

41A.3 Provides for the parties to agree a replacement adjudicator if the appointed one dies, becomes ill or otherwise unavailable and therefore cannot act. Alternatively either party may apply to the nominator for a replacement. Again the JCT Adjudication Agreement is to be used.

41A.4.1 Provides that either party can give a notice of intention to refer to adjudication, that the notice shall give brief details of the dispute in the notice. Within seven days of the date of the notice, or the execution of the JCT agreement by the adjudicator if later, the party giving the notice shall refer (called 'the referral') the dispute to the adjudicator for his decision. With the 'referral' the party must give particulars of the dispute, a summary of the contentions on which he relies, a statement of the relief or remedy which is sought and any material he wishes the adjudicator to consider. A copy of all this must be sent to the other party.

41A.4.2 Provides that 'the referral' shall be by actual delivery or by fax or by registered post or by recorded delivery. All supporting documents must be sent with the 'referral'. If fax is used, then for record purposes, it should also be sent by actual delivery or by first-class post.

41A.5 Deals with the conduct of the arbitrator.

41A.5.1 Provides that the adjudicator shall immediately upon receipt of 'the referral' confirm receipt to the parties.

41A.5.2 Provides that the party not making 'the referral' may within seven days of the date of 'the referral' send to the adjudicator and the other party a written statement of the contentions on which he relies and any material he wishes the adjudicator to consider.

41A.5.3 Provides that the adjudicator shall within twenty-eight days of his receipt of 'the referral' act as an adjudicator, not as an expert or an arbitrator, and send his decision in writing to the parties. If the referring party agrees, the period of twenty-eight days may be extended by fourteen days; or by agreement of the parties, the period may be extended as agreed.

41A.5.4 Provides that the adjudicator is not required to give reasons for his decision.

41A.5.5 Provides that the adjudicator shall act impartially, set his own procedure and has absolute discretion to take the initiative in ascertaining the facts and the law. Eight subclauses provide that the adjudicator can

> use his own knowledge and/or experience; open up, review and revise any certificate, opinion, decision, requirement or notice issued given or made under the contract as if no such certificate, opinion, decision, requirement or notice had been issued or made; require from the parties further information from that they submit; require the parties to carry out tests or open up work or further open up work; visit the site or any workshop; obtain such information from any employee or representative of the parties; obtain from others such information and advice as he considers necessary on legal matters, subject to giving notice to the parties together with a statement or estimate of the cost involved; deciding the circumstances or period for which a simple rate of interest shall be paid.

41A.5.6 Provides that if either party fails to enter into the JCT adjudication agreement then the decision of the adjudicator is not invalidated.

41A.5.7 Provides that each party shall meet its own costs except that the adjudicator may order who should pay the cost of any testing he requires.

41A.6.1 Provides that the adjudicator shall state how payment of his fee and reasonable expenses shall be apportioned between the parties. In default of this, the fee and expenses shall be equally apportioned.

41A.6.2 Provides that the parties are jointly and severally liable to the adjudicator for his fee and expenses.

41A.7.1 Provides that the adjudicator's decision is binding on the parties until the dispute is finally determined by arbitration or legal proceedings or by an agreement in writing by the parties made after the decision of the adjudicator was given. A footnote in the contract states that arbitration or legal proceedings are not an appeal against the decision of the adjudicator but are a consideration of the dispute as if no decision had been made by him.

41A.7.2 Provides that the parties shall comply with the decisions of the adjudicator and ensure they are given effect.

41A.7.3 Provides that if either party does not comply with a decision of the adjudicator then the other party shall be able to take legal proceedings to secure such compliance. It remains unclear how enforceable some decisions of adjudication will be and case law does not suggest that enforceability will be as simple as Latham intended.

41A.8 Provides that the adjudicator is immune from anything done or omitted to be done when acting as an adjudicator unless resulting from something done in bad faith.

There is a JCT Adjudication Agreement that will apply either if the parties have named an adjudicator in the contract or have one appointed by applying to one of the nominating bodies (the Royal Institute of British Architects, the Royal Institution of Chartered Surveyors, the Construction Confederation or the National Specialist Contractors Council) for appointment of an adjudicator.

Clause 41B: arbitration

41B Provides for arbitration either if the parties choose to have article 7A in their contract by deleting the alternative of article 7B (legal proceedings). It provides for the Construction Industry Model Arbitration Rules (CIMAR) current at the base date to be used in any arbitration (unless the parties were to delete their use from the contract).

41B.1.1 Provides for a party to serve on the other a notice of arbitration in accordance with CIMAR rule 2.1. The notice should be written identifying the dispute and requiring the other party to agree to the appointment of an arbitrator. CIMAR rule 2.3 requires that the parties agree within fourteen days after the notice has been served (or any agreed extension) the name of an arbitrator, failing which either party may apply to the nominating person in the contract for an appointment. The Royal Institute of British Architects, the Royal Institution of Chartered Surveyors or the Chartered Institute of Arbitrators may appoint under the contract provisions.

Clauses follow that contain some of the Arbitration Act 1996 provisions and refer to the CIMAR Rules. The JCT reproduces the CIMAR Rules as the JCT 1998 edition of CIMAR.

Clause 41C: legal procedings

This new clause provides that the parties may include within their contract that any dispute may be determined by legal proceedings. Clause 41C.1 provides that the issues that can go to a court are the same as those that can go to arbitration as provided in article 7A and clause 41B. It is unclear how enforceable some decisions of adjudication will be, and case law does not suggest that enforceability will be as simple as Latham intended.

JCT INTERMEDIATE FORM OF BUILDING CONTRACT 1984

Amendment 12 to IFC 84 has provided for adjudication and revised provisions for arbitration. There is no provision for contractual reference to legal proceedings, as in JCT 80 and the Agreement for Minor Building Works. Presumably, if legal proceedings are the preference of the parties at contract stage, then this would require deletion of the arbitration provision, thereby making recourse to the courts compulsory.

A new article 5 provides for disputes to go to adjudication and article 7 for arbitration to be used either in the first case (by agreement of the parties to omit adjudication) or after adjudication has not determined a dispute. Article 7 provides for any matter to go to arbitration except in connection with enforcement of any decision of an adjudicator. Arbitration can be during the works or after completion, except for VAT matters and the statutory tax deduction scheme.

Section 9 provides for similar arrangements for adjudication (clause 9A) and arbitration (clause 9B) to those contained in JCT 80.

Clause 9A: adjudication

9A.1 Provides for application of the clause when either party 'refers' to adjudication.

9A.2 Provides for the identity of the adjudicator to be as named in the contract or as nominated by 'the nominator' (by one of the listed bodies set out under JCT 80).

9A.2.1 Requires a 'nominated' adjudicator to contract on the JCT Adjudication Agreement.

9A.2.2 Provides for referral of the dispute within seven days of the date of intention to refer, as done in clause 9A.1.

9A.2.3 Provides that the parties will execute the JCT Adjudication Agreement with the adjudicator.

9A.3 Provides a mechanism for replacement of the adjudicator through death or other disability to act.

9A.4.1 Requires one party to give 'notice of intention to refer' a dispute to adjudication, together with brief details of the dispute in the notice. Within seven days of that notice or, if later, within seven days of executing the JCT Adjudication Agreement, the party giving 'notice of intention to refer' shall refer (called 'the referral') to the adjudicator for his decision. It is accompanied by a summary of the issue and the remedy or relief sought, with a copy to other party.

9A.4.2 Provides for methods of communication.

9A.5.1 to 9A.5.3 Provide a timetable as JCT 80.

9A.5.4 States that the adjudicator is not obliged to give reasons.

9A.5.5.1 to 9A.5.5.8 Allows the adjudicator to act as set out under JCT 80 clause 41A.5.5.

9A.5.6 Provides that failure of either party to enter into the JCT agreement or to comply with the requirements of the adjudicator shall not invalidate the adjudicator's decision.

9A.5.7 Provides for parties to meet their own costs but for the adjudicator to direct on who should pay the cost of any test or opening up required, under clause 9A.5.5.4.

9A.6.1 Provides for the adjudicator to apportion payment of his fee and reasonable expenses between the parties.

9A.6.2 Makes the parties jointly and severally liable to the adjudicator for his fees.

9A.7.1 Makes the decision of the adjudicator binding until the dispute is finally determined by arbitration, legal proceedings or agreement of the parties. Any new tribunal shall not be considered an appeal but a consideration as if no adjudication had taken place.

9A.7.2 Requires the parties to comply with decisions of the adjudicator and ensure that decisions are given effect.

9A.7.3 Provides that if a party does not comply then the other can take legal proceedings.

9A.8 Allows immunity to the adjudicator.

Clause 9B: arbitration

9B.1.1 Provides that either party, in accordance with article 7, may serve on the other party a written notice of arbitration requiring him to agree to the appointment of the arbitrator. If the parties fail to agree within fourteen days from this notice (or within any agreed extension) then the appointment process takes place. CIMAR rule 2.5 provides that the arbitrator is appointed upon his agreement to act or his appointment under rule 2.3 whether or not his terms have been accepted.

9B.1.2 Provides for joinder of proceedings and that a party may give a further notice for the same arbitrator to act in consolidation of proceedings with a further dispute.

9B.2 Allows the arbitrator to decide on a similar set of issues as outlined in JCT 80.

9B.3 Provides that the arbitrator's award is final and binding, subject to appeal on any question of law arising in the course of the reference (this follows sections 45 and 69 of the Arbitration Act)

9B.4.1 and 9B.4.2 Allow for appeals to the courts.

9B.5 Refers to the Arbitration Act.

9B.6 Provides for the JCT edition of the CIMAR Rules current at the base date to apply to the conduct of arbitration.

JCT AGREEMENT FOR MINOR BUILDING WORKS 1980

Amendment MW11 to the JCT Minor Works form (article 4, part 4.1) provides that an adjudicator must be appointed either by agreement of parties or by appointment by one of the nominating bodies as listed under IFC 84. If there remains a dispute after adjudication then an arbitrator shall be appointed by agreement or by appointment from one of the three bodies listed in JCT 80.

This form has a supplementary memorandum, of which Part D deals with adjudication and Part E with arbitration. Article 4.2 provides for appointment of an arbitrator, subject to article 4.1, by agreement or by one of the appointing bodies. As mentioned in the guidance notes is the possibility that where the employer is a 'consumer' as defined in the Unfair Terms in Consumer Contracts Regulations 1994, then by article 4.2A, legal proceedings may be preferred to arbitration because of the possibility of a potential breach of those regulations. This is not considered to be applicable where the Minor Works form will have been used and drafted by an employer's professional team.

Adjudication article 4.1 and Part D of supplementary memorandum

Clauses D1 to D8 provide similar requirements to those in JCT 80 and IFC 84. Use of the JCT Adjudication Agreement is part of the provisions.

Arbitration article 4.2 and Part E of supplementary memorandum

Clauses E1 to E7 provide similar requirements to those in JCT 80 and IFC 84.

Legal proceedings article 4.2A and Part E of supplementary memorandum

Clause 4.2A.2 provides by contract the ability of the court to open up certificates; this is to avoid the *North Regional Health Authority* v. *Derek Crouch* (1984) restriction that had been adopted by the courts until *Beaufort House* v. *Gilbert Ash* (1998).

GC/WORKS/1 CONTRACT WITH QUANTITIES (1998)

Material changes have been introduced from the previous form GC/Works/Edition 3 to strengthen the procedures. All disputes can now be adjudicated and the adjudicator is to be a fully independent person. No provisions are now subject to the 'final and conclusive' resolution by the employer, that was in the previous form. GC/Works/1 Contract (1998) details contract provisions for the procedures to be followed in order to adjudicate disputes.

In the commentary to the contract, it is stated that the procedures in condition 59 (adjudication) are to be followed in order to adjudicate disputes and it 'is intended to avoid festering disputes harming the working relationships on which the project depends for success'. If adjudication is not successful then condition 60 (arbitration and choice of law) comes into the procedures for dispute resolution. The commentary states that the adjudicator and the named substitute adjudicator should, wherever possible, be appointed before, or simultaneously with, entry into the contract. It is stated that some decisions of the employer, although necessarily open to adjudication, cannot in practice be reversed, and any adjustments must therefore be made by way of financial compensation, provided the adjudicator finds the employer was at fault.

The commentary states that the same adjudicator should be named in all contracts on the same project – whether with contractors, subcontractors or consultants – in order to avoid diverse decisions being given by different adjudicators. It suggests that a main contractor may wish to ensure the same person is named in subcontracts.

Adjudication (condition 59)

Condition 59 reflects section 108 of the Act. The adjudicator is to be named in the Abstract of Particulars. He has power to award costs and his decisions are binding until varied or overruled in arbitration or litigation; they must be complied with forthwith. The adjudicator is given express power to vary or overrule decisions made by or on behalf of the employer, but the effect is limited to financial compensation in some cases. Condition 59 has eleven subsections:

59(1) Both employer and contractor can give notice of intention to refer to adjudication. There is a model form 20 in the commentary for this notice. Within seven days of such notice a further notice of referral, model form 21, actually refers the dispute to adjudication.

59(2) The notice of referral must set out the principal facts and arguments relating to the dispute. Copies of all relevant documents in the possession of the party giving the notice shall be enclosed with the notice. A copy of the notice shall be sent to the other party, the project manager (PM) and the quantity surveyor (QS). The task of assembling the documents may be considerable and will combine the steps which, in an arbitration would be a claimant's points of claim and documents that would be produced in discovery. (Chapter 27 explains the meaning of 'discovery'.) Part of the criticism of adjudication is that the referring party will have relative leisure and time to prepare his notice or referral and assemble his ace while the other party will have a strictly limited period to respond. During the construction period, the referring party is much more likely to be the contractor than the employer, because disputes will more than likely arise from the contractor wanting more money than the employer or his consultants may consider justified.

59(3a) If the named adjudicator is unable to act or ceases to be independent of the employer, the contractor, the PM and/or the QS, he is then substituted by the alternate named at contract stage.

59(3b) It is a condition precedent to appointment that the adjudicator shall agree to abide by the time limits in this condition.

59(3c) This sets out provisions for appointment of an adjudicator if one has not been agreed at contract stage.

59(4) The replying party, the PM and the QS may submit representations to the adjudicator no later than seven days from the receipt of the notice of referral. This is a very limited period and will not allow elaboration or great sophistication in the material produced. The other party can subsequently submit more material to the adjudicator during this period.

59(5) The adjudicator must notify his decision no earlier than ten days and no later than twenty-eight days from receipt of the notice to refer or such longer period as the parties may agree after the dispute has been referred. The adjudicator may extend the period of twenty-eight days by up to fourteen days with the agreement of the referring party. The adjudicator's decision is valid if issued after the time allowed. The time limits arise from section 108 of the Act. The adjudicator's decision shall state how the cost of his fees and expenses (or salary with overheads) is to be apportioned between the parties and whether or not one party shall bear the whole or a part of the legal and other costs of the other party.

59(6) The adjudicator does not have to hold a hearing but he may do if he so wishes. He can take the initiative in ascertaining the facts and the law and the parties shall enable him to do so. He is to act as an 'expert adjudicator' not as an arbitrator but he shall have the powers of an arbitrator acting in accordance with condition 60 (arbitration and the law) of the GC/Works/1 (1998) form. He has very full powers to assess and award damages, legal and other costs and expenses. Notwithstanding condition 47 (finance charges) of this form to award interest, the adjudicator can award simple or compound interest from such dates, at such rates and with rests as he considers on:
(a) The whole or part of his award (this is the term used in this contract, although many commentators use this term for arbitrator's decisions and use 'decision' for adjudicator's decisions) for any period up to the date of his award.
(b) The whole or part of any amount claimed during the proceedings and outstanding at the start of the proceedings but paid before the award was made. This can include for an award of interest and damages, legal and other costs and expenses.

59(7) Subject to condition 60 (arbitration and choice of law) the adjudicator's decision is binding until the dispute is finally settled by legal proceedings, by arbitration (if the parties agree to arbitration or the contract provides for arbitration) or by other agreement. The parties could, of course, agree to accept the decision as final and binding to settle their dispute.

59(8) The adjudicator also has powers to vary or overrule any decisions previously made by the employer, the PM or the QS other than
(a) under condition 26 (site admittance)
(b) under condition 27 (passes)
(c, d, e) certain parts of condition 56 (determination and notice of determination) and condition 58(d) (decisions to determine the contract)
(f) decisions to assign (condition 61)

59(9) Notwithstanding condition 60 (arbitration and the choice of law), the parties must forthwith comply with any decision of the adjudicator and shall submit to summary judgment and enforcement of those decisions.

59(10) If one of the parties requests, the adjudicator shall give reasons within fourteen days of the decision being notified to the requesting party.

59(11) The adjudicator, and any employee or agent of him, is not liable for anything done or omitted by his actions as the adjudicator, unless caused by an act or omission in bad faith.

Arbitration and the choice of law (condition 60)

If adjudication has not brought a final resolution of a dispute, the contract provides a second level of dispute resolution. These procedures can only be used after completion of the works. The commentary to this form advises that legal advice should be taken before beginning and throughout any proceedings. An arbitrator has additional express powers to rectify the contract, order inspections, measurements or valuations, consolidate arbitral proceedings and the same power as the adjudicator to vary or overrule decisions. The arbitrator is named in the Abstract of Particulars.

The arbitrator will determine disputes within the framework of the Arbitration Act 1996, which provides for recourse to the courts to determine a point of law during an arbitration or after an arbitrator's award on a question of law.

ICE MAJOR WORKS CONTRACT 1991

The ICE Conditions of Contract, Sixth Edition, January 1991 have been amended to take account of settlement of disputes.

Clause 66(2) provides that if at any time the employer of the contractor is dissatisfied with any decision, opinion, instruction, direction, certificate or valuation of the engineer or with any other matter in connection with or arising out of the contractor carrying out the works, then the matter has to be referred to and settled by the engineer. If either party is dissatisfied with his decision or if a decision has not been taken within the time allowed, then either party may serve a notice of dispute.

ICE now contains options for conciliation and adjudication before there is reference to arbitration. It is intended that most disputes will be dealt with as soon as possible after they arise. Clause 66 has a number of subsections:

66(1) Provides that a procedure as laid down in the contract is to be followed in order to avoid, or failing that, to settle disputes.

66(2) Provides that if at any time (a) the contractor is dissatisfied with any act or inaction of the engineer or (b) if either the contractor or the employer is dissatisfied with any decision, opinion and so on, as stated above, then the matter is referred to the engineer who shall notify his written decision to both parties within one month of the referral to him.

66(3) Provides that both parties have agreed (in the contract) that there is no dispute until (a) the time for the engineer to give his decision under 66(2) has passed or the decision is unacceptable or it has not been implemented such that one party has served a notice on the other and on the engineer in writing or (b) an adjudicator has given a decision on a dispute under clause 66(6) and one party is not giving effect to it such that the other party has served on him and on the engineer a 'notice of dispute' then the dispute shall be as stated in that notice. As with other contracts, the word 'dispute' shall be held to include 'difference'.

66(4) Provides the parties shall continue to perform their obligations as if there were no dispute and the parties shall give effect to every decision of the engineer on a matter of dissatisfaction and to the adjudicator on a dispute under clause 66(6).

Conciliation

Clause 66(5) provides that (a) either party may at any time before service of a 'notice to refer to arbitration' under clause 66(9) by a written notice seek the agreement of the other party to having the dispute considered under the Civil Engineers' Conciliation Procedure (1994) or as amended subsequently and (b) if the other party agrees to this procedure then any recommendation of the conciliator shall be deemed to have been accepted as finally determining the dispute by agreement unless a notice of adjudication or a notice to refer to arbitration is issued not later than one month after receipt of the recommendation by the dissenting party. There is no contractual provision for legal proceedings to be an alternative to arbitration, unlike the JCT 80 and GC/Works/1 forms.

Adjudication

As an alternative to conciliation, adjudication may be activated, by reason of the Act and as provided in the ICE contract. This is provided for in clause 66:

66 (6)(a) Provides that either party has the right to refer a dispute to adjudication by giving a written 'notice of adjudication' to the other party at any time of his intention to go to the adjudicator. The adjudication shall be carried out in accordance with the ICE Adjudication Procedure (1997) or later amendment.

66 (6)(b) Provides that the adjudicator, unless already appointed, shall be appointed within seven days of the notice of referral to him.

66 (6)(c) Requires that the adjudicator reaches his decision within twenty-eight days of the referral notice.

66 (6)(d) Provides that the period of twenty-eight days may be extended by up to fourteen days with the consent of the referring party.

66 (6)(e) Requires the adjudicator to act impartially.

66 (6)(f) Requires the adjudicator to take the initiative in ascertaining the facts and the law.

66 (7) Provides that the decision of the adjudicator shall be binding until a dispute is finally determined by agreement of the parties, by legal proceedings or by arbitration if the contract provides for this or the parties subsequently agree to arbitrate.

Arbitration

Clause 66 (9a) provides that all disputes arising under or in connection with carrying out the works, other than failure to give effect to a decision of an adjudicator, shall be finally determined by reference to arbitration. A party seeking arbitration shall serve a written notice on the other party, called a 'notice to refer', to go to arbitration. Subsections to clause 66(9) are referred to below:

66 (9b) Provides that where an adjudicator has given a decision under clause 66(6) on a dispute then a notice to refer must be served within three months of the giving of the adjudication decision otherwise it shall become final as well as binding.

66 (10a) Provides that the arbitrator shall be a person agreed by the parties.

66 (10b) Provides that when the parties fail to reach agreement within one month of either party serving a written 'notice to concur' in the appointment then the dispute shall be referred to a person to be appointed on application to the president of the ICE.

66 (10c) Provides a procedure for appointing a new arbitrator if the first one, for reasons listed, fails to be able to act.

66 (10d) Provides that unless the parties agree otherwise, any arbitration can proceed before the works are complete. Of course, they can agree to arbitrate after completion.

66 (11) Provides the usual matters for procedure in an arbitration.

66 (12a) Provides for the engineer to be called as a witness before a conciliator, adjudicator or arbitrator and presumably, though not stated in legal proceedings, on any matter relevant to the dispute.

66 (12b) Provides for all matters placed before a conciliator to be done without prejudice and that the conciliator cannot be called as a witness by anyone claiming in connection with the matters referred to him.

NEW ENGINEERING CONTRACT 1994

Following the first edition of the NEC documentation, extensive comments were made on the provisions for settlement of disputes. Provisions for dispute resolution were not in keeping with the overall philosophy of the documentation and provided very little opportunity for disputes to be resolved until the construction works were completed and serious disputes were left to be resolved by litigation. The current edition of the ECC has been amended to take account of the Housing Grants, Construction and Regeneration Act (1996) and is committed to a two-stage process of resolution with adjudication being tried first and then, if the adjudicator's decision is contested, the second stage is by 'review by a tribunal'.

The nature of the tribunal has to be identified in the contract data and agreed by the parties, presumably before they enter into contract. The tribunal can be arbitration, expert determination, litigation or a dispute resolution panel. The ECC provisions remain unclear and critics of the ECC argue their comments accordingly.

Adjudication

The NEC has been amended in 1998 to take account of the Act by the issue of an addendum, under Option Y(UK)2: The Housing Grants Construction and Regeneration Act 1996. Clause 90.1 provides that the parties and the project manager (PM) have to follow the contract procedure for disputes. Subsections of clause 90 are referred to below:

90.2 Provides that if the contractor is dissatisfied with an action or inaction by the PM he notifies this to PM no later than four weeks of becoming aware of this; within two weeks of this, the contractor and PM attend a meeting to seek resolution of the dispute.

90.3 Provides if either party is dissatisfied with any other matter, he similarly notifies the PM and the other party within four weeks, and within two weeks a meeting is held between PM and the parties.

90.4 Provides that no matter shall be a dispute until a notice of dissatisfaction has been given and the matter has not been resolved within four weeks. The word 'dispute' (which includes 'difference') has that meaning.

90.5 Provides that either party may give notice to the other party at any time of his intention to refer a dispute to adjudication. The notifying party may then refer the dispute to the adjudicator within seven days of the notice.

90.6 Provides that the referral must contain details of the dispute to be considered. Any further information from a party must be provided within fourteen days of referral.

90.7 Provides that the parties carry on with their required actions under the contract during the period of the adjudication as though no referral had taken place and there had been no dispute.

90.8 Provides that the adjudicator shall act impartially and may take the initiative in ascertaining the facts and the law.

90.9 Requires the adjudicator to reach a decision within twenty-eight days of the referral or such longer period as the parties may agree after the dispute has been referred. The adjudicator may extend the twenty-eight days by up to fourteen days with the consent of the notifying party.

90.10 Requires the PM to provide reasons to the parties and to the PM with his decision.

90.11 Provides that the decision is binding until the dispute is finally determined by agreement or by the tribunal.

90.12 Provides immunity to the adjudicator for his actions or inactions when acting as the adjudicator unless he has acted or not in bad faith. His employees are similarly protected.

Combining procedures

- Clause 91 is amended to be called 'Combining procedures' (rather than 'the adjudication') and relates to combining main and subcontract disputes.
- Clause 91.1 provides that a subcontractor may attend a meeting between the parties and the PM to seek dispute resolution.
- Clause 92.1 states that the decision is enforceable as a matter of contractual obligation. It has been stated that there may well be defences to stop enforcement, perhaps arising from the content of the decision and/or the conduct of the adjudicator or if legal issues have been decided.
- Clause 92.1 provides for how the adjudicator is to act. The person appointed as the adjudicator is named in the contract data and it is the intention of the NEC documentation that the appointment is made under the NEC 'Adjudicator's Contract'. The adjudicator's fees are shared equally between the parties regardless of his decision. If for whatever reason the adjudicator does not or cannot act, ECC clause 92.2 provides a process for appointing a replacement adjudicator.
  (a) The adjudicator is to act as an independent adjudicator not as an arbitrator.
  (b) The adjudicator's decision is enforceable as a contractual obligation not as an arbitral award.
  (c) The adjudicator's powers include power to review and revise any action or inaction of the PM and supervisor.
  (d) The adjudicator is to make his assessment in the same way that a compensation event is assessed.
- Clause 93.1 provides that tribunal proceedings, that is reference to a court, arbitration or other tribunal, may not be started before completion of the whole of the works or earlier termination. The adjudicator's decision is final and binding unless and until revised by a tribunal. The tribunal has not been defined but should be inserted in the contract data by the employer. A dispute cannot be referred to the tribunal unless it has first been referred to the adjudicator.
- Clause 93.2 provides that evidence can be given to a tribunal that was not given to the adjudicator. The tribunal can review and revise any action or inaction of the PM or the supervisor and any decision of the adjudicator.

PART 8

CASE STUDIES ON CLAIMS

CHAPTER 29

INTRODUCTION TO CASE STUDIES

The following chapters give case studies of three hypothetical projects (whatever realities they may incorporate) to illustrate some of the principles discussed in earlier chapters, followed by a short discussion of some principles for drawing up an omnibus claim document. If such case studies are to be kept within reasonable length, they are bound to suffer from weaknesses, even if hopefully they have the strength of putting flesh upon theoretical discussion.

One weakness is that they can illustrate only in outline, because of their brevity and the constraints of being part of an inanimate book, without the complex interactions of life. Many of the ramifications of earlier chapters are thus left unexplored, although sometimes new avenues are uncovered. As a result, the chosen project has to be delineated in a rather dehydrated form, with many of the realistic details shrivelled up or missing altogether. This leaves a great deal to the imagination of the reader, but also isolates critical features more readily. It is just this isolation which can prove difficult in real life, when there is much detail and perhaps a few red herrings as well.

Furthermore, the projects have to be made artificial in some respects, again to keep treatment within bounds. What may really be a mass of calculations or narrative becomes much more akin to a formula into which just a few figures may be dropped. In the process, the formula distorts reality in places. All of this reads like an apologia, as it is!

More positively, the case studies may be seen as imaginary constructs which embody the essence of elements from various actual projects, chosen for their problematic nature, as well as a number of quite fictitious elements. The attempt has been made to follow through to their inevitable conclusions the elements woven in like this, subject to them illustrating some useful point. This means that in places the elements written into the scenario have turned out to have unexpected consequences in later development, but that these have been accepted and faced, rather than that the 'plot' has been rewritten to avoid them.

A frequent approach to this type of exercise is to set out a full claim document, or whatever is in question, presented from the contractor's point of view. This is then subjected to analysis from the employer's side, with the result that the time or money claimed is substantially reduced. This can give the impression that the contractually naive (or worse) contractor has been beaten, and rightly so it seems, but it also gives an unbalanced view of where right may lie, while having the virtue of setting alternative concepts down together for comparison. In that many claim documents do originate from the contractor in finished form, this approach is obviously useful and often to be commended.

However, with most current forms of contract, the potential for progressive presentation and agreement is high. If necessary, mediation, adjudication and further tribunal determination could be considered. Facilities for negotiation and options of mediatory and tribunal activity are growing (Chapters 27 and 28). There is now the contractual right of the parties to seek adjudication on an issue at any stage before or after completion, followed by contractual ability to seek arbitration or legal proceedings. It has not been possible to divert the course of case studies into these processes – they would have to be the subject of a separate book, perhaps at another time. The progressive approach of settling 'claims' is much more in keeping with the spirit under the JCT terminology of ascertainment by the architect or the quantity surveyor on the basis of data provided by the contractor, rather than the big bang at the far end of the contractual universe. It also tends to a less adversarial attitude between the parties. The distinctive pattern followed in the present studies is therefore to describe progressive discussion and perhaps agreement, at least of the direct elements of cost or time, while leaving it unclear just how soon this agreement was reached. This means that there is less emphasis on setting up a contractor's Aunt Sally for demolition, although it is still possible to construct one from the implications of the more consensus models laid out. A contractor looking for throwaway points to include in claims can still find them! An example of the full-claim approach is given in Chapter 26, based upon the case study in Chapter 32.

This methodology is related to a stronger narrative form than is common in sample claims, to show how situations develop and, by inference, can be at least partly controlled. It emphasises the value of progressive information while memories are fresh and of taking mitigating action by both parties. The illustrative value of the 'full claim at the end' method is limited in what it allows here, because all is fossilised in the records. The present method even allows the parties to change their minds on the way along.

The case studies proceed with different levels of detail. Thus the first, related to quite a large scheme of new work, goes into most detail through two chapters about the procedures and calculations when problems are arising. The other cases deal with smaller schemes of new and alteration work, and give no calculative detail in support of the narrative, which covers most of the relevant differences without resorting to repetition of essentially similar detail from the first study. Although these two studies provide self-contained discussion and explanation, the reader more concerned with the type of work set out in them should therefore still read the other study first, or at least go back to it for detailed amplification of what is his primary interest.

Even within the first study, the method and level of detail in calculations varies quite inconsistently and without apology, whereas precise procedures over notices etc are ignored. The aim is not to set out model approaches into which anyone may drop actual events or figures in place of those given. It is rather to highlight the key principles at least once, but not with slavish repetition, and to use whatever method of calculation best fits this aim, often one which could not be used in the given form in practice. Precise and fully detailed calculations can best be derived from working experience or from standard texts on estimating practice. No attempt is made to relate prices and other figures given to any particular period of time, as this can so easily become dated. In short, let the reader beware over the detail. It is the principles which are critical. For this reason, reference back to earlier and more digressive chapters is vital to fill out the significance of the illustrative material presented here.

CHAPTER 30

MAJOR CASE STUDY: LABORATORY PROJECT, PART 1

- Particulars of works
- Main events and their results in outline
- Extra payments in early stages
- Extra payments in later stages
- Some principles emerging over primary loss and expense

This case study concerns the execution of fairly large works by a regional contractor. The works covered new construction on a greenfield site, across the public road from another site of the employer. The problems stemmed primarily from a number of late decisions by the employer while work was under way, 'assisted' by some unexpected characteristics of such an innocent site. The main events are described briefly together below, before being explored separately.

PARTICULARS OF WORKS

Contract details

The contract form was the JCT private edition 1980 with quantities. It was on a fixed price basis, apart from statutory fluctuations. The sectional completion supplement was used.

Scope of works

The general layout of the contract works, with the major modifications introduced during progress, is shown in Figure 30.1. An analysis of the contract bills is given in Table 30.1 at the end of this chapter, but may be consulted here to give a feel for the scale of the works. The original works may be divided into three main portions:

(a) A three-storey in situ concrete-framed laboratory, containing also the administration offices for the site (which was one of several operated by the same group).
(b) A single-storey steel-framed process shop, of mostly open-plan layout, with some office space and toilets.
(c) Extensive bulk excavations to cope with a sloping site, site roads, car parks and pavings, drains and services and other incidental external works.

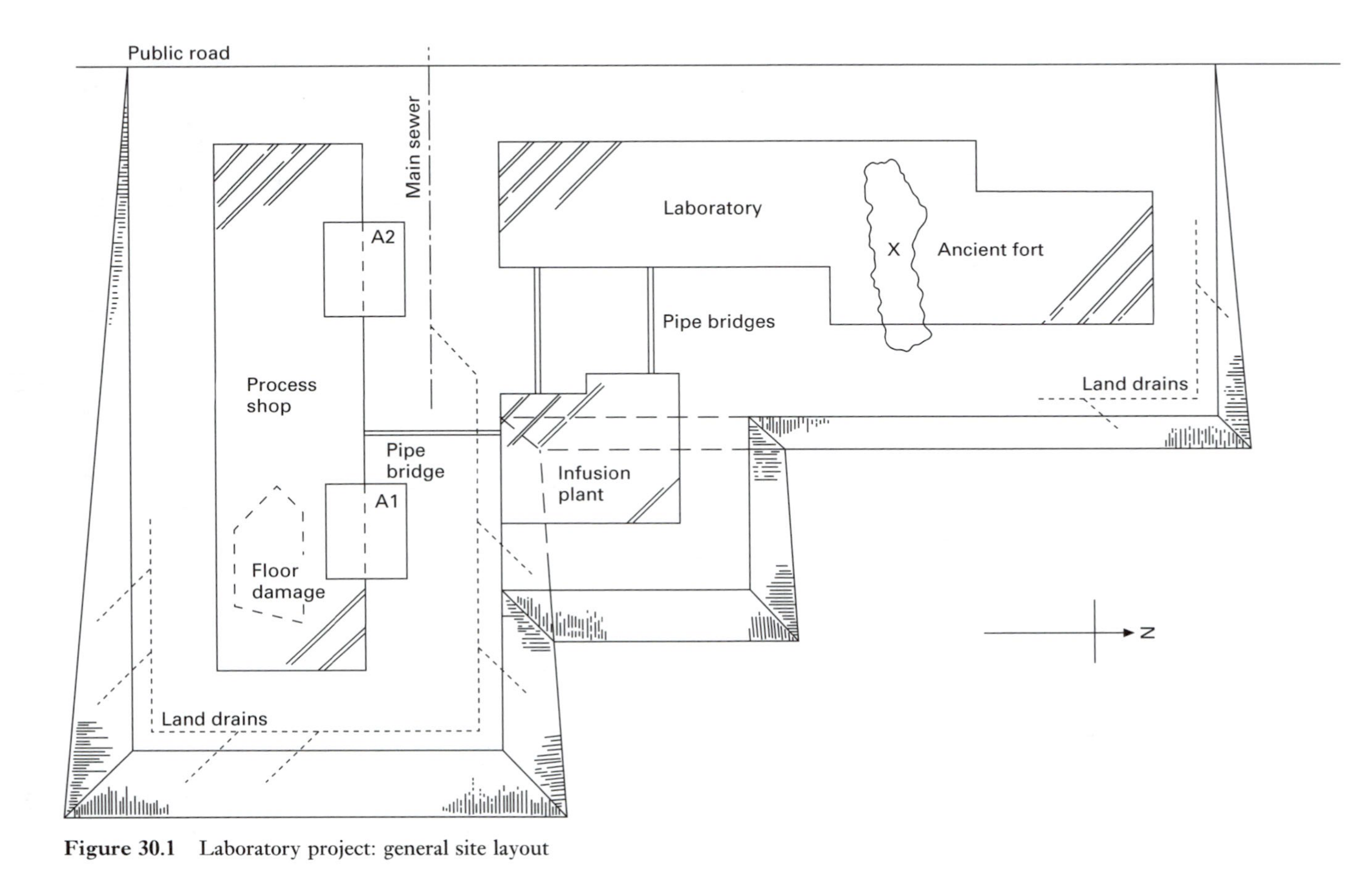

Figure 30.1 Laboratory project: general site layout

Among the nominated subcontractors were those for heating and electrical installations and laboratory fittings. Process installations were by direct contractors, coming in late in the construction period, so they had a clear run on the main floor of the process shop.

Programme of works

The requirement for sectional completion was for the process shop to be completed within thirty-six weeks, with the laboratory completed within forty-eight weeks. Completion of the external works was allocated between these periods to allow each building to come into use accordingly.

The contractor produced a network analysis of his programme soon after work started on-site. The key structure of this is shown in simplified form by the bar chart given in Figure 30.2. It indicates that the major part of the bulk excavations were to be performed at once, as the process shop (required first) was at the more

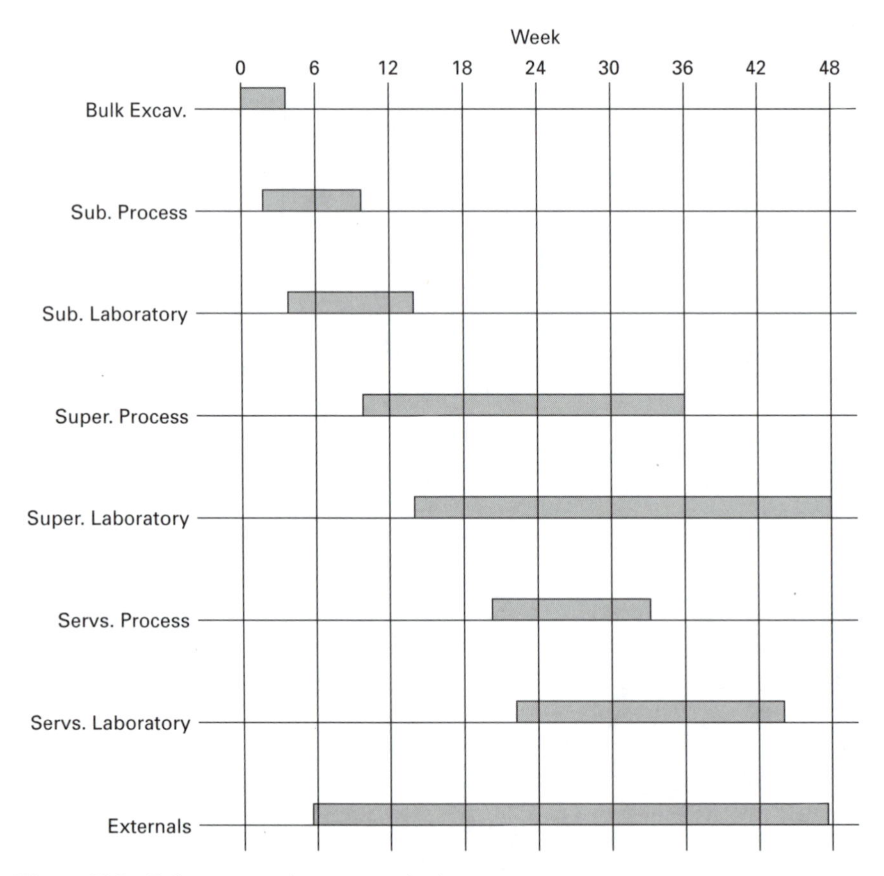

Figure 30.2 Laboratory project: summarised programme

deeply cut end of the site. The two buildings would then be staggered in their demands for comparable work, to permit gangs to move between them. Site works would be phased in detail to suit.

The programme was somewhat optimistic in places and generous in others. It reserved very little by way of float time for the project as a whole. The network at various stages is shown in Figures 30.3 to 30.7 at the end of this chapter. It follows conventional notation, but modifies this in later stages by recording the achieved programme, rather than the intended programme.

MAIN EVENTS AND THEIR RESULTS IN OUTLINE

Groundworks and annexes initially

In view of the programme, the bulk excavations started at the deep end of the site and were complete there, with foundation excavations started, when the architect issued an instruction postponing work on the foundations for the south side of the process shop, pending redesign. The contractor confirmed that he would hold all further foundation work to the shop, as he considered work would be uneconomical carried out piecemeal. He stated that special protection of the exposed formation would definitely be needed, as this was already showing signs of wetness and threatening deterioration. He indicated that he might be incurring further extra expense and delay, due to compression of the relative programmes for the two buildings.

As surface excavation moved along the site of the laboratory, the walls of an ancient fort were discovered, straddling the building at X in Figure 30.1. The contractor stated that he intended to cease work altogether at this, but the architect informed him that he must make such progress as he could, to comply with the contract. He then continued excavation from the south extreme of the site, up to the area affected by the fort, duly adjudged an antiquity and engulfed with archaeologists. The contractor received further instructions to provide assistance to the investigators to speed up their work.

Four weeks after the postponement instruction on the process shop, revised drawings were issued, showing two large staff annexes, A1 and A2, and some reduction in staff accommodation within the main area. Advanced drawings had been made available to the steelwork subcontractor for fabrication, but further delay beyond the four weeks was inevitable. Foundation work therefore proceeded, leaving out work to take steelwork bases later.

For the laboratory, foundation work was performed from both ends, involving extra setting out, slower working and a number of errors, until the investigation was completed and the remains of the fort removed. Throughout, the contractor was informally coaxed by the architect to take measures to reduce delay, rather than just stop. He also lost track of giving adequate notices over problems when the fort had been discovered.

Introduction of additional work

When the superstructures of both buildings were completing and internal work was in hand, the employer changed his plans for some cycles of his own processes on the site. The architect therefore instructed the introduction of an infusion plant, as shown in Figure 30.1.

The contractor considered challenging this on the grounds that it constituted a variation of the contract (see 'Variations and variation of the contract', in Chapter 6) by extending its scope substantially and at a late stage when comparable work was not being performed. In view of the programme and cost effects, he might well have succeeded, if needs be by adjudication which would have taken place at once. Instead, he settled for taking on the extra work, but protected himself by seeking both extension of time and loss and expense for the original works, as well as suitable rates for the extra building and its external works performed out of sequence in a restricted area.

Involved in this change of mind, were pipe bridges to each of the original buildings from the infusion plant, interfering with plant used in the main external works. Changes in the nature of services and delay in providing fresh drawings and specifications affected the work of nominated subcontractors and of the contractor inside the existing buildings, by then at the stage of finishings.

A late mishap

All of these additions and changes had been absorbed and were under way, and machines and equipment were being installed in the process shop, when the site itself again took a hand. An area of the main floor and some associated pits adjoining annex A1 started to lift and crack severely. This was traced to water pressure due to a spring, now constricted by the construction work. It was necessary to interfere with work on the site roads and other finishings and lay a system of land drains, as shown in Figure 30.1, to remove the immediate problem and to guard against it cropping up elsewhere. Only when some of this work was advanced could remedial work inside the process shop be taken far. After this, the work moved to completion without incident!

EXTRA PAYMENTS IN EARLY STAGES

Extra payments arose in several ways. There were separate disturbances to each building during performance of substructure work, followed by the interaction or knock-on effect of the two as work moved into the superstructure phase. These affected the original programme in Figures 30.3 and 30.4, and moved matters to what is shown in Figures 30.5 and 30.6.

Postponement for the process shop annexes

The instruction over this came only two weeks into the contract period, as bulk excavations at the north of the site were almost complete and foundations were being taken out at the deeper end. This represented event 1 on the network and the initial phase of loss and expense for the contractor extended into the bases (activity 1/3) and foundation walls (activity 3/5). It also affected the nominated steelwork subcontractor (activities 3/5 and 5/6).

The contractor's extra expense

Work stopped in the area, apart from bringing up foundation construction in already open excavations (which were then filled back) where no changes were indicated, and propping excavation sides elsewhere. When foundations were constructed, it was necessary to leave out work to accommodate steelwork, which was held up completely. Protective hardcore to a minimum thickness was laid over much of the exposed formation.

The contractor's initial claim was twofold:

(a) All extra work and work delayed and likely to be carried out with disturbance should be paid for as daywork.
(b) The total amount for preliminaries should be adjusted pro rata to the extension of time granted.

The architect referred the whole question of payment to the quantity surveyor, while granting an extension of four weeks on an interim basis, as the time he expected he would spend in producing enough information to allow the annexes to be incorporated into the site work.

The quantity surveyor disagreed over the rather sweeping payment suggestions made, on the grounds that to substitute daywork so radically would mean that the contractor was being paid the difference between his tendered prices and his costs, rather than his loss and expense (Chapter 11). He would also be receiving profit on the difference, not a loss and expense principle. In passing, he pointed out to the contractor that his preliminaries pricing appeared low in parts (as had been raised during examination of the tender). If, therefore, the measured prices included an element of preliminaries cost, the contractor could be losing to that extent.

So far as preliminaries adjustment was concerned, he explained that an extension of time did not entitle the contractor to a pro rata increase, even though the work affected was on the critical path of the network. The preliminaries included allowances for items not directly related to time, or even not related to it at all, such as the provisions for beginning and ending the project. Again, what was included in the contract bills was not the measure of ascertainment; whether high or low, this had to be actual loss and expense. By this time the problem of the ancient fort in the laboratory area had appeared, so the prolongation question was obviously going to be more complex and needed looking at as a whole. This is taken in the next chapter.

The outline of the agreement reached after some while was therefore to pay the following items:

(a) Costs as recorded for transferring some labour and plant to the laboratory site to install bases there, until work became available on the process shop site, and then returning it. These would have been quite small in any case, but became almost academic when that site gave trouble.

Total £670.00 (based on site observations)

(b) Half of the measured cost of the protective hardcore, on the basis that rather more hardcore would have been required for the steel erection at a later date anyway and as the base for the main floor. All that was happening was that the bed was being provided somewhat earlier and would need some extra topping up to restore levels, before actual thickening. The significance of the condition of the subsoil was lost on everyone at this stage.

Total 5,540 m^2 at £2.50 per m^2 = £13,850

This includes using less than economical plant at the time required.

(c) Standing time of plant not able to be redeployed elsewhere on the site. The principal item here was the main batching plant for site-produced concrete. This eventually stood idle for five weeks continuously (it was being erected late in the period of bulk excavation and for a week following), and intermittently for a total of a further three weeks during the time of piecemeal working. It was possible to transfer some of its attendant operators elsewhere, but not all. Some minor plant was also affected.

Total 8 weeks at £740.00 per week = £ 5,920

(d) Extra setting out for the annexes at the cost of the engineer already full-time on-site was included, pending review over the effect on site staff generally when the several disturbance matters were clarified. It was established that the net addition due to variations would not itself cover this excess setting out.

Total 3 days at £240.00 = £720.00

(e) Extra cost of executing foundation walls in short sections piecemeal between the gaps left for steelwork to avoid excessive delay, an element of mitigation at the contractor's discretion. (This is an instance of how a trade-off should be looked for, without the optimum solution necessarily being clear in advance.) Three figures were discussed:

Allowed productivity in bill rates = 2.10 h/m^2
Achieved productivity on similar work in unaffected sections of foundations = 1.90 h/m^2
Achieved productivity in the affected sections = 2.60 h/m^2

The latter two figures were relevant, not what had been assumed for tendering purposes (Chapter 11). In this instance they gave a higher reimbursement than using the tender figure, and so ascertained the actual loss the contractor was suffering and preserved his higher than anticipated profit. The result could have been the other way round.

Craftsman at £10.00 and 50% labourer at £8.00 = £14.00 per hour
Total 320(2.60 – 1.90) × £14.00 = £3,135

(f) Extra cost of filling in foundation walls between these sections and around steelwork when erected. This was assessed on similar principles to (e), but on a cost per occurrence, taking account of such features as double-handling of materials to the scattered points of use. Each occurrence involved an average of 4 m^2, with 40% productivity at the point of work, that is an excess expense of 1.5 times, and additional handling expense up to the point.

Total	37 × 4 × 1.90 × 1.5 × £14.00 =	5,905
	37 × £16.00 =	590
		£5,995

(g) Additional attendance on the steelwork subcontractor, mainly involving special areas of hardstanding for him to manoeuvre his mobile crane.

Total 22 × £40.00 (average) = £880

These items gave a total for the contractor of £31,670. It may be questioned whether some of these items should be treated as variations, rather than loss and expense, in so far as a measured basis is used. None of them is due to a change in design etc as defined by clause 13 producing a difference in the finished works, but all are due to a disturbance by postponement of regular progress. Even the walls come within this criterion, as they are not *varied* work performed in dissimilar conditions from what was envisaged in the contract bills. The changed annexes themselves (the cause of the upset) fall to be valued by the variation rules. Had they been introduced later and out of sequence, they might have qualified for enhanced rate treatment, as happened with the infusion plant.

Subcontractor's extra expense

A separate ascertainment was performed for the steelwork subcontractor (activities 3/5 and 3/6), in view of his nominated status. This covered extra expense at his works, rather than at site, where the provision of extra hardstandings eased matters and where the erected walls did not interfere with movement. Described in outline, the items were as follows:

(a) Fabricated steelwork not used, less scrap value.

Total 15 tonnes at £1,300.00 = £19,500

(b) Altering fabricated steelwork on a time basis.

Total for labour and equipment = £4,750

(c) Storing steelwork, before and after fabrication, while the architect's design and that of the subcontractor were being revised.

Total 6 weeks at £550.00 = £3,300

(d) Design time in redesigning work, allowed as the excess not covered by the inclusive rates in the subcontract bills.

Total 12 days at £200.00 = £2,400

These items gave a total for the subcontractor of £29,950. No prolongation costs fell to the subcontractor, other than those of storage, in the circumstances prevailing off-site. The whole incident was too short for him to substantiate any secondary loss and expense (Chapter 12), which possibility remained open for the contractor on his work.

In the subcontractor's case it is more cogent to argue that some of the items were by way of variations, so that all but (c) were technically included in the account accordingly. These amounts were all included in the total set against the prime cost sum. The contractor then obtained his profit margin on the part given as loss and expense, quite correctly under the wording of clause 30.6.2.9. This contrasts with his own part of the loss and expense and with what happens when there is a domestic subcontractor (as there was among some of the amounts already discussed).

Effect of ancient fort

The employer could hardly be blamed for this happening, and probably not even his own ancestors were involved! Excavation moved to the laboratory in activity 1/2 and uncovered the fort. It was the contractor's responsibility under clause 34.1 to do all he could not to disturb in any way what he had found and to notify the architect of the discovery.

The architect instructed the contractor to work round the fort and to allow archaeologists to take records, as a preliminary to removing the remains for safe keeping. These were comparatively diffuse and so not to be preserved where they were, but well worth taking away. The contractor wanted to suspend work altogether, no doubt daunted by this second catastrophe (from his point of view) within the early weeks on-site. The architect reminded him of his obligation under clause 25.3.4.1 to use 'his best endeavours to prevent delay'. Work therefore continued on what was most of the substructure, and the central portion was filled in later. The present consideration takes matters up to event 9 of the network.

The response of the contractor was not too precise at this stage. This second delay to an extent helped to bring the two buildings back towards the phasing intended by the programme (at event 9) – a slightly perverse benefit! He did not give proper notices about disturbance of regular progress, but told the architect when this was queried that he was under clause 34, which said nothing about notices (or properly 'applications'), as did clause 26. Even so, it is helpful to all concerned if some similar pattern of actions is established. In the end, the architect instructed the quantity surveyor to ascertain the loss and expense which undoubtedly was taking place, and which he was looking at meanwhile.

The contractor put forward a claim for several items:

(a) Extra payment for attendance on the archaeologists, who ranked effectively as direct contractors under clause 29.
(b) Extra setting-out costs, due to the position of the fort.
(c) Extra excavation and filling incurred by the problems of levels on the two sides of the fort not marrying up.

(d) An overall drop in productivity in the central third of the substructure for all trades up to floor level, and especially in the basement, which was close to the fort area.
(e) Daywork for carrying out all work in joining the building when the fort was removed.
(f) Extension of time for the delay occasioned to the substructure, with the position reserved beyond that, in view of what was happening to the process shop.

The last of these may be cleared first, as it related to time and not payment, which is immediately being discussed. It was pointed out that none of the activities were upon the original critical path and any question of extension would have to be set against whether their late performance affected the superstructure, which *was* critical.

The other questions were decided as follows:

(a) Payment for attendance as instructed could only be dealt with as daywork. It thus effectively became a variation by its method of valuation, even though there is nothing express in the contract allowing the architect to instruct the contractor to provide attendance, only to allow third parties to work on the site. The alternative is for the contractor to work directly for them and become, for those purposes, a third party himself. This might lead to numbers of legal problems if any injury occurred.
(b) As the fort was entirely below ground, in the nature of things, and largely below reduced level excavation, the claim for additional setting out costs was rejected, apart from engineer's time in setting out central detail as a filler operation, when the fort had gone.

    Total 1 day at £240.00 = £240

(c) The same view was taken initially of the claim for extra reduced level excavation carried out in error, leading to extra disposal and filling. After extended discussion, it was agreed that there had been difficulties in maintaining progress, over which the architect had pressed quite strenuously to counterbalance the other delay which the employer had pitched in so suddenly. The claim was therefore met 'without prejudice', as is sometimes politic in a situation of mixed pressures.

    Total 40 m^3 of excavation, removal and filling at £40.00 = £1,600

(d) The productivity drop was held to have been stated too broadly across the types of work concerned, which would be affected differently, and proper evidence of actual drops had not been given. The situation was unclear, even though there was an instinctive feeling that there had been drops. As the contractor had not given notice of what he now alleged had happened, he had weakened his own position, but not necessarily precluded a right to payment, remembering that this was clause 34. A straight comparison with the actual cost of similar unaffected work was difficult, as there were differences between work in the several areas and as, when it was similar, it had not been performed in separate parcels, but by the same gangs continually moving about between the areas. This pattern was not itself due to the disturbance. The result was that the quantity surveyor had to exercise judgment (based upon the bill rates and an assessment of how they compared generally with productivity) in consultation with the architect, the clerk of works and the contractor – a common state of affairs. An outline of the settlement was as follows:

(i) Bases (actually performed ahead of those for the process shop), little effect after delay, labour content 30%, allow 10% of this content for one-quarter:

50,500 × 30% × 10% × ¼ = £379

(ii) Walls, more effect, labour content 45%; allow 15% of this content for one-third:

76,100 × 45% × 15% × ⅓ = £1,712

(iii) Basement, affected by delay leading to standing time, assessed as six tradesmen and four labourers for one week:

6 × 40 × £5.50 =	1,320
4 × 40 × £4.50 =	720
	£2,040

(iv) Plant (minor items) drops in production and standing time associated with these labour losses (not shown in detail) = £344.

Total for this head of claim = £6,115

(e) The recorded daywork was examined against the measured values for the work concerned. The comparison showed an excess of labour and plant costs (again related to bill rates and an assessment of likely productivity) calculated at 17%. This was considered reasonable and was used as the basis of ascertainment.

Total £1,335

The problem of how to 'ascertain' when no comparable data for undisturbed work are available, is a constantly recurring one. Despite all that may be maintained on the subject, there comes a point at which judgment has to be exercised, as when calculating a difficult 'star rate'.

In this instance, work on the walls (activity 9/11) was due to wait for the gangs to become available from the process shop, itself delayed. A wait had been allowed in the original network, while the activity was not critical anyway. The amended network shows that, in fact, labour moved across to the laboratory during a waiting time on the process shop, and then back again to allow the progress achieved to happen. This may again have led to some of the drop in productivity, but overall was beneficial in reducing the cost of delay for both parties. The effect of this reappears when looking at the next stage of the works.

The loss and expense total of items (b) to (e) was therefore £9,290, assuming payment for the erroneous excavation to be for loss and expense.

Early superstructure for both buildings

This section concerns the knock-on or ripple effect of what had already happened on activities which were not put out of rhythm by the immediate disturbances, but which suffered because their own interrelations were changed.

The main effects up to events 6, 7 and 11 have been described and evaluated. The broad position thus far is shown in Figure 30.5, which gives a revision of the first part of the network to allow for what happened, so that only times achieved are entered. For some preceding and subsequent events, two times are shown: achieved times and (in parentheses) times which might have been justified under the contract. The achieved times take account of what the contractor did by way of mitigation, broadly saving programme time and overheads and adding to direct site expenditure.

The extra time incurred owing to the increased work content in the process shop annexes has also been included, largely in elements A and G (activities 7/10 and 15/20). The effect was to move much of the more complex office and toilet accommodation out into the projections, leaving more clear floor space in the main run of the building, so that most of the net additional construction (as distinct from disturbance) was quite simple work. This aspect is returned to when extension of time is discussed later. The work itself was valued under clause 13 and need not be discussed here.

With the commencement of superstructure work, the contractor faced the consequences of the earlier disruption working their way through. It had been intended in the original programme that the same gangs should work first on the process shop and then transfer to the laboratory. This allowed for a one-week pause on the laboratory between the two elements A, whereupon the laboratory also came on to the critical path, related to its own completion date. This saved gang confusion and switching. There were several gangs in most elements and sometimes the same gang in several elements, but this aspect has been omitted to simplify the example.

As affairs had turned out, the contractor now faced a six-week pause at this point, if he maintained the same diagram logic. Although a contractor under the JCT contract is not obliged to go to extra and irrecoverable expense to regain lost time, neither may he simply hold woodenly to what he planned when things go awry. Often, as here, it does not pay him either. When there is loss and expense in prospect, he has the mitigation issue at stake as well.

The pause was enough to raise the possibility of rescheduling a whole string of elements to put the laboratory ahead, but this would have thrown the whole sectional completion sequence into difficulty later. What happened was that the gangs were shifted to and fro between elements A to C on the buildings, at the expense of productivity, but to reinstate the basic sequence as soon as practicable. This was agreed to be a reasonable course of action, so that the loss and expense were reimbursed. It is an example of the sort of area where the policy should be examined carefully, rather than just whether a particular amount of loss and expense was actually incurred, before an agreement is reached.

It is still the case that benefit of the doubt for lack of the advantage of hindsight should often be allowed. As usual over the contractor's programme and methods, the architect cannot instruct or insist, but consultation (perhaps implication of others) at an awkward stage can save disagreement later.

The result is shown on the network in Figure 30.5, where the original logic is retained for ease of reference, except over the steelwork, and where elements A, B and C are shown linked to their counterparts in the laboratory by two-way arrows (activities 7/10, 10/12 and 12/14). For the laboratory, the times which would have resulted but for the reorganisation are given in parentheses. Their effects would then have rippled through the rest of the network. Element A of the process shop is lengthened only because of its increased work content, whereas element B and those following become just subcritical.

The contractor argued for a 15% drop in productivity for labour and all plant involved on both buildings in these elements. This was purely impressionistic – as might be said for some of the allowances made earlier – but it would not stand up to any rigorous comparison with the appropriate contents of measured rates, which were taken as a first datum level to see whether the percentage had any possible validity before detailed work went ahead. Work then reverted to an analysis of what had happened, followed by a comparison of productivity on the two buildings. Agreement was as follows:

(a) The crane and hoist and all attendant labour used for the laboratory superstructure were both tied up for ten weeks instead of eight (longer than claimed in percentage terms). No such effect occurred on the process shop, other than that due to the extra work content and covered by variation valuation, which could allow for any adjustment for plant caused by quantity variation (although in this instance no adjustment resulted).

Crane at £530.00 per week × 2	=	1,060
Hoist at £140.00 per week × 2	=	320
Driver at £10.00 per hour × 40 × 2	=	800
Attendants at £8.00 per hour × 40 × 2	=	640
	Total =	£2,820

(b) Moving equipment to and fro between and sometimes within buildings, and setting up to accommodate shorter runs of activity. These were assessed by reference to those involved directly and those who observed what happened (such as the clerk of works), and covered labour, transport and lack of productivity of the equipment itself. There was a total of 24 moves, averaging 6 persons for 3 hours, several types of transport and 'standing' time for moved equipment of 5 hours:

£10.00 × 6 × 3 = £180
£10.00 × 3 = £30
£32.00 × 5 = £160

which leads to

£370.00 × 24 = £8,880

(c) Dislocation of gangs producing work, including refamiliarisation time. A total of 24 breaks, averaging 8 persons per gang:

£10.00 × 6 × 5 = £300
£8.00 × 2 × 5 = £80

which leads to

£380.00 × 24 = £9,120

(d) Drop in productivity in the laboratory. This was an arguable concept, when the foregoing items had been taken into account. Comparison with the process shop during periods when switching had been overcome was rather inconclusive, as there was a restricted amount of really similar work. When evidence was available, it suggested quite a marginal drop. This was agreed, on a conciliatory basis, at 2% of the labour expenditure of £101,500.

Total £2,030

(e) Additional overtime. To keep pace with the process shop, overtime at 10% of the normal week was worked. Labour expenditure £101,500, exclusive of foregoing amounts. Overtime was one-eleventh of the total time and attracted a 25% higher rate, less standing payments for overheads on labour, transport, importation etc, a net increase of 13% per hour. For 100 hours basic, therefore, 10 net hours overtime were worked at an added cost of 1.3 hours non-productive time. This constituted 1.17% of the labour bill (1.3 out of 111.3 hours).

Total £1,190

The relative smallness of this item may be noted. It was not a 'compensation' to the workforce, an aspect discussed in Chapter 11.

These items gave a total for the contractor of £24,040, including a small non-labour element. This is about 22% on the unadjusted labour content for both buildings, rather than 30% as originally put forward, but includes items not put forward explicitly by the contractor.

Review of position reached

The previous sections have brought the question of the early disturbances to a tidy level. It is useful to review the section on client's risk analysis in Chapter 1, to consider how far the employer might have avoided disturbance and whether this would have been a net advantage. Ascertainment of amounts would in practice probably have taken quite a time beyond the construction programme to achieve. But the figures used are very rounded and they are given for illustrative purposes only. More detail is given in some instances than in others, simply to make a point, but in all cases there would be more detail based on normal estimating practices. The total loss and expense amounts come to £94,950, of which £24,040 are ripple effects from the two substructure disruptions, occurring in the superstructures.

What has not been done so far is to deal with the effect of prolongation on site overheads, or to look at secondary loss and expense. This would be premature for two reasons: the overall progress of the project might not warrant any such reimbursement but, if it does, it is likely to become more complex, rather than remain unchanged.

Equally, the matter of extension of time is also hanging. In week 20 the architect upgraded his original four-week extension for the whole project to six weeks for the process shop and kept it at four weeks for the laboratory. This was towards the end of elements A for each building. The actual position at the ends of the two elements C was one week later for the process shop and one week earlier for the laboratory. This is acceptably close to reality, as the architect has the power to review the extensions granted up to twelve weeks after practical completion and there may be other causes leading to the differences, some qualifying and others not.

Some alternative possibilities

Before passing to the later stages of the project, some further options may be taken briefly. These are treated as alternatives, to avoid making the main example too complex.

Subcontracts

When being urged to keep up progress, the contractor might have considered introducing a second gang to double his workforce on an element. If he had sublet this element, he might have found himself in the position that the firm concerned could not provide the extra labour, so that he had to seek a second firm or drop behind on-site. He would be entitled to drop behind, if he had explored all reasonable ways to maintain progress in the face of his problems. According to the precise circumstances, he would not necessarily be obliged to bring on this extra resource. Presumably, the first subcontractor was the most satisfactory over price, quality etc, available when subtenders were sought. It is quite possible, although needing to be demonstrated, that any further firm is not so satisfactory, especially over price.

Given such a situation, the contractor has several points to consider:

(a) Obtaining consent to subletting as usual, unless the firm appears on the list of domestic subcontractors given in the contract bills.
(b) The difference in price he now faces. In obtaining consent to sublet, he is advised to obtain agreement at least in principle to reimbursement of the extra expense he has to meet. Usually in a domestic arrangement, the price is entirely at the contractor's risk, whether he has to choose from a list or wishes to sublet at his own initiative, and whether or not his choice drops out (Chapter 7).
(c) The primary reimbursable difference is that between the two subcontract prices, measured or otherwise, without adjustment for profit or attendance. As ever, it is the contractor's net loss and expense which is in view.
(d) There could be a difference due to the performance efficiency of the second firm being lower than that of the first, perhaps over the level of defects or other

delay factors. Usually, this is entirely the responsibility of the contractor, however the subletting comes about. Here it comes into play as part of an endeavour to remedy a situation which is not the contractor's fault. If he is forced to look at a firm which he suspects is going to be less satisfactory for himself, he should protect his position in advance by warning the architect and obtaining his endorsement. Actual reimbursement will still be subject to the test of fact.

(e) In the situation of work split between two buildings and being performed in parallel, rather than in series, the contractor may incur extra supervision or coordination with his own direct activities.

The other option which is very likely to arise, is that of a domestic subcontractor's loss and expense, without any question of a change of subcontractor. Here the subcontractor's position is governed by the form of subcontract applying, and there is no obligation for the standard form DOM/1 to be used with a JCT form, desirable though this is. If it is used, it gives the subcontractor rights to reimbursement on the same lines as those in the main contract. If this happens, the contractor alone has to negotiate with the subcontractor and come to a settlement, and the architect and the quantity surveyor are not involved directly.

However, they are due to settle with the contractor on the basis of his direct and net loss and expense. When there is a domestic subcontractor involved, this should mean that what he receives from the contractor is what the contractor receives from the employer. The difference in price or prices between main contract and subcontract is not relevant. This does not mean that the contractor can agree anything in scope, if not cost level, with the subcontractor and then simply pass it up to the employer to pay. In view of this, there is often a need for the architect or the quantity surveyor to become involved in a monitoring way in subcontract loss and expense settlements, or at least to reopen their basis on occasions. As earlier discussion in the present case study has shown, it may be necessary to use judgment to a degree in ascertaining loss and expense, when what is agreed at main contract level may differ in principle from what is agreed at subcontract level.

None of these points apply directly to nominated subcontracts, where the architect and the quantity surveyor are brought in via the obligatory subcontract form. This is illustrated by the instance of the steelwork subcontractor earlier in the present case study, whose amounts were settled direct and on which the contractor received profit.

The other situation is where either type of subcontractor, domestic or nominated (or for that matter 'named' under the IFC contract), is involved in a 'three-cornered' dispute – a dispute in which it appears that some part of his loss and expense falls to be met by the employer ultimately, and some part by the contractor alone. Here it is inevitable that the employer's consultants will be brought in to clarify the position and draw the lines. Their part is then to settle what concerns the employer for payment – and keep out of the rest! This is illustrated in outline by what happens over steelwork in the case study in Chapter 32, but in practice it is inevitably complicated.

Absence of ancient fort

Although the fort caused extra expense in its own right, it did take the pressure off the utter distortion of the construction programme. If it had been absent, the contractor would have been faced with retardation of the process shop programme, while the laboratory programme was still pressing ahead. In these circumstances, he would have had to rearrange his whole programme to put the laboratory ahead, if the system of work gangs passing from one building to the other was to have been maintained (an alternative being for twin gangs to have been employed, as mentioned above). Here are some of the types of loss and expense that would have emerged:

(a) Costs of reorganisation, such as those of supervisory staff engaged in preparing the new programme and seeing the gangs into their amended workstations.
(b) For the earlier gangs, the costs of removing equipment and plant from one building to the other, according to what had been set up already.
(c) Standing or waiting time associated with the last operation.

Against these features, the contractor would have had the possibility of switching his main concrete batching plant to the laboratory, rather than having it stand mostly idle as actually happened. The possibility of apparently retrieving the situation by putting the process shop back in front would not have been present, except at considerable expense and actual delay. None of this would have affected extension of time, as it actually needed to be granted, the laboratory being subcritical at this stage.

EXTRA PAYMENTS IN LATER STAGES

Introduction of infusion plant

The programme settled down after the ripple effects of the substructure disturbances had worked through the project, that is by about week 25. This illustrates the possibility of a contractor regaining momentum after a disruption, given peace and quiet for long enough, and the broad philosophy of fading effects set out under 'Client's risk analysis' in Chapter 1. For simplicity, the common occurrence of a mass of minor delays due to late instructions by the architect has been omitted. On this unique contract, it must have been a great consolation to the contractor – not to mention the architect! If the architect's instructions had been late, the result would have been dealt with by calculations (known as 'ascertainment') similar to those outlined earlier in this case study.

The introduction of the quite major requirement of the infusion plant on-site in week 32 posed several questions:

(a) It was seen as a variation of the contract itself, by changing the scope of works in a major way. The contractor could therefore have declined to take on the extra work, whatever terms he might have been offered as an inducement.

Strictly, the fact that it was proposed to introduce it so late did not affect the basic objection, as he could have declined the day after signing the contract. However, the timing did make matters worse in practical terms, and it was also likely to extend the total construction period in a more significant manner than envisaged by clause 25, so changing another fundamental term of the contract, that of time.

(b) Given this possibility, the employer could have been placed in a very difficult position, as the new facility had become a vital feature of his operations for technical reasons which are not relevant to this narrative. This is the sort of thing which any employer should try to avoid or limit by forethought (see 'Client's risk analysis' in Chapter 1). If he cannot, he has the choice of paying the sitting contractor, perhaps over the odds, or of introducing another contractor. The latter is contingent upon the stage of the current works and the configuration of the site permitting it. In itself, it varies the present contract by changing the site availability, but could be done as a species of non-fundamental breach. Clearly, neither of these solutions is at all favourable in the present instance.

(c) If the contractor had not cooperated, the employer could have done nothing, unless he had been prepared to breach the contract by termination and bring in another contractor to perform all the works. This would have been very expensive (Chapters 9 and 18) and have extended the time for the existing works. It would probably have taken longer than finishing off and then constructing the extra plant.

(d) As the contractor decided to accept the extra work, he was entitled to suitable preliminaries costs and to measurement of what he did, both being priced at rates to take account of the circumstances in which he was working. He was also entitled to a suitable time for the work in its own right.

(e) With this done, there was the disturbance factor for the existing works, given that the new plant itself was to be treated as a very large variation instruction. There were two possibilities: to include the extra expense in the figure for the new plant, or to cope with it as it occurred under clause 26. The former had its virtues by being clear-cut in settlement, but had all the problems associated with prior estimation of any disturbance element (that it would have meant ignoring the 'loss and expense' principle would not have been critical in the circumstances of negotiation at the time).

(f) It was therefore decided to follow the second course on loss and expense, especially as there were some doubts at the time of agreement over the precise programme of the new building, and so over its impact on the original works still going ahead.

(g) A supplementary agreement was entered into, setting out the points covered above and all other details to give contractual completeness.

None of the detailed negotiations over the infusion plant itself need to be described here, although the agreed figures for preliminaries and extension of time are of interest later.

Loss and expense flowing from infusion plant

The installations provided in the new plant had an impact on the services in each building, because they linked to them and also changed them in a number of ways. These services were all critical activities, affected by the alteration of installed work and delays over redesign in elements (b) (activity 20/24 of the process shop) and (f) (activity 33/34 of the laboratory), with further redesign causing delay in following elements. The likelihood of time and cost both being involved was clearly high.

But some of the effects of the extra work are a clear example of how there can be loss and expense without necessarily extension of time, a situation which may be reversed. There were some relatively minor effects on the building elements running alongside the services inside the buildings, which were subcritical, and quite marked effects on the roads and other such work proceeding as a non-critical activity in the rear parts of the site.

There was some debate over the fact that these areas had been started quite early within the time span allocated. The employer complained that, had they been programmed later, there would not have been a disturbance effect. The contractor's reply was that, had he left the work until late, he might have hit unforeseen snags, whereas late introduction of delaying variations could have jeopardised completion on time. Furthermore, this was a fixed price contract, so he had every incentive to work ahead of inflation. Rightly, he won the day, as he was simply proceeding 'regularly and diligently'; see *Greater London Council* v. *Cleveland Bridge* in Chapter 4.

The key items in the settlement were as follows:

(a) Alterations to the building elements were all dealt with by daywork covering removal and some of the reinstatement, whereas larger sections of work were measured and priced pro rata to what was done originally. As this work went ahead continuously with that before and after, no loss and expense due to standing time occurred and the variation reimbursement covered everything else.

(b) In the case of the services (all by nominated subcontractors), runs of pipework and cabling were extensively affected by the introduction of different specifications to suit the output from the new plant. Removals were dealt with by daywork and replacement by measured items priced pro rata to the bill items. This pricing did not take account of the piecemeal working inherent in the switching of systems and the waiting that beset the work, as design ran alongside the site work and as some plant and equipment was delayed by ordering times. Installation of items of plant and equipment within the existing buildings was also changed, but this was paid for adequately by daywork for removals and items pro rata to those in the bills for the replacements. An example of the numerous calculations for various types of pipes and cables (inclusive of fittings etc) will suffice:

Allowed productivity in bill rate	= 0.35 h/m
Allowed productivity in new rate	= 0.45 h/m
Achieved productivity for original work	= 0.40 h/m
Achieved productivity for revised work	= 0.65 h/m

Calculated productivity for revised work in
undisturbed conditions (0.40 × 45 ÷ 35) = 0.51 h/m
Drop in productivity (0.65 – 0.51) = 0.14 h/m
Cost for 100 m (100 × 0.14 × £10.00) = £140

This approach was necessary in the absence of comparable undisturbed work and to allow for the generally lower than budgeted productivity.

Total loss and expense in services
Process shop = £7,120
Laboratory = £4,530

The high amount of measured work and daywork helped to keep the loss and expense down here, indicating the narrow border often existing between the elements.

(c) Attendance by the contractor on the subcontractors was out of proportion to the contract allowances. This consisted of a number of strands, all of which had to be ascertained by reference to what had happened up to the beginning of the disturbance and by the exercise of judgment over the allocation of what happened thereafter, including when work had reverted to a more normal rhythm. The main strands were as follows:

- Supervisory staff coordinating main and subcontract work.
- Attendant labour moving materials and equipment as needed.
- Skilled labour performing alterations to permanent work, and not covered by the main variation valuation.
- Scaffolding, staging, hoists, chutes and similar facilities, both in extra provision and in adaptation.
- Special attendance items were also affected, but could be isolated more readily.

Total loss and expense in attendance:
Process shop = £8,720
Laboratory = £5,955

(d) Waiting time was incurred by labour and plant performing the roads etc while additional drains and services were laid for the infusion plant and while two pipe bridges were erected across them to serve the existing buildings. Some alteration to the road layout was needed near the new building, but this was covered by daywork. The pipe bridges also caused some hindrance to the taller items of plant when roadworks resumed under them, so leading to some work in small areas. Here are the totals:

(i) Waiting time of labour and plant:

45 hours at £65.00 = £2,925

(ii) Work in small areas:

Production should have cost	11,400
Actual cost	14,700
Loss	£3,300

(e) Cleaning costs arose because of the removal of spoil across the partly completed areas of surfacings; this also meant filling in of lifted material. Both items were covered entirely by daywork.

As before, these items exclude the costs of prolongation. The net result was a delay of three weeks in the process shop and four weeks in the laboratory, with a total for loss and expense of £32,550.

Damage to process shop floor

Although the work just accounted for produced a total delay of three weeks in the process shop, its effects took much longer to work their way through, so the affected elements in fact proceeded alongside those following. This was also true of the laboratory, but this falls out of the picture for the present.

When all the elements from events 20 and 21 onwards were going on together (except the final decorating and clearing up in element U), the final blow was struck at progress by the site itself. This came in the form of the rupture of areas of the main shop floor and associated pits and ducts adjoining annex A1, that is the annex further into the site and so lower in relation to the original ground levels. The cause turned out to be a spring, itself outside the site, but feeding considerable quantities of water under the process shop. The floor was redesigned as a variation, with other effects covered as loss and expense, in the absence of a successful insurance claim (see next section). The financial results were several, in this order, but subject to some overlap:

(a) The clearance of the affected area and the suspension of work in and around it at lower levels in the building. Installation of services in the roof space was able to continue with some rescheduling.

Clearance work on daywork (not valued here)		
Standing time for labour and equipment of the contractor and several subcontractors	=	16,240
Extra supervision of services work by subcontractors	=	1,720
Extra attendance by the contractor on services work	=	1,430
Total	=	£18,370

(b) A special site investigation to check the exact problem and its extent. This led to the suspension of external works on the north side of the shop in expectation of additional work there as a result.

Site investigation (not included in this final account)		
Total loss and expense: standing time of labour and plant (reduced by transfers to other areas of the site)	=	£5,620

(c) An extensive system of land drains at the feet of the banks all round the site was installed, causing disturbance of executed work in more areas than where work had initially been stopped.

Removal of already executed work was paid for as daywork (not valued here), whereas the drainage system was measured and priced at special rates

Total loss and expense: further piecemeal working and standing time while the drains were being installed = £6,510

(d) Repair and modifications to the damaged parts of the process shop when the land drainage system was sufficiently advanced to prevent further damage.

All of this work was paid for by daywork

(e) Resumption of other work held up by the incident.

This was welcomed with relief, but no extra expense was incurred by way of champagne!

These items gave a total loss and expense amount for the floor damage of £30,440. The financial effects of this story have been listed without much detail, as the principles are those which have been used already. The time effect was to place completion as follows:

(a) Process shop in week 58, owing to the effects of the floor damage on top of the introduction of the infusion plant already discussed, rather than week 45 as in the revised network in Figure 30.6, with the addition of three weeks to give week 48.
(b) The external works finished in week 62.
(c) The laboratory held to week 55, that is in accordance with the revised network plus four weeks.
(d) The infusion plant almost held to its own time of finishing in week 57, by finishing in week 59.

These effects are shown in the final network in Figure 30.7.

Some further aspects of the damage to the floor

The financial effects have so far been listed without comment on where liability for the lateness of the incident might lie. This question was put when it was realised that the early need to protect the formation while the annexes were being designed should have raised signals in someone's mind. Several possibilities were canvassed:

(a) Whether the precontract site investigation was deficient in scope or execution. If it was deficient in scope, who had supplied an inadequate brief?
(b) Whether the architect or consulting engineer should have noticed the significance of the early problem and taken action then.
(c) Whether the contractor had failed to exercise a duty to warn over a matter within his province as a competent and skilled contractor (Chapter 5).
(d) Whether the employer had so pressed everyone in those early days to minimise the delay which he had introduced by his own change of mind, that none of them had had sufficient opportunity to notice what was in evidence in the formation.

(e) Whether in fact the formation had been bad enough for the correct conclusions to be drawn.
(f) Whether the contract insurances could be called upon to meet the costs.

These options, and there might be more, pass the blame around among all the main characters in the story or give it to none. The whole matter is far too complex to portray in a chapter devoted to other issues, or in many ways adequately within a book and away from the actual site, but should be noted as an example of what leads up to allocation of responsibility in a disputed instance. Only in possibility (c) would the contractor have to bear the cost. It is assumed this one did not apply, so he was reimbursed for his loss and expense and other amounts and did not face a substantial counter-charge, even including liquidated damages.

The insurance option is complex, the contractor having insured the works. The insurer is liable for all risks, but excluding inter alia 'any work . . . lost or damaged as a result of its own defect in design'. The floor was clearly inadequate for the stresses; whether it was defective in design depended on the view taken of (a), (b), (d) and (e) above. The insurer could probably have avoided liability upon any of them being found to favour him. He in fact did so, leaving reimbursement of the contractor to follow the variations method, as it would have done had clause 22B or clause 22C (for an extension) applied.

Had a counter-charge arisen against the contractor, there could have been excess costs to be met by the employer and incurred by direct contractors, working in the process shop on plant installations when the disturbance occurred. These were already running late, but the responsibility was the employer's, so no argument could arise that the contractor had caused them to slip into the present disturbance. Unless a contractor is made expressly responsible by a sectional completion arrangement for meeting particular dates or otherwise directly facilitating the work of such persons, it is difficult to see where his contractual responsibility to the employer would lie, other than in the matters covered by the indemnity provisions. These relate to 'the carrying out of the Works' and are unlikely to be construed as relating to a duty-to-warn situation. If the contractor had *negligently* failed to warn, an action in tort might have lain.

Clause 29 simply provides that 'the Contractor shall permit the execution of such work' and that the employer is responsible for these persons, who are not subcontractors. It appears likely that the employer would have to rely on any liability of the contractor for liquidated damages to recoup his expenses due to delay to and disturbance of these persons, just as he would over direct contractors due to move in on completion of the building contract.

The narrative has carefully kept matters just far enough apart to avoid the problem of overlapping causes of loss and expense, so that individual matters can be taken as such. A re-reading of the account, with some compression of what has been given, will begin to introduce extra complications, which could not be properly dealt with in an example. The situation is beginning to be there by implication in the process shop in the overlap of the work due to the infusion plant and the consequences of the floor being damaged, but is so minor in context that it may be ignored when all the loss and expense is reimbursable anyway. Affairs would have been different if even part responsibility for the floor damage had been taken by the contractor.

SOME PRINCIPLES EMERGING OVER PRIMARY LOSS AND EXPENSE

All the items of loss and expense have been given as they occurred. They come to the following total:

Annexes, contractor's work	31,670
Annexes, subcontractor's work	29,950
Ancient fort	9,290
Early superstructure	24,040
Infusion plant effects	32,550
Floor damage and drains	30,440
	£157,940

The total is relatively unimportant, as the point is to demonstrate the principles and practicalities involved. Here are the more important considerations:

(a) In situations of complete or nearly complete cessation, it is usually easy to arrive at the costs of suspension, waiting or standing time.
(b) When there is comparable work (what SMM7 calls 'similar work') performed, it may be possible to compare productivity of disturbed work with undisturbed. This is easier to say than to achieve, unless observations can be set up in advance and provided they can be adequately controlled to secure reliability. Often it may be suspected that this approach is held to have occurred as a means of covering the use of adjusted prime cost and judgment (perhaps the same thing) to arrive at an answer.
(c) When work is not comparable, what is included in the variation account and what in the loss and expense element is not always easy to divide. The narrative as given is not beyond criticism, and various competent practitioners could well arrive at different allocations. In general, it is as well to include as much as possible in the variation or remeasurement account, both to secure a more certain profit and overheads margin and because the point of clause 26 is to cover what 'would not be reimbursed by a payment under any other provision'.
(d) When several sets of disturbance overlap, it can become very difficult to segregate their effects. This does not matter greatly when all the responsibility lies with one party, as the contract does not require each cause to be accounted for separately when this is impracticable (Chapter 17). It does matter when there is divided responsibility, as it becomes more difficult to apportion the financial effects.
(e) When numerous sets of small disturbances occur, they often merge into a constant climate of disturbance, as when information is constantly late in reaching the contractor. Then there is no problem with segregating effects, as

they are individually too small to warrant or even permit it; the difficulty lies in identifying the total effect when this is the sum of many almost insignificant effects. Ascertainment becomes something of a guess, hopefully educated. It may rely on past experience, if only of what has been acceptable before. This reads cynically, but probably there is a tendency to settle for something near to this reality in many cases.

(f) There is no direct correlation between the timing of a disturbance within the total programme or the scale of, for instance, extra work introduced on the one hand, and the level of loss and expense on the other. (Compare loss and expense due to the annexes including ripple effects, totalling £52,775, with the net addition of work of £67,890, excluding preliminaries.) This tells against 'rule of thumb' settlements. But the discussion under 'Client's risk analysis' in Chapter 1 is still relevant.

(g) Situations do arise in which the decision as to who has to bear the cost or the loss and expense should precede the actual ascertainment decision. Examples are those where there is negligence by the contractor or when a three-way dispute involves a subcontractor as well.

These points may also be borne in mind when reading the next chapter, which they partly anticipate.

Table 30.1 Laboratory project: analysis of contract bills

Preliminaries

Item	Amount	Total
Setting up site	20,500	
Supervision	78,000	
Major plant	90,500	
Accommodation	58,500	
Sundry items	30,200	
Final clearance	7,400	285,100

External works

Item	Amount	Total
Bulk excavations	91,400	
Drains and mains	176,800	
Roads, car parks	109,200	
Site finishings	29,700	407,100

Process shop

Item	Amount	Total
Bulk excavation (in externals)		
Substructure		
Bases	116,400	
Walls	159,300	275,700
Superstructure		
Steel 1	53,200	
Steel 2	29,700	
Element A	74,600	
Element B	36,400	
Element C	34,200	
Element D	38,700	
Element E	83,700	
Element G	54,800	
Element H	74,500	
Element K	40,600	
Element L	31,200	
Element M	36,400	
Element N	33,300	
Element U	28,400	649,700
Services		
Element a	55,700	
Element b	201,400	
Element c	118,400	375,500
Fittings		
None		
		£1,300,900

Laboratory

Item	Amount	Total
Bulk excavation (in externals)		
Substructure		
Bases	50,500	
Walls	76,100	
Basement	92,200	218,800
Superstructure		
Element A	120,400	
Element B	48,300	
Element C	45,900	
Element F	68,600	
Element D	75,900	
Element H	117,600R	
Element J	74,000	
Element K	89,700	
Element P	83,800	
Element Q	69,600	
Element R	135,900	
Element S	114,100	
Element T	50,100	
Element U	71,200	1,165,100
Services		
Element d	76,500	
Element e	91,200	
Element f	127,600	
Element h	178,000	473,300
Fittings		
Element g	293,300	
Element j	243,100	536,400
		£2,393,600

Summary

Preliminaries	285,100
Process shop	1,300,900
Laboratory	2,393,600
External works	407,100
	4,386,700
Insurances etc	64,800
Contingencies	300,000
Total	£4,751,500

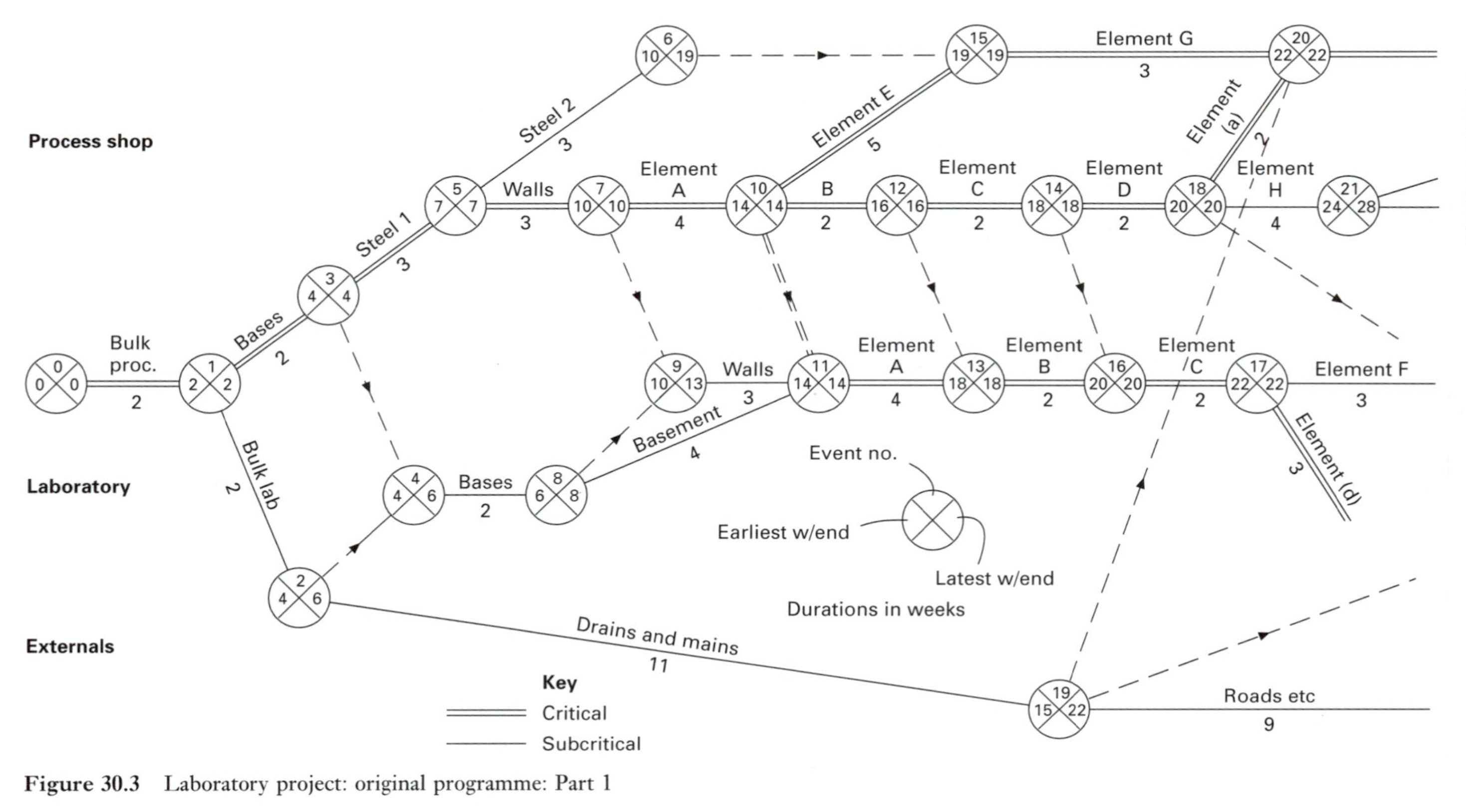

Figure 30.3 Laboratory project: original programme: Part 1

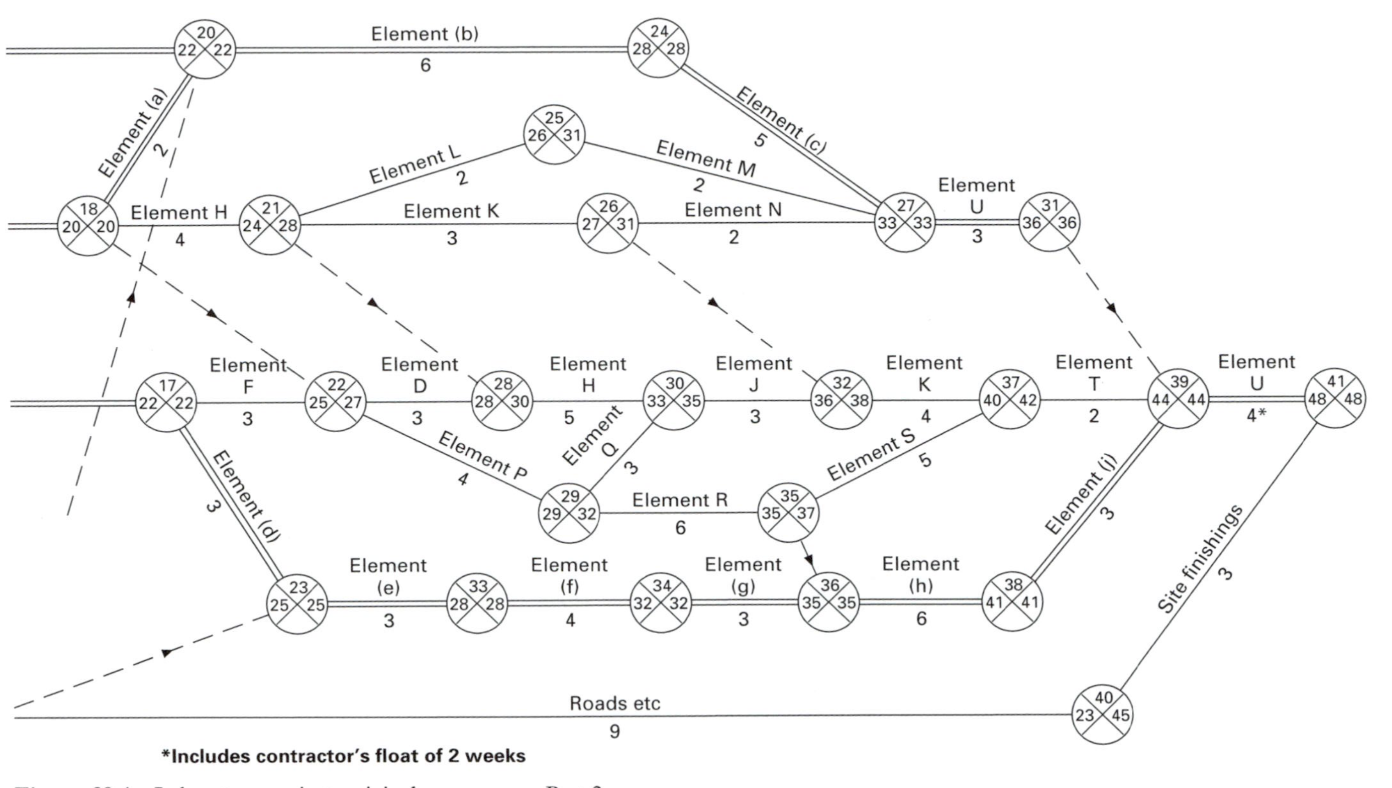

***Includes contractor's float of 2 weeks**

Figure 30.4 Laboratory project: original programme: Part 2

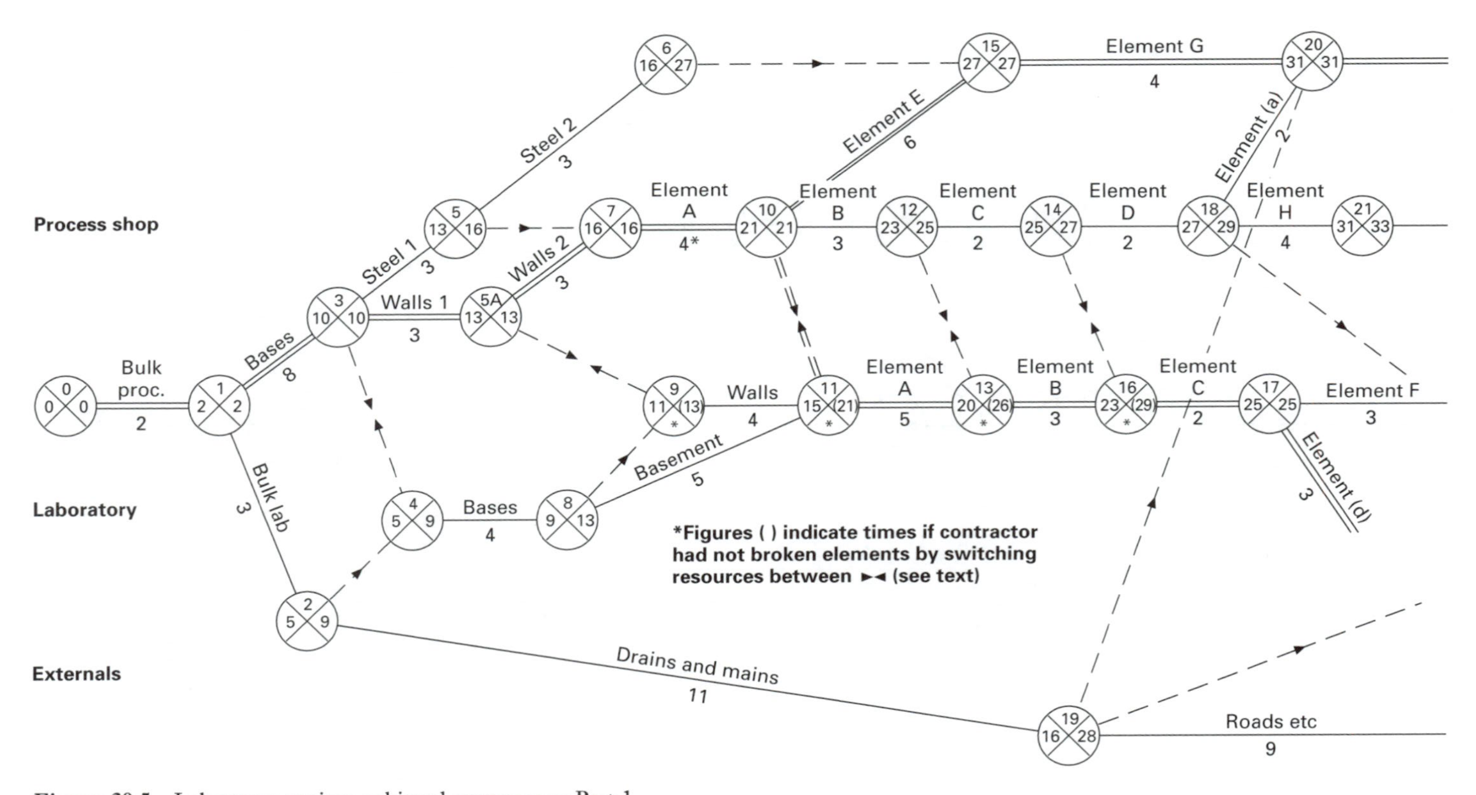

Figure 30.5 Laboratory project: achieved programme: Part 1

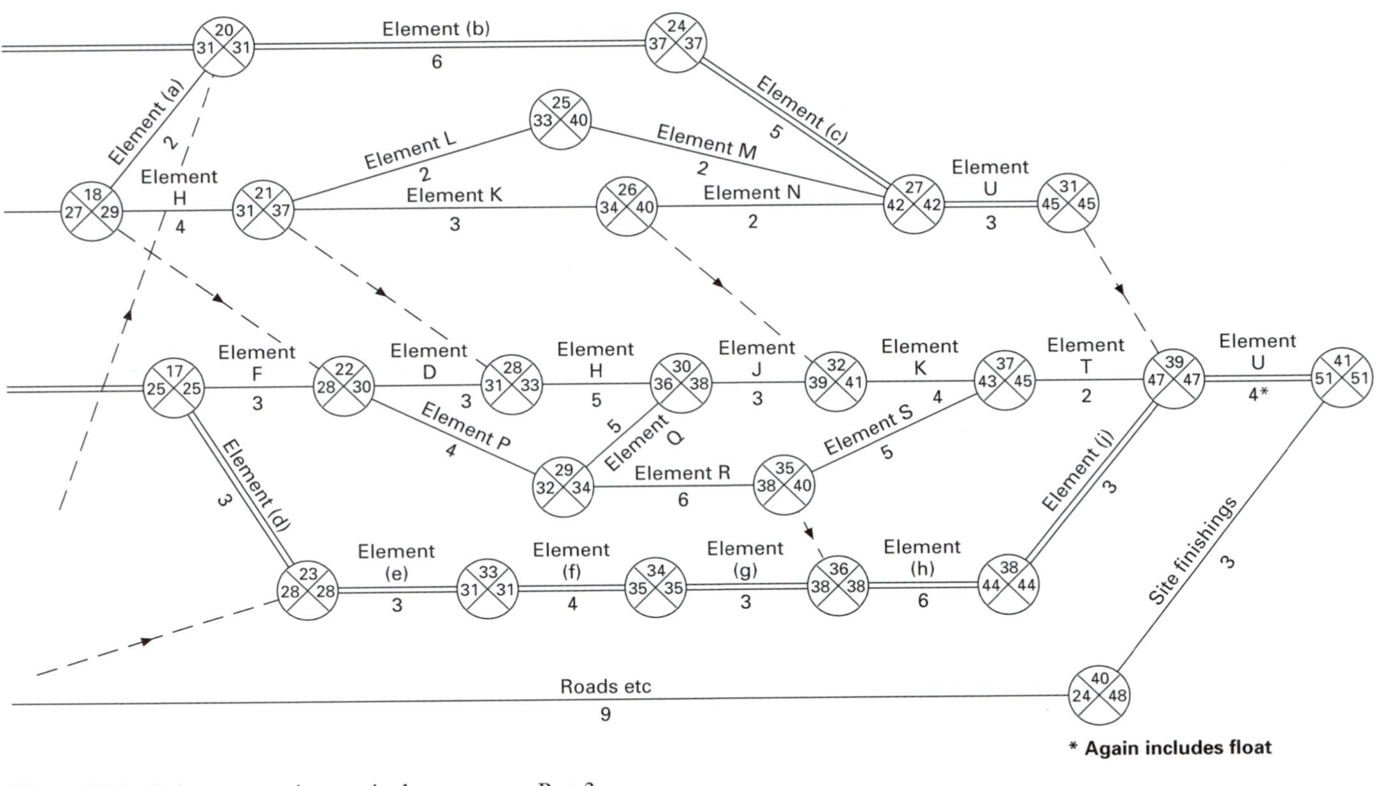

Figure 30.6 Laboratory project: revised programme: Part 2

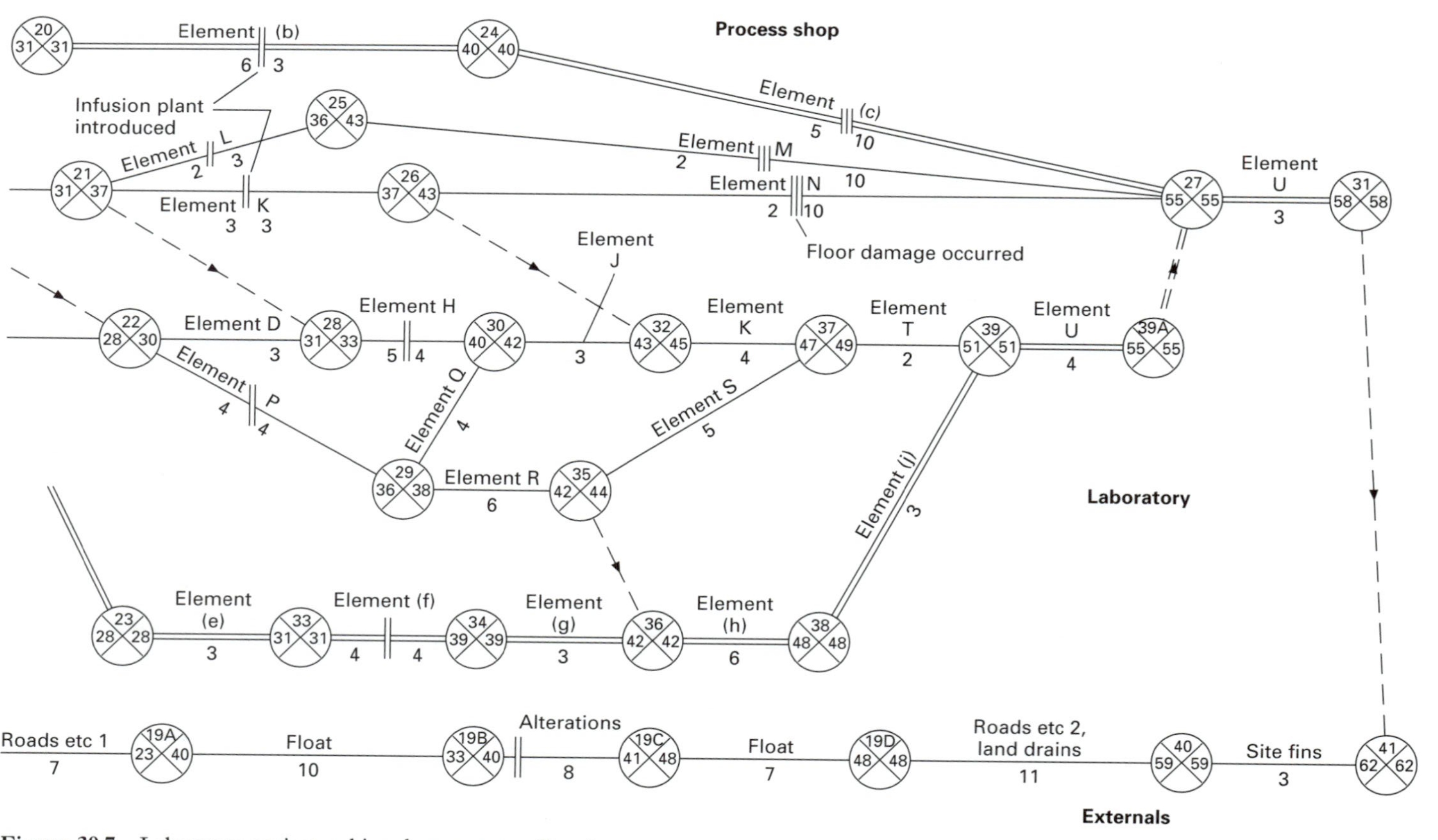

Figure 30.7 Laboratory project: achieved programme: Part 2

CHAPTER 31

MAJOR CASE STUDY: LABORATORY PROJECT, PART 2

- Extension of time
- Preliminaries adjustment due to scope of work
- Loss and expense on site overheads
- Inflation on fixed price contract
- General drop in productivity of labour
- Loss and expense on head office overheads
- Some final comments

With the individual loss and expense amounts dealt with for each phase of progress, there remained the amounts for supervision and other site overheads and prolongation and, distinctly, extension of time. These, and items in the 'secondary' category (Chapter 12), interlocked in various ways in the final settlement and are taken in the following order:

(a) Extension of time
(b) Preliminaries adjustment, due to the increased scope of the works
(c) Loss and expense items
    (i) Site overheads
    (ii) Inflation on this fixed price contract
    (iii) General drop in productivity
    (iv) Head office overheads
    (v) Profit and financing charges

Each of them had either cropped up during progress or was raised by the contractor during settlement. Several may be seen as having a 'global' flavour, as noted in the introductory part of Chapter 12. Throughout this chapter, Table 30.1 may be referred to for original contract values.

EXTENSION OF TIME

The various delays to progress which were connected with loss and expense by sharing the same cause (not that the one cause automatically justified extension of time, or vice versa) have been mentioned during the sequential narrative and the cumulative effect stated. The delays on the critical path were as follows.

	Process Shop	Laboratory	Externals
Original time (weeks)	36	48	48
Delay (weeks)			
Annexes	6	1	–
Fort	–	–	–
Superstructures	3	2	–
Infusion plant	3	4	3
Floor damage	10	–	11
Total	58	55	62

In addition, there were four weeks of subcritical delays to the laboratory, due to the fort. The infusion plant itself was two weeks behind the agreed date at week 59.

Most of these times have been explained in relation to the events which occurred. Here and there, the contractor asked for more time or for time in relation to subcritical activities. The value of the network was that it showed these features up quite clearly. When there is not a network, it is necessary to go through calculations which are equivalent to at least partly setting one up.

The sequence of events given has ignored any question of overlap in causes of delay (Chapter 10). These are broadly of two types:

(a) *Other causes given as relevant events in clause 25*: Some of them may lead only to extension of time and others may (in appropriate cases, but not automatically) lead also or instead to payment for loss and expense. The fort was an instance of loss and expense without extension of time, as that part of the laboratory programme was subcritical. If a right to extension does arise, it is calculated by the overall duration of the two delays, not by the sum of their individual durations. The sum of their individual durations would obviously exaggerate the effect.

(b) *Other causes which are always the contractor's responsibility*: Examples are normal adverse weather or failure to obtain materials in time, when no shortage exists. In these cases the contractor is entitled only to such part, if any, of the delay due to the relevant event as stretches beyond that due to the other cause.

One or two of the times were adjusted by the architect during his periodic reviews. The contractor did argue for longer extensions in places to give him a little more leeway than the one week of float built into elements U for each building. Properly, he should have been granted extensions which were abated by these float periods, as clause 25.3.1.2 refers to delay 'beyond the Completion Date'. This is a frequent cause of debate, as the contractor thereby loses the buffer which he has included to cover possible slips on his part, which is a perfectly reasonable precaution. It occurs in exaggerated form when the contractor allows for finishing well ahead of the contract date (Chapter 4). It is suggested that a fair approach to this problem is to allow extensions in the earlier stages which preserve the contractor's float to cover his own needs, but to tail it off towards the end as it becomes less necessary.

In this project, the contractor used up his float to ease the problems of tight working in the confusion which had set in, and no one objected. Calculation of extension of time within a one-week margin becomes quite debatable in circumstances of the nature of those suffered here. The origin of the circumstances also would have weighed in any doubtful area against the employer in an appeal to arbitration or proceedings.

The only times needing further comment are those due to the floor damage in the process shop. Within the shop itself, affairs were at a highly complex stage of working towards finality when the damage came to light. There was work involving services by subcontractors, the contractor's own finishings and extensive direct work for the employer. On top of the delay while the remedy was being sought and during which no remedial work could be started, there followed a period when all concerned were likely to get in one another's ways, with no one in convincing overall charge of the programme. There was little option but to accept whatever finishing time was achieved, in the absence of gross mismanagement, and this was done.

The other aspect of this lay in the external works, which finished after the buildings and not with them. As often happens, the sectional completion requirements had not been terribly explicit over the precise areas to be completed with each building. This could well have undermined any putative argument about areas which were needed to go with the process shop, and so any claim that full liquidated damages should apply for that building until all was handed over. Although the external works were definitely non-critical in the original programme and had been progressing quite well, the incidence of the land drains on top of the work caused by the infusion plant threw them well out. Any argument for some proportion of liquidated damages for the external works alone was very likely to fail and was not pursued. *Bramall & Ogden* v. *Sheffield* (Chapter 9) is illustrative of the results of lack of precision here.

The usefulness of a network analysis for establishing controlling extensions of time can be seen, even from the limited evidence of a book exercise. In its absence, something equivalent should be set up to isolate critical activities from subcritical and non-critical activities, so the genuine areas of delay can be seen. Only delays affecting the end date are relevant, although subcritical activities can easily be made critical by other delays.

PRELIMINARIES ADJUSTMENT DUE TO SCOPE OF WORK

This is a straightforward matter, arising from normal estimating principles and mentioned in clause 13.5.3.3 as a matter for variation valuation 'where appropriate'. The procedure is discussed in Chapter 6. Although not a question of loss and expense, it needs to be cleared here because loss and expense matters hinge on it.

Reasons for adjustment

In this contract, several issues had emerged to suggest an adjustment of preliminaries. Everything relating to loss and expense due to extra supervision or prolongation had to be ignored for clause 13 purposes and, conversely, clause 13 matters had to be excluded for loss and expense ascertainment. This is not mere pedantry, but relates to the differing methods of calculation and their results.

There had been a lifting of the value of measured and cognate work, main and subcontract, for these reasons:

(a) Extra work in the annexes of the process shop, totalling £67,890.
(b) Introduction of the infusion plant with its own variations, totalling £346,700, excluding preliminaries.
(c) Remedial and additional work in the process shop, owing to the spring, totalling £38,600.
(d) Land drains and related work in the external areas of the site, totalling £43,230.
(e) A general incremental edging up of the value of other work by a series of architect's instructions, totalling £145,820.

The total value of all this work was thus £642,240, or £295,540 without the infusion plant.

The contractor had negotiated for the infusion plant on the basis that he would set up a distinct subsidiary site organisation to cope with it in the short time available, which also fell within the closing stages of the original works, when the emphasis for organisation was changing with the changing character of the work. Here are his agreed additional preliminaries, set beside those in the original contract sum:

	Original	**Additional**
Setting up site	20,500	1,200
Supervision	78,000	6,700
Major plant	90,500	8,400
Accommodation	58,500	3,400
Sundry items	30,200	6,900
Final clearance	7,400	1,000
	£285,100	£27,600

These additional figures were high when the work content of the two parts were compared. They reflected the fact of a much smaller 'contract within a contract', but also the strong bargaining position of the contractor in the circumstances.

It was necessary to establish what would have happened, had all the other extra work been instructed in a regular way, so as not to disturb regular progress.

The contractor's first proposal was that there should be a straight adjustment of all the original preliminaries amounts in proportion to the adjusted amounts for measured and other work, other than the amount for setting up site, when the extra work had not been instructed, but including final clearance when it had. This would have given an increase of some £20,000 and ignored economies of scale (otherwise the effect of marginal costing) and also took no account of the change in contract duration or of features such as those illustrated by the S-curve of Figure 12.1(a). It was established in ensuing discussion that major plant and accommodation had not been increased on-site, but had been used more intensively, extendedly or efficiently, according to the various circumstances. Some 60% of the amounts for sundry items had been similarly affected, but the rest not at all. Final clearance had been a little more piecemeal than expected, because of the staggered finish to the works.

Apart from the effects of disturbance of regular progress, there would have been an extension of time of two weeks for the annexes and general extra work, this applying to each building. There would also have been an extension of three weeks for the extra work inside the process shop to remedy the floor, had this been instructed and introduced evenly. The land drains themselves would have added four weeks on a similar basis to the external works. As the results of the floor damage did *not* occur in this steady manner, it is open to debate whether *any* of the extra preliminaries should be excluded from the loss and expense amounts, so affecting the levels of them. Practitioners differ over items such as these, but it is submitted that the present pattern is the more correct. The resulting completions would have been

Process shop	Week 41
Laboratory	Week 50
External works	Week 52

These were all due to additional work, valued as variations, not to 'neutral' causes such as weather. They therefore constituted elements in the valuation of extra preliminaries, in accordance with clause 13.5.3.3. That there was no separate original completion date for the external works may be ignored in the present context. Equally, it was decided not to argue over whether the contractor might have finished one week early all round, by maintaining his initial float (see above) and assuming no variations at all, that is by weeks 35 and 47.

The calculation of additional preliminaries, affecting only the original amounts, without the infusion plant, eventually was as given under the following headings.

Supervision

The extra work in the annexes had had little effect, remembering that additional setting out had been covered so far as loss and expense (but see later under 'Site overheads'). It was also not possible to isolate individual costs for the other extra work, other than the land drains. It was, however, agreed that some extra due to general coordination on-site and paperwork was inevitable, while there was a prolongation, quite apart from the disturbance element. This would have been less on subcontract work than on the contractor's direct work, but the proportion of these was about the same between the original and the additional work. Some element of supervision value (about one-third plus all forepersons, from precontract discussions) was agreed to be in the measured rates, and so already reflected in the variations net addition.

Original work, without preliminaries = £4,101,600

The staffing covered in the various figures was:

Contract manager (visiting)
Agent and two sub-agents/engineers
Site surveyor
Bonus, costing, wages and administration

These were to be available according to the build-up and rundown of the project. To maintain simplicity, the identities of these persons are ignored and they are allocated for illustrative purposes in the following manner (see also the histograms in Figure 31.1) out of the amount of £78,000 in the preliminaries, while not affecting the principles of adjustment. The split between buildings (including share of external works) was as follows:

Process shop 25% = £19,500
Laboratory 60% = £46,800
Common to both 15% = £11,700

The intensity of use of the various categories is reduced to 'units' throughout the original programme:

Process shop
Weeks 1–5 at 1 unit = 5 units
Weeks 6–10 at 2 units = 10 units
Weeks 11–20 at 3 units = 30 units
Weeks 21–36 at 2 units = 32 units
£19,500 ÷ 77 units = £253.25 per unit

Laboratory
Weeks 1–5 at 1 unit = 5 units
Weeks 6–10 at 2 units = 10 units
Weeks 11–22 at 3 units = 36 units
Weeks 23–48 at 2 units = 52 units
£46,800 ÷ 103 units = £454.40 per unit

Common to both
Weeks 1–48 at 1 unit = 48 units
£11,700 ÷ 48 units = £243.75

From these figures, increased 'allowances' were made, based on the timing of the qualifying extensions (see the delay times given under 'Loss and expense on site overheads'). These all assumed regular working and attendance as at other times, so that no mitigation factor was introduced:

Process shop
1 week at 3 units and 4 weeks at 2 units = £253.25 × 11 = £2,786

Laboratory
1 week at 3 units and 1 week at 2 units = £454.40 × 5 = £2,272

Common to both
2 weeks at 1 unit = £243.75 × 2 = £488

There was also special supervision because of the floor repair work and land drains (had both been performed regularly, but in their 'special conditions', as clause 13.5.1.2 has it), in each case two extra staff for part of the time:

(£100 + £70) × 30 days = £5,100

Total additional allowance = £10,646

The straight expression of the supervision amount as a percentage of measured and other work would have given £4,970. The higher amount (exclusive of the last element for repair etc) is due to the time element, abated by the effect of economies of scale. The contractor would have lost out on his original proposal.

Major plant

The original allowance of £90,500 was split:

Transportation, setting up and removal (not in general items)	= £4,350
Hire charges or equivalent	= £33,650
Operating labour etc	= £52,500

For even greater simplicity than was used for supervision, plant is taken as being entirely available during the first two-thirds of the programme for each building; again, it would actually be calculated in considerably more detail. Plant costs were higher during the earlier stages of each building and slightly higher in proportion for the laboratory, owing to its concrete frame and multi-storey nature. Half of the variations came in the finishings stage of each building, when less use of major plant occurred, so that only one week's extension was relevant. The exception was the work to the process shop floor, which needed a variety of plant, but most of this was brought in specially because of the timing and covered by daywork. The agreement was based on the following in principle, but worked out in more detail:

Hire and operating total	= £86,150	
Allocation to process shop, nearly one-third (28,720 – 2,200)	= £26,520	
Allocation to laboratory, just over two-thirds (57,430 + 2,200)	= £59,630	
Allow for process shop, 1 week extra on 24 weeks (26,520 ÷ 24)		= 1,110
Allow for laboratory, 1 week extra on 32 weeks (59, 630 ÷ 32)		= 1,860
Extra major plant allowance for 3 weeks of floor repair in process shop (420 × 3)		= 1,260
		Total = £4,230

Accommodation

Accommodation was allowed for the whole period of 48 weeks originally, on the basis that it was not reasonable to remove sections early, when the laboratory was more complex and larger. Setting-up and clearing amounts in this case were in the respective general site items. It was agreed that the two-week overall extension was relevant and that the extra extension to the process shop caused no extra cost. The accommodation was not needed later during completion of the external works, when the employer made a room available in the process shop and when disturbance applied anyway. A simple proportioning therefore covered the elements of hire, maintenance and heating and lighting:

Allow for whole project, 2 weeks extra on 48 weeks (58,500 ÷ 24) = £2,440

Sundry items and final clearance

The sundry items were affected in assorted ways, and for some 60% of their value as noted above. Without giving detail to repeat principles already illustrated, the result was:

Original sundry items	=	£30,200
Allow 60% for 2 weeks extra on		
48 weeks (30,200 × 60% ÷ 24)	=	£760

The slight spreading of final clearance was compensated by an addition of £354, based (and this could be questioned) on costs produced.

A review

These various extra preliminaries amounts totalled £18,430, less than proportionate to the increase in work executed, when allowance has been made for the special features included. However, they take no account of loss and expense, all of which follows, apart from small elements of supervision included earlier.

All the calculations ignore the sum for contingencies, not just because it is not part of preliminaries (where it is sometimes misleadingly placed), but because the contractor cannot allow any preliminaries in relation to it, as he is quite unaware of whether it will be used, and if so how. This singles it out from other provisional sums, which usually state their scope, so that preliminaries can be allowed in addition (see 'Subsidiary provisions over valuation' and discussion of defined and undefined provisional sums in Chapter 6).

If the contractor had allowed most of his preliminaries in his rates, he would have obtained an excessive extra, if the rates had been applied to the increased quantities without question and in particular in the absence of any programme extension. This gives point to the reference in clause 13.5.1.2 to instructions which 'significantly change the quantity', and also to 'allowance . . . made for any addition to or reduction of preliminary items' in clause 13.5.3.3. A reduction of preliminary amounts, even when they have been priced as nil, may be the best way of reflecting this position, effectively reversing what has been done above.

Having prepared the way with the extended digression in this preliminaries section, the discussion returns to loss and expense.

LOSS AND EXPENSE ON SITE OVERHEADS

Relationship to preliminaries adjustment

Ascertainment here has to proceed from the base established of adjusted preliminaries related to the works as varied, but with the works performed in a regular manner and with costs taken into account, rather than estimated figures in the preliminaries. In principle it deals with both disruption and prolongation. In this project, it is prolongation which predominates, because its sheer extent swallows up the effects of disruption itself. Staff generally will have plenty of spare time to deal with extra activity!

It is immediately evident that the statement just given introduces a conceptual problem which has been ignored so far in this case study for ease of working, but which can conveniently be highlighted now. Because the contractor's original intentions have been considerably modified, there is now an additional set of figures of consequence to add to those relevant:

(a) Expenditure which would have been incurred with the works completed unaffected by variations etc or disturbance.
(b) Expenditure which would have been incurred with variations etc taken into the reckoning.
(c) Expenditure as actually incurred.

The second of these figures is not the same as that resulting from the adjustment of preliminaries just performed, although related to it, any more than the first is the figure included in the contract bills. Its calculation requires a hypothetical exercise (as so many are) arriving at a revision of the contractor's programmed plant use and costs. The third figure can then be used in conjunction with it to reach the amount of loss and expense, although this needs analysis and adjustment to ascertain which part of the actual cost is caused by the disturbance factor – it is not the simple difference. When one theoretically calculated figure is used to ascertain another which involves a number of conjectures, as this must, the result begins to look rather suspect.

There are some possible ways round this problem of what may be termed the original method:

(a) To calculate all extra costs as part of the loss and expense and not include any in the preliminaries adjustment. This is not what the contract requires (consider a contract with preliminaries adjustment and no loss and expense) and, in any case, it gives a different result.
(b) To do the reverse. This is open to the same objections.
(c) To calculate a global allowance in a suitable form for all extra elements and then to divide this for preliminaries and loss and expense. The figures can then be adjusted to reflect the correct levels of pricing. This may well be the point of departure for many calculations, by supplying a controlling framework.
(d) To perform the two calculations in parallel, so that both relate to the original programmed intention. This avoids the multi-layer approach of the original

method. It does mean that all loss and expense is related to the contract works, without variations, and this may be unrealistic in concept.

In the present case and for illustrative purposes, all calculations follow the last method, ignoring the effect of disturbance on variations. This avoids tracing the value of variations through a labyrinth of figures for very little purpose. There were, however, enough general variations (e.g. those not due to the damage to the floor) in this case to be affected by disturbance of regular progress to a significant degree, so in practice some of that work would need to be examined.

The extent of delay covered in relation to variation instructions, including all work which also led to disturbance, was as follows.

	Process shop	Laboratory	Externals
Original time (weeks)	36	48	48
Original delay (weeks)			
Annexes	1	–	–
Superstructures	1	2	–
Floor damage	3	–	4
Total	41	50	52
Further delay (weeks)	17	5	10

The infusion plant started later than the rest, but also lost two weeks because the floor damage affected work around the plant. The detailed allocation of the total of each delay is shown under 'Extension of time' above.

The great distinction here is that the contractor has to show actual loss and expense, rather than secure an adjustment of preliminaries, even though this in itself is far from a straight pro rata. This can be quite difficult for him, as some expenses creep up gradually. He really has to show that he employed extra personnel and incurred related expenditure, specifically because of disturbance of regular progress. In this case the contractor again asked for a pro rata adjustment, as he had done over his preliminaries; in fact, he originally lumped the two together. With the preliminaries out of the way, progress was made on the other element. It became evident quite early in negotiations that the question of prolongation was sufficiently large to outweigh any suggestion that extra staff had been introduced

It was convenient to examine the problem under the items given in the preliminaries, as these had been priced to highlight a fairly logical split, even if the amounts were a little uneven, with some parts included in the prices for measured work. This meant looking at both the original works and the added infusion plant.

Supervision

The approach here was to calculate the cost of staff, and then allow the additional time spent due to prolongation beyond what had already been covered in the adjustment of preliminaries, because varied work necessitated extra time.

To aid illustration, these cost figures are shown as though they were adjustments of what was allowed in the preliminaries – a method *not* to be used as such in reality. In practice the actual costs inclusive of direct on-costs should be used. The present method allows a comparison to be made with the preliminaries to show differences in values and scope.

The reference is therefore to the set of figures used above in the preliminaries calculations, and also to Figure 31.1 for the periods now involved. This assumes for simplicity that the originally estimated periods and levels were appropriate in reality. This they may well not be, and always the actual figures without disturbance should form the basis from which loss and expense is ascertained. As they occurred, the various elements of prolongation affected original periods when there were differing levels of staffing units as have been defined. The prolongation of the original periods is as follows:

Process shop

Weeks 6–10	6 weeks at 2 units =	12 units
Weeks 11–20	1 week at 3 units =	3 units
Weeks 21–36	10 weeks at 2 units =	20 units
	Total =	35 units

Laboratory

Weeks 6–10	1 week at 2 units =	2 units
Weeks 11–22	1 week at 3 units =	3 units
Weeks 23–48	3 weeks at 2 units =	6 units
	Total =	11 units

Common to both

Weeks 1–48 5 weeks at 1 unit = 5 units

Using the preliminaries-based unit values, the total prolongation amount due to disturbance would be calculated like this:

Process shop	35 units at £253.25 =	8,863.75
Laboratory	11 units at £454.40 =	4,998.40
Common to both	5 units at £243.75 =	1,218.75
Total rounded down to nearest pound	=	£15,080

It has been noted when dealing with preliminaries that about one-third of the values anticipated for the staff listed there were included in measured rates, as were all forepersons. None of these have been covered for loss and expense purposes in the individual disturbance calculations so far. All should therefore be included in the present calculations.

Staff values as above	15,080.00
Allow say 50% for portion in measured rates	7,540.00
Allow forepersons as in measured rates	5,100.00
	£27,720.00

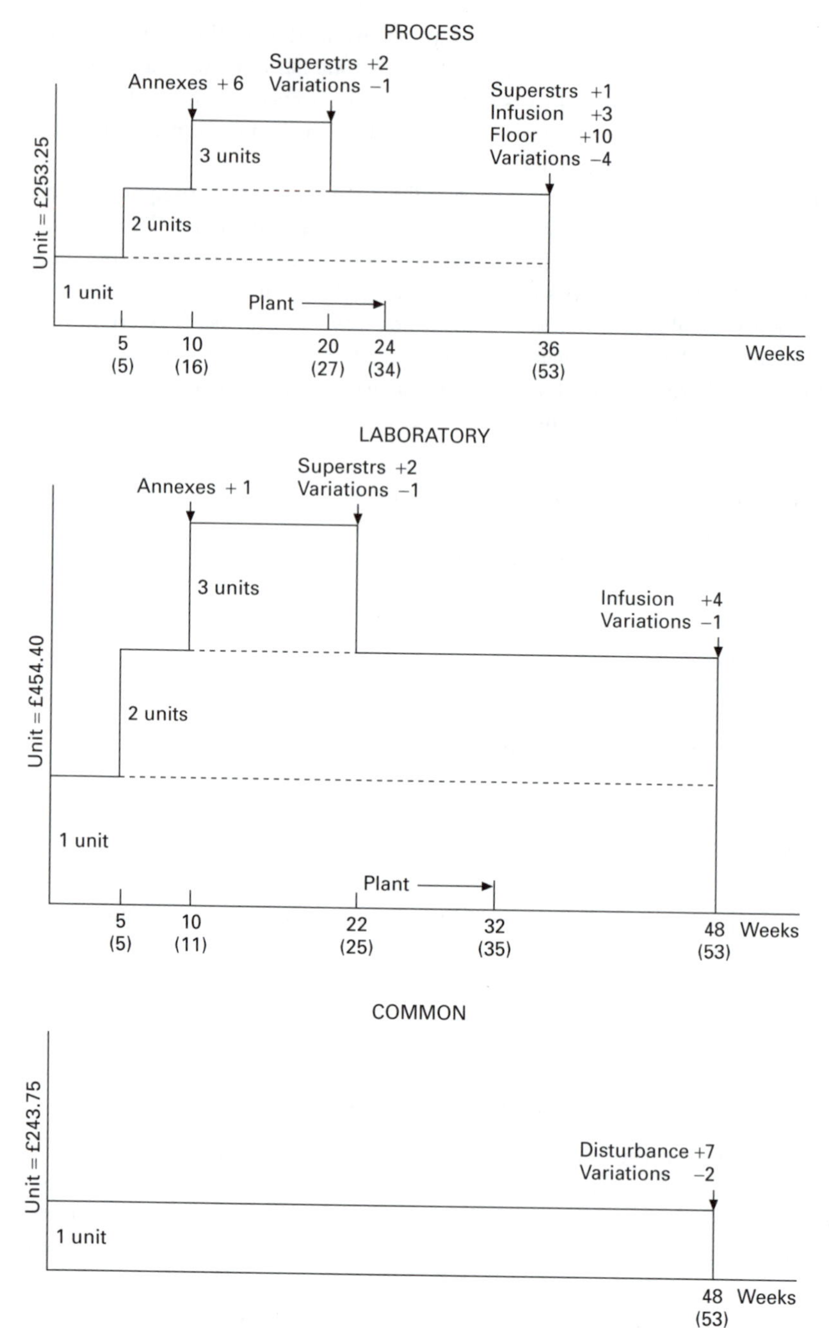

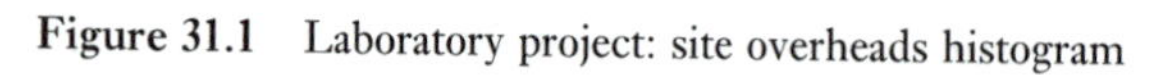

Figure 31.1 Laboratory project: site overheads histogram

Less head office and profit amounts, adjusted up or down for actual levels paid, say 20%	5,540.00
	£22,180.00

This gives the direct total for prolongation, and this is considerably different in result from any direct proportioning such as the contractor suggested. An extension of the 48-week project period by five weeks (the overall effect on the main completion, ignoring extension of time as granted) gives only £8,125 when applied to the contract amount of £78,000. The calculation of the sum £15,080 shows the fallacy of such an approach, which ignores the effect in either direction of subsidiary prolongations within the total programme (a cognate fallacy may be seen in a crude formula approach to calculating head office overheads, as shown hereafter and mentioned in Chapter 12). The other weaknesses show in the calculation that leads to £22,180.

It bears repeating that ascertainment should be done as such by building up from actual costs as they would have been and then were, but the foregoing paragraphs more clearly isolate the potential traps in what is being done.

Some qualifications remain under this heading to the flat rate calculation performed. This ignores the further programme extension of four weeks due for the external works alone. At this stage, only a skeleton supervisory staff was needed, partly on a visiting basis, so the allowable cost became

4 weeks at £420.00 = £1,680.00

The earlier amounts agreed for loss and expense included some elements of a supervisory nature. When the overall ascertainment of prolongation occurred, it became clear that some of the earlier amounts were duplicated therein, as staff were able to perform the extra work while waiting to carry on with their routine tasks. For tidiness of presentation in the final account, these items were deleted from the earlier sections:

Extra setting out for the annexes	360.00
Extra setting out due to the fort	120.00
Part of extra supervision coordinating services work in both buildings after introduction of the infusion plant	740.00
	£1,220.00

The earlier supervision under the last item had occurred during the weeks preceding the disturbance caused by the process shop floor breaking up, and so required the introduction of additional staff from outside the project to cope with the temporary workload. When the floor delayed matters, regular staff had enough time to cope, once the first crisis was over and they were waiting for fresh instructions from the architect.

There was also the question of whether the contractor had been able to make any saving in cost by redeploying staff, or even whether he should have been able to do so when he did not. The staff working on the laboratory had not been hindered for any single long period which would have justified their physical redeployment. Those engaged in common work between the buildings, such as wages and administration, were similarly placed as there was always some relevant work to occupy them. If anything, they had had to work harder during the closing stages, when the process shop had dragged on, but no extra staff or overtime payments had resulted. They had just not had the rather easier time which they had expected! The contractor was unable to demonstrate that he would otherwise have brought in extra work from other sites for them to perform during their slack period. If he had been able to do so, he would have had a valid claim for displacement.

There was one more subtle argument lying in the background. The supplementary agreement for the introduction of the infusion plant had included its own amounts for preliminaries, but had excluded any element of consequential loss and expense on the rest of the works so that, for instance, extra amounts for labour and plant have already been considered here. The amounts for preliminaries for the plant had been pitched fairly high, in view of the contractor's strong bargaining position. But was it not the case that, as a result of the introduction of the plant, an element in the loss and expense was an adjusted overall expenditure for supervision on the rest of the scheme? This was argued to be a net reduction, needless to say from the employer's side, on the basis that the common elements of wages etc in the supervisory staff were not increased, so there was a 'negative loss and expense'. This argument could be extended to other elements of supervision, and to accommodation etc yet to be considered in this chapter.

Two points may be made about this argument. Firstly, it is one thing to exclude from an agreement the effects on other sections of the physical work, as these do not overlap in cost terms with the new portion introduced, but quite another to exclude the effects on clearly overlapping site overheads. It must be held that the amount for preliminaries in the infusion plant agreement was based on a 'net extra' cost basis, that is it took into account any spreading of supervision without extra staff. The saving was to be seen against the provisions of clause 13 and not clause 26.

Secondly, clause 26 of the JCT form deals with positive loss and expense incurred and does not recognise any abatement due to savings made elsewhere in the project's execution or on another, as distinct from balancing extra expenditure in an area of cost against savings in the same area in the same period. It is at least possible, even if not probable, that a contractor in mitigating his loss by transferring resources elsewhere could also be enhancing his efficiency there by bringing in resources which he would otherwise have to do without. This is a somewhat different point from that raised on the present project, but there is still no right to secure an abatement of the amount ascertained.

In this instance the contractor agreed not to pursue his claim strongly for extra expense on the infusion plant due to delay caused by the land drain work. He settled for a 'nominal allowance' of £100. Perhaps he was conscience-stricken about his previous deal!

Major plant

It is possible to deal here with this element rather on the lines followed under the adjustment of preliminaries, and so more briefly than site overheads, as the guiding principles are similar. Although the total allowed in the contract preliminaries, exclusive of setting up and removal, was £86,150, the actual net costs incurred for the contract period were less (the contract period only is used for the reasons given at the beginning of this site overheads section). These were less because of the exclusion of head office overheads and profit and also the variation up and down of payments made for plant and attendant labour. The actual figures for the original programme were as follows.

	Hire cost	Operating cost	Time used (weeks)
Process shop	11,260	13,150	24
Laboratory	20,310	31,690	32
	£31,570	£44,840	

As with the preliminaries, regular use over the time is assumed here in presentation and proportionate extension of use for the total periods indicated in Figure 31.1. This would need checking in practice, as usual. The results with any adjustment were as follows:

Process shop, hire
£11,260 for 24 weeks
£15,950 for 34 weeks
£1,420 reduction for partial removal
£ 3,270 net extra cost

Laboratory, hire
£20,310 for 32 weeks
£22,210 for 35 weeks
No reduction possible
£1,900 extra cost

Process shop, operating
£13,150 for 24 weeks
£18,630 for 34 weeks
£1,770 reduction for partial redeployment
£3,710 net extra cost

Laboratory, operating
£31,690 for 32 weeks
£34,660 for 35 weeks
No reduction possible
£2,970 extra cost

These figures gave a total extra cost of £11,850. When ascertaining the loss and expense due to postponement of work when introducing the process shop annexes, an amount of £4,320 had been included for plant standing idle during that period. This was now extracted from the earlier figures when the principle of an overall ascertainment was agreed (as happened with supervision), to keep all overhead costs in one place.

Suspension of hire or redeployment of operatives became possible when the longer delays occurred on the process shop, but was not reasonable with the shorter delays for the laboratory.

Accommodation

Here no reduction was possible by way of removing items for short periods, for obvious reasons. The basic calculation for ascertainment was therefore to take the actual net cost for the original contract period and extend it:

Amount in preliminaries	£58,500
Actual original net cost	£49,400
Extension for 5 weeks beyond 48 weeks	£5,150
Reduction for less attendance when reduced occupancy	£470
Net extra cost	£4,680

Nothing was needed for the further period of the external works, because of the provision of enough accommodation by the employer, which included cleaning etc without charge.

Sundry items

These fell out rather as they did for preliminaries adjustment, so that similar calculations suffice:

Amount in preliminaries	£30,200
Actual original net cost	£27,100
Proportion affected 60%	£16,260
Extension for 5 weeks beyond 48 weeks	£1,690

Total for site overheads

The foregoing figures may be brought together as follows:

Supervision prolongation	22,180
Supervision, external works	1,680
Major plant	11,850
Accommodation	4,680
Sundry items	1,690
	£42,080

INFLATION ON FIXED PRICE CONTRACT

When delay occurs, there is a reasonable presupposition that the contractor will incur greater costs, which are not reimbursed in the absence of a sufficiently embracing fluctuations clause. This was the case in the project considered. The sensible approach to this question to give an adequately close answer is to use some form of indexation. If this is not done, the alternative is to perform calculations based upon inspection of time sheets and invoices in detail, as happens when fluctuations are calculated under the cost-related JCT clauses 38 and 39 or their equivalents. Admittedly, it is cost which is in question to arrive at loss and expense, but it would be asking too much to extend this rigorously to the ascertainment of the comparatively small differential between what the contractor should have incurred and what he actually did incur. No court is likely to expect quite this amount of detail.

According to what is covered by the contract fluctuations clause, this may be done by some adaptation of the standard figures provided for the operation of JCT clause 40 and its relatives. The principle involved may be illustrated by making some exclusions from and simplifications to the total operation which would actually take place:

(a) Only the process shop is considered, without its associated external works.
(b) Only the contractor's own workforce and that of domestic subcontractors are considered, it being assumed that the steelwork and the services alone were by nominated subcontractors.
(c) Only the original contract work involved is considered, so that variations etc are excluded.
(d) Preliminaries are ignored.
(e) The indexation figures are represented by percentage changes to keep them 'timeless' and are given in four-week steps, rather than monthly.
(f) One indexation percentage is given to cover all categories of work as usually defined. The various work elements shown in the programme would themselves consist of several categories in a number of instances.
(g) It is assumed that the works started in step with an indexation period.
(h) The same percentages are used for what must be deemed to have been included in the contractor's rates to allow for inflation and what needed to be allowed for the time shift which occurred. This is marginally inaccurate mathematically, as the first value is entirely within the total rate, whereas the second value is partly within and partly additional.

The necessary ascertainment calculations are set out in Table 31.1. The 'original' figures are derived from the contractor's programme and the contract bills and the 'achieved' figures from the actual programme and the bills. The application of the same percentages to the differing values performed lead to totals in each case, giving a simple difference, which is the loss and expense figure.

The exclusion of nominated subcontractors is not just a concession to simplicity. Their tenders are usually obtained at some time different from that of the

Table 31.1 Laboratory project: extra inflation due to delay (process shop only)

Weeks	Index (%)	Original				Achieved			
		Elements (%)	Amounts	Totals	Inflation	Elements (%)	Amounts	Totals	Inflation
1–4	100	Bases 100	116,400	116,400	–	–	–	–	–
5–8	100.4	Walls 30	47,790	47,790	190	Bases 50	58,200	58,200	230
9–12	100.7	Walls 70	111,510			Bases 50	58,200		
		A 50	37,300	148,810	1,040	Walls 35	55,760	113,960	800
13–16	101.0	A 50	37,300			Walls 65	103,540	103,540	1,040
		B 100	36,400	73,700	740				
17–20	101.4	C 100	34,200			A 80	59,680	59,680	840
		D 100	38,700						
		G 40	21,920	94,820	1,330				
21–24	101.8	E 50	41,850			A 20	14,920		
		G 60	32,880			B 100	36,400		
		H 100	74,500	149,230	2,690	C 50	17,100		
						E 50	41,850	110,270	1,980
25–28	102.1	E 50	41,850			C 50	17,100		
		K 100	40,600			D 100	38,700		
		L 100	31,200	113,650	2,390	E 50	41,850		
						G 40	21,920		
						H 25	18,620	138,190	2,900
29–32	102.6	M 100	36,400			G 60	32,880		
		N 100	33,300	69,700	1,810	H 75	55,880		
						K 30	12,180		
						L 50	15,600	116,540	3,030
33–36	102,7	U 100	28,400	28,400	770	K 70	28,420		
						L 50	15,600	44,020	1,190
37–40	103.0					M 50	18,200		
						N 30	9,990	28,190	850
41–44	103.1					N 10	3,330	3,330	100
45–48	103.7					M 10	3,640	3,640	130
49–52	104.1					M 10	3,640		
						N 10	3,330	6,970	290
53–56	104.4					M 30	10,920		
						N 50	16,650	27,570	1,210
57–60	104.9					U 100	28,400	28,400	1,390
			842,500	842,500	10,960		842,500	842,500	15,980
					Less anticipated amount:				10,960
						Extra inflation:			5,020

contractor. When this time is later, it is possible for the tenders to be based upon the programme then current, so that any delay inflation which already exists can be allowed in the subcontract sums. If the question of loss and expense due to inflation arises for these persons, it will be calculated similarly to the present example, but with distinct indexation levels.

Domestic subcontractors are different, in that the contractor has to give firm prices for their work when he himself tenders. He is held to these prices whatever the levels he actually has to accept and whenever he accepts them, subject to the considerations over lists of permitted persons mentioned in Chapter 7. Within the simplifications of the example, they are therefore subject to the same indexation as the contractor himself, so far as what the contractor recovers. He may well pay different amounts (probably less) because of the timing of later subtenders, although also in reality what happens will depend on the character of their work and so on the appropriate figures.

Within the original programme there were some elements that were subcritical (as, furthermore, were the external works not taken into the reckoning here), but all of them are shown as happening at the earliest time possible, thus preserving the contractor's float and minimising his inflation exposure. This is reasonable practice, but gives the maximum shift to the achieved time, and so the maximum inflation differential. Equally, none of the achieved times is queried as to whether it represents an unduly retarded execution, usually a very difficult thing to do in circumstances of disturbance. Both assumptions should be examined in practice before calculations proceed.

The achieved programme takes account of all extensions of time granted, which include those arising out of variations and not having a relation to prolongation as here intended. No extensions due to 'neutral' relevant events, such as weather or strikes, occurred in this project. Had the project not been subject to disturbance of regular progress, but had there still been extensions of time, the contractor would not have received payment for the effect of inflation. This represents the distribution of risk between employer and contractor over delays, and is unexceptionable over items like weather or strikes (where the granting of an extension may be the sticking point), but can be criticised over employer-generated matters like variations.

The contractor needs to be watching when there is a heavy crop of variations which extend the programme in inflationary times, but which do so in a tidy way as far as progress is concerned, so there is no loss and expense sector in which to allow for inflation. The problem is that variations usually accrete over time, rather than appear in one lump. The only resting place seems to be in the reference to 'conditions' in, for instance, JCT clause 13.5. When this refers to 'similar conditions' in clause 13.5.1, it certainly covers actual varied work performed regularly, but later than expected, and so leading to costs such as unusual supervision or plant. For work not varied, but delayed by variations, clause 13.5.5 may be useful, by its reference to a variation that 'substantially changes the conditions under which any other work is executed'. The term 'conditions' has a primary and intended reference to physical matters, it must be assumed, but what may have been intended may not determine the interpretation which the courts, ultimately, may supply. It is suggested that a wider interpretation is reasonable in the proper circumstances.

In this case there were delays because of variations as such, and further delays because of the effects of the variations on other work. It is earlier argued, when discussing extension of time, that these two should be distinguished for such purposes as preliminaries adjustment. This distinction is equally valid here, although the financial outcome is unaffected, accepting the preceding argument about 'conditions'. Strictly, the calculations in Table 31.1 need to be abated and reallocated appropriately, with the attendant difficulties of the variations and the disturbances being so closely linked. The line of least resistance is to leave what is properly a variation valuation in with a loss and expense ascertainment. There is a limit to hair-splitting!

With the process shop at £5,020 as shown in detail, the laboratory came out on the same basis at £2,780. This gave a total for inflation of £7,800.

GENERAL DROP IN PRODUCTIVITY OF LABOUR

The contractor raised this issue more in hope than anger, and was not too surprised when it was turned down. It is always very difficult to substantiate such an item, or at least to demonstrate how extensive are the loss and expense. This is because it consists of the diffuse effects of disturbance of regular progress, which cannot be clearly attached to a particular part of the programme, or to a particular part of the contractor's costs. This does not mean that nothing is there, just that it is difficult to quantify because it is the frictional result of lack of confidence, shortage of long-term work in view, or the other causes which are broad and rather vague.

The major pointer to the possibility of a general drop is that a string of disturbances has congealed into one protracted disturbance. When the main isolated effects have been evaluated, there remains the continuous effect. In this case there were several high points of disturbance:

(a) The postponement of work on the process shop pending revised annex drawings.
(b) Discovery of the ancient fort on the laboratory site.
(c) Introduction of the infusion plant.
(d) The breaking up of the floor in the process shop, followed by remedial work and the land drainage system.

The first two of these were dealt with in a fairly self-contained way. The first had some tapering consequential effects, partly a general drop of productivity in their following time bracket, by unsettling the sequence of work elements in both buildings. All these items of extra expense were ascertained, the disturbance worked its way out and work settled down into a rhythm again. The infusion plant occurred as a distinct feature, and again its effects were reasonably compact, in that they impinged on services and related work in the two buildings, when each was clear of major structural work. The plant also affected the external works, mainly by introducing specific constraints on working.

The final blow of the floor problem did overlap with the effects on services in the process shop, due to the infusion plant. This was so much so that the two sets of disturbance ran into one and, effectively, all the operations in the building were under review once the extent of the floor damage became apparent.

As the settlements for the separate disturbances had been cleared so carefully as they went along, with all the effects monitored closely, there was little left to pick up. This the contractor agreed without much show of reluctance.

LOSS AND EXPENSE ON HEAD OFFICE OVERHEADS

The state of the rest of the final account

The narrative now passes into the area of secondary loss and expense, following the terminology of Chapters 11 and 12. As a prelude, here is a summary of the total position of the final account adjusted for everything except secondary loss and expense:

Contract sum		4,751,500
Less contingency		300,000
		4,451,500
Variations net addition		
Measured and daywork	295,540	
Preliminaries	18,430	313,970
Primary loss and expense		
Annexes	61,620	
Fort	9,290	
Ripple effects of these	24,040	
Infusion plant effects	32,550	
Floor damage	30,440	
Site overheads	42,080	
Inflation	7,800	207,820
		£4,973,290

Any assumptions or blatant inaccuracies in these figures may be traced in earlier discussion, as for instance the exclusion of preliminaries and variations in the inflation allowance for the process shop and the assumed amount for the laboratory. Strict continuity of detail is not of the essence in the overall study.

Use of standard formulae

The results produced by the standard formulae described and critically assessed in Chapter 12 may be given first, to act as a comparison; remembering that they tend to give high results in most instances.

To keep the formulae on an equal footing, a head office figure of 6½% is used in each case, applying to the contractor's own work and the work of all types of subcontractor (this is a variable factor, but is assumed for simplicity in tracing the figures, and is how the formulae approach matters). The Hudson formula uses the percentage in the tender and the Emden formula uses the actual percentage. This is the essential difference between them, and here it means that they give the same result. The Eichleay formula does not use a percentage, but the total head office amount and turnover; the figures used equate to 6½%.

As there was sectional completion in the project, it becomes necessary to split the totals and use the formula concerned twice each time. The extended programme duration is the total, less the original and the extensions not related to disturbance of regular progress.

Hudson or Emden

Process shop $\left(\frac{6.5}{100}\right) \times 1{,}617{,}500 \times \left(\frac{58 - 41}{36}\right) = 49{,}650$

Laboratory $\left(\frac{6.5}{100}\right) \times 2{,}834{,}000 \times \left(\frac{55 - 50}{48}\right) = 19{,}190$

Total = £68,840

Eichleay

Process shop $1{,}950{,}000 \times \left(\frac{1{,}861{,}000}{30{,}000{,}000}\right) \times \left(\frac{17}{53}\right) = 38{,}800$

Laboratory $1{,}950{,}000 \times \left(\frac{3{,}061{,}250}{30{,}000{,}000}\right) \times \left(\frac{5}{53}\right) = 18{,}770$

Total = £57,570

The Hudson and Emden formulae are related to the contract sum; the Eichleay formula is related to the final account, taken as all amounts other than head office overheads. The difference between the two results is to be traced mostly to the process shop, with the laboratory contributing in a much smaller way, and is due to the use in Eichleay of the final programme time in place of the original. Indeed, the use of the final money amounts increases what would otherwise be there, unless inclusion of the total of all contemporary accounts has cancelled this out. This illustrates the tendency of the Eichleay formula to come rather closer to reality, as the main criticism of the other formulae in particular is that they are too inclusive and so too high in outcome. But subject to these comparisons, all formulae relate in essence to the prolongation position, rather than what disruption occurred, with or without prolongation.

There has already been a recovery of head office overheads in the items for variations etc, including extra preliminaries. These total £313,970, so that the presumed inclusion of 6½% in the rates and amounts yields £20,407. Given the coarseness of the formulae, it must be assumed that this amount is being covered twice, by the Eichleay formula at least. The higher the final account, the more this is so.

The figures derived from this survey are referred to later. Meanwhile, discussion can return to examination from the other and more piecemeal end of the problem.

Directly identifiable costs

Directly identifiable costs are usually in the minority when loss and expense is in question, for two reasons: they may be small in amount compared with the figures just calculated, but also they often have not been noted as problems have developed on the faraway site. The following were identified in relation to the times of maximum disturbance on-site, although they did not all occur at once or in complete weeks:

Director	1,500 per week
Administration	700 per week
Buying	500 per week
	£2,700 per week

With a five-week total, this gave £13,500 in all. Each figure included for overheads and expenses on the member concerned, itself a way of paying amounts by formula.

Rate of progress and cash flow costs

Although the standard formulae all use broad-brush overall figures, they ignore the effects of varied speeds of work, expenditure and payment during a disturbed programme – or, for that matter, the undisturbed one used as the datum. To illustrate some possibilities, the cash flow situation for the process shop, the more disturbed building, is shown in Table 31.2. This compares the expected rate of payment with the achieved rate of payment, using only the contract values to keep the comparison direct. No preliminaries or external works are allowed; the figures are those used for the preceding inflation calculation. They therefore also exclude nominated subcontractors, who were paid early (in the case of steelwork) or who performed late, so avoiding delay in payment.

It is clear that the original programme gave a rather flat payment rate, the normally expected S-curve flattening between weeks 13 and 20. The achieved rate started late, as weeks 1–4 were blank. Thereafter, the flat period was squeezed somewhat and payment generally attenuated. The period beyond, when completion should have occurred, showed severely reduced payments. Although the extra direct costs of disturbance need to be looked at for additional overhead expense, the pattern being described also suggests the possibility of a relative loss in relation to the original work, because of the flattened S-curve even without disturbance.

The simpler case is that of financing charges on the retention held until and after practical completion. Retention was at the rate of 3% until that event and then

Table 31.2 Laboratory project: extra interest on retention due to delay (process shop only) for retention = 3%, interest = 9%

Weeks	Original				Achieved			
	Flow	Cumulative	Blocks	Interest	Flow	Cumulative	Blocks	Interest
1–4	116,400	116,400		193.40	–	–		–
5–8	47,790	164,190		69.48	58,200	58,200		157.14
9–12	148,810	313,000	313,000	185.44	113,960	172,160	172,160	284.03
13–16	73,700	386,700		76.53	103,540	275,700		236.55
17–20	94,820	481.520		78.77	59,680	335,380		123.95
21–24	149,230	630,750	317,750	92.99	110,270	445,650	273,490	206.12
25–28	113,650	744,400		47.21	138,190	583,840		229.61
29–32	69,700	814,100		14.48	116,540	700,380		169.43
33–36	28,400	842,500	211,750	–	44,020	744,400	298,750	54.86
37–40					28,190	772,590		29.27
41–44					3,330	775,920		2.77
45–48					3,640	779,560	35,160	2.26
49–52					6,970	786,530		2.90
53–56					27,570	814,100		5.72
57–60					28,400	842,500	62,940	–
				758.30				1,504.61
						Less original:		758.30
						Extra interest:		746.31

1½%, the standard JCT arrangement. The formula used to calculate interest before practical completion, both original and achieved, was

$$\begin{pmatrix}\text{Gross payment}\\ \text{for period}\end{pmatrix} \times 3\% \times 9\% \times \frac{1}{52} \times \begin{pmatrix}\text{Number of weeks to}\\ \text{practical completion}\end{pmatrix} = \text{Amount}$$

The effect of the prolonged holding of the full level is shown in the figure. Although the fund did not start to accumulate until one period late and then built up more slowly than intended, it was then held for six extra periods, during which a comparatively small share of the fund accumulated. The complete position over extra financing of retention was as follows:

Process shop		
Before practical completion, as Table 31.2	746	
Less allowance for non-qualifying weeks, i.e. those subject to extension only	96	
	650	
After practical completion, retention of 1½% on £842,500 = £12,638 held for an extra period of 17 weeks at 9% per annum interest	372	1,022

Laboratory		
Before practical completion	152	
After practical completion	152	304
Preliminaries		
Before practical completion	67	
After practical completion	67	134
		£1,460

The much smaller excess expense was to be expected for the laboratory, where the total extension was less and the redistribution of payments was less radical. As an alternative, a redistribution can reduce a contractor's costs under this head by pushing payments into the closing periods. This could be offset against a relevant gross loss, but not deducted in the absence of such a loss. A relevant gross loss is not easy to define, but broadly it is a loss due to overall prolongation, rather than due to a localised disturbance, not having wider repercussions.

The more complex and significant case is that of capital utilisation. The contractor, as is necessary practice, had made some form of allocation of working capital to the project, based upon his rate of expenditure. This in turn may be assumed to bear a close relationship to the rate of payment, subject to the considerations mentioned in Chapter 3, such as front-loading the pricing structure. How he makes the allocation is another question.

No contractor actually deposits the required amount in a piggy bank on the mantelpiece. In most cases there will not be an absolutely specific allocation of funds within the organisation. Broadly, what is happening is that a contractor decides about work he might take on by reference to his overall capitalisation, cash flow prospects and other forecast workload. It is generally considered that somewhere between 20% and 30% of annual turnover is needed as working capital to finance operations. This takes account of what is needed to pay bills ahead of being paid (not all bills, but especially site wages and including the costs of obtaining work and many head office charges), to fund such matters as retention and to cover the unevennesses which are bound to occur. These include claims amounts and the late recovery of profit!

Given a definite and maintained programme for any project, the contractor can calculate fairly closely where he is going. Given the circumstances which he encountered on the present project, he is faced with an attenuation of his programme and a need to engage his capital over an extended period. It is not adequate to argue that, say, £1,000,000 tied up over six months is the same as £500,000 tied up over twelve months. The question is, Did the contractor know in advance that he would be tying up his capital over a longer period, but would actually need rather less of it? If not, he could not reduce his allocation and release some for other purposes.

Within the present project, the comparatively minor delay to the laboratory did not allow the contractor to reassess his overall financial position in such a way as to reallocate his capital within his broader strategy. In the case of the process shop, there was considerably more delay, but it still cropped up in instalments. Only in the closing stages, when the floor went wrong, might it even be suggested there was a reasonable opportunity to reduce what was being allowed within head office philosophy. Again, this is not a matter of precise shuffling of funds, but of taking a total view over prospects.

A possible level of ascertainment, and that adopted, was to use the original contract amounts only, as follows:

Process shop (1,617,500 at 22½% capital and 9% interest = 32,755 pa)

Contract period	36 weeks	
Disturbance period	53 weeks	
Reduction for final rundown period of floor	4 weeks	
Net excess period	13 weeks	
	Allow last 32,755 × 13/52 =	8,189

Laboratory (2,834,000 at 22½% capital and 9% interest = 57,389 pa)

Contract period	48 weeks	
Disturbance period	53 weeks	
(no reduction)		
Excess period	5 weeks	
	Allow last 57,389 × 5/52 =	5,518
		£13,707

The use of the original amounts allows for the inclusion of overhead recovery in the variations etc in the final account. The extended time for these is not included in the above calculations. The present amounts ascertained represent what is needed to fund the loss and expense elements in their effect on progress.

It will depend on how the percentage levels are ascertained from the contractor's records as to whether the amounts already given for extra interest on retention should be allowed separately as they have been, or included in the general extra financing charges. In all these areas there is room for some difference of approach to the goal, so that calculations would be structured differently, with the final figure hopefully close in each method.

Other head office costs

There remain several costs at head office which virtually defy segregation between projects, such as rent on premises (or equivalent capital charges) and insurances of various kinds. These costs can only be ascertained by going back over several years' figures, and so following the judicial exhortation given in *Peak Construction* v. *McKinney* (Chapter 9). Examination produced a reasonable assessment for present purposes of 2.35% on site-based costs:

Primary loss and expense items £207,820 at 2.35% = £4,883

Loss of profit on other work forgone

As explained in Chapter 12, this is not the addition of profit on loss and expense amounts, as these are intended only to put the contractor back in the position he would have been in had disturbance of regular progress not taken place. Rather, it is profit forgone elsewhere because of the intrusion of the extra volume of work and delays into the contractor's planned business activities. As such it was dependent here, as always, upon several contingencies.

The first was the availability of extra work and the second the credibility of the disturbance encountered actually leading to a probability of its loss from the order book. The contractor was able to demonstrate that, during the progress of the present project, he was running with a quite full load and there was enough work about of a suitable variety for him to expect to obtain some of it. In fact, he was obtaining work steadily. This met the first criterion sufficiently, as only *probability* is required under it.

The second was more awkward. It was very difficult to argue that the addition of less than 1% to the contractor's workload over the period, taking the figure of £30,000,000 used in the Eichleay calculation, could in itself have deflected other work. But if this position be taken to its rigid conclusion, it would mean that no contractor suffering proportionately small losses on a number of concurrent contracts could ever recoup anything. Any large contractor would regularly be financing the misdeeds (from his viewpoint) of employers. It was necessary to look at the position in relation to what was happening generally, and this was that several contracts were being afflicted by disturbance of regular progress to an extensive degree. On this basis, the present project had to take its share, and so an allowance was made. (Without this, the case study would of course be lacking!) The principle is one which receives differences of interpretation, although the present view is that reimbursement is frequently justified.

Arriving at an amount presents two further questions: the sum on which it is based and the percentage to apply. For the sum, the whole of the loss and expense was taken, as the contractor's order book was full. The only really arguable item was that for extra financing charges, on the basis that this could be interpreted as including reward for capital which was by way of profit. It was demonstrated that the level of 9% did not actually reward those running the enterprise, but simply dealt with borrowing as such.

The percentage aspect allowed of several possibilities:

(a) The percentage included in the tender, a level of 3.3%. This was just an aim, and not necessarily a reality. It therefore could not be taken alone.
(b) The percentage achieved, a level of 3.1%. This had more claim to be used, as it was real. It did, however, represent what had happened against the background of the tender as embodied in the contract sum. This could well differ from the climate in which work was being obtained during the currency of the project, and so of work which was arguably being forgone.
(c) The percentage being included in current tenders, a level of 3.6%. This was due to some upturn in the work coming into the market allowing more advantageous tenders. It was still conjectural at the stage of settlement.

(d) The percentage which would result, being actual loss. It would have been possible to wait and then average a representative selection of contracts (without their loss and expense amounts!). It was decided to assume some shortfall in profit realisation, as was happening in the actual project, but to pitch this at less than 0.2%.

A figure of 3.5% was therefore agreed, based on a shortfall of 0.1%. The ascertainment of profit forgone and the final total for all loss and expense became

Primary loss and expense		207,820
Secondary loss and expense		
Head office staff	13,500	
Interest on retention	1,460	
Extra financing costs	13,707	
Rent, insurance etc	4,883	33,550
		£241,370
Profit on other work at 3.5%		8,450
		£249,820

The amount for secondary loss and expense may be compared with the results of the standard formulae given before:

Loss and expense amounts, as above	33,550
Amount assessed as in variations	20,410
	£53,960

Hudson and Emden gave £68,840 and Eichleay gave £57,570. The first pair make no deduction for the margin for overheads in variations etc. In the case of Eichleay, it depends on how the figures used for final account values are extracted from the accounts, that is whether variations on any excessive scale are isolated and allowance made in using the formula. As used in this study, they are reasonable. Nevertheless, the relative closeness of the three results is a matter of some surprise and was *not* produced by reworking the arithmetic of the case study until it emerged! Formulae can often be much further out.

SOME FINAL COMMENTS

Here the study proper ends, but a few notes may be added. In most places throughout, the distinction between contractor and subcontractor has been ignored, because it has been irrelevant. It has also been ignored in the interests of simplicity in cases where there might actually be a difference in the figures, but not in the principles involved. The need for a two-tier system of negotiation and calculation should be apparent, as is discussed in Chapters 7 and 13 on subcontractors and multiple contracts. It is also possible to take the principles illustrated and transfer them into the dispute situations which crop up between contractor and subcontractor without the employer being involved at all, or even those in which there is a triangular negotiation going on. Each depends on its own set of contract documents and procedures, but is essentially the same.

Numerous subsidiary features can appear in a dispute situation, which have not been covered. Thus there may be effects stemming from the shift in timing following delay, such as seasonally different working conditions (for better or for worse) or running continuous activities so that they straddle holiday periods or similar occasions. The most common cause of delay and expense, it is widely believed, is delayed or changed instructions. Except for the really major cases of change, these too have been excluded, because they add nothing new. When they consist of a string of small instructions, it is not practicable to illustrate them meaningfully in a book. Then there are the 'neutral' causes, such as weather and government activity, which have to be extracted from the reckoning. All that could be said in a theoretical example would be 'extract them', which becomes tedious and so has not been said.

What has been said at the risk of tedium is that the methods of calculation have regularly been simplified or sometimes distorted in the cause of explanation, rather than wooden, repetitive instruction. Here it is being said again – so may the point be taken and excused! The other case studies do illustrate one or two of the more interesting points not taken here, but by way of compensation give no calculations at all.

How the employer felt about the outcome must be conjectured. It might be wondered, for instance, whether he would have wanted the process shop annexes, had he realised the disturbance costs he would be facing compared with the costs of straight variations. The facilities might have been invaluable, of course. The section on client's risk analysis over such issues in Chapter 1 may be considered, as well as the virtues of thinking things through before starting on-site. The annexes did appear as a change of mind quite early, but then such things frequently do.

Be all of this as it may, and excluding the infusion plant supplementary contract, the financial outcome was as follows:

Contract sum, less contingencies	4,451,500
Net addition for variations	313,970
Loss and expense amounts	249,820
	£5,015,290

There have been many worse.

CHAPTER 32

MINOR CASE STUDY: FACTORY PROJECT

- Particulars of works
- Main events and their results
- Approach to settlement

This case study concerns the execution of moderately sized works by a moderately sized local contractor. The works were performed in fairly cramped conditions at the works of an existing and fully operative manufacturing business, and covered new works and alterations, with some phasing. Each of these features led to the problems encountered. By comparison with the case study in Chapters 30 and 31, there is much more technical detail given here but, in compensation, no figures at all. The procedures followed, once problems have arisen, are quite good – maybe someone had read a sound book or two! The material in this chapter is, however, also used as the basis for discussion of the comprehensive claim approach in Chapter 26 and may be read with this in mind.

PARTICULARS OF WORKS

Contract details

The contract form was the IFC 1984 contract, using the 'with quantities' option within it. It was on a fixed price basis, except for statutory fluctuations. A sectional completion supplementary condition was not used – despite the phasing required.

Scope of works

The general layout is shown in Figure 32.1, with the following main items:

(a) Construction of a research and development bay on an area of the employer's site which was virtually unused and easily accessible from the highway. The building was a steel-framed cladded shed with a free-standing structure inside, containing a large installation (the 'maxi-machine') and with a control facility adjacent at upper level. There was a substation internally and a system of ducts in the concrete ground floor.

(b) Alterations at first floor level to area X of existing offices over a stores building, removal of a corner of the building at both floor levels and construction of a first-floor bridge linking to the control facility in the R&D building.

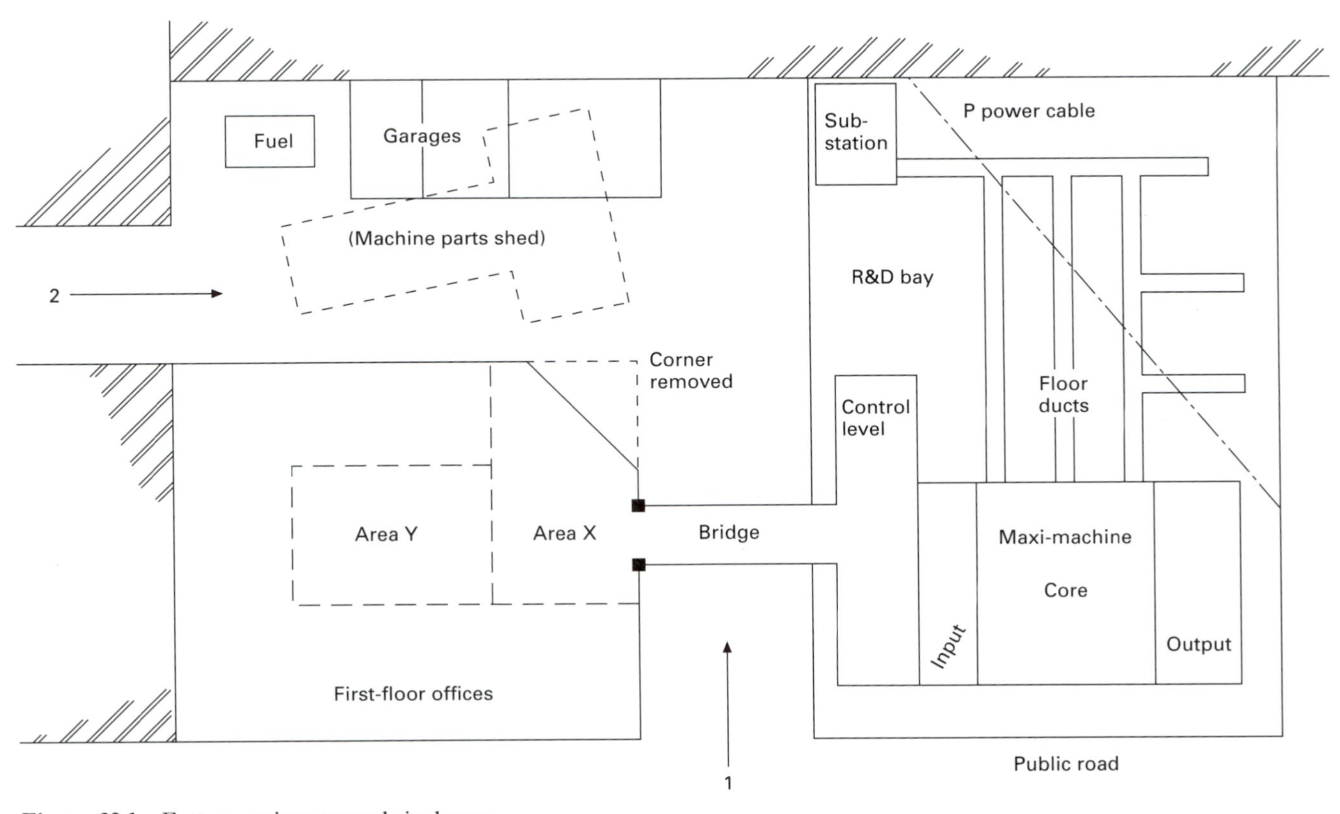

Figure 32.1 Factory project: general site layout

(c) Demolition of a single-storey machine parts shed adjoining the stores building and construction of a range of maintenance garages and a fuel pit.

Within the whole, the steel frame of the R&D bay and electrical work generally were by named subcontractors, as were other elements of work. The provision and installation of the maxi-machine was by a direct contractor and was to proceed during the contract works. Other services included in the contract were essentially simple, as the employer was installing special elements himself, after the construction period.

Programme of works

A simplified bar chart of what was intended is given in Figure 32.2(a). This was produced by the contractor from the phasing requirements set out in the

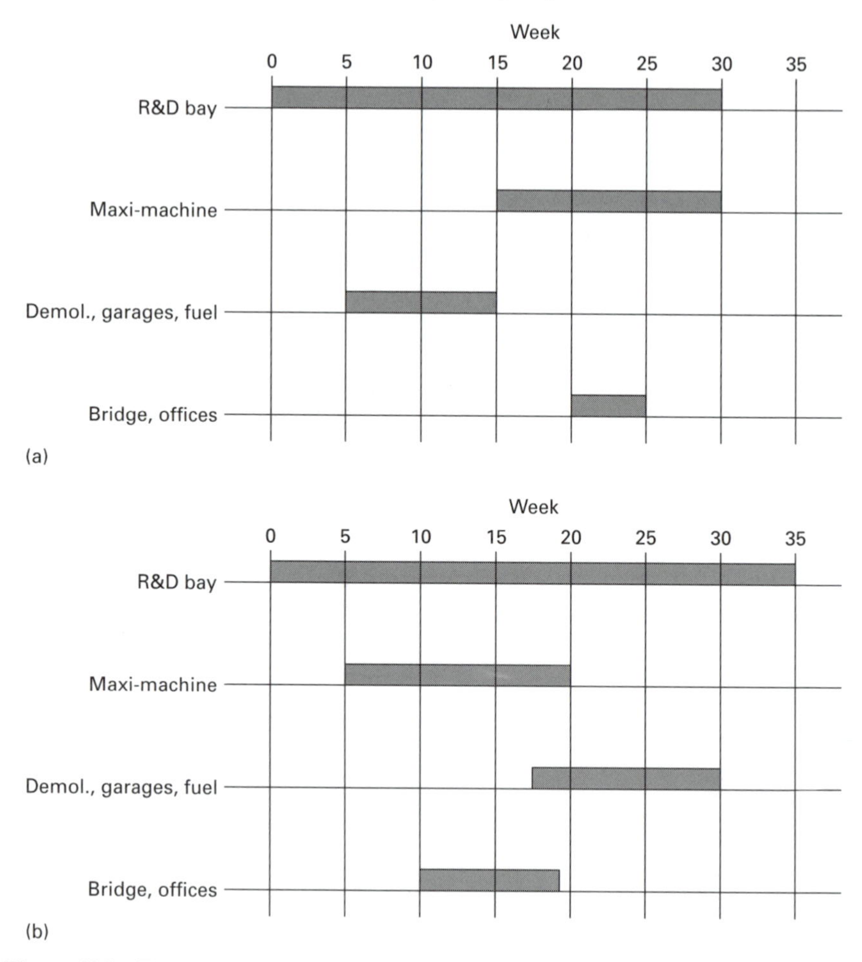

Figure 32.2 Factory project: summarised programme: (a) contract programme and (b) achieved programme

preliminaries of the contract bills, which showed a sequence and gave a single completion date, with a statement that the contractor should give adequate notice in consultation with the works engineer of when he would require to enter particular portions of the site. The contractor provided information during precontract discussions as a master programme, which the IFC contract does not actually require; here are the key points:

(a) Construct the R&D bay, with access 1 available (week 1 to week 30).
(b) Maxi-machine, proceeding within last and involving integration of contractor's supporting and enclosing work (week 15 to week 30).
(c) Demolition of machine parts shed and construction of garages and fuel pit, with access 2 becoming available as a result (week 5 to week 15).
(d) Construct bridge and perform alterations to existing offices etc, so restricting height of access 1 (week 20 to week 25).

The contract bills stated that obligations and restrictions would apply, as referred to in clause 3.6.2, over the following:

(a) Access, due to the need for prior demolition work, and to give notice before access 2 could be cleared by factory users.
(b) Working space, due to the same reasons.
(c) Order of working, due to the need to enter the offices at a late stage of the works.

There *was* only one completion date, despite the mandatory nature of the period for altering the offices. This could have proved difficult (see *Bramall & Ogden* v. *Sheffield* in Chapter 9), but was accepted throughout by the parties.

MAIN EVENTS AND THEIR RESULTS

Changes during progress

The employer found that he could obtain a very favourable price for the maxi-machine, provided he had this work done earlier than he had planned. He decided to accept this proposition, leading to several consequences:

(a) It became necessary to erect the steelwork and cladding at the end of the building housing the machine ahead of the rest (rather than the reverse) and earlier than had been anticipated. This threw out the contractor's order of providing foundation work and necessitated extra work by way of temporary protection for the maxi-machine, which cluttered the main working area for the bay.
(b) Revised floor duct layouts and details arose from the technical nature of the machine chosen and from afterthoughts about the layout of subsidiary machines within the bay. This led to working inside the newly constructed building envelope, with uneconomical plant and the need to provide clean conditions during concurrent installation of the main machine. During excavation for the

ducts, an existing power cable at P, serving another part of the works, was unexpectedly encountered crossing the area.

(c) The bridge from the existing offices was needed earlier, due to commissioning of the machine. This restricted movement from access 1 and led to complications in altering the offices adjoining, and these were exacerbated by a requirement to alter more offices than had been foreseen (area Y). As these were areas still occupied by staff, weekend working and other special measures were introduced by agreement.

In addition to these changes, it became necessary to postpone the demolition of the machine parts shed, so this could remain in use during the early stages of the maxi-machine installation. This had several effects:

(a) Access 2 was not available on time, affecting the intention to circulate construction vehicles from one access to the other as work built up, also complicated by the early erection of the bridge.
(b) Construction of the garages was deferred, so the contractor could not use them for temporary storage of materials. His intention to do so had not been made clear in the tender or during precontract discussions.
(c) Construction of the fuel pit became difficult at the planned time and was delayed by the contractor. It was then subject to late changes of design in moderately important respects.

The revised programme actually followed is set out in Figure 32.2(b), showing the rearranged sequence. An overall prolongation of five weeks occurred, spread evenly from the end of week 5 onwards for the R&D bay.

Timing of changes

Although the changes were quite drastic, by way of variations etc, had they all been decided upon and instructed at the beginning, it would have been possible for the contractor to have revised his programme to accommodate them and for the contract sum to have been adjusted by valuing the effects as variations. This could even have been tied up entirely in advance of work proceeding.

With events as they actually occurred, the first line of enquiry is how far agreement of variations is still suitable. The basic work of, for instance, placing concrete in the foundations, floor and ducts of the R&D bay still ranks for measurement and valuation, even if it is performed in different conditions from those envisaged in the contract. They may be distinctly more awkward than would have applied had the contractor been able to carry out a considered replanning, but still be performed without disturbance in the sense of clause 4.11, which is the last provision of the contract to be invoked.

This becomes a major consideration for the contractor, who is required to give notice as soon as he can of expected disturbance, but not of changed conditions for variation purposes. In this instance, he was able to warn quite early over the revision due to the advancement of installing the maxi-machine, although the full effects of it in terms of clean conditions and other secondary effects were not immediately evident. He therefore did so, even though he could not clearly discern the extent of disturbance. On the other hand, he found the delay in demolition of the machine parts building on him at the last moment, when he was about to implement his programme.

When a contractor is in any doubt about whether a loss and expense situation is developing, he must prudently give notice as soon as he has sensible grounds to believe so. This enables all concerned to be warned and take any possible avoiding action, and at the least be ready to keep records rather than to try to 'invent facts' afterwards. In the present case, had the employer sounded out the contractor through the architect, he might have decided that the likely extra cost to the contractor would outweigh the saving on the price of the machine (see the example of risk analysis in Chapter 1). This sort of action is quite different from the 'cry wolf' attitude which reacts every time a revised bending schedule for one beam is issued.

An architect also has to decide how to react. He cannot reject any reasonable approach from the contractor out of hand, but should not rise to every bait by as good as agreeing that something is due. Although it is for him to decide whether there is a case and then whether to ascertain the amount himself, in practice he often needs to discuss with the quantity surveyor just where the boundary is to be drawn. This time there was room for manoeuvre over what were to be variations and what were loss and expense. A quantity surveyor is entitled to deal and should deal with the variations on contract terms under clause 3.7, simply because an instruction has been issued, whether the contractor or the architect acts over some element in any way or not under clause 4.11. Duplication or a gap (if the contractor is not alert) can therefore arise by lack of communication. This is a good reason for the quantity surveyor to do the ascertainment in any case.

APPROACH TO SETTLEMENT

Contractor's overall analysis

Although the disturbance factor here was quite considerable, what the contractor had to do first of all was to decide whether there were clearly separable elements in his claim. From these, some progress could then be made towards isolating the effects, rather than using the unacceptable method of 'cost, less valuation'. The narration of events above both helps and hinders here, because it divides affairs up geographically on the site. Geography, time and operational relations vary in their importance. When things go wrong one step at a time, this progression may help proper segregation or merely lead to an incremental or overlapping build-up of total confusion. In this case, he decided to apply under several heads for loss and expense. This he did promptly for the first six items, while giving broad warning of the last two:

(a) Disturbance of sequence and timing of the superstructure of the R&D bay, as stated separately above: the primary effect.
(b) Revision of the floor ducts, also as stated separately above: the secondary effect.
(c) Work inside the existing offices, both due to the bridge being early (although itself constructed without problems) and to working partly in occupied spaces.
(d) Access problems, due to the bridge and the late removal of the machine parts shed, but not to the order of erecting the R&D superstructure.
(e) Storage problems off-site owing to the garages not being available.
(f) Localised problem of the fuel pit.
(g) General loss of productivity on other work, due to the sheer level of disturbance caused by the specific matters.
(h) General question of prolongation, affecting supervision and preliminaries.

This division had much to commend it in its logic, although it meant care in practical segregation. The first three items had most risk of confusion with the valuation of variations. They also involved loss and expense claims from named subcontractors for steel and services, which had to be taken separately by the contractor, but included in the whole. Although sufficient division is needed, an excess is better avoided, as there comes a point at which overlapping relevant matters do not allow such finesse. The contractor may find that he lacks flexibility in negotiations as a result.

He proceeded to put forward separate claims for each of these, in two main groups: those flowing from the maxi-machine decision, and those flowing from the delay over releasing the existing shed. The architect immediately referred them to the quantity surveyor, who established some baselines with the contractor over what would be included in his variation account. The great virtue of this cooperative approach all round contrasts with the drawn battle lines approach so often encountered. Some comment on presenting a consolidated, single-claim document at completion is made in Chapter 26.

Each element of loss and expense is discussed separately. The total amount of extension of time was assessed after practical completion, following several intermediate extensions. It was made on the lines considered in Chapters 30 and 31.

Primary effect on R&D bay

This was related to the erection of the superstructure. The contractor was concerned to reduce the total delay effect, in his own interests as well as the employer's. It may be wondered whether a larger contractor might have found it easier to mitigate the loss and expense in the process, by a concentration of resources at particular points, so that work went on in tidier parcels. It is suggested that the extent of mitigation which is reasonable will vary with the managerial capabilities and resources of the actual contractor: the employer probably gained initially on price by engaging a smaller contractor with less overheads and must expect to lose when trouble comes. Whereas a contractor should mitigate, it is the employer who causes the disturbance.

The subsidiary items in the loss and expense settlement were as follows:

(a) Extra costs of delivering ready-mixed concrete for the area of the floor beyond the section required for the maxi-machine. These included problems of manoeuvring lorries and receiving concrete in the restricted area available between the bay and the still-standing machine parts shed. It was agreed that all costs of placing concrete, once it was shot from the lorries, were part of the revised measured prices for working in conditions dissimilar from those envisaged in the bill prices.
(b) Steelwork subcontractor's loss and expense working in phases, especially setting out work, extra temporary bracing and the difficulty of handling steel and using a smaller crane inside the area of the second phase (see next section for a further three-way item arising from this).
(c) Standing time associated with the provision of temporary protection for the maxi-machine, while other construction work was going on. It was recognised that most of the standing time due to the bringing in of large and delicate parts of the machine had been allowed in the contract sum, but that an element of excess arose out of this occurring earlier when more, dirtier carcassing work was in progress.
(d) Additional supervisory time in dealing with the foregoing by reorganising work, setting out and cooperating with the machine installer.

Some arbitrariness may be detected about what has gone into variations and what has gone into loss and expense. For example, more loss and expense could have gone into (a) and less into (b). These are problems associated with a proper interpretation of the contract, as distinct from a sensible reimbursement of the contractor and husbanding of the employer's resources.

Subsidiary problem with steelwork

A relatively small problem arose over the steelwork mentioned under (b) above, due to the need for extra temporary bracing. The nominated subcontractor presented his shop drawings to the architect, as a development of the architect's general arrangement drawings, which simply gave the main layout of structural work. These drawings were a revision of what the subcontractor had prepared when submitting his original tender. The architect checked them over in relation to the permanent members shown (there were variations) and passed them back through the contractor as 'approved'. In passing, it may be noted that the IFC 84 contract has no provisions for design by the contractor or any of his subcontractors, and it is widely held to be a 'construct only' contract, like the JCT 80 form. Perhaps more critically, when the architect unequivocally approves a design, he is assuming a responsibility for it, a point taken up in detail in *Design and Build Contract Practice.*

Such a burden may be considered fair in the contractual circumstances, but all that the architect had actually done was to check that the scheme fitted into his own design. He did not check the members structurally. As it happened, none of this caused any trouble over the permanent work. What did cause trouble was the positioning of some of the temporary bracing that was in the way of some of the contractor's own permanent work, which had to be constructed before the bracing was removed. The contractor did not notice this aspect, but simply passed the approved drawings on to the subcontractor without comment.

The steelwork was erected before the discrepancy was discovered. The subcontractor presented an item for the costs of adaptation to the contractor, who in turn presented them with his own costs of delay to the architect, as being loss and expense. The architect rejected them, stating that the contractor should have noticed and was responsible for coordinating information and warning of discrepancies. As a supplementary point, he stated that the subcontractor should have been aware of the permanent work which was affected. The subcontractor's answer was that the architect should have been at least as aware. The scene was set for a dispute which might have revolved for the duration of settlement of the final account. Clearly, the contractor had to pay the subcontractor, whatever happened, unless the subcontractor accepted the architect's point about him. The architect was equally clearly following a safe line by refusing to consider the item initially. Until the position was cleared, he could only compromise the employer by admitting anything.

Eventually, it was agreed that the architect had given the contractor the impression by his actions that he had approved everything on the subcontractor's drawings. Furthermore, in the tight circumstances existing when the installation of the maxi-machine was being shifted to the early stages of the programme, it was agreed that any question of the contractor being able to coordinate every piece of information in advance was uncertain. He was therefore given the benefit of the doubt and paid for the costs incurred.

This is an instance in which the precise lines of liability arise not so much from abstract contractual analysis, as from the pattern of events interpreted with some leniency when the moderately sized contractor had been placed under strain by the changes introduced into the scheme.

Secondary effect on R&D bay

Once the steelwork and cladding were erected, the contractor faced the subsidiary effects. These were twofold: the constricting effects of the surrounding structure on his work within it, mainly more concreting, and the extra care needed and protective measures involved in dealing with the concurrent work on the maxi-machine itself – the cause of the disturbance. He could, of course, have delayed the erection of the steelwork surrounding the area until all the concreting had been completed, but this he judged less desirable in his own interests of overall progress on the various trades engaged on the works.

Whether this was also less desirable here in the interests of the employer is a distinct question, although the answer is often the same in many cases, as the employer is seeking to maintain completion. There is a financial break-even point for each party between cost and return, and the two points are unlikely to be absolutely identical. The architect has no power to interfere in what the contractor actually does, only to judge it afterwards, so that it usually becomes a rather academic question. The contractor has an obligation over delay to 'use his best endeavours' and to 'do all that may reasonably be required' by the architect. These mean 'best' and 'all', so the contractor must act decisively, but they are still fairly loose and are usually held do not oblige him to take much positive action without extra payment (and this the contract does not cover).

The position was complicated, as work proceeded, by the discovery of the power cable P, which presumably the employer should have known about (unless a previous owner had had it laid and not said anything). At least, the contractor could not be expected to know. At such a point, it becomes very difficult to segregate the effects of the several matters, and really unnecessary, as it is possible to rely on the blanket expression of clause 4.11 'any one or more of the matters' to which it refers.

The resulting settlement involved the following:

(a) Continuing extra costs of concreting, which included working with special care once the maxi-machine was being installed, all paid for as variations.
(b) Breaking up ducts and the surrounding floor prior to relaying, paid for as variations by daywork.
(c) Standing and reorganisation costs while the daywork was going on, paid for as loss and expense, involving several trades which could not work in the disturbed conditions or had to move materials and equipment to allow it to proceed.
(d) Further loss and expense (mainly waiting and reorganisation) due to the discovery of the cable, checking by the employer of its significance and arranging for the electricity board to deal with it. Additional excavation and attendance by the contractor was paid for as a variation by daywork.
(e) Suspension of the work of trades causing dirt and vibration during critical periods of the installation of the maxi-machine. This was dealt with as loss and expense and covered some wastage of materials which deteriorated due to the delay.
(f) Additional supervision, again due to disorganisation and cooperation with the machine installer.

Again, the problem of achieving clear demarcation between variations and loss and expense may be noted.

Work inside offices

The instruction to carry out additional work in area Y of the offices was a straightforward variation in a rather small special section, and not of sufficient magnitude to constitute an unacceptable enlargement of the contract (this contrasts with the infusion plant in Chapter 30). Even with the requirement to work at weekends, valuation was quite normal to perform.

Two difficulties in negotiations arose early on:

(a) Both the original and extra areas of offices were to be partly in use during the alterations, requiring careful operations, carried out cleanly and quietly, and subject to unpredictable delays and piecemeal working.
(b) The major work concerned was to have been sublet by the contractor and consent had been obtained before the variation was issued. It may be added that the position would probably have been resolved similarly had consent not been obtained so early. The subcontractor in question was unable to perform the work at the revised date and the best alternative firm was substantially dearer in quoting for the original scope of work.

In principle, all of these effects were susceptible to assessment by the loss and expense approach, although whether they were able to be 'ascertained' is doubtful. It was eventually agreed, before the work was carried out, to place the whole within clause 3.7 and have it valued by daywork, with that of the actual subcontractor based upon special terms. On top of this valuation, a special allowance was added to cover the intensive supervision required.

The erection of the bridge, breaking through into the existing building and removal of the corner and rebuilding on the splay were all performed at bill rates at the same time as the alterations, allowance being made in protection items for the extra care needed. The main extra requirement was for clear segregation of costs of measured work and daywork on-site.

Access problems

There had been special provisions in the contract for using access 1 initially and then access 2 as well, once the machine parts shed was cleared away. This was what happened, but the shed remained in place much longer than stated. This effectively introduced a variation of obligations and restrictions under clause 3.6.2. Alternatively, it could be seen as a 'failure . . . to give . . . ingress' under clause 4.12.6, so leading to reimbursement of loss and expense. It was actually given as an instruction to postpone removal of the shed, from which the other consequences flowed, so again giving loss and expense as the correct outcome.

In this instance there were no ordinary measured or lump sum items in the contract bills to vary in isolation. This is typical of such an occurrence, as is the lack of easily verifiable cost data for what would have happened but for the change. Whether the adjustment of the contract sum is dealt with by variation or ascertainment is largely a matter of words: calculation starts from a narrow base and is similar. If it is from a base of variation, clause 3.7.6 allows for 'any addition to or reduction of' preliminaries items, which is not a pro rata adjustment. It was necessary to do this, to take into account adjustment of original amounts concealed within measured and preliminaries items generally. The parallel instance of preliminaries adjustment in Chapter 31 fills out this paragraph with calculations.

The main elements in the present instance were the extra time spent by every vehicle which entered the site until it left again, with the associated slower unloading and extra attendance, and the need to keep the small areas adjoining clear of whatever might otherwise have been dumped there. To the extent that vehicles belonged to merchants, it was not possible for the contractor to demonstrate that he incurred higher quotations for materials delivered in these conditions. Any general and wider drop in productivity was put forward separately.

Storage problems

The contractor had not stated when tendering that he intended to use the employer's new buildings for storage of materials. He won the day on the basis that he had shown early construction in his master programme, there was no provision in the contract for partial possession by the employer leading to their early surrender, and the lack of other space on-site made their use a reasonable thing to posit.

With this established, it became fairly easy to deal with loss and expense on the basis of provision of extra space at the contractor's yard at a time-related charge, reflecting its non-availability for other purposes and costs due to congestion, with associated extra security and double-handling costs from the yard to site. These were clear and separate elements without a problem of segregation from costs which would have been incurred in the absence of the disturbance.

Fuel pit

There was a simple case of late information while the fuel pit was under construction, leading to a straight allowance of standing time for labour and plant.

Had the information been delayed before work started, the contractor might have been expected to hold back until it was available, perhaps diverting his resources elsewhere meantime. This could have involved putting them on another project, and would have depended on what he had available as a smaller contractor. Properly, for him to do this, a postponement instruction would have been required. In its absence, he would have been entitled to go on working until the information failed to appear by the date on which he had requested it, and then have incurred his costs. He was not entitled or required to stop ahead of cast-iron certainty (meaning something proper and in writing), until he was out of work. If he had, he might even have been risking action for failing to proceed regularly.

General loss of productivity

Much of the loss of productivity had been covered under the items isolated as disruptive. The contractor's contention was that the areas of work between them had suffered from the 'shadow' effect produced. A review of the programmes, proposed and actual, in Figure 32.2 shows that the programme was considerably changed, and most of its major components were affected in detail. No one seriously considered that there had been no loss: the only question was how to ascertain it. In the nature of things, it was not possible in any certain way to compare what should have happened with what had happened. 'As-certain-ment' was not practicable. A somewhat parallel situation occurred in *J. Crosby & Sons Ltd* v. *Portland Urban District Council* (1967), when it was held that a lump sum might be awarded for overall delay and disorganisation, provided this introduced no duplication with other amounts and no inadmissible profit allowance.

It was agreed that the site records showed discontinuities in the areas of work, with some information provided piecemeal and a tendency to smaller parcels of work than might have been expected. Against this, an inspection of costs, so far as any segregation could be achieved, suggested that nothing catastrophic had occurred. It was then agreed to treat most of the question as one of evaluation of work under varied conditions, as the other major changes had been introduced as variations. This had two advantages in aiding settlement: ascertainment was not used, and adjustment of measured rates could be introduced. The contractor was also happy, in that he could secure an overheads and profit margin. It is not always entirely clear in circumstances of this kind what basis should be used, and a doctrinaire split could have been engineered here: that chosen should generally be considered first before resorting to loss and expense. This time the parties came back to it!

The contractor proposed an allowance of 10% of the actual labour and plant costs incurred for all relevant work. Eventually, it was agreed to allow 7½% of all work not accounted for in the other major items, except for the garages and the internal linings of the R&D bay envelope with all painting work. The following reasons were given:

(a) The internal linings of the R&D bay envelope and all painting work had proceeded at approximately the right time and were sublet as entities, as intended.
(b) The garages had changed position in the programme, but had been performed in an orderly sequence, somewhat late. This had caused some extra expense in concreting work, which was valued in its own right as a lump sum.

Prolongation and preliminaries

There had been a total extension of time of five weeks, none of it relating to the first five weeks of the construction period. Extra supervision costs had been allowed for:

(a) The R&D bay in respect of the primary and secondary effects, as given above. Here extra major plant allowances had also been made.
(b) Alteration work in the existing offices.

It was agreed that these amounts effectively dealt with most of the prolongation effects of supervision (as well as the intensification effects), as it was largely a case of using the same personnel over longer periods. An allowance of one member of staff for three weeks was made to cover the extra supervision needed on the garages, due to their displacement into the closing weeks.

Other preliminaries costs were the access and storage costs. Both of these had already received separate treatment, covering most aspects. It was agreed to pay extra costs in moving the small site offices due to the sequencing of work. These included an allowance for staff disturbance and working for a short time from the contractor's head office.

There was a mild flirtation with the elements of financing charges and inflation due to prolongation, before the suggestion was dropped. These elements are worked out in Chapter 31, over a much larger contract. Here a contract of twenty-five weeks was extended by five weeks due to disturbance of regular progress, and this appears quite small, as it was in a period of low interest rates and inflation. The effects would have looked more persuasive for an ascertainment, given a contract of twenty-five months and five months extension in more adverse conditions. Certainly, it is easier to locate amounts when they are larger. As is also suggested in the next chapter, the simple question of there being a smaller contractor can affect the outcome unfairly. He may be performing smaller contracts, but suffering an accumulation of small losses which in total are significant. He should fight for his case, often another difficulty given his size.

CHAPTER 33

MINOR CASE STUDY: FLATS PROJECT

- Particulars of works
- Main events and their results
- Comments on narrative

This case study concerns rehabilitation of privately owned, rented blocks of flats and is given as narrative only without calculations, as for the study immediately preceding. However, in this instance, comments on the implications are excluded from the narrative and given in the latter part of the chapter, referenced by bracketed numbers following paragraphs in the narrative. This allows the reader to form an opinion, if so wished, before comparing it with the suggestions offered.

PARTICULARS OF WORKS

The contract form was in the IFC 1984 form, with a priced specification and with a sectional completion supplement based upon that available for use with the JCT main form. (1)

The works covered six blocks of four-storey flats, arranged semi-detached in three buildings. A longitudinal section through blocks A and B, as typical, is shown in Figure 33.1(a). The main elements of work were as follows:

(a) New partitions and doors to revise internal planning.
(b) Replacement windows.
(c) Refitting kitchens and bathrooms, and renewing services and plumbing. The employer was to supply the following for fixing by the contractor: kitchen units, refrigerators, sinks and bathroom fittings.
(d) Hot water central heating.
(e) Making good internal finishings and redecorating.
(f) Sundry minor structural repairs, varying with the flat.
(g) New handrails and finishings to common stairs.

Named subcontractors were included in the contract for three types of work:

(a) Services, plumbing and central heating.
(b) Electrical installations.
(c) Stair finishings.

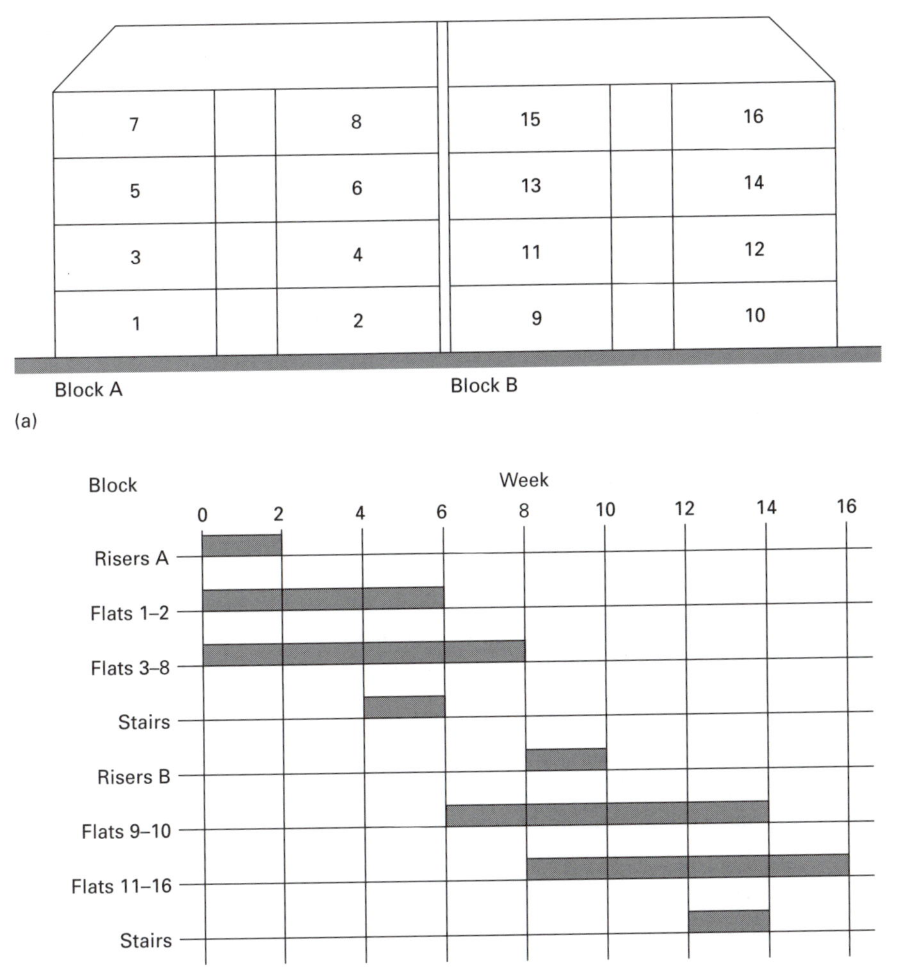

Figure 33.1 Flats project: summarised programme: (a) typical section and (b) proposed first part of programme

The flats in each block were to be vacated successively and rehabilitated before the next block was released to the contractor. The tenants from the first block were to move to alternative accommodation, the succeeding groups were to move into renovated flats and, finally, the first group was to move into the last block handed back. There was thus no overlap between blocks. With a period of eight weeks for each block, this gave a total contract period of forty-eight weeks. (2)

The contractor's programme for the first two blocks is shown in Figure 33.1(b), and it continued similarly for the remaining blocks. This indicated some aim to overlap minor elements between blocks, and so stretch beyond the exact wording of the contract, but was acceptable to the architect and the employer.

MAIN EVENTS AND THEIR RESULTS

Block A

Two tenants in the first block were two weeks late in moving out. The contractor asked for a week's extension of time on the basis of postponement of possession of the site. This the architect disputed, as only part of the initial site entitlement was affected. The contractor indicated that he would reserve his position. (3)

A consignment of kitchen units for two complete blocks was delivered in week 5, in accordance with the employer's schedule made available to tenderers. These did not fit and needed minor, but awkward, alterations in most cases, partly because the site dimensions of the flats varied. The contractor was instructed to make the alterations to the block A units, to order the supplier to make the remaining alterations and to check following batches before delivery. He required to be paid for the work involved and sought a further week's extension, which was granted. (4)

The plumbing and heating subcontractor was sent off-site for a two-week period while these problems were being resolved, under an instruction issued to him by the architect. This was too long for the contractor, who contended that he was delayed a further week, which the architect disputed. The electrical subcontractor was not affected. Other trades employed by the contractor directly or as domestic subcontractors were not seriously delayed, but had to work around what was going on, or failing to go on. (5)

The first two flats were completed by the end of week 8 and the whole block by the end of week 10. The contractor argued that he was justifiably behind by the two weeks extension he was claiming on all of these flats, while the architect maintained that he was one week behind on the whole of the block. (6)

Blocks B and C

These two blocks ran to schedule, subject to starting late as a follow-on from the previous block. Again, two flats were handed over two weeks ahead of the rest for block B, but not block C. (7)

Block D

As work began here on the main consignment of flats, following two weeks of advanced work on two flats, the plumbing and heating subcontractor failed to appear to begin preparatory work. By the beginning of week 3 of the contract programme for the block, he was insolvent. The architect hesitated for all that week and the next over what to do, as he explored various options. Early in week 5 on the block, he instructed the omission of the work from the contract under clause 3.3.3(c) and notified the contractor that the work would be performed by a person engaged directly by the employer. The contractor objected to the omission of work on which he had included profit in his tender and argued that the architect could not act in this way.

As the contractor was unable to sustain his case that the architect could not issue such an instruction and that it was unreasonable for the employer to introduce someone to perform absolutely necessary work, the introduction of the direct contractor went ahead. The contractor switched his attack to claiming extension of time and loss and expense. Again, he sought loss of profit on the omitted work as part of the loss and expense. (8)

The block was finished six weeks later than was acceptable under the already extended time flowing from earlier events. When it became clear that this delay was developing, the contractor claimed an extension of this duration, while also noting that the float of two weeks which he had built up by finishing flats in each of the earlier blocks two weeks ahead was now lost, so prejudicing his completion even more. (9)

Block E

By what means it need not be asked, the employer persuaded four of the tenants of this block to move out into alternative accommodation four weeks earlier than was needed if the contractor were to have the programmed, unimpeded access to the whole block, given his actual state of progress. The architect then instructed the contractor to take possession of these flats at once and to proceed with work in them. The contractor did so after one week, but proceeded fairly slowly with the earlier elements of work. The architect suggested that he should be working faster, whereupon the contractor made one or two unrepeatable suggestions, but also pointed out that the trades needed for the following elements could not be released any sooner from block D. (10)

At this juncture, the contractor asked for adjudication but decided not to proceed with his application (this option came into being under the amendments to standard forms brought about under the Housing Grants, Construction and Regeneration Act 1996). This concentrated the minds of all participants, but did not produce any instant solutions. (11)

Work on the remaining four flats in the block began on completion of work to block D. The contractor made application for loss and expense due to work being out of phase. The architect contended that the employer had acted to mitigate the contractor's loss by bringing forward flats in block E and that the availability of flats in block F further helped. There was therefore, he concluded, little substance in the contractor's application. (12)

Four of the flats were handed over three weeks ahead of the programmed date, so enabling the employer to move tenants. Before the other four were completed, weaknesses in the existing roof were discovered. These caused delay to the service risers and tanks (installed by the direct contractor) for the remaining flats, as well as additional work to the roof itself. The flats were not handed over until four weeks after the programmed date and the roof work went on for a further six weeks. The architect contended that the contractor had been responsible for two weeks of the delay to the flats themselves, but accepted that the roof work had led to extra time for delivery of components and engaging suitable craftsmen. In reply, the contractor stated that the direct contractor had delayed beyond what was necessary, to the tune of 1½ weeks. (13)

Block F

In the case of block F, the employer again managed to move tenants out early, so making four flats available to the contractor three weeks ahead of due time. This was still somewhat ahead of what the contractor could absorb, quite apart from what he argued he was contractually bound to do.

It also accentuated the degree of disturbance which the contractor maintained he faced. He continued to state that he was put to extra expense by having to work in two blocks, even if they were in one building. The architect countered by insisting that work was still running at the same rate, it was just that it had been changed in its exact content at any one time during these closing stages. He conceded that there would be a more attenuated end to the project than had been planned, so extending some costs.

Owing to the need to work between blocks and in relation to the delays to the roof of block E, some confusion crept into the coordination of the working of the contractor and the direct plumbing and heating contractor, despite the repetitive nature of the blocks. The architect held that the contractor had failed to provide programme detail as needed, whereas the contractor held that he was not responsible for the other person's programme, only for seeing that there was no clash. At any rate, there was a further delay of two weeks in finishing the final block. (14)

COMMENTS ON NARRATIVE

These comments have been separated from the account of events etc to allow the reader to form an independent opinion, if so wished. In the rather confused occurrences described, and briefly at that, there is room for difference of opinion. The comments given here aim to light up certain aspects of affairs, perhaps by placing a particular interpretation on some of them. The numbers in parentheses match the comments with the relevant section of the narrative.

Particulars given

(1) The important matter contractually is the use of the sectional completion supplement. It is needed to cover the question of liquidated damages satisfactorily when there are several dates for handing parts of the works to the employer. Although the IFC contract does have a partial possession clause available as an optional extra, this clause is not adequate to protect the right of the employer to claim liquidated damages if there is an overrun on one or more sections, but they still finish ahead of the final (and only) completion date. This was demonstrated in the case of *Bramall and Ogden Ltd* v. *Sheffield City Council* considered in Chapter 9.

(2) The question of phased possession is not considered in any standard building contract (as distinct from civil engineering) except GC/Works/1 and, oddly, the JCT sectional *completion* supplement. This arrangement occurs frequently enough, as in the present project. It only receives attention by implication in the provisions over obligations and restrictions imposed by the employer through the term 'in any specific order', as mentioned in Chapter 6 in connection with variations to them in IFC clause 3.6.3 and correspondingly in the JCT contract.

Block A

(3) IFC clause 2.2 mentions simply 'the giving of possession', following on clause 2.1 which specifically mentions, 'Possession of the site shall be given.' This appears to envisage complete possession or none at all, giving the advantage to the contractor in this first exchange. Had he just accepted the position by moving in and starting work, he would probably have lost his advantage by inferred acceptance of what had happened. He professed to have reserved his position, and perhaps he did, but the ultimate question (as ever) will be, Even if the event is a valid postponement of possession, did it actually cause delay sufficient to extend completion of the works?

(4) These units were items ranking under clauses 2.4.9 and 4.12.4 as a supply by the employer, where deficiency in that supply could lead to the contractor obtaining extension of time or reimbursement for loss and expense. The extension granted was clearly in order, and later on, an extra payment might be in order if disturbance resulted. The contractor should have applied for such payment 'within a reasonable time of it becoming apparent' and he may well have lost his right at any rate to adequate reimbursement by not applying soon enough for records to be kept. There is little which could have been done to reduce the effect by counteraction from the architect: the units were there and the later batches were still a long time off.

It was reasonable enough for the contractor to accept an instruction to remedy the units, so long as he did nothing which made him responsible for them as units. It could have turned out less than reasonable for the employer, as he had two contractors who had dealt with the same units, so leaving doubt over ultimate suitability. On the other hand, the contractor should not have accepted responsibility for dealing with the supplier over altering the remaining units on-site or for any checking of those yet to come. He might even have found himself having to visit the works. Although he might be paid for time expended, possible liability over what was supplied could return to him in the wrong circumstances. He should firmly leave these matters with the employer or his agent, presumably the architect or a clerk of works.

(5) The plumbing and heating subcontractor was of the named variety, but this did not give the architect any more right to deal direct with him than with any other type of subcontractor. There is no such right even with a nominated subcontractor under the JCT contract. Furthermore, the nature of the dealing was to interfere with the contractor's site organisation and control, again something which the architect has no authority to do. Provided the contractor was not asking

something entirely out of reason by a week's extension, the architect was well advised to grant it and keep quiet! The subcontractor himself might have complained had he not gone into liquidation so soon after.

When a relatively small disturbance effect is produced on workpeople in a small contract, the interpretation of 'regular progress . . . materially affected' in clause 4.11 needs care. In proportion to the size of the project, the disruption suffered here may have been as severe relatively as an intrinsically much larger one on a large project. It is its effect on *progress*, a relative matter, which is in question. There is a tendency in small projects to play down the effects, especially if the contractor is also small and perhaps not so well versed in contractual matters as his larger brethren. In an alterations contract, such as here, there is also the problem of identifying disturbance in what may be an apparently less controlled-looking scene at the best of times. More tolerance is often due.

(6) The contractor was working to his own programme, with two flats ahead of the rest, which had been tacitly accepted by the architect on behalf of the employer. However, it was not contractual in that it did not affect when the contractor was due to finish the block. All flats were due together, but could be earlier (see observations in Chapter 4 that the contractor is entitled to finish early). If his claim to two weeks extension was correct, he was clear, as that was how he had finished. If he was entitled to only one, as the architect maintained, then he was a week behind, even though two flats were ahead. No measure of trade-off was strictly possible. The two of them remained apart on the issue.

Blocks B and C

(7) No new problem existed here, work was still two weeks behind overall. As the availability of a block was dependent on completion of its predecessor, there could be no question of holding the final completion date at the original position, despite the delay on the first block. The failure to be early on part of the third block was just that: the contractor was not early, but still running as contract.

Block D

(8) It is established, by the case of *Percy Bilton* v. *Greater London Council* (see Chapter 7 for details of this and related cases), that delay reasonably resulting from the withdrawal of a nominated subcontractor is the responsibility of the contractor and does not lead to extension of time, but that excess delay due to the architect unreasonably failing to renominate does lead to extension of time. It would appear that the position is similar with a named subcontractor under the present contract.

In the immediate instance, there was a delay by the architect, but hardly an unreasonable one in the circumstances. He was entitled to weigh the options in his client's interests: a rash decision could even have opened him to a charge of professional negligence. On top of this would have been the time required for the direct contractor to prepare and come on-site. The reference in clause 3.3.4(b) to extension of time and loss and expense flowing from an instruction must be read in the light of the relevant clauses on those matters. These hold out remedies for the contractor when an instruction is not issued timeously, but not simply because there is an instruction. It is difficult to see that the mere omission of the work would lead to delay for the contractor.

Against this, there was the introduction of a direct contractor. If he did not fit in with the contractor's programme when the period of reasonable delay had passed, the contractor would have cause to seek his remedies. These would arise under clause 3.11, which is stated to apply, and from this under the usual clauses 2.3 and 4.11.

The contractor raised the question of loss of profit as part of his loss and expense. As there was not disturbance of regular progress involved, he was on the wrong track: his entitlement to recompense was there, but arising out of the variation introduced. Part of the valuation, perhaps all of it, would be concerned with this element.

(9) In view of these remarks, the contractor was not entitled to the whole of the six weeks extension represented by the overall delay. In fact, he was fortunate that the total delay was not longer. He was given two weeks, with the result that he had quite a problem to regain time in the period left to him. He had noted the loss of his two weeks float, but this could not be adduced as further delay: it was there to help guard against a contingency such as had occurred, as well as his own domestic lapses. In any case, it did not give him a clear float, but just some margin on two flats in each block.

Block E

(10) It was helpful of the employer to try to move matters on by making some flats available early, no doubt with an eye to his own affairs rather than the contractor's problems. As it was half a block that was involved, it was not actually going to help the total finish, but might reduce the liability (if any) to liquidated damages. But the architect had no authority to instruct the contractor to take possession early, and the contractor need not have budged, once he was there. He was also beyond contractual reproach in not hurrying over the work, in view of the limited progress he could foresee, presumably with some justification.

This is one of those places where most of the standard contracts do nothing very well about regaining lost time. There is no power to instruct acceleration or anything which rearranges the programme to serve it. There is only the obligation of the contractor in clause 2.3 (in this contract) to 'use constantly his best endeavours to prevent delay' and to 'do all that may be reasonably required [by the architect] . . . to proceed'. The contractor could easily have challenged on his authority to instruct, but he would have been ill-advised not to do his best by cooperating. Because it was an instruction, the contractor stood possibly to gain in the end, as discussed below.

(11) The contractor was quite correct contractually, but not particularly wise practically. Adjudication could be commenced to a quick timescale but the contractor was not confident that an adjudicator would find favourably. Given that several moderately complex arguments were in prospect and that nothing really critical hung on speed, the contractor did not seek adjudication. All this may have done would have been to clarify the extension of time question, so removing doubt

over what the contractor might do about his programme. He appeared to act out of frustration and without weighing the matter that deeply.

(12) The contractor's application would have fallen completely, had he gone ahead with the early batch of flats without an instruction. He would then have been seen as trying to regain time, and so mitigating everyone's losses, but at his own cost. As it was, he had obeyed an instruction, even if one not authorised by the contract. There was therefore an implied term that he would be paid for the costs of meeting the instruction. There was no variation to the works, but consequences which fell into the loss and expense category.

These centred around working in two blocks at once, so necessitating extra movements of the hoist and the gangs between blocks D and E, which were not under one roof. So far as the gangs were concerned, this meant shifting equipment more often and further than was necessary when adjoining flats were being altered. The effect for which the contractor sought recompense was the net extra expense, so any saving (as mentioned by the architect) on loss and expense already building up would be automatic. The benefit suggested by the advancing of work in flats in block F was negative, by putting more work out of phase.

Agreement over loss and expense eventually went ahead broadly on this basis, although the amounts which the contractor put forward for disruption of the regularity of gang working were substantially reduced. They were shown to include amounts that were caused by him trying to make up the lost time which he now realised would not be covered by an extension. A claim for extra supervision was rejected on lack of evidence: the foreperson could get around just as well!

The staggered working in these blocks also led to difficulties for the named subcontractor carrying out stair finishings. He had to work in smaller sections, suspend operations at short notice and arrange for the contractor to provide additional temporary protection. This led to loss and expense on his part, although he worked into the evenings to keep up with the programme, again at agreed extra cost.

Some delay to the direct plumbing and heating contractor also arose. Agreeing this was the concern of the employer through the architect, without the contractor. There were elements of what was put forward which the architect contended were the fault of the contractor, and over which he referred the direct contractor to the present contractor. The specialist steadfastly refused to take this route, arguing that he had no contract with anyone but the employer. In the end, the various contestants met and resolved liabilities, as was sensible on such a small project. In principle the direct contractor was correct over liabilities: whether the employer recovered less, or more, from the contractor was not his concern. The main problem was evidential, and a meeting was the quickest way to establish facts.

(13) The weakness of the roof was an unexpected blow to all concerned. It was in no way the responsibility of the contractor, although he could not very well avoid performing the work or engaging those who did. It would have been intolerable to have had another contractor working so closely with those already present, especially with the tensions which had built up! Extra payment and time were inevitable in a situation in which there was no immediate yardstick for comparison.

The argument over delay did not blow up until the flats concerned were just being finished. None of those involved had kept any records of consequence about what had happened, and how far it tied in with the roof problem. The architect could well have warned the contractor that he was not satisfied, whereas the contractor could have protected himself by warning the plumbing and heating contractor. This contractor might have seen trouble coming, but then again he might have been selected as a scapegoat. These areas are always difficult. In the end, each one concerned was absolved from liability to any other and the time and cost effects were left where they lay. The lessons are clear, but not always easy to implement.

Block F

(14) The arguments over this block present little which is new, but do show affairs settling down into the general acrimony which can cause molehills to grow into their big brothers. There were the phasing arguments, the triangular arguments between parties and the lack of detail over programming disputes.

The architect's argument that work could proceed at the same overall rate in the block, although embracing different proportions of gangs, was true, but of little consolation to the contractor over the uneven working within the totality. It is just this unevenness which causes extra expense within seemingly good progress. On the other hand, the architect appeared to have some ground for complaint about the lack of information from the contractor to programme the direct contractor. It was not the contractor's responsibility to supervise the other, but he should give reasonable data to the architect, as a matter of course. For the contractor to 'permit the execution of such work', he must do more than just sit and watch.

Some final matters

It is not necessary to give a complete summary of the outcome. The essential points have been made, so the result in principle can be deduced with moderate accuracy.

The secondary loss and expense aspects may be mentioned. Given the scale of the works and of the contractor, it may seem that little is due. Again, the scale question must not detract from the reality of head office costs and financing charges for the smaller contractor. It is often the case that a protracted contract can affect such a contractor far more seriously than a larger one, just because it is a bigger share of turnover and clogs his endeavours more effectively.

It is also the smaller contractor who tends to be less well organised to take up his own interests here and protect his position in advance or retrieve it afterwards. This perhaps calls for more tolerance between the parties, bearing in mind that the employer may be small as well. If matters do reach reference to a tribunal (adjudication, arbitration or legal proceedings), everyone must remember that tolerance then has to bow before the application of the law.

In this instance the weakness of records and procedures on both sides no doubt led to compromises and loss of positions all round.

TABLE OF CASES

The following abbreviations are used for the main report series:

AC	Appeal Cases	IR	Irish Reports
All ER	All England Law Reports	KB	King's Bench
BLISS	Building Law Information Subscription Service	LGR	Local Government Reports
BLR	Building Law Reports	LJ	Law Journal
CA	Court of Appeal	LJKB	Law Journal King's Bench
Ch	Chancery Reports	LJR	Law Journal Reports
CILL	Construction Industry Law Letter	Lloyd's Rep	Lloyd's Law Reports
CLR	Construction Law Reports	LT	Law Times
Const LJ	Construction Law Journal	NSW	New South Wales
EG	Estates Gazette	ORB	Official Referees Business
Ex	Exchequer	QB	Queen's Bench
FT	Financial Times	SJ	Solicitor's Journal
HL	House of Lords	TLR	Times Law Reports
		WLR	Weekly Law Reports

Consultation of the full law report is desirable to obtain full details of the decision and principles. Different decisions based upon apparently similar situations may arise out of the precise contractual terms and circumstances. Before decisions of consequence are taken in a legal matter professional advice should be obtained. References in heavy type are to page numbers where some details of the case are given and references in normal type indicate briefer text.

Following this alphabetical table the cases are listed on pages 561–570, as they occur under the topics in each chapter.

TABLE OF CASES UNDER CHAPTER/TOPIC HEADINGS

This table allows groups of cases and individual cases to be traced both under chapter headings and under topics within a chapter. Case titles are abbreviated without law report details, which are given in the preceding table of cases. Heavy type indicates the main page reference that generally contains some detail of the case. Before decisions of consequence are taken in a legal matter professional advice should be obtained.

Chapter 3 – Tendering: the contractor's policy, examination and the contract

Examination and acceptance of the tender

Offers, statements and representations

Resulting contract (or not)

Chapter 4 – Physical and time problems

Programme and method statements

Site possession, access, ground and other conditions

Chapter 5 – Information, instructions and approvals

Mistakes in contract and other documents

Approval of the works

Dissatisfaction over certificates

Deductions from certified amounts

Ownership of materials and goods

Chapter 9 – Remedies for breach of contract

Damages for breach of contract at common law

Quantum meruit

Liquidated damages

Costs of preparing claims for extension of time and loss and expense

Chapter 10 – Delay and disturbance of progress

Concurrency of causes of delay and disturbance to programme

Chapter 11 – Ascertainment of primary loss and expense

Labour and plant

Chapter 12 – Ascertainment of secondary loss and expense

Head office overheads

Loss of profit

Interest and financing charges

Costs of preparing claims for extension of time and loss and expense

Chapter 14 – Global claims

Chapter 15 – Causes justifying extension of time and/or reimbursement of loss and expense

Some principles

Causes leading to extension of time only

Causes leading to extension of time and/or loss and expense

Chapter 16 – Delay and extension of time

Late completion

Procedures for extension of time

Chapter 17 – Contract procedures over loss and expense

Action by the contractor

Response by the architect

Chapter 18 – Damage to the works and determination of employment

Damage to the works, and related insurance

Determination of contractual employment

Chapter 20 – GC/Works/1 with quantities contract

Chapter 21 – ICE Major Works contract

Chapter 22 – The New Engineering Contract

Chapter 23 – JCT Agreement for minor work

Chapter 25 – Evidence and negotiation

Chapter 27 – Methods of dispute resolution

Chapter 28 – Provision for dispute resolution in standard contracts

Chapter 30 – Major case study: laboratory project part 1

Chapter 31 – Major case study: laboratory project part 2

Chapter 32 – Minor case study: factory project

Chapter 33 – Minor case study: flats project

INDEX OF CONTRACT CLAUSES

This index covers references to main and sub-contract clauses, other than cross-references within the text. No distinction is made between the variant editions of JCT 80 contracts.

The contract forms are listed in the following order:

JCT 63
JCT 80
IFC 84
JCT Minor Works
DOM/1
DOM/2
IN/SC
NAM/SC
NAM/T
NSC/C
NSC/T
NSW/W
GC/Works/1
ICE Fifth Edition
ICE Sixth Edition
NEC

INDEX OF SUBJECTS

This index groups subsidiary topics extensively under main topics, numbers of which are key topics throughout the book, while making considerable use of cross-references. This enables these topics to be compared and researched systematically from the index, according to the particular interest of the reader.

Except when a specific term is needed, some terms used in the JCT contracts, such as 'architect', 'instruction', 'relevant event' and 'variation' are used generally to cover cognate terms in other contracts to avoid undue repetition of entries and should be read accordingly.